Josef Börcsök

Functional Safety

Stay informed!

Keep yourself up to date and subscribe to our newsletter!

Our product and seminar newsletters inform you regularly about new knowledge for your profession.

free of charge – without obligation – unsubscribe at any time

VDE VERLAG
Technik. Wissen. Weiterwissen.

www.vde-verlag.de/newsletter

Buy this title as an e-book too and save 60%!

Buyers of this book are entitled to a special offer: in addition to the printed book, you can purchase the e-book for only 40% of the regular price.

Additional benefits:

- Full searchability of the content for fast research
- with bookmarks and links directly to the desired information
- PDF format means it can be used anywhere

Download your personal e-book now at this link:

- www.vde-verlag.de/ebook_en
- Enter your personal, one-time-use e-book code:

> 483337UX5FEXH1VX

- Add the e-book to your shopping cart and order it at the offer price

Note: The e-book code was created especially for you and may not be shared with third parties. If the book is withdrawn then the associated e-book code will be invalid.

Josef Börcsök

Functional Safety

Basic Principles of
Safety-related Systems

2nd rev. Ed.

VDE VERLAG GMBH

ICS 13.110; 13.310; 25.040

All rights reserved. Any utilization in breach of the strict limits of copyright law, without the prior approval of the publisher, is prohibited. Reproductions of common names, brand names, trademarks etc. in this publication are not subject to the acceptance that these names could be regarded as free or could be used by anyone, even without particular marking, in the sense of the trademark and brand protection legislation. Publication does not imply that the solutions described are not protected by intellectual property rights (e.g. patents and utility models). The publisher assumes no liability for the correctness and practicability of the programs, circuits, and any other arrangements and instructions published, nor for the correctness of the technical content of this publication. The up-to-date valid versions of the relevant statutory and official regulations and technical regulations (e.g. VDE body of regulations) have to be respected.

The author and the publishing house have taken great care in preparing the text and illustrations. Nevertheless, complete freedom from errors cannot bei guaranteed. For this reason, neither the author nor the publishing house can assume any responsibility for the information given in this book. On no account can the author or the publishing house be made responsible for any damage, direct or indirect, arising from the use of this information.

Bibliographic information published by the Deutsche Nationalbibliothek
The Deutsche Nationalbibliothek lists this publication in the Deutsche Nationalbibliografie; detailed bibliographic data are available on the Internet at *https://portal.dnb.de*.

ISBN 978-3-8007-3337-8 (print)
ISBN 978-3-8007-5361-1 (e-book)

All rights reserved.

© 2025   VDE VERLAG GMBH · Berlin · Offenbach
         Bismarckstr. 33, 10625 Berlin
         buchverlag@vde-verlag.de

Cover photo: HIMA Paul Hildebrandt GmbH + Co KG, Brühl

Printed in Germany by Elanders GmbH, Waiblingen   2025-03

Preface to the 1st Edition

With the introduction of electronic programmable systems in safety-relevant applications, "Functional Safety" has become a central concept. The term "Functional Safety" appears in titles of the international standards IEC 61508 and IEC 61511, which were published a number of years ago. These standards define how functional safety can be attained in safety-relevant applications using electrical, electronic and electronic programmable components and systems.

In general, functional safety means that a component or a system performs its safety-relevant task correctly and in accordance with the risk to be managed. The system either performs this function, even if internal faults or failures occur, or will assume a predefined safe state.

To fulfil this requirement, an understanding of safety engineering and a comprehensive knowledge of the existing standards are required. This begins with examining a safety system's lifecycle, performing hazard and risk analysis, specifying the requirements of safety-related components and systems, developing and implementing the systems, and the process ends with the system's operation and maintenance.

The work "Functional Safety" examines all relevant topics in detail. The reader gains an overview on the historical development of safety systems and of the standards related to programmable engineering. This includes the discussion of different application areas as well as national and international standards. The book presents basic concepts such as risk and reliability analysis, faults, root cause analysis and failures as well as all necessary safety-related parameters, their definition and evaluation. This includes both qualitative and quantitative factors.

Hardware and software requirements of a safety-related system are given appropriate consideration as well as the potential approaches and measures for achieving the required quality and safety. The comprehensive section on software is a true highlight as it is often difficult to obtain a thorough overview of this subject. Practical examples of how these concepts are applied round off this section.

IEC 61508, as a basic standard, and IEC 61511, as a sector standard for the process industry, are the currently most important standards on functional safety. Both are comprehensive and complex. The reader gains a clear overview, useful explanations and helpful procedures for applying the standards. Safety management and the required documentation are covered, as are measures for avoiding and controlling faults in a system's hardware and software.

The last section provides terminology and definitions for all relevant concepts and depicts quantitative or statistical parameters.

In my every-day project work and numerous discussions during conferences on safety-relevant systems, I am often asked where appropriate literature can be found about this topic. In my opinion, the author has provided a thorough introduction to the subject matter while simultaneously offering the experienced user a valuable reference work.

I have known the author for many years as an expert who has participated in safety-relevant projects and standardization work. I would underscore his dedication as a contributor to the "TÜV Rheinland Functional Safety Programs", where he provided valuable proposals in interest groups and as an expert for functional safety.

October 2006

Dipl.-Ing. Heinz Gall
TÜV Rheinland Industrie Service GmbH
Business Segment Manager: Automation, Software and Information Technology

Acknowledgment

To write a book on safety technology, as most experts in this area surely know, is a long and difficult undertaking. Within the scope of research and development projects, many companies and universities around the world are working to achieve a deeper understanding of the processes or the failure reaction of components and systems. As such, a constant stream of innovation would normally have to be added to the book during its drafting phase. For this reason, elaborations must be limited in their scope, even with the knowledge that substantial progress in safety related technology can be expected in the year to come.

Irrespective of new knowledge that will be gained in the coming years, this book's intended purpose is to reflect the current state-of-the-art of science and technology. While this objective cannot be achieved in its entirety, all relevant topics about functional safety should be outlined and, where possible, extensively explained and described.

The selection of topics that should be covered in a book about functional safety is also not easy. Although a broad range of subjects are completely indisputable, additional topics, depending on the point of view of an expert, must also be considered.

Thought provoking discussions with national and international functional safety experts led to numerous proposals and suggestions for completing this book. The support and constructive participation of certifying bodies committed to functional safety also contributed to finalizing this book.

I would like to sincerely thank all of my colleagues in the industrial sector and the university environment, as well as all postgraduates, graduates and students I worked with and who have contributed in various ways to making this book a reality.

Particular thanks are due to my wife Susanne and our daughters, Jasmin and Julia, who motivated me during this period, offering their patience and support while sacrificing our time together.

Josef Börcsök

Contents

Contents 9

1   Introduction 21

2   Historic Developments of Safety Systems and Standards 23

3   **Standards and guidelines** 27
3.1   Standard committees 27
3.2   Standards 30
    3.2.1 DIN V 19250 32
    3.2.2 DIN V VDE 0801 32
    3.2.3 IEC 61508 35
    3.2.4 IEC 61511 38
    3.2.5 EN ISO 13849 41
    3.2.6 IEC 61131 43
    3.2.7 ISA TR 84.02 44
    3.2.8 RTCA DO 178B 45
3.3   Definitions around the term of safety 47
3.4   State of the Art 50
    3.4.1 Automobile area – ISO 26262 50
    3.4.2 Aviation 55
    3.4.3 Automation technology 55

4   **Faults, Fault Causes and Failures** 57
4.1   Failure rates 57
4.2   Fault-Failure-Deviations 62
4.3   Failure sources 63
4.4   Failure tolerance 64
4.5   Common cause failures 65

5   **Parameter of Risk- and Reliability Analysis** 67
5.1   Reliability Parameters 68
5.2   Probability of failure 70
5.3   Average Lifetime 70

5.4	Average Repair-Time	71
5.5	Average Duration of Usefulness	71
5.6	Availability	72
5.7	Failure Rate	72
5.8	SFF	74
5.9	DC	74
	5.9.1 Tests	75
5.10	MTTF	76
	5.10.1 MTTF – Spurious Trip Rate	76
5.11	PFD	76

6	**Measures for a Risk Analysis**	**81**
6.1	Basic Concepts	81
6.2	Methods of Danger Analysis	82
	6.2.1 Forward- and Backward-Search	82
	6.2.2 Top-Down and Bottom-Up Search	83
6.3	Probability Analysis	83
	6.3.1 Statistical Analysis	83
	6.3.2 Fault Propagation Model	84

7	**Risk Matrix**	**85**

8	**Risk Graph**	**89**
8.1	Risk Graph according to DIN V 19250	89
	8.1.1 Correlation between Risk, Acceptable Risk, Residual Risk and the Risk Reduction	90
	8.1.2 Risk Parameter	91
	8.1.3 Further Risk Parameters	94
	8.1.4 Risk Graph	94
	8.1.5 Requirement Classes	96
8.2	Risk Graph according to IEC 61508-5 and IEC 61511-3	97
8.3	Risk Graph according to DIN EN 954-1	98

9	**Fault tree analysis**	**103**
9.1	Field of application and purpose Fault Tree Analysis	103
9.2	Terms	104
9.3	Graphical representation	106
9.4	Analysis procedure	107
	9.4.1 Analysis steps	107
	9.4.2 System analysis	108
	9.4.3 Undesirable event and failure criteria	109
	9.4.4 Relevant reliability parameters and time intervals	109
	9.4.5 Component failure modes	109

		9.4.6 Fault tree creation 109
		9.4.7 Evaluation of the fault tree 113
9.5	Fault tree analysis 120

10 Event tree analysis 121
10.1 Components Event Tree Analysis 122

11 LOPA 127
11.1 Layers of Protection 128
11.2 LOPA Valuation 131
11.3 Typical Protection Levels 132
        11.3.1 Basic Process Control System 133
        11.3.2 Physical equipment 134
        11.3.3 External systems to reduce the risk 135
11.4 Several actuating events 135

12 Reliability Block Diagram Analysis 137
12.1 Reliability models 142
        12.1.1 Systems without Redundancy 142
        12.1.2 Systems with Redundancy 144
        12.1.3 Mixed systems 148
12.2 Redundant Systems with Different Failure Rates 159
12.3 Substitution of Redundant System Components through
     Single System Components 164

13 Markov Model 167
13.1 Introduction 167
13.2 Possibilities with Markov Models 168
13.3 Theoretical Principals of the Markov Models 168
13.4 Time dependent Markov Models 173
13.5 Implementation of a Markov Calculation for a Safety Related System 173
        13.5.1 Transition Matrix P for System Model 176

14 Lifecycle Analysis of a Safety System 183
14.1 Hazard and Risk Analysis 183
14.2 Execution of a Risk Evaluation Analysis 183
14.3 Life Cycle Phases 185
        14.3.1 Development of a safety-instrumented function 185
        14.3.2 Failure models and PFD calculation 187
        14.3.3 System Architecture 189
14.4 Overall Planning 193
14.5 Realization of a SIS 193

14.6	Installation, Startup and Validation	195
14.7	Operation, Maintenance and Repair	195
14.8	Modification and Retrofit	196
14.9	Summary	196

15   Common Cause Failure 199
- 15.1 General 199
- 15.2 Common cause failures 200
  - 15.2.1 Analysis of Common Cause Failures 201
- 15.3 Common Mode Failure 204
- 15.4 Examples for Failures through a Common Cause 205
- 15.5 Technologies for the Evaluation of SIS Designs for CCF 206
  - 15.5.1 Industrial Standards 206
  - 15.5.2 Technical organization-specific Guidelines and Standards ... 206
  - 15.5.3 Qualitative hazard identification methods 207
  - 15.5.4 Qualitative Valuation 207
  - 15.5.5 Checklists 208
- 15.6 Quantitative Evaluation of Common Cause Failures 208
  - 15.6.1 Explicit Methods 209
  - 15.6.2 Implicit Methods of Common Cause Failures 216
    - 15.6.2.1 Basic-Parameter-Model 217
    - 15.6.2.2 Beta-Factor-Model 217
    - 15.6.2.3 Multy Greek Letter Model 218
    - 15.6.2.4 α-Factor Model 218
    - 15.6.2.5 Binomial Failure Rate Model (BFR) 219
- 15.7 β-Factor 220
  - 15.7.1 The Effect of the β-factor on safety 221
  - 15.7.2 Assessment of the β-factor 223
- 15.8 1oo2 System 225
  - 15.8.1 Probability of Failure with Common Cause Failures 225
- 15.9 Measures against Failures through Common Cause 227

16   Proof Test 229
- 16.1 Monitoring and Conducting of Proof Tests 229
- 16.2 Types of Proof Tests 230
- 16.3 Reliability Function and MTTF 231
  - 16.3.1 Failure Probability 231
  - 16.3.2 Probability of Failure on Demand 232
  - 16.3.3 Proof Test Interval T_1 232
- 16.4 Definition of the Proof Test according to IEC/EN 61508 233
- 16.5 Consequences of an Insufficient Proof Test 233
- 16.6 Differences between Diagnostic Test and Proof Test 234
  - 16.6.1 Definition of Diagnostic and Proof Test 234

	16.6.2 Performance Indicators	235
	16.6.3 Results of Calculations with or without Diagnosis	236
	16.6.4 PFD-Calculation with Variable Proof Test Coverage	237
16.7	Influence of Proof Test Interval on PFD_{avg}-Value	238
16.8	Risk Reduction	240
	16.8.1 Risk Rate and Average Failure Probability	241
	16.8.2 Proof Test Frequency	242
	16.8.3 Proof Test Expansion Factor	244
17	**Hardware of Safety-Related Systems**	**247**
17.1	Normative Architectural Specifications	247
	17.1.1 Quality in Safety for Users of Safety-Critical Systems	247
	17.1.2 Implementing Safety for Manufacturers of Safety-Critical Systems	248
17.2	Hardware Safety Life Cycle	249
	17.2.1 Safety Requirements Specification	249
	17.2.2 Safety Validation Planning	251
	17.2.3 Design and Development of the E/E/PES	251
17.3	Hardware Fault Tolerance	251
17.4	Constraints	253
	17.4.1 Architectural Constraints	253
	17.4.2 General Concepts of Risk Reduction	253
17.5	1oo1 System	256
	17.5.1 *PFD*-Fault Tree in 1oo1-Architecture	256
	17.5.2 Markov Model of 1oo1-Architecture	258
	17.5.3 Calculation of the MTTF-Value of a 1oo1-Architecture	259
17.6	Additional Architectures	261
18	**Software requirements for a system with functional safety**	**277**
18.1	Software in systems with functional safety	277
	18.1.1 Software requirements	281
	18.1.2 Non-functional requirements	281
	18.1.2.1 Goal setting	281
	18.1.2.2 Goal control	282
	18.1.3 Categories of non-functional requirements	282
18.2	Software development	284
	18.2.1 Models of software development	286
	18.2.1.1 Waterfall model	286
	18.2.1.2 Spiral model	287
	18.2.1.3 V-Model	288
	18.2.1.4 Project planning	289
	18.2.2 Specification of requirements	289
	18.2.2.1 Characteristics of a specification	290
	18.2.2.2 Description of requirements	291
	18.2.2.3 Formality of requirements	291
	18.2.2.4 Customer requirement specifications	292

18.2.3 Software architecture 292
18.2.3.1 Breakdown into components 293
18.2.3.2 Intersections 293
18.2.3.3 Communication within the system.. 294
18.2.3.4 Ability to test components 294
18.2.3.5 Additional quality characteristics... 294
18.2.3.6 Resources... 295
18.2.3.7 Quality of the solution 295
18.2.4. Possible architectural styles 296
18.2.4.1 Functional orientation 296
18.2.4.2 Object orientation 297
18.2.5 Reusable architectural structures 297
18.2.5.1 Design patterns 297
18.2.5.2 Frames.. 298
18.2.5.3 Architectural design... 298
18.2.6 Programming convention... 298
18.2.6.1 Documentation and appearance of source text 298
18.2.6.2 Naming convention.. 299
18.2.7 Software development with UML 300
18.2.7.1 Object-oriented analysis 300
18.2.8 Object-oriented design... 301
18.2.8.1 Architecture 302
18.2.8.2 Assigning procedural structures... 302
18.2.8.3 Developing design classes 303
18.2.8.4 Describing component intersections 303
18.2.8.5 Specializing status models 303
18.2.8.6 Object flow of activity models... 303
18.2.8.7 Modeling interaction models 303
18.2.8.8 Developing tests... 304
18.2.8.9 Specifying attributes 304
18.2.9 The use of CASE tools... 304
18.2.9.1 Round trip engineering with CASE tools 305
18.2.9.2 MDA 306
18.2.9.3 Comparison of UML CASE tools.. 306
18.2.10 Software quality 307
18.2.10.1 Quality plan... 309
18.2.11 Software reliability 310
18.2.11.1 Measurements of reliability 310
18.2.11.2 Differences between hardware and software reliability..................... 311
18.2.11.3 Increase in reliability by verification and validation.......................... 313
18.2.11.4 Validation of reliability.. 314
18.2.11.5 Proof of reliability.. 315
18.2.12 Measuring software quality.. 315
18.2.12.1 Lines of code (LoC)... 317
18.2.12.2 McCabe measure.. 317
18.2.12.3 Halstead measures.. 318
18.2.12.4 Usefulness of formulas 319

18.2.13 Failures in software systems 319
18.2.14 Testing procedure 321
18.2.14.1 Testing procedure 321
18.2.14.2 Black box test methods 322
18.2.14.3 White box test methods 323
18.2.14.4 Intuitive test case determination 324
18.2.15 Testing in practice 325
18.2.16 Integration 325
18.2.16.1 Top-down integration 325
18.2.16.2 Bottom-up integration 326
18.2.16.3 Outside-in integration 326
18.2.17 System and certification test 327

19 Application examples 329
19.1 Practical Implementation of the IEC 61508 Safety Standard 329
    19.1.1 IEC 61508 Norm 330
    19.1.1.1 Functional Safety Management 332
    19.1.1.2 Pipe to Pipe Approach 334
    19.1.1.3 Quantitative Safety Evaluation 335
19.2 Determining the SIL of a Processor Based System 335
    19.2.1 SIL Requirements 336
    19.2.2 Determining the SIL of a Processor Unit with Processor Periphery 337
    19.2.3 DC-Measures for a Processor Unit with Processor Periphery 337
    19.2.3.1 Processor Units 337
    19.2.3.2 Read-Only Memory 338
    19.2.3.3 Alterable Memory 339
19.3 Determining the SIL of a Safety Function 340
    19.3.1 Determining the SIL of a Safety Function 341
    19.3.2 Modification of the Architecture of a Safety Function 342
    19.3.3 Determination of the SIL of a Modified Safety Function 344
    19.3.4 Modification of the Safety Function 345
    19.3.5 Determining the SIL of a Safety Function with Diagnosis 347
19.4 Determining the SIL of a Safety Loop 348
    19.4.1 Determining the SIL of the Safety Loop 350
19.5 Examples of Reliability Analysis 353
    19.5.1 Example 1 (Chemical Installation) 353
    19.5.1.1 Risk Graph 353
    19.5.1.2 Event Tree 355
    19.5.1.3 Error Tree Analysis 356
    19.5.1.4 Reliability Block Diagram 357
    19.5.2 Example 2 (Driver-Side Airbag) 358
    19.5.2.1 Risk graph 358
    19.5.2.2 Event Tree 360
    19.5.2.3 Error Tree Analysis 360
    19.5.2.4 Reliability Block Diagram 361
    19.5.3 Example 3 (Airplane) 362

    19.5.3.1 Risk Graph 362
    19.5.3.2 Event Tree 364
    19.5.3.3 Fault Tree Analysis 364
   19.5.4 Example 4 (Pipeline) 367
    19.5.4.1 Risk Graph 367
    19.5.4.2 Event Tree 368
    19.5.4.3 Error Tree Analysis 369
   19.5.5 Example 5 (Coliseum) 370
    19.5.5.1 Risk Graph 371
    19.5.5.2 Event Tree 371
    19.5.5.3 Error Tree Analysis 372

20   IEC/EN 61508 373

20.1 IEC/EN 61508-1 374
   20.1.1 Outline and Field of Application 374
   20.1.2 Compliance with this Standard 376
   20.1.3 Documentation 376
   20.1.4 Safety Management 376
   20.1.5 The Complete Safety Lifecycle 378
   20.1.6 Verification 380
   20.1.7 Evaluation of Functional Safety 380

20.2 IEC/EN 61508-2 380
   20.2.1 Field of Application 380
   20.2.2 The E/E/PES Safety Lifecycle 381
   20.2.3 Techniques and Measures for Control of Failures during Operation 383
   20.2.4 Methods for Avoiding Systematic Errors during
      Different Phases of the Lifecycle 383

20.3 IEC/EN 61508-3 383
   20.3.1 Field of Application 383
   20.3.2 Quality Management System of Software 383
   20.3.3 Software Safety Lifecycle 383
   20.3.4 Evaluation of Functional Safety 385
   20.3.5 Appendix A – Guidelines for the Selection of Techniques and Methods. 385

20.4 IEC/EN 61508-4 385
   20.4.1 Terms Regarding Safety 385
   20.4.2 Terms relating to Devices and Equipment 386
   20.4.3 System Terms 386
   20.4.4 Terms relating to Safety Functions and Safety Integrity 388
   20.4.5 Terms relating to Errors, Failure, and Deviation 389
   20.4.6 Terms relating to Lifecycle 389
   20.4.7 Terms relating to Verification of Safety Measures 389

20.5 IEC/EN 61508-5 390
   20.5.1 Field of Application 390
   20.5.2 Appendix A – Underlying Concepts 390
   20.5.3 Appendix B – ALARP and the Concept of Tolerable Risk 391

		20.5.4 Appendix C – Quantitative Methods for Determining the Safety Integrity Level 393

        20.5.4 Appendix C – Quantitative Methods for Determining the
Safety Integrity Level 393
        20.5.5 Appendix D – Qualitative Methods for Determining the
Safety Integrity Level (Risk Graph) 394
        20.5.6 Appendix E – Specification of the Safety Integrity Level A
Qualitative Procedure – Matrix of the Extent of a Dangerous Event 395

20.6 IEC/EN 61508-6 396
        20.6.1 Field of Application 396
        20.6.2 Appendix A – Application of IEC/EN 61508-2 and -3 396
        20.6.3 Appendix B – Exemplary Procedure for Determining Hardware Failures 396
        20.6.4 Appendix D – Methods for Quantifying the Consequences of
Hardware Failures due to the Same Cause in E/E/PES 401

20.7 IEC/EN 61508-7 401
        20.7.1 Field of Application 401
        20.7.2 Appendix A – Overview of Procedures and Measures for E/E/PES:
Control of Accidental Hardware Failures 401
        20.7.3 Appendix B – Overview of Techniques and Measures for
Prevention of Systematic Failures 403
        20.7.4 Appendix C – Overview of Techniques and Measures for
Achieving Safety Integrity of Software 404

21 IEC 61511 405

21.1 Scope of Application 405
21.2 Subdivision of Standard IEC 61511 407
21.3 Terms and Abbreviations 410
        21.3.1 Abbreviations 410
        21.3.2 Terms 411
21.4 Management of Functional Safety 422
        21.4.1 Goal 422
        21.4.2 Requirements 422
        21.4.3 Evaluation, Auditing, and Revisions 422
        21.4.4 SIS Configuration Management 423
21.5 Safety Lifecycle Requirements 423
21.6 Verification 426
        21.6.1 Goal 426
        21.6.2 Requirements 426
21.7 Hazard Analysis and Risk Evaluation 426
        21.7.1 Goal 426
        21.7.2 Requirements 426
21.8 Allocation of Safety Functions to Protection Layers 427
        21.8.1 Goal 427
        21.8.2 Allocation Requirements 427
        21.8.3 Safety Integrity Level 4 Requirements 427
        21.8.4 Demands on Factory Devices Used as Protective Layers 428
        21.8.5 Requirements for Failure Avoidance 428

21.9	Safety Specification of the SIS	429
	21.9.1 Goal	429
	21.9.2 SIS Safety Requirements	429
21.10	SIS Design and Planning	429
	21.10.1 Goal	429
	21.10.2 General Requirements	429
	21.10.3 Demands on Safety Behavior upon Error Detection	430
	21.10.4 Demands on Hardware Error Tolerance	430
	21.10.5 Demands on the Selection of Components and Subsystems	431
	21.10.6 Field Devices	431
	21.10.7 Interfaces	431
	21.10.8 Maintenance and Test Device Requirements	432
	21.10.9 Failure Probability of Safety-technical Functions	432
21.11	Application Software Requirements	432
	21.11.1 Demands on the Safety Lifecycle of Application Software	433
	21.11.2 Specification of Application Software Safety Requirements	437
	21.11.3 Validation Planning for Application Software Safety	438
	21.11.4 Design and Construction of Application Software	438
	21.11.5 Integration of the Application Software into the SIS Subsystem	439
	21.11.6 Procedure for Modification of Application Software	440
	21.11.7 Verification of Application Software	440
21.12	Final Inspection	440
	21.12.1 Goals	440
	21.12.2 Recommendations	440
21.13	SIS Assembly and Implementation	440
21.14	SIS Safety Validation	441
21.15	Operation and Maintenance of the SIS	441
	21.15.1 Goals	441
	21.15.2 Requirements	441
	21.15.3 Re-examination and Inspection	442
21.16	SIS Modifications	442
	21.16.1 Goals	442
	21.16.2 Requirements	442
21.17	Decommissioning of the SIS	442
21.18	Documentation Requirements	443
	21.18.1 Goal	443
	21.18.2 Requirements	443
22	**Terms and Definitions**	**445**
22.1	Safety Systems	445
	22.1.1 Risk	445
	22.1.2 Partial Risk	445
	22.1.3 Risk Limit	445
	22.1.4 Risk Parameters	445
	22.1.5 Requirement Class	445

		22.1.6 Measures	446
		22.1.7 Protection	446
		22.1.8 Measurement and Control Protection Measures	446
		22.1.9 MSR Protection Installation	446
		22.1.10 Undesired Event	446
		22.1.11 Error	446
		22.1.12 Redundancy	446
		22.1.13 Diverse Redundancy	446
		22.1.14 Failsafe	447
	22.2.	Dependability	447
		22.2.1 Reliability	448
		22.2.2 Availability	449
		22.2.3 Safety	450
		22.2.4 Maintainability	450
	22.3	Documentation of Failure Behavior	450
		22.3.1 Density Function, resp. Failure Density $f(t)$	450
		22.3.2 Failure Probability, resp. Distribution Function $F(t)$	453
		22.3.3 Reliability, resp. Survival Probability $R(t)$	456
		22.3.4 Failure rate $\lambda(t)$	458
		22.3.5 Description of Failure Behavior by Examples	460
		22.3.6 Boolean Theory	464
	22.4	Time Factor	465
		22.4.1 MTTF	466
		22.4.2 $MTTF_{spurious}$	466
		22.4.3 MTBF	466
		22.4.4 MTTR	467
		22.4.5 Example for Calculation of MTTF	467
		22.4.6 Continuous Availability	467
		22.4.7 Downtime DT	469
		22.4.8 Uptime UT	470
		22.4.9 Mean Down Time MDT	470
	22.5	General Thoughts About Terms and Standards	470
		22.5.1 Degree of Diagnostic Coverage DC	472
		22.5.2 Common Cause Failure CCF	473
		22.5.3 Probability of Failure on Demand PFD	474
		22.5.4 Failure Rates	475
		22.5.5 Risk, Damage, and Danger	478
		22.5.6 Hazard Rate	479
		22.5.7 Safety Integrity Level SIL	479
	22.6	Process Control Technique PLT	482
	22.7	Performance Level PL	483
Literature			485
Index			517

1 Introduction

Modern technical systems, which control safety critical processes, are getting more and more complex. There are several reasons for this. On one hand, the requirements of the systems are getting more while at the same time the systems are getting smaller. On the other hand, it is necessary, in order to stay globally competitive, to offer systems that technically can perform more and in the mean time that are safer.

This applies even more in the field of digital safe automation technology, which utilizes complex digital circuits. Normally microprocessors are used which solve problems via programs. These "intelligent" systems can be found nowadays, due to the rapid development in microelectronics, in safety related control and automation systems of any kind or in general anywhere were momentous difficult tasks need to be performed.

Digital computer systems of any kind are especially utilized for safety related task. Tasks can include the supervision or control of automobiles, trains, airplanes or power stations and chemicals plants. Also the medical world is a growing application area. In the future robots carrying out operations and video transmission systems for endoscopes will be used whose information will be transmitted via networks and which can be controlled and monitored by doctors. In each of these cases a failure of the system could lead to serious damage or even lead to casualties. These dangerous situations results, besides the academic interest to develop and designs safe systems and processes, also in social interest in such processes and systems. It is the fear for dangerous events, which is among others depends on the capability to control a certain, chosen situation.

In general, modern complex control and applications systems for safety critical purposes utilize different subsystems. These subsystems can themselves be complex systems. These subsystems can be realized in hardware as well as software. The analysis of such systems can in principle be performed from two different view points. First of all they can be viewed from a management point of view. With this kind of view the aspects of how to reach reliability, for example how it can be planned, examined or steered, are moved into the foreground. Here, the reliability is analyzed as one of the many quality properties. The second point of view is based on a technical, academic analysis. Here aspects are analyzed, as for example the reliability or the risk determination of a complex system or the generation of an applicable model. One should differentiate in this case between analyzing a software model and a hardware model. Unfortunately, hardware models cannot immediately be projected on software models and vice versa.

Reliability means the functioning under all circumstances. Especially due to the complex software in modern systems it is not possible to fully exclude all failures. Safety means here that even in the case of an occurring failure the system cannot go into a critical state.

Safety relevant applications and automation systems are divided into systems with and without safe states. Here it is assumed that either the technical process has a safe state or it is a process which does not have a safe state.

A process state is considered safe state when no danger can occur. These state needs thus to be reached in case of a failure of the application or automation system which controls this process. Safety related systems have, compared to the control systems, additional requirements concerning safety related aspects, e.g., failure correction or safety integrity. Due to the importance of these considerations concerning system safety they are generally drawn up in their own document, in order to ensure appropriate safety of the system.

The different functional, non-functional and safety related technical requirements of the system result, together with the general system characteristics, in a list of systems specific attributes. Among others these are:

- Reliability
- Availability
- Fail safe operation
- System integrity
- Data integrity
- System repair
- Maintenance
- Dependability

MTTF and PFD values were already presented in the past to have a tangible measurement for the analysis of safety critical systems. The PFD value represents the performance of the system in terms of freedom from failures. The smaller the PFD value, the better the system. General system aspects must be considered in most industrial applications as safety applies to all elements of a loop. A loop is defined as a system consisting of input modules (digital inputs), logic solvers (components like CPU's processing logic), and output modules (digital outputs), as well as sensors and actuators.

2 Historic Developments of Safety Systems and Standards

The introduction referred to the ever increasing use of technical, computer controlled systems. With this use, additional problems occurred, which were then not yet regarded in their extent and effect.

Looking closer to the historic development of computer-aided safety systems, you can see that forecasts and prognoses are very difficult to accomplish. With the fast development of the information and computer technologies, see Figure 2.1, a forecast, which concerns a period of more than 5-7 years, are only possible with an extreme uncertainty.

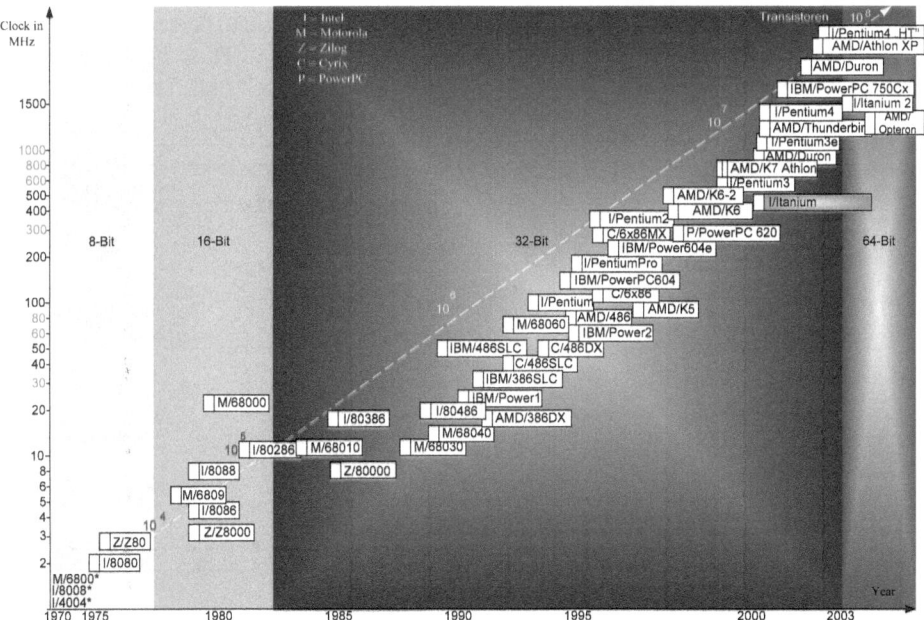

Figure 2.1: Computer development since 1970

As an example of these difficult forecast possibilities, next are some statements of specialists from the past. Thomas Watson, IBM-CEO 1956 -1971, made the following statement around 1943: „*I think, there is a market for perhaps five computers world-wide*". Or around the year 1968, when an engineer from the IBM computer department delivered the following statement for the invention of the microchip: „*But what ... should this be good for?"*, or from Ken Olson, president and founder of DEC, 1977, when he made the statement: *"There is no reason, why someone should have a computer at home"* or, as last example, an utterance made by the world-wide well-known Bill Gates around 1981: *"640 Kbyte should be sufficient for everyone"*.

As described, computer systems become ever more complex and their application areas expand continuously over almost all well-known areas of technology and the personal environment. Here are mentioned exemplary only the mobile radio and medical technology as well as automation and automobile industry. Not only the hardware but also the software developed itself, in some application areas, for example the safety engineering, even more complex than the hardware. Ever higher reliability and security is demanded, because computer errors and losses can cause accidents and damage and as consequence high claims for damages. Therefore, new and better solutions and systems must be developed, which possess a very high reliability and security, and which are economically profitable. The chances, which are offered by the use of new technologies, can be used both within the range of the manufacturing and the development. The theoretical and the practically tested reliability and safety methods must be systematically applied. The software must be, as already applied with the hardware, test and controllable, in order to reduce hardware errors/component losses (technical errors) or operating errors (human errors in the handling) as they appear with the use of conventional technology regularly. Of course errors arise with the use of new technology. These errors can be based in the software and lead as consequence to crashes of programs. Errors can occur by compatibility problems with a software update or express themselves as interface problems, if different program products cooperate.

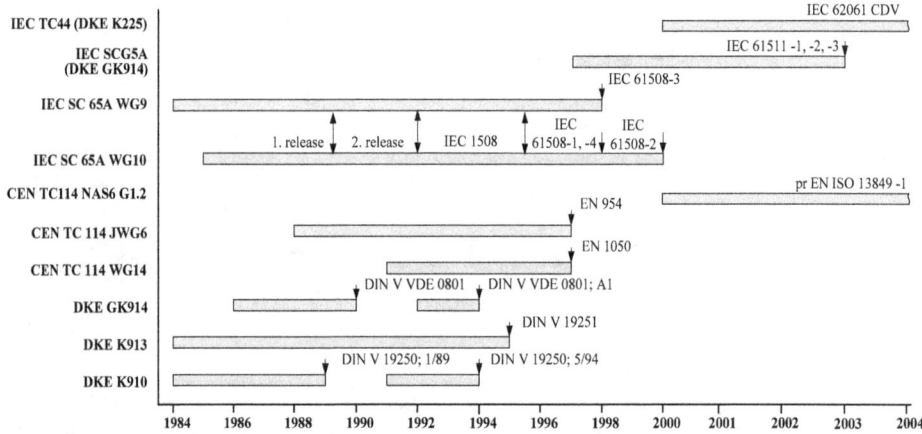

Figure 2.2: Development of standards and guidelines since 1984

Also operating errors with the manual mode/emergency mode can lead to serious consequences. With new and particular further development, already well-known sources of errors shift again and again. Therefore the reliability/safety methods must be adapted and used systematically. The standards represent an important source of information for the developer. Standards can fulfill different important purposes. Guidelines for draft and development techniques are made available by standards. The user has the certainty that a product was developed according to methods, which are well-known and effective. A uniform approximation between different teams is promoted by working with standards. Standards represent an assistance that a product reaches a certain quality level and they offer a defined decision basis in case of an argument.

As shown in Figure 2.2, the completion of a standard can take several years before, after usual several versions, a final version comes on the market. Background is usually that the authors of a standard represent different interests and a common consent needs to be found first.

Safety related systems must fulfill certain standards for their intended use, which specify the development methods and the system requirements. The certification of the safety system and the final total process, in which the safety system is integrated, depends on keeping these standards and criteria, which define them. Not within all industrial ranges the observance of standards is as strictly regulated as with the application in the safety related measuring and control engineering or in the emergency shut down technology.

Apart from the formal standards, additionally different guidelines exist, from which necessities at the system development can be derived. Examples are industrial specific standards and guidelines as they are used in the chemical, petrochemical, aviation, atomic as well as mining industry. They describe in each case the main dangers of those industries. The remaining risks, which accompany with the requirements to the reliability and the availability, are defined and regulated by considerations of the supervisory authorities and the industry. Out of this, the standards for safety requirements to systems of the specific industrial sector result. More general standards and guidelines refer against it to all industrial ranges.

3 Standards and guidelines

The use of safety-related computer systems within different industrial areas led to a wealth of experiences and knowledge within this range. Today, compressed proven knowledge is present in the form of standards and guidelines, which are to help the development engineer to achieve a uniformly high level of quality and safety when developing new systems. Not only with the development, but also the operation of a safety-related system requires certification of the system in accordance with the standards and for the application they were developed. Certifying should take place via the responsible certification authorities. Therefore many different committees are world-wide busy with the development of standards for development, verification and operations of safety systems.

Next, some of these organizations which develop standards and guidelines, or who actively participate in the development, are briefly introduced.

3.1 Standard committees

RTCA: The RTCA (US Radio Technical Commission for Aeronautics) develops and publishes, as an advisory committee, guidelines and minimum requirements for technologies used in aviation. The committee SC-167 is responsible for standards for software certifying.

The US aviation authority FAA (Federal Aviation Administration) specified in 1993 that the guidelines, particularly the document of DO-178B[1] published by the committee SC-167, are basis for the approval of new software by the FAA.

DoD: The US Department of Defense (DoD) is endeavoured to use standards already established if they are present in the desired form. Their own standards exist particularly within the range of nuclear weapons, like for example „the Nuclear Surety Design Program for to Nuclear Weapon system software and firmware" (AFR 122-9[2]).

DGQ: The goal of the German society for quality registered association (DGQ) is it to develop quality techniques and to promote the know-how to quality management.

Through close co-operation with institutes for standardization it made an important contribution for the development of the guidelines belonging to the ISO 9000[3]. The latter is not appropriate for safety related systems or software. Rather they are occupied with the general quality of the safety device. For this reason they are applicable for nearly everything, and thus also the basis for safety engineering, industry and economics, manufactur-

---

[1] [DO-92] DO178B, RTCA, Inc. 1992
[2] [AFR-87] AFR 122-9 Report, The Nuclear Security Design Certification Program for Nuclear Weapon System Software and Firmware
[3] [EN-90] EN ISO 9000, *Quality management systems*, s. a. [ISOa05], [ISOa00], [ISOb00]

ing and production, development and research, furnishing services and much more besides.

An enterprise, which wants to be certified according to ISO 9000, must compile and specify their own procedures for quality assurance according to ISO 9000. Independent auditors regularly audit whether these procedures comply with ISO 9000 and whether they are implemented accordingly in operations.

NASA: The US space agency NASA (national Aeronautics and space administration) developed many own standards. For new programs often also new standards are specified, like for example for the space station „internationally space station alpha". The requirements made there are substantially more detailed than in other standards.

MISRA: The MISRA (Motor Industry software Reliability Association) is a US organization of manufacturers in the area of the automotive engineering.

This organization has developed some guidelines for the development and use of safety software considering specially requirements for the use in vehicle technology.

In addition applies that for example drivers of different experience show different behaviours or that the response times of drivers can be very different.

Also a compensation factor has to be set for those cases where a system used for improving safety can lead to more risks taking and to more dangerous behaviour of the user.

IEC: The IEC (International Electrotechnical Commission) develops and administers standards in co-operation with national committees.

To achieve this working groups are used, which deal in each case with a central topic of interest. Take for example „Technical Committee 65", which is responsible for „Measurement and control in industrial processes". The document „IEC 61508: Functional Safety: Safety related system"[4] resulted directly from this".

The IEC 61508 divides systems regarding their hazard potential into one of four risk or requirement classes („Safety Integrity level" or SIL) and prescribed according to the respective class requirements for the development and implementation of the systems. Table 3.1 shows the connections in detail.

Another IEC document dealing with safety critical software is „IEC 880: Software for Computers in the Safety System of Nuclear Power Stations "[5].

DIN: The German Institute for standardization (DIN) is with its subsidiary organizations the responsible institution for standardisation works in Germany and represents the German interests in world-wide and European standardization organizations. In the 80's, IEC and DIN investigated the basic requirements of safety systems for use in measurement and automated control.

In Germany, various standards have been developed over the last two decades in almost all technical areas relating to safety-critical systems.

---

[4] [IECa]   IEC 61508, International Standard 61508 *Functional Safety: Safety-Related System*
[5] [IEC-86]   DIN IEC 880, Software for computers in the safety systems of nuclear power stations

3.1 Standard committees

Table 3.1: Specifications and requirements for the different SIL levels according to IEC 61508

Specifications	SIL 4	SIL 3	SIL 2	SIL 1	Applicability: Hardware (H) Software (S)
Requirements and Design Specifications	Formally (mathematically)	Semi-Formal (e.g. natural speech)	Informal (e.g. natural speech)	Informal (e.g. natural speech)	H/S
Configuration management	Full (automatically for development & production)	Full (automatically for development & production)	Yes	Manual	H/S
Prototype	yes	yes	Optional	Optional	H/S
Structured Design methods (e.g. Flow charts, relationship or transfer diagrams)	yes	yes	Preferred	Optional	H/S
Design examination	Yes (project team)	Yes (project team)	Yes (project team)	Test (experts)	H/S
Project-Management	Yes	Yes	Yes	preferred	H/S
Independent technical evaluation	Yes	preferred	optional	Optional	H/S
Data evaluation analyses and correcting actions	Yes	Yes	Yes	Yes	H/S
Statistic analyses	Yes	Yes	Optional	Optional	H/S
Dynamic analyses (e.g. automatic testing)	Yes	Yes	Yes	Yes	S
Independent testing	Yes (Accomplished by an external organization)	Yes (Accomplished by an external office)	Yes (Preferred, if accomplished by an external office)	Optional	H/S
Environmental / serviceable testing	Yes	Yes	Preferred	Optional	H
Computer produced simulations for construction unit tolerances	Yes	Yes	Preferred	Optional	H
Additional product monitoring (e.g. independent examinations)	Yes (Accomplished by an external organization)	Yes (Accomplished by an external organization)	Yes (Preferred, if accomplished by an external organization)	Optional	H/S
ISO 9001	Yes	Yes	Yes	Yes	H/S

Of course, not all safety standards can be presented and explained here. However, the most important ones, including the historical ones that form the basis for the current safety standards, will be briefly explained. The most important historical safety standards are counted the DIN V 19250 „Fundamental safety aspects to be considered for measurement and control equipment"[6] and the DIN V VDE of 0801 „Principles for computers in safety-related systems"[7].

---

[6] [DINa98] DIN V 19250, Grundlegende Sicherheitsbetrachtungen für MSR-Einrichtungen
[7] [DINc98] DIN V VDE 0801, Grundsätze für Rechner in Systemen mit Sicherheitsaufgaben

In DIN V 19250, similar to the current generic safety standard IEC 61508, an application is classified regarding its hazard potential according to a risk or requirement class.

Also the need for quantitative evaluation is pointed out. Parameters like possible extent of the damage, damage frequency, probability density and -duration of persons within the range of the system, possibility of the danger warning, probability of occurring certain events and others are used.

According to DIN V 19250 eight requirement classes are differentiated. Figure 3.1 presents the four safety integrity levels (SIL) according to IEC compared to the AK levels.

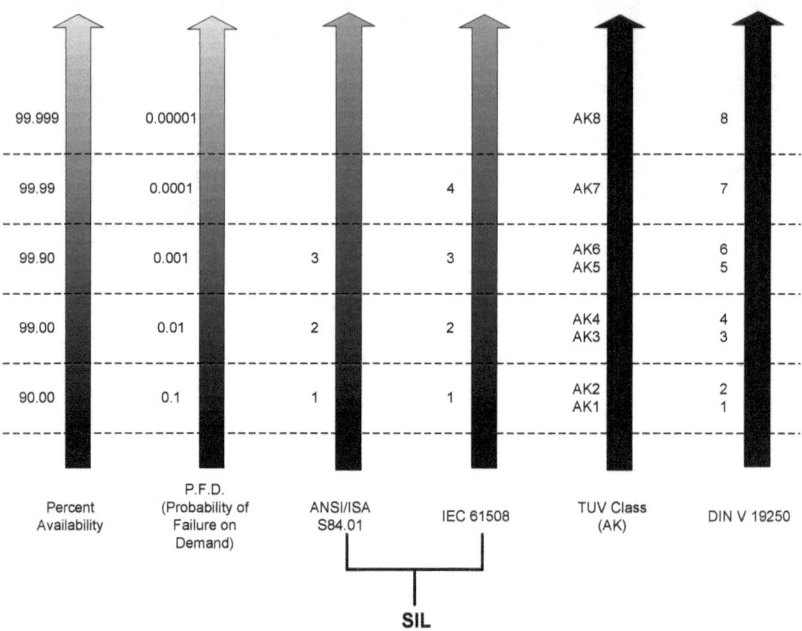

Figure 3.1: Comparison of the requirements classes with the SILs

3.2 Standards

Safety-related systems (SIS – Safety instrumented system) are used already for years for carrying out safety-related functions in the process industry. The safety systems should bring the supervised process into a safe state in case of a demand. The required safety functions depend on many application specific factors and form a part of the overall safety, which the equipment under control, and the process control system requires if they possess inherent risk. Safety, which is to be ensured by the safety function, can be achieved by a multiplicity of protective systems, which are based on the most different technologies, like mechanics, pneumatics or programmable electronics. These safety functions serve functional safety and reduce the risk of a functional loss with possible serious consequences. Which safety function is used and where, is determined by a Process Hazards analysis (PHA)

3.2 Standards

The term „safety system" is replaced in the different standards also by the designations „safety critical" or „safety-related system". The general understanding of a "safety, safety-critical or safety-related system" is a system, which consists of one or more safety functions and which, if they fail, can lead to dangerous consequences. Therefore, it is necessary to develop safety critical systems with a certain minimum standard level, so that the demands on functional safety are fulfilled and can be reviewed. Applications of such systems reach from transportation (to air, on and under the water or on the road or course), to energy production (nuclear power stations, solar power stations), to general industrial processes (chemical, petrochemical, metallurgical), up to medical supply, like life-supporting systems or patient monitors.

In the standards numerous methods are found to carry out safety analysis, which are used during a whole life cycle of a safety system. Thus it is guaranteed that the safety systems fulfil the requirements – during safety engineering for example the requirements of the functional safety – of the appropriate standards. All standards, which are concerned in general or specifically with functional safety, stress the value of possible failure reduction in a SIS by good design and by the employment of proven engineer technology. The standards are concerned thereby both with random hardware and/or software failures as well as systematic failures and with common cause failures, thus failures with a common root.

For the use of safety-related systems, which consist of hard- and software, national and international standards, guidelines and regulations need to be met. In order to get an overview of these various documents are briefly described next with their name as well as their applicability. This representation does not claim to be complete. In the case of doubt the validity of the appropriate standards and regulations should be inquired at the responsible institutions. Over the past two decades, standards dealing with "functional safety" have become established in almost all technical areas, and the number is increasing year on year. Only a few of these standards are mentioned here to illustrate the wide range of "safety standards".

- **EN IEC 61508:** Functional safety of electrical/electronic/programmable electronic safety-related systems
- **EN IEC 61511:** Functional safety – Safety instrumented systems for the process industry
- **DIN IEC 61513:** Nuclear power plants – Instrumentation and control of safety-related systems – General system requirements
- **EN ISO 13849:** Safety of machinery – Safety-related parts of control systems
- **EN IEC 62061:** Safety of machinery – Functional safety of safety-related electrical, electronic and programmable electronic control systems Introduction
- **ISO 26262:** Road vehicles – Functional safety
- **ISO 25119:** Functional Safety for Tractors and Machinery for Agriculture and Forestry
- **DIN EN 50129:** Railway applications – Telecommunications, signaling and data processing systems – Safety-related electronic systems for signaling
- **DIN EN 50156:** Electrical equipment for combustion plants and associated equipment
- **IEC 62304:** Medical device software – Software life cycle processes

- **IEC 60601:** Medical electrical equipment – Part 1: General requirements for basic safety and essential performance
- **DIN EN ISO 10128:** Robots and robotic systems – Safety requirements for industrial robots

3.2.1 DIN V 19250

This already historic standard carries the title „Fundamental safety aspects to be considered for measurement and control equipment"[8]. It describes a qualitative procedure for a risk estimation, whereby the result is represented in the form of application classes. Since the procedure may not be applied to the entire system, only the risk is determined, which is covered by the protection system. The procedure is independent of the application and independent of the technology of the protection device.

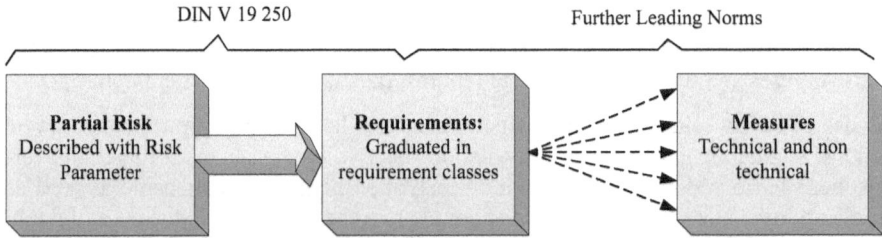

Figure 3.2: Connection between the partial risk which can be covered by a protection system, the requirements and the measures for a unit.[9]

3.2.2 DIN V VDE 0801

DIN V VDE 801 is also one of the historic safety standards and is entitled „Principles for computers in safety-related systems" and supplements the DIN V 19250 „Fundamental safety aspects to be considered for measurement and control equipment". The DIN V VDE 0801[10] describes the possible measures in safe computer systems, in order to reach the requirements in DIN V 19251 „Instrumentation protection device – requirements and measures for secured function"[11] defined requirement classes.

Generally the safety-relevant requirements of computers, and/or field bus systems, are application independent. The level of the requirements depends on the risk and the hazard potential, which result from respective application.

When comparing the requirement classes according to DIN V 19250 and DIN V VDE 0801 with safety integrity the level according to IEC 61508[12] the following aspects are to

---

[8] [DIN-94]   DIN V 19250: 1994-05
[9] [DINb94] DIN V 19250, Leittechnik: Grundlegende Sicherheitsbetrachtungen für MSR-Schutzeinrichtungen
[10] [DINc98] DIN V VDE 0801, Funktionale Sicherheit sicherheitsbezogener elektrischer/elektronischer/programmierbarer elektronischer Systeme (E/E/PES)
[11] [DINa95] DIN V 19251, Leittechnik; MSR-Schutzeinrichtungen; Anforderungen und Maßnahmen zur gesicherten Funktion
[12] [IECa]   IEC 61508, International Standard 61508 Functional Safety: Safety-Related System

be particularly considered. In DIN V VDE 0801 a computer system is defined as safety system. There the computer system is regarded by the input of the system up to its output. In principle only demands are made against the computer system. The other components for connection to the plant or the process (field instrumentation) are not considered.

IEC 61508 considers the safety-relevant function, i.e., it will always be an entire function chain, e.g. from sensor – to computer (logical/programmable unit) – to actuator. Figure 3.3 shows the different allocations.

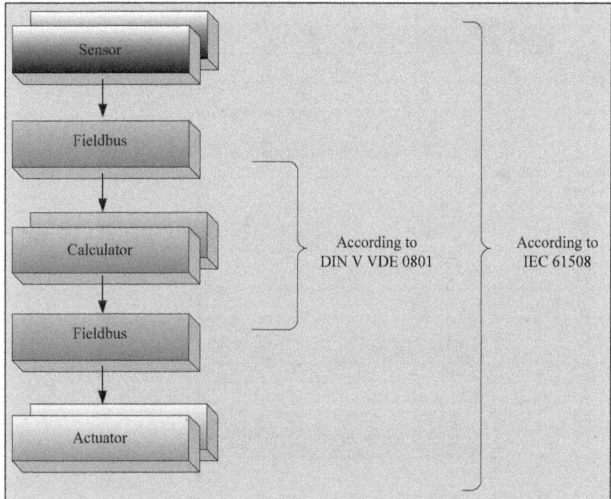

Figure 3.3: Function chain sensor-computer-actuator with the view according to DIN V VDE 0801[13] and IEC 61508[14]

The represented function chain must completely fulfil the requirements of the respective safety integrity level after IEC 61508. Thus also the field bus systems are to be included into the view.

With the definition of the safety integrity level result the range and effectiveness of the safety-relevant measures which must be realized. Which extent and which effectiveness the individual measures to avoiding and control failures must have, is fixed in the application-independent standard IEC 61508.

The effectiveness of the measures from the DIN V VDE 0801, which can be met, in order to reach the requirement classes according to DIN V 19250, is represented in Table 3.2

Table 3.2 lists different failure causes. Measures to avoid and control failures are assigned according to their effectiveness to the requirement classes, and/or SIL as minimum requirements. From the table it can be derived whether measures to avoid and/or control failures are to be met. Here the combinations of measures to avoid and control failures must be selected according to the requirement class and/or SIL in such a way that the probability of a dangerous failure becomes sufficiently small.

---

[13] [DINc98] DIN V VDE 0801, Funktionale Sicherheit sicherheitsbezogener elektrischer/elektronischer/programmierbarer elektronischer Systeme (E/E/PES)
[14] [IECa] IEC 61508, International Standard 61508 Functional Safety: Safety-Related System

Table 3.2: Measures from DIN V VDE 0801[15], in order to reach the requirement classes according to DIN V 19250[16]

Malfunction due to Random Failures in the Hardware	Requirements according to DIN V 19250							
	1	2	3	4	5	6	7	8
	Measures to manage Failures							
1 Single failure								
2 Several failure								
3 Systematic failure in the Hardware	Measures to prevent Failures							
	Measures to manage Failures							
4 Systematic failure in the Software	Measures to prevent Failures							
	Measures to manage Failures							
5 Application and Manipulation Failures	Measures to prevent Failures							
	Measures to manage Failures							
6 Failures caused by environment and application influences	Measures to prevent Failures							
	Measures to manage Failures							

The distinct grey tones mark the differently demanded effectiveness' of the safety-relevant measures:

---

[15] [DINc98] DIN V VDE 0801, Funktionale Sicherheit sicherheitsbezogener elektrischer/elektronischer/programmierbarer elektronischer Systeme (E/E/PES)
[16] [DINb94] DIN V 19250, Leittechnik: Grundlegende Sicherheitsbetrachtungen für MSR-Schutzeinrichtungen

3.2.3 IEC 61508

This standard, which is called basic safety standard, describes the fundamental, complete life cycle of safety-related systems and is available in its version IEC 61508:2010.

IEC 61508 describes and identifies necessary activities, defines the responsibilities for different phases and describes the necessary or required expertise or experience/competence for each life cycle phase.

It also identifies the requirements for the necessary documentation and provides assistance in dealing with FSM (Functional Safety Management), verification and validation. The scope of IEC 61508 is the consideration of electrical/electronic/programmable electronic (E/E/PE) systems that are used to perform safety functions.

The standard specifically states (Part 1.1.b)[17] that this standard is generally valid and applicable to all safety-related E/E/PE systems, regardless of the application.

It is divided into seven parts, which can be subdivided as follows:

Part 1:[18] This part carries the title „ general requirements" and describes the fundamental requirements for the development of safety applications (concept, application area, definition, hazard and risk analysis).

Part 2:[19] This part has the title „requirements for electrical/electronics/programmable electronics safety related systems" and is applicable to each safety related system/subsystem and their units, which contains at least one electrical, electronic or programmable electronic unit. This part specifies how the safety requirements and their allocation to the safety related E/E/PE systems are refined and converted into functional requirements.

Part 3:[20] This part has the heading „software requirements" and applies to the development of the software, which is part of a safety-related system. The development of the software results thereby in defined sections. Each section of the software safety life cycle must be divided into elementary activities, whereby range of application, inputs and outputs have to be specified.

Part 4:[21] This part has the title „definitions and abbreviations ". The different terms and abbreviations are used in the parts of 1 to 7 of the IEC 61508.

Part 5:[22] This part has the heading „Examples of methods for the determination of safety integrity level ". It contains information about different risk concepts in dependence of the required safety function and beyond that this part of the standard contains statements about the connection between risk and safety integrity. A further emphasis of this part is the indication of different methods to the determination of the safety integrity level for safety referred E/E/PE systems, safety referred systems of other technology and external mechanisms for minimizing the risk.

---

[17] [IECa]   IEC 61508, International Standard 61508 *Functional Safety: Safety-Related System*
[18] [IECa99], [EN-a01]
[19] [IECb01], [EN-b01]
[20] [IECc99], [EN-c01]
[21] [IECd99], [EN-d01]
[22] [IECe99]   IEC 61508-5:1998 + Corrigendum 1999, International Standard 61508, Functional safety of electrical/electronic/programmable electronic safety-related systems, Part 5

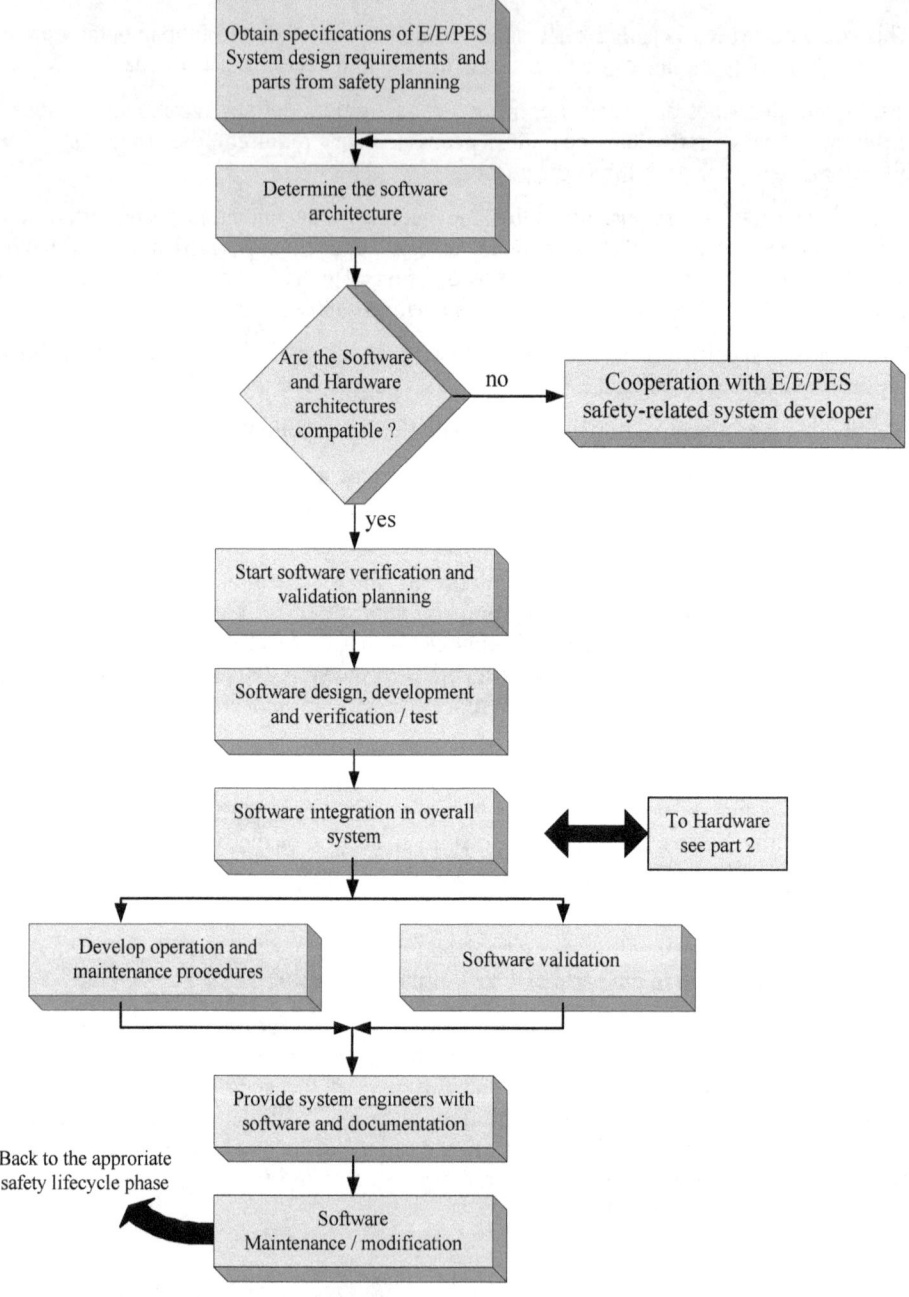

Figure 3.4: Structural software production for safe systems, after IEC 61508[23]

---

[23] [IECa]   IEC 61508, International Standard 61508 *Functional Safety: Safety-Related System*

3.2 Standards

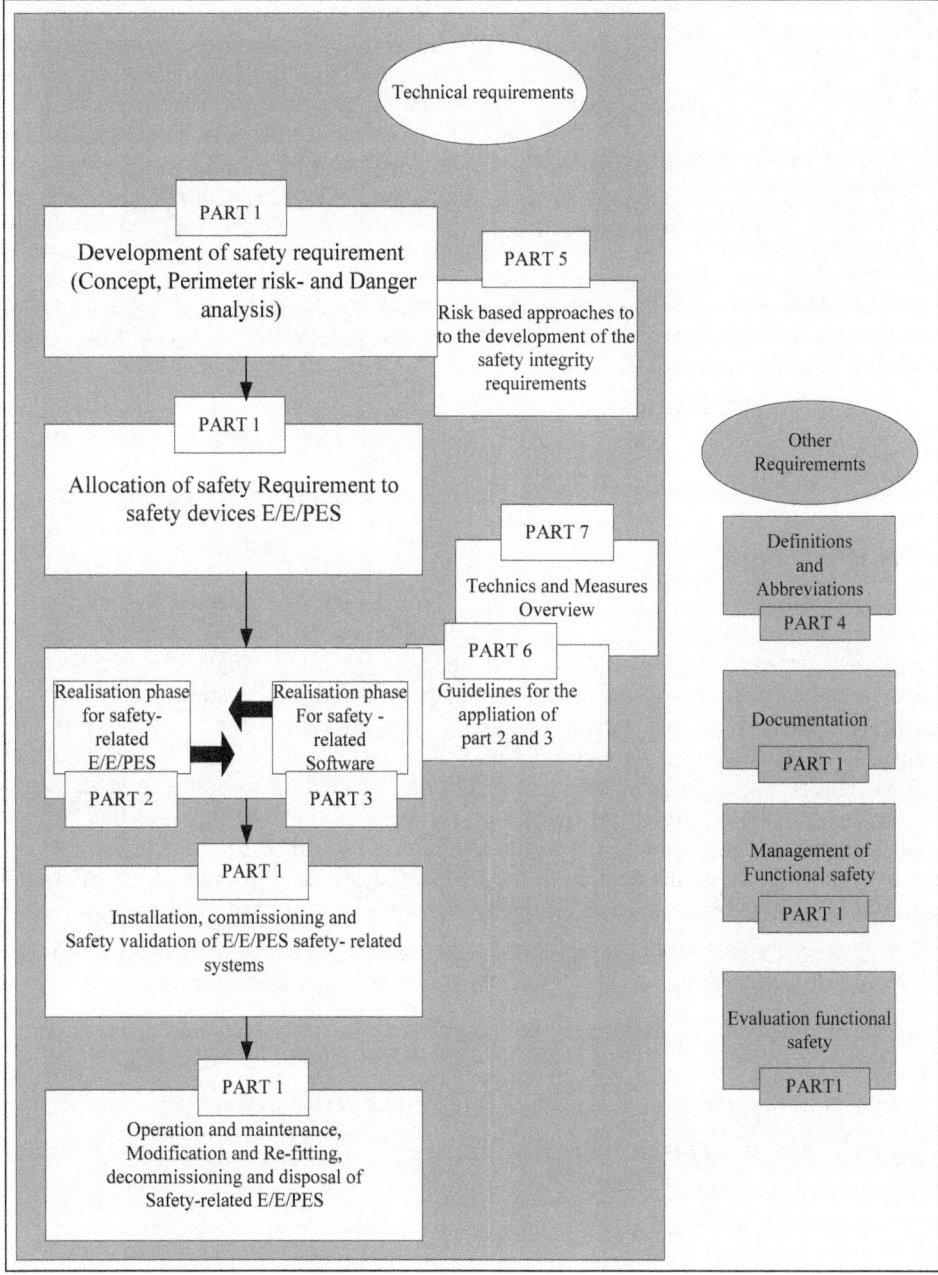

Figure 3.5: Overall structure of the IEC/EN 61508[24]

---

[24] [IECa] IEC 61508, International Standard 61508 *Functional Safety: Safety-Related System*

Part 6:[25] This part of the standard carries the title „Guidelines on the application IEC 61508-2 and IEC 61508-3". In appendix A an overview of the requirements specified in part 2 and 3 is given as well as the functional steps of their implementation. In appendix B procedures are presented for the computation of the probabilities of hardware losses. Appendix C contains an example for the calculation of the diagnostic covering factor. In appendix D a method is presented for the determination of the probability of failure, under the condition that failures with a common cause exist. In appendix E different examples are presented for the application of the software tables according to the safety integrity of level.

Part 7:[26] This part of the standard has the heading „Overview of techniques and measures" and offers an overview of the different safety procedures and -measures for the use of part 2 and 3 of the IEC 61508.

The parts 1, 2, 3 and 4 are also used as basic safety standard and are consulted by technical committees when creating standards according to IEC Guide 104 and ISO/IEC Guide 51.

3.2.4 IEC 61511

The title of this standard reads: „Functional safety: Safety Instrumented System for the process industry sector"[27] – in German „Functional safety: Safety PLT safety devices for the process industry"[28] and is available with the frame IEC 61511:2019. This international standard applies to safety-related systems in the process industry. It applies IEC 61508 to the process industry and demands the execution of a hazard and risk analysis. From this analysis a specification for safety-relevant systems can be provided. The safety-related system consists of those components and subsystems, starting from sensor over Logic Solver up to actuator, which are necessary for the execution of the specified safety-relevant function. This standard fits within the framework of IEC 61508 applicable to the process industry. The terms defined in the IEC 61508 „safety life cycle" and „safety integrity level" (SIL) forms the basis for the application of this international standard.

The IEC 61511:2019 exists under the title „Functional safety: Safety-relevant systems for the process industry" and consists of the following parts (see Figure 3.8):

- Part 1: Framework, definitions, system, hardware and software requirements (normative);
- Part 2: Guidelines for the application of IEC 61511-1 (informative);
- Part 3: Guidance for the determination of the required safety integrity levels (informative).

---

[25] [IECf00] IEC 61508-6:2000, International Standard 61508, Functional safety of electrical/electronic/programmable electronic safety-related systems, Part 6
[26] [IECg00] IEC 61508-7:2000, International Standard 61508, Functional safety of electrical/electronic/programmable electronic safety-related systems, Part 7
[27] [IEC-02] IEC 65A/324/FDIS 2002, Functional safety: Safety Instrumented Systems for the process Industry sector
[28] [IEC-02] DIN IEC 61511, Parts 1 to 3, (VDE 0810 Part 1), Functional safety: Safety Instrumented Systems for the process Industry sector

3.2 Standards

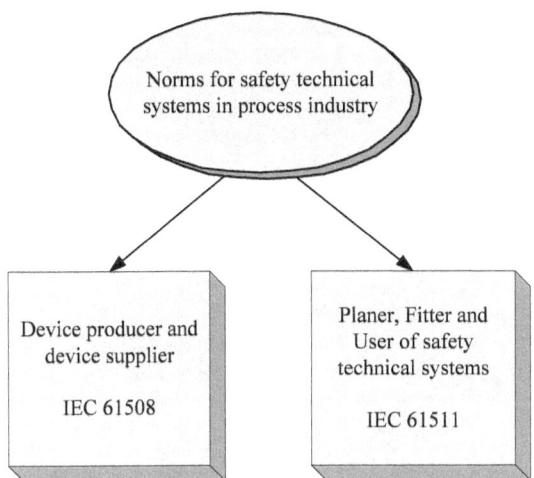

Figure 3.6: Relation between IEC 61508 and IEC 61511[29]

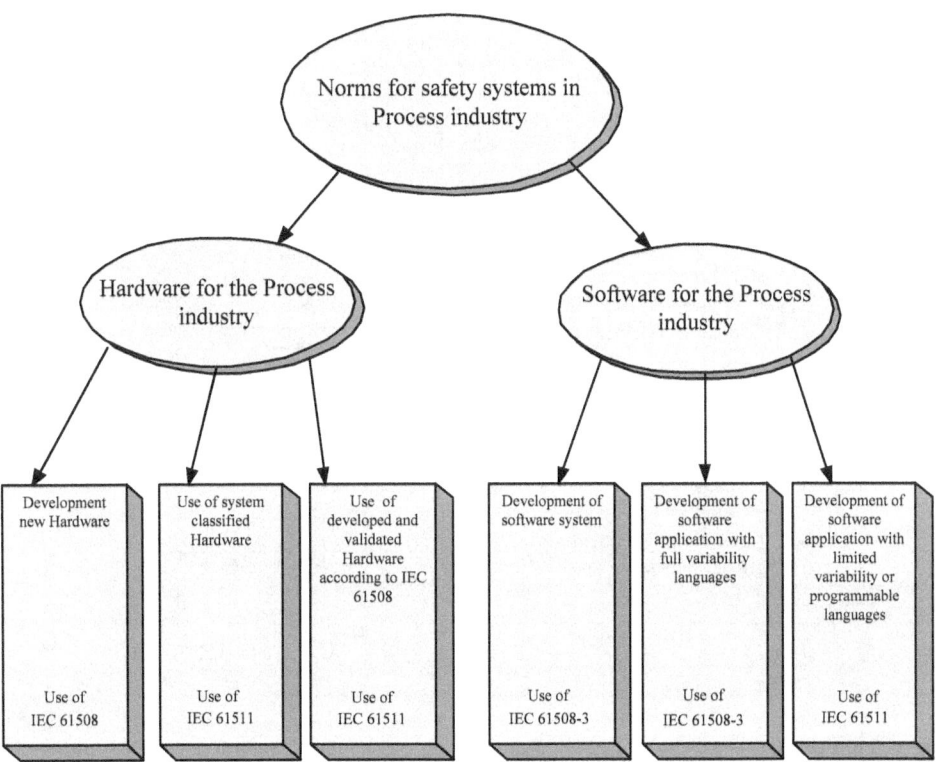

Figure 3.7: Relation between IEC 61508 and IEC 61511 applicable in the process industry [30]

---

[29] [DINa03] DIN IEC 61511, Teil 1 bis 3 (VDE 0810 Teil 1), Funktionale Sicherheit: Sicherheitstechnische Systeme für die Prozessindustrie

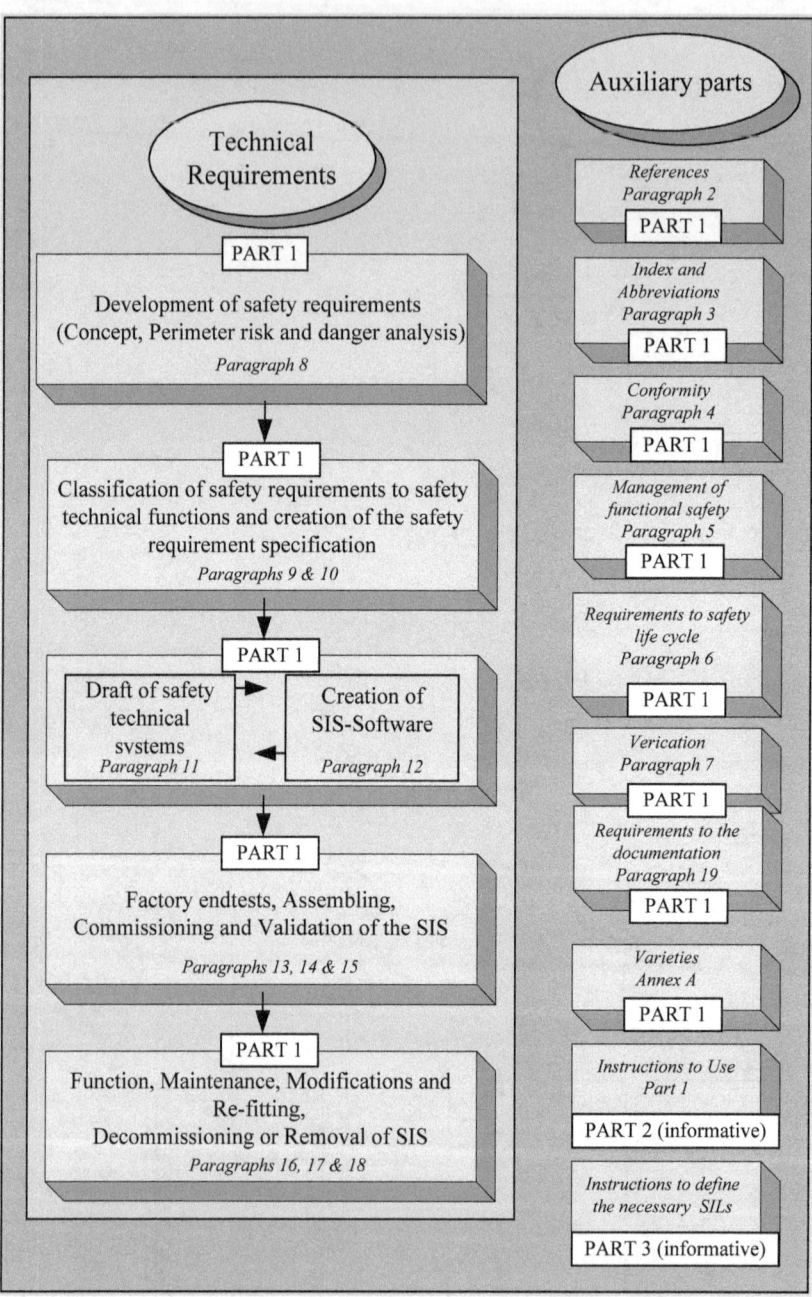

Figure 3.8: The three parts of the IEC 61511[31]

---

[30] a. a. O.
[31] [DINa03] DIN IEC 61511, Teil 1 bis 3, (VDE 0810 Teil 1), Funktionale Sicherheit: Sicherheitstechnische Systeme für die Prozessindustrie

3.2.5 EN ISO 13849

The European Union and national regulations and laws state that machines must fulfil specified requirements of the health and safety guidelines (EHSR). Until the 2000s, the basis for functional safety in the area of "Safety of machinery" was EN 954-1, but this standard did not cover the use of electronic, programmable systems for safety functions.

This standard (EN 954-1) was replaced in 2006 by the standards EN ISO 13849 (Safety of machinery – Safety-related parts of control systems) and EN IEC 62061 (Safety of machinery – Functional safety of safety-related electrical, electronic and programmable electronic control systems).

The safety standard DIN EN ISO 13849-1 contains recommendations for system design, integration and validation and combines the complex probabilistic approach of IEC 61508 with the deterministic concept presented in EN 954 on the basis of risk analyses.

The DIN EN ISO 13849 standard consists of two parts.

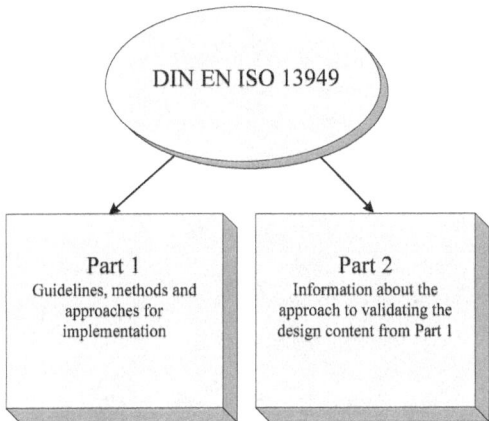

Figure 3.9: DIN EN ISO 13849

Part 1[32] contains general information about the guidelines, as well as methods and approaches for implementation. The second part contains detailed information on the approach to validation of the design content from Part 1. The standard can be applied to the safety-related parts of machine control systems. It also contains special requirements for safety-related parts with programmable electronic systems.

Part 1 of DIN EN ISO 13849 is divided into 11 normative chapters 1 to 11 and 9 informative chapters.

Safety integrity levels (SIL) are specified in IEC 61508 and EN IEC 62061 as a classification scheme for functional safety. These safety levels are a measure of the safety-related reliability of the system under consideration. These are failure limit values (PFD, PFH), each of which covers a specified time period (T1, proof test interval).

---

[32] Source: DIN EN ISO 13849:2016

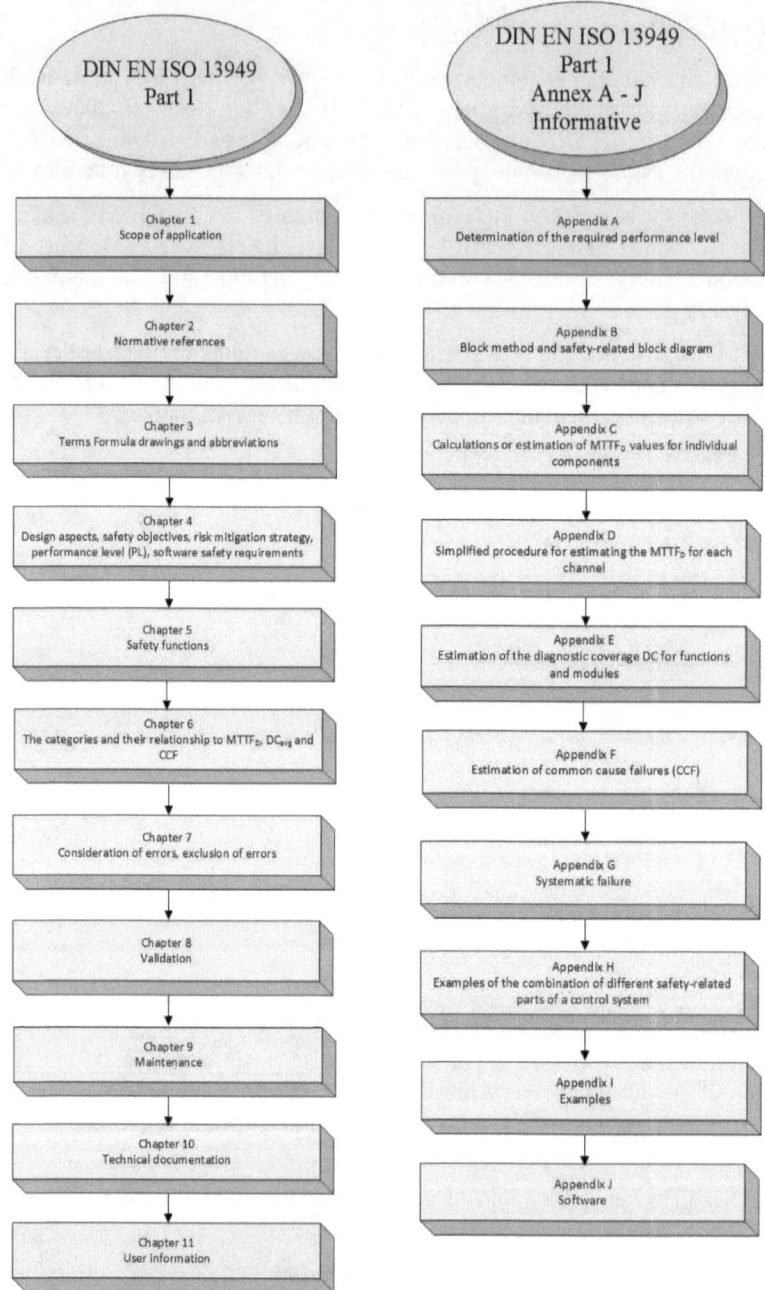

Figure 3.10: Part 1 of DIN EN ISO 13849[33]

The safety standard DIN EN ISO 13849 considers so-called performance levels (PL). Five different performance levels (PL a to PL e) are defined in DIN EN ISO 13849. Dur-

---

[33] Source: DIN EN ISO 13849:2016

ing processing, the required PL for each safety function is first determined by the given requirements and risk considerations. This risk analysis is generally determined using a risk graph. The severity of the injury S (severity), the frequency or duration of exposure to the hazard F (frequency), and the possibility of avoiding the hazard P (probability), are used to determine the performance level (PL).

When analysing a safety-related system, factors such as the MTTFD (Mean Time To Dangerous Failure), the DC (Diagnostic Coverage), the CCF (Common Cause Failure) and the structural properties (divided into categories) are also included and evaluated.

3.2.6 IEC 61131

This international standard carries the title „Programmable controllers"[34] and describes completely the necessary requirements for programmable controllers (PLC) and the applicable peripherals. It is arranged into seven parts.

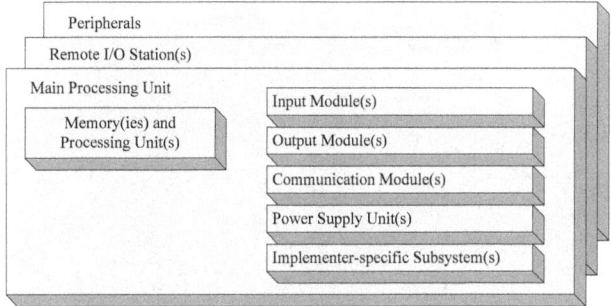

Figure 3.11: PLC-hardware model[35]

Part 1: General information

Part 2: Equipment requirement and test (requirements for PLC and the tests necessary for it). This part of the standard describes the requirements for PLC's and the necessary tests also in relation with the peripherals in the form of „Programming and debugging tools (PADTs) and man-machine interface.

Part 3: Programming languages: This part defines the minimum requirements concerning the fundamental programming elements, the syntactic and semantic rules, which are valid for the most frequently used programming languages, including the graphic languages with „Ladder Logic "and functional blocks.

Part 4: User guidelines: This part gives a general overview of the standard as well as application guidelines for the final user of PLCs.

Part 5: Communication (specifications for the data communication): This part defines data communication between the PLCs and other electronic systems applying Manufacturing Message Specification (MMS) in accordance with the international standard ISO/IEC 9506.

---

[34] [IEC-03] IEC 61131, Programmable Controllers
[35] [IEC-03] IEC 61131, *Programmable Controllers,* part 5

Part 6: Functional Safety: This part specifies the requirements for PLCs and the associated peripheral devices. A programmable controller and peripheral devices that fulfil these requirements are considered suitable for use in a safety system.

Part 7: Fuzzy control programming: This part defines fundamental programming elements of the Fuzzy logic for controllers, how they are used for example in PLCs.

Part 8: Guidelines for the application and implementation of programming languages: This part gives guidance to the hand of the software developer, in order to use the programming languages specified in part 3.

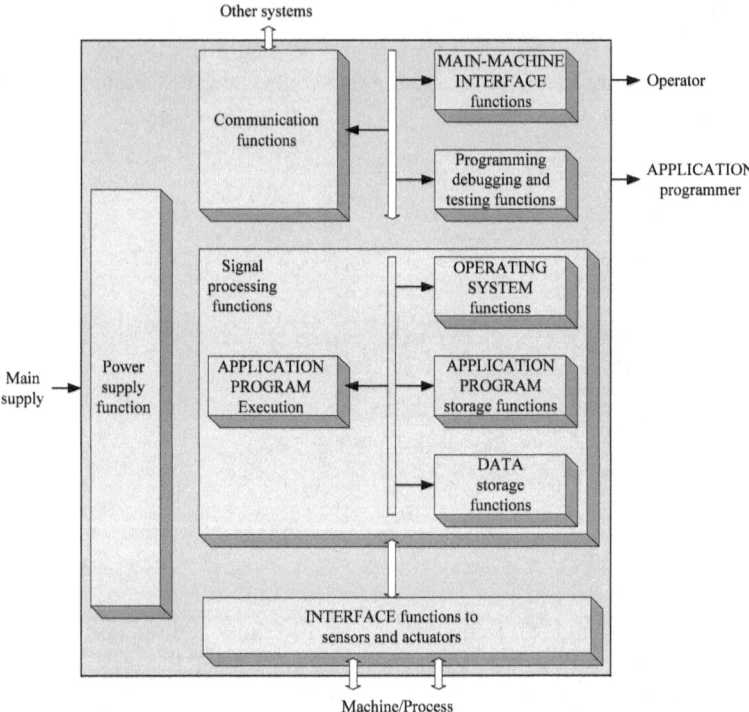

Figure 3.12: Basic Functional Structure of a PLC System[36]

3.2.7 ISA TR 84.02

The standard ISA TR84.0.02[37] shows how to model a complete SIS, which contains sensors, logic solver and the actuating elements. To a certain extent the system analysis techniques permit that these elements can be analyzed independently. This permits a pre-selection to the developer of the safety system with the installation configuration, in order to obtain the necessary safety integrity level. The ISA TR84.0.02 is divided into five parts, as is represented in Figure 3.13.

---

[36] [PLC-03] www.plcopen.org
[37] [ISA-02] ISA-TR84.00.02-2002, Parts 1-5. Safety Instrumented Functions (SIF) – Safety Integrity Level (SIL) Evaluation Techniques

3.2 Standards

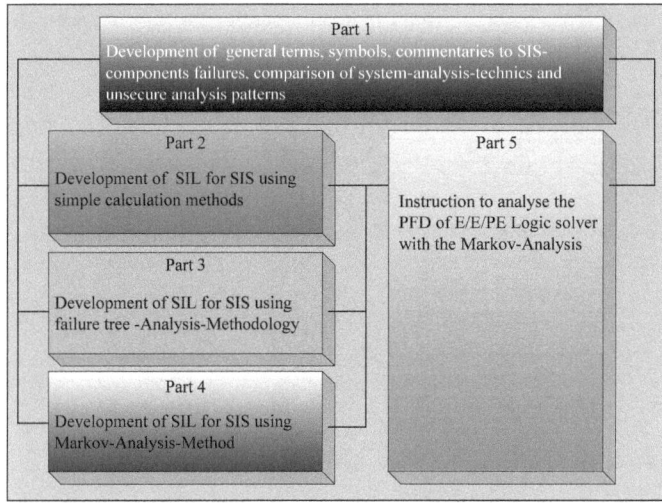

Figure 3.13: The five parts of the standard TR84.0.02[38]

Part 1 contains a detailed listing of all definitions and their parts, which were used in this document. These are alike with the ANSI/ISA S84.01-1996[39] and the IEC 61508[40] standard. This part contains likewise background information, how the elements or components are to be modelled, as well as a closer view of the hardware components.

Part 2 offers equations for the computation of the SIL values for the requirement to safety-relevant systems (SIS) according to the ANSI/ISA S84.01-1996 „application type of safety-relevant systems in the process industry".

Part 3 represents a fault tree analysis for the computation of the SIL values for SIS according to the ANSI/ISA S84.01-1996.

In part 4 Markov techniques are presented for SIL value calculation.

Part 5 computes the PFD of E/E/PE Logic solver only by using the Markov model.

3.2.8 RTCA DO 178

This standard, which is now available in the form RTCA DO178C, is entitled „software considerations in airborne systems and equipment Certification "[41]. The guidelines for the development of aviation software in the USA are defined by the standard DO-178B[42]. The standard, developed from the radios Technical Commission for Aeronautics (RTCA), covers guidelines for the production of software for the employment on board of aircrafts. DO-178C serves as method for the release of components for numerous critical environ-

---

[38] a. a. O.
[39] [ISA-96] ISA-84.01-1996. Application of Safety Instrumented Systems for the Process Industries
[40] [IECa00] IEC 61508, International Standard 61508: Functional safety of electrical/electronic/programmable electronic safety-related systems
[41] [RTCA92] RTCA/EUROCAE 1992; Software Considerations in Airborn Systems and Equipment Certification
[42] [DO-92] DO178B, RTCA, Inc. 1992, Fehlerliste für elektrische Bauelemente – Bei der Prüfung unterstellte Fehlerarten, BIA-Handbuch 7

ments in air, space travel and defense area, military and nuclear technology as well as medicine and communication.

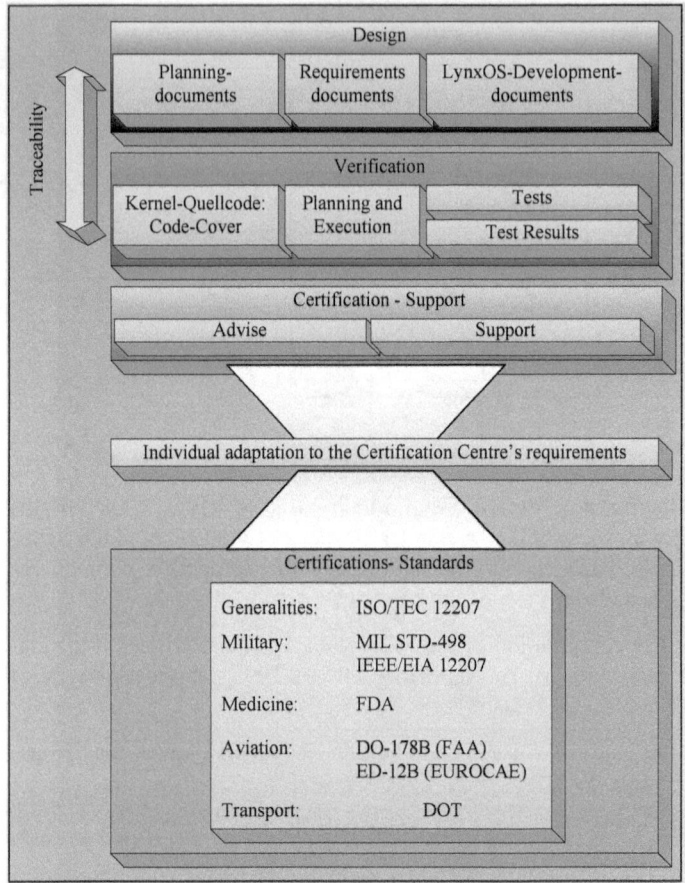

Figure 3.14: Fields, which cover the different standards, here in particular the DO-178B

The standard DO-178C mainly refers to processes in connection with system and software life cycles, as represented in Figure 3.14. The desired certification level is decisive for which requirements must be fulfilled, and decides in some cases even how accurately the requirements are to be kept.

The DO-178C-Certification levels are divided into A, B, C, D or E, whereby the organization of the level takes place according to software failure consequences, as represented in Table 3.3. The system certifying takes place according to a certain certification level. Seized are here the hardware, the application and the system software, which implement a certain function as a unit. The availability of a well-known certifiable operating system can drastically reduce the cost and time for a system certification.

Table 3.3: DO-178B Certification Levels[43]

Level	Consequences of a loss
A	**Catastrophic**
	A loss prevents the safe continuation of the flight and/or the safe landing. Loss means the loss of human lives.
B	**Dangerous/serious**
	A loss limits the ability of the airplane or the crew to deal with adverse operating conditions. Possible consequences are a substantial decrease of the safety tolerances or the operability, a physical impairment or increased working load of the crew, which possibly prevents them from the reliable or complete completion of their tasks or has negative effects on the passengers (with serious or potentially fatal injuries for a small number of these passengers).
C	**Substantial**
	A loss limits the ability of the airplane or the crew to become finished with the adverse operating conditions. Possible consequences are a substantial decrease of the safety tolerances or the operability, a substantially increased working load for the crew or conditions, which impair the efficiency of the crew as well as adversities for the passengers (perhaps also injuries).
D	**Slightly**
	A loss on this level is not crucial for the safety of the aircraft. The necessary remedies can be mastered by the crew without a problem. Slight losses are for example a low decrease of the safety tolerances or the operability, a low increase of the working load for the crew (e.g. Change of the routine flight plan) as well as certain inconveniences for the passengers.
E	**No effect**
	A loss has no effects to the operating ability of the aircraft and doesn't increase the working load of the crew.

3.3 Definitions around the term of safety

Next an overview of different terms and their definitions in the context of safety are given. It becomes evident that there exists no uniform standard for terms and definitions, but within different areas definitions and terms are used.

According DIN EN 61508-4[44]

Harm: Physical injury or damage of the health of humans, either directly or indirectly as a result of damage of goods or the environment [ISO/IEC Guide 51:1990 (modified)].

Hazard: Potential source of damage [ISO/IEC Guide 51:1990].

NOTE: The term includes the endangerments of persons, those within a short time interval develops (for example by fires and explosion) and also those, which has a long-term effect on the health of a person (for example by release of a poisonous substance).

Hazardous situation: Circumstance, by which a person is exposed to an endangerment.

Hazardous event: Endangerment situation, which leads to a damage.

---

[43] [DO-92] DO178B, RTCA, Inc. 1992, Fehlerliste für elektrische Bauelemente – Bei der Prüfung unterstellte Fehlerarten, BIA-Handbuch 7
[44] [IECd99] IEC 61508-4:1998 + Corrigendum 1999, International Standard: 61508 Functional safety of electrical/electronic/programmable electronic safety-related systems, Part 4: Definitions and abbreviations

Risk: Combination of the probability, with which a damage arises, and which extent of this damage [ISO/IEC Guide 51:1990 (modified)].

NOTE: For the further discussion of this term see appendix A of IEC 61508-5[45].

Tolerable risk: Risk, which is portable, based on the current social value conceptions in a given connection.

NOTE: See attachment B of the IEC 61508-5[46].

Residual risk: The risk remaining despite preventive measures.

Safety: Freedom of unacceptable risks.

Functional safety: Part of total safety, related to the EUC and the EUC leading or control system, which depends on the correct function of the E/E/PE safety related systems, systems of other technology and external mechanisms for risk reduction.

Safe state: State of the EUC, in which safety is reached.

NOTE: During the transition from a potentially dangerous condition to the final safe condition the EUC can go through a number of safety intermediate conditions. For some situations a safe condition exists only so long, as the EUC is subject to a continuous control. Such a continuous control can take place for a short or an indefinite period.

Reasonably foreseeable misuse: Use of a product, procedure or a service under conditions or for purposes, which are not intended by a supplier, but those, can happen caused by the product, the procedures or the service in connection with or as a result of usual human behaviours.

According to DIN EN 14971[47]

Damage: Physical injury or damage of the health of humans or damage of goods or the environment [ISO/IEC Guide 51:1999, Definition 3.3[48]].

Hazard: Potential source of damage [ISO/IEC Guide 51:1999, Definition 3.5].

Hazardous situation: Condition, in which humans, goods or the environment are exposed to one or more endangerments [ISO/IEC Guide 51:1999, Definition 3.6].

Intended use / purpose: Application of a product, a procedure or an achievement according to the specifications, instructions and data supplied by the manufacturer.

Proof: Information, whose correctness can be proven, and which is based on the fact, which by observation, measurement, investigation or are won by other preliminary investigations [ISO 8402:1994, Definition 2.19[49]].

Procedure: Fixed way to implement an activity [ISO 8402:1994, Definition 1.3].

---

[45] [IECe99] IEC 61508-5:1998 + Corrigendum 1999, International Standard: 61508 Functional safety of electrical/electronic/programmable electronic safety-related systems, Part 5: Examples of methods for the determination of safety integrity level

[46] a. a. O.

[47] [DINa00] DIN EN ISO 14971, Medicine products – application of the risk management to medicine of products. [DINb03]

[48] [ISOa99] ISO/IEC Guide 51:1999, Safety aspects – Guidelines for their inclusion in standards

[49] [DINd94] DIN EN ISO 8402:1994, Quality management and quality assurance of terms

Process: Sentence of in interrelations standing means and activities, which transform inputs into results [ISO 8402:1994, Definition 1.2].

Residual risk: Risk, which remains after the application of safety measures [ISO/IEC Guide 51:1999, Definition 3.9].

Risk: Combination of the probability of the occurrence of a DAMAGE and the SEVERITY LEVEL of this DAMAGE [ISO/IEC Guide 51:1999, Definition 3.2].

Risk analysis: Systematic evaluation of available information, in order to identify endangerments and measure RISKS [ISO/IEC Guide 51:1999, Definition 3.10].

Risk evaluation: Whole of the procedure, which covers RISK ANALYSIS and RISK EVALUATION [ISO/IEC Guide 51:1999, Definition 3.12].

Risk control: Process, by which decisions are caused and preventive measures are implemented, in order to reduce risks or hold them within fixed borders.

Risk evaluation: Evaluation on the basis of a RISK ANALYSIS whether on the basis of the values recognized by the society a justifiable RISK in a given connection was reached.

Note: on the basis of the ISO/IEC Guide 51:1999, Definition 3.11 and 3.7.

Risk management: Systematic use of management principles, PROCEDURES and practices for the analysis, evaluation and control of RISKS.

Risk management document: Composition of RECORDINGS and other documents, which are produced by the risk management process and not necessarily have to be in a place.

Safety: Freedom of untenable RISKS [ISO/IEC Guide 51:1999, Definition 3.1].

Severity level: Measure of the possible consequences of a HAZARD.

Verification: Confirmation due to an investigation and by supply of a PROOF that fixed demands were fulfilled.

NOTE: In the Design the VERIFICATION refers to the PROCESS of the investigation of the result of a given activity, by which the conformity with the fixed requirement for this activity is determined. (ISO 8402:1994, Definition 2.17).

According to DIN 40041[50]

Dependability: Condition of a unit concerning their suitability to fulfil the reliability requirements during or after given time intervals with given application conditions.

Reliability: Suitability of a unit to fulfil a demanded function on given application conditions.

Nonconformity: Non compliance with a requirement

Failure: Completion of the operability of a material unit in the context of the certified demand.

---

[50] [DINb90] DIN 40041:1990-12, *Zuverlässigkeit, Begriffe*

According to DIN 9000[51]

Dependability: Recapitulatory expression for the description of the availability and its factors of influence operability, maintainability and maintenance readiness.

NOTE: Reliability is used only for general descriptions in not-quantitative sense. [IEC 60050-191:1990[52]].

Verification: Confirmation by supply of an objective proof that fixed requirements were fulfilled.

NOTE: The designation „verified" is used for the designation of the appropriate status.

Validation: Confirmation by supply of an objective proof that the requirements for a specific intended use or a specific intended application were fulfilled.

NOTE 1: The designation „validated" is used for the designation of the appropriate status.

NOTE 2: The application conditions for validating can be genuine or simulated.

According to VDE 31000[53]

Damage: Damage is a disadvantage by injury of right goods due to a certain technical procedure or condition.

Risk: The risk, which is connected with a certain technical procedure or condition, is described in summary by a probability statement, which considers the frequency of the entrance of an event leading to the damage and the extent of the damage which can be expected with the event entrance.

Threshold risk: Threshold risk is the largest still justifiable risk of a certain technical procedure or condition. Generally the threshold risk cannot be seized quantitatively.

Danger: Danger is a state of affairs, with which the risk is larger than the threshold risk.

Safety: Safety is a state of affairs, with which the risk is not larger than the threshold risk.

Protection: Protection is the decrease of the risk by measures, which limit either the entrance frequency or the extent of the damage or two.

3.4 State of the Art

3.4.1 Automobile area – ISO 26262

Compared to other areas the automobile industry produces relatively large quantities of systems. In return this means however that changes of characteristics of a special system are sooner profitable as with small numbers of systems. The result is a large variety of safety-related electronic systems. The principle of the V-model became generally accepted within the automobile industry for the development of safety-related electronic systems and has manifested itself in the ISO 26262 standard since 2011. Large parts of the

---

[51] [EN-90] EN ISO 9000, Quality management systems
[52] [IECa90] IEC 60050-191 Ed. 1.0 b: 1990, International Electrotechnical Vocabulary. Chapter 191: Dependability and quality of service
[53] [DINa87] DIN VDE 31000 Teil 2:1987-12, Allgemeine Leitsätze für das sicherheitsgerechte Gestalten technischer Erzeugnisse – Begriffe der Sicherheitstechnik – Grundbegriffe

development are supported by appropriately qualified tools and development tools (tool qualification). Often also the „rapid prototyping" is used. Another aspect is the requirement for the composition of electronic systems, which play an ever increasing role for the co-operation between vehicle manufacturers and suppliers.

A major focus during the development of vehicles and vehicle systems is on the system test. Over several months first the subsystems and afterwards the overall system are intensively tested. The ISO 26262 standard also applies here. Figure 3.15 shows as an example a development process which is related to the V-model. In addition to test cases also rapid prototyping can be recognized. This basic model can also be found in the ISO 26262 standard (in 4: Product development at the System Level; in 5: Product development at the Hardware Level; and in 6: Product development at the Software Level).

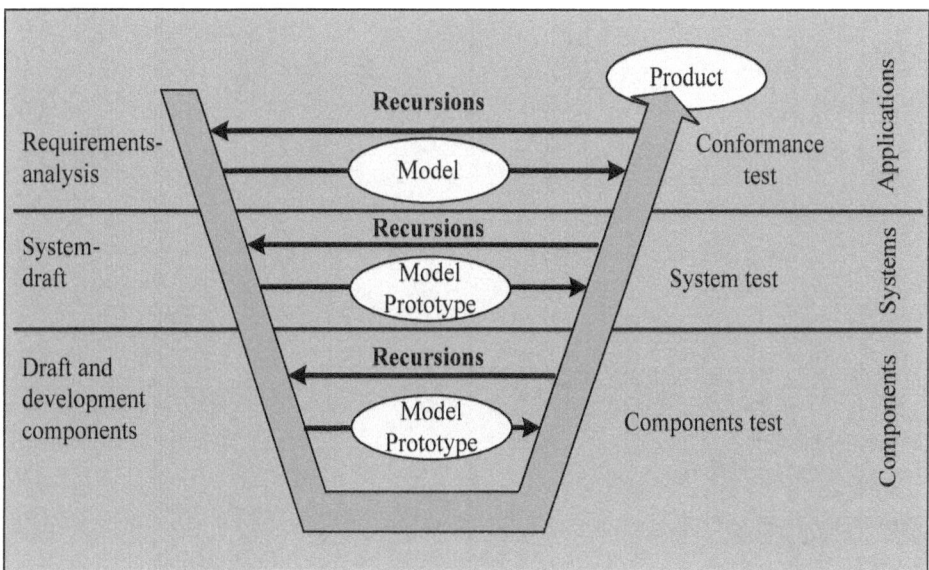

Figure 3.15: The V-model

Due to the increasingly innovative use of electrical and electronic components in the automotive sector, which also control and monitor safety-relevant systems, the ISO 26262:2011[54] standard ("Road vehicles – Functional safety") was introduced in 2011.

The standard has since been published in a second version ISO 26262:2018[55] and has been extended in this current version by two additional parts: Part 11 (Guidelines on application of ISO 26262 to semiconductor) and Part 12 (Adaption of ISO 26262) for motorcycles).

The ISO 26262 standard is a standard for safety-related electrical/electronic systems within road vehicles. Application of the standard is voluntary, but in practice more and more car manufacturers are demanding that their suppliers apply it to new projects.

---

[54] International Standard ISO 26262:2011
[55] International Standard ISO 26262:2018

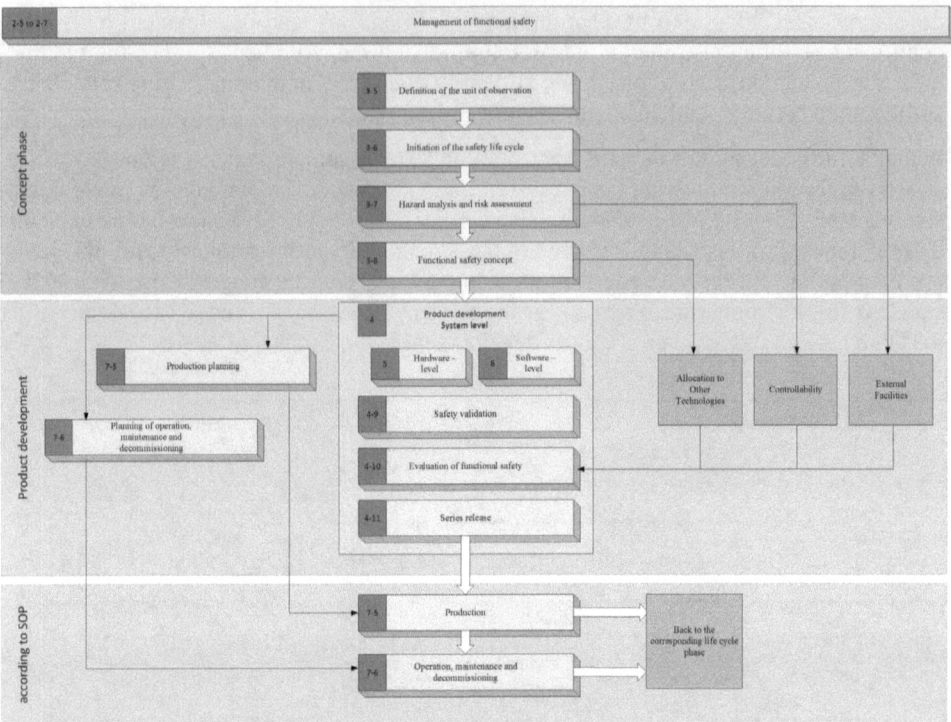

Figure 3.16: ISO 26262 Automotive Safety Lifecycle[56]

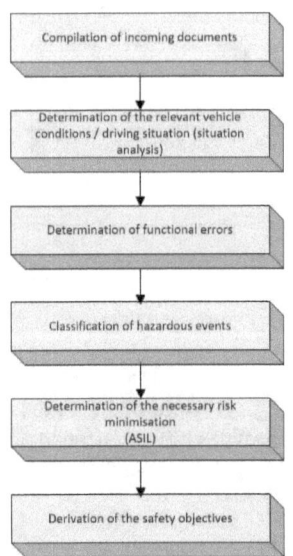

Figure 3.17: ISO 26262 Hazard analysis and risk assessment (G+R) according to ISO 26262[57]

---

[56] Source: International Standard ISO 26262:2018
[57] Source: International Standard ISO 26262:2018

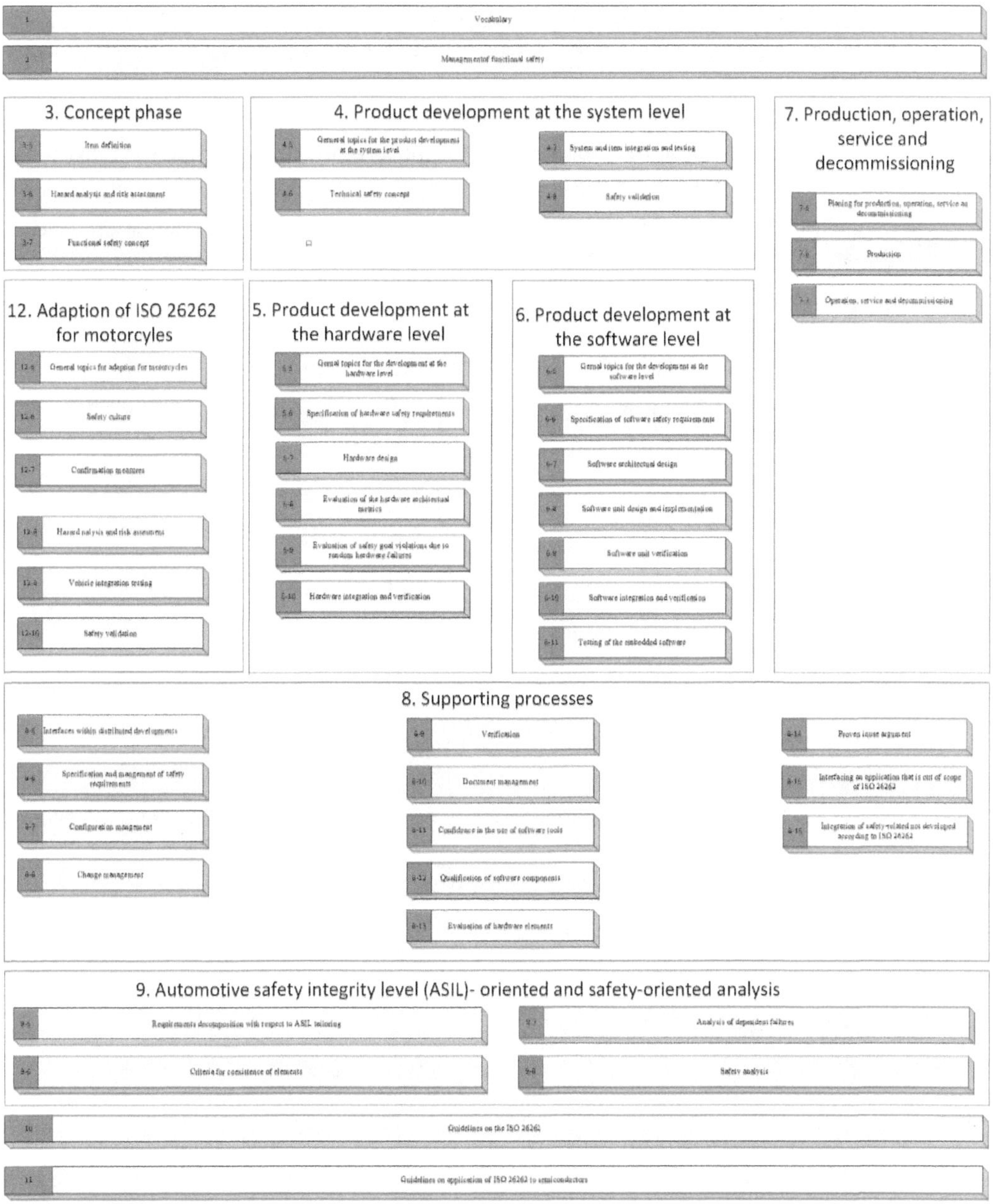

3.4 State of the Art

Figure 3.18: Overview of ISO 26262[58]

As with IEC 61508 as a generic safety standard, ISO 26262 also describes a safety life cycle. The ISO 26262 standard specifies a safety life cycle. It describes methods, activities and work products (documentation, project planning, etc.) that must be carried out during the development, production, operation and maintenance of safety-related systems

---
[58] Source: International Standard ISO 26262:2018

in motor vehicles. In this context, functional safety means the prevention of damage to persons or the system environment caused by the failure or incorrect behaviour of a system.

According to ISO 26262, a hazard analysis and risk assessment must therefore be carried out. For this purpose, ISO 26262 defines a qualitative procedure for analysis that is tailored to the automotive sector.

As with IEC 61508, the effort and requirements of ISO 26262 are based on the hazards and risks directly associated with the safety-related system under consideration.

The ISO 26262 standard only covers the safety of electrical, electronic and programmable components in vehicles and does not provide methods for the safety considerations of mechanical, hydraulic or pneumatic components.

ISO 26262 is divided into twelve parts. Parts 3 to 7 relate to the individual phases of the life cycle, the remaining parts are phase-independent. The ISO 26262 parts cover the following content:

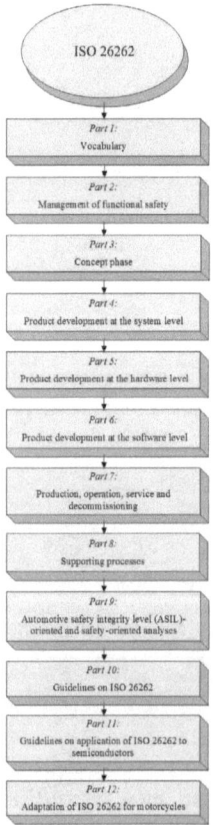

Figure 3.19: Overview of ISO 26262[59]

---

[59] Source: International Standard ISO 26262:2018

3.4 State of the Art

The core processes are represented by parts 3 (concept phase), 4 (product development system level), 5 (product development hardware) and 6 (product development software). The other parts can be regarded as support processes and notes on implementation.

In IEC 61508, the systems under consideration are categorised by assigning them to a safety integrity level (SIL) according to the hazards and risks involved.

ISO 26262 adapts this procedure and the classification for specific use in the automotive sector. Here, the systems are assigned to the ASIL (Automotive Safety Integrity Levels). These result from the severity of the possible damage S (Severity), the frequency of occurrence of the hazardous situation E (Probability of exposure) and the controllability by the driver C (Controllability).

ASILs, like the SIL levels in IEC 61508, are classes for specifying the necessary safety requirements of the system. These classes are associated with measures and techniques for minimising risk. The highest class defined in ISO 26262 is ASIL D and ASIL A is the lowest. It is therefore easy to recognise that the ASIL D class is the most demanding and requires the most effective measures.

3.4.2 Aviation

Besides many other, specially militarily driven standards, the standards SAE ARP 4754[60] and SAE ARP 4761[61] are wide spread within the development of civilian and military electronic systems in the aviation. In these standards a process is described, which puts a special focus, apart from the functional development of the system, on the view of safety and reliability of the system which can be developed. Substantial elements are a hazard analysis and the guarantee of the safety requirements. In particular quantitative analysis methods are used. The process is very documentation intensive, like most development processes from the aviation industry. Since the boundary conditions strongly differ in the aviation industry from those in the automobile industry it is not possible to transfer the process concepts directly. The core content is usable, but must be adapted to the boundary conditions.

3.4.3 Automation technology

In the automation industry several standards are established, such as IEC 61508, IEC 61511, ISO 13849 and IEC 62061. These are characterised by the life cycle model and the certification of system components in order to create a certified overall system. A characteristic is here the categorization of the system in so-called „safety integrity level "and based on this the adjustment of the development process. Due to the concrete application on the automation and power plant industry the aforementioned process concepts are however only conditionally suitable for automobile development. Also the concept of certified components is not convertible to the automobile industry.

---

[60] [ARPa96] SAE-ARP4754, Certification Considerations for Highly-Integrated Or Complex Aircraft Systems
[61] [ARPb96] SAE-ARP4761, Guidelines and Methods for Conducting the Safety Assessment Process on Civil Airborne Systems and Equipment

4 Faults, Fault Causes and Failures

All technical systems are due to their structure subject to a certain probability of failure. That is why these systems, and particularly all safety related systems, need to be developed, operated and maintained according to national and international regulations and standards. The IEC/EN 61508 standard is applicable to systems where the safety related functions and applications are carried out with:

- Electrical systems (E)
- Electronic systems (E)
- Programmable electronic systems (PES)

It can be applied in principle to all E/E/PE safety-related systems as well as in particular in those application areas where there are no specific safety standards. This standard considers a systematic and risk-based approach of the safety-relevant problems.

4.1 Failure rates

The failure rate $\lambda(t)$ is one of the most important items when examining safety related systems. In particular the individual components, of which the safety-related computer system is comprised, are examined individually for their failure rates. The failure rate of the given item equals basically the probability, based on δt within the interval $(t, t + \delta t)$, given the condition that it did not fail at time t. The distribution consists of the periods of early failures, the failures which approach constant failure rates, and the wear out failures. Industrially almost all systems, which operate, control or monitor safety related processes, are treated so that early failures do occur which makes it possible to start the actual usage phase with a constant, or in the strict academic sense assuming an approximate constant failure rate:

$$\lambda(t) = \lambda .\qquad\qquad(4.1)$$

The failure rate of a new part can in general only be determined experimentally. For existing parts it is possible to find appropriate values in specific failure rate catalogs.[55] In principle the statement can be made that the failure rate $\lambda(t)$ equals to the negative value of the derivative at the time t differential logarithmic reliability function:

---

[55] [IECc97] IEC 61709, Electronic Components Reliability
    [CNET93]   CNET-Ausfallratenkatalog
    [MIL 217]   MIL-HDBK-217
    [SIEM91] Siemens, SN 29500-2; *Ausfallraten Bauelemente*

$$\lambda(t) = -\frac{d}{dt} \cdot \{\ln R(t)\}$$
$$= -\frac{1}{R(t)} \cdot \frac{d R(t)}{dt}. \qquad (4.2)$$

Equation 4.2 shows that the failure rate λ(t) fully determines the probability function R(t). With $R(0) = 1$ the following equation results:

$$R(t) = e^{-\int_0^t \lambda(\tau)\, dt}. \qquad (4.3)$$

For a constant failure rate (approximately time independent) $\lambda(t) = \lambda$. This results in the equation:

$$R(t) = e^{-\lambda t}. \qquad (4.4)$$

Eq. 4.4 shows that the probability function *R(t)* is a falling exponential function $e^{-\lambda t}$ for a constant failure rate λ, as shown in figure 4.1.

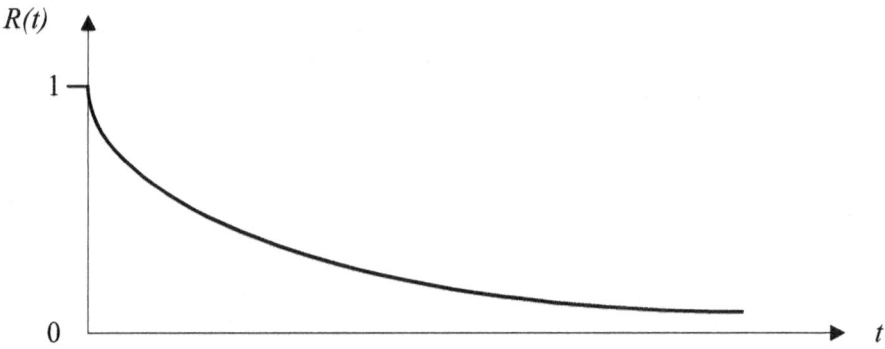

Figure 4.1: Probability function R(t) as function of time

Here, λ is called the average failure rate. It describes the average number of failures that can be expected in a defined time interval:

$$\lambda = \frac{1}{MTBF}. \qquad (4.5)$$

The so called "bathtub curve" applies, as seen in Figure 4.2, if the failure rate cannot be assumed to be constant.

4.1 Failure rates

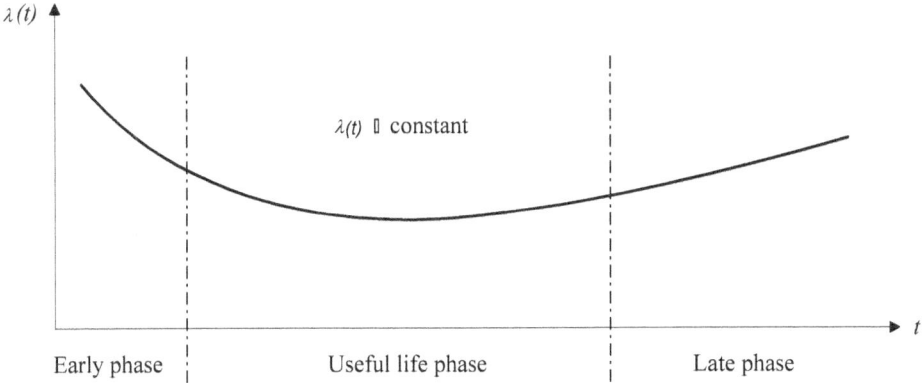

Figure 4.2: „Bath Tub Curve" describing the failure rate as function of time

The Weibull distribution, also known under the name „Bath Tub Curve", is divided into three areas:

- Early phase: The failure rate is high at the beginning due to early failures of parts (e.g., bad parts, soldering failures, etc.).
- Operational use: The failure rate $\lambda(t)$ reaches a minimum during the operational phase (the bottom of the bath tub curve). The failure rate can be assumed to be constant during this phase.
- Wear out phase: The failure rate increases during the wear out phase because the parts reach their end-off life.

As already mentioned the failures of parts can only be determined experimentally. Failure rate catalogs are established for existing parts. The calculations performed in this book are based on failure rates as supplied via the datasheets of the manufacturers or by utilizing the Siemens standard. The data is presented as so called FIT-Values (Failure In Time), where *1 FIT* equals the following unit:

$$\lambda = 1\,Fit = 1 \cdot 10^{-9}\frac{1}{h}. \tag{4.6}$$

Special characteristic values of the failure rates are used to validate safety related applications. Within a failure analysis these are differentiated as follows:

- Safe failure
- Dangerous failure

The safe failures, whether detected or undetected, have no influence on the safety function of the system. Dangerous failures in the safety function lead on the other hand to a dangerous state of the system. Dangerous failures are furthermore divided into:

- Dangerous detected failures
- Dangerous undetected failures

The failure rates and their distribution are shown graphically in Figure 4.3.

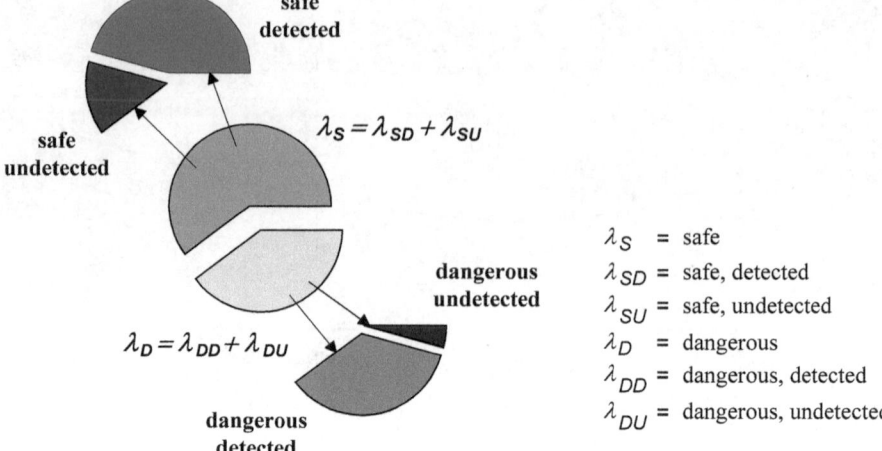

Figure 4.3: Failure rates

It can be stated that in principle there are no systems without a remaining failure rate. The connection between the individual failure rates is given by the following equations.

The overall failure rate, described as λ_B, within a certain time unit is equal to the sum of safe and dangerous failure per time unit (λ_S und λ_D) and is described by Eq. 4.7.

$$\lambda_B = \lambda_S + \lambda_D \tag{4.7}$$

The connection between λ_S and λ_D is described by the safety related factor S, see Eq. 4.8. The S factor gives the ratio between all dangerous failures and the total of all possible failures.

$$S = \frac{\lambda_D}{\lambda_S + \lambda_D} \tag{4.8}$$

The correlation between the safe failure rate λ_S and the base failure rate λ_B as well as between the dangerous failure rate λ_D and the base failure rate λ_B can be expressed by the S factor, see Eq. 4.9 and Eq. 4.10.

$$\lambda_S = \lambda_B \cdot (1-S) \tag{4.9}$$

$$\lambda_D = \lambda_B \cdot S \tag{4.10}$$

The amount of dangerous failures within a certain period of time, described by λ_D, is the sum of all dangerous detected and dangerous undetected failures per time unit (λ_{DD} and λ_{DU}) and is described by Eq. 4.11.

$$\lambda_D = \lambda_{DD} + \lambda_{DU} \tag{4.11}$$

The DC parameter describes the relationship between the failure rate for dangerous detected failures and the failure rate for dangerous failures. This parameter represents the

4.1 Failure rates

ratios between the failure rates which can be detected by diagnostics and the failure rate for dangerous failures.

$$DC = \frac{\lambda_{DD}}{\lambda_D} \tag{4.12}$$

The detected dangerous failures per time unit as well as the undetected dangerous failures per time unit can be expressed by the base failure rate λ_B, the diagnostic parameter DC and the S factor, see Eq. 4.13 and Eq. 4.14.

$$\lambda_{DD} = \lambda_B \cdot S \cdot DC \tag{4.13}$$

$$\lambda_{DU} = \lambda_B \cdot S \cdot (1 - DC) \tag{4.14}$$

The mentioned DC parameter is established via diagnostic measures which are described in Part 2 of the IEC/EN 61508[56].

The safe failure fraction (SFF) is another factor which is often presented in relationship with the failure rates. The SFF parameter determines the portion of safety relevant failures. It is formed out of the ratio of all safe and dangerous detected failures and the sum of all safe and dangerous failures, see Eq. 4.15.

$$SFF = \frac{\sum \lambda_S + \sum \lambda_{DD}}{\sum \lambda_S + \sum \lambda_{DD} + \sum \lambda_{DU}} \tag{4.15}$$

Eq. 4.16 is applied if one desires to determine the diagnostic coverage (DC) for the total system.

$$DC = \frac{\sum \lambda_{DD}}{\sum \lambda_D} \tag{4.16}$$

This diagnostic parameter gives the ratios between all dangerous failures that are detected by diagnostics and the total dangerous failures.

The relationship between the SFF and DC factor for a given S factor of a system is presented in the equations Eq. 4.17 and Eq 4.18.

$$DC = \frac{SFF - 1 + S}{S} \tag{4.17}$$

$$SFF = 1 + S \cdot (DC - 1) \tag{4.18}$$

In order to carry out a meaningful probability of failure analysis on a safety related systems it is necessary to analyze all possible failure modes of the applicable design.

---

[56] [IECb01] IEC 61508-2:2001, International Standard: 61508 Functional safety of electrical/electronic/programmable electronic safety-related systems, Part 2: Requirements for electrical/electronic/programmable electronic safety-related systems

4.2 Fault-Failure-Deviations

The DIN 40041[57] standards has the following definition:

- A fault occurs when the characteristic value does not reach the given demand. A failure is therefore a condition.
- A failure occurs when a system stops to perform a particular process (including the first failure). Therefore a failure is an event.

As a basic rule counts that all faults and failures can be described by basic mechanisms:

- Physical or chemical failure mechanisms and effects
- Human error like logical lack, understanding, interpretation or careless deviations.

The following key factors can be relevant in the case of physical or chemical failure mechanisms:

- Kind of production (e.g., the allowed failure tolerance)
- Environmental conditions (e.g., aggressive environment)
- Load and overstressing, which leads to aging and usage.

A human error during specification, design and documentation of al components of a safety system lead to a failure. A basic understand of all these aspects is normally required to carry out a reliability analysis of the system. The reliability concept is also for developers important, as an increase in reliability (e.g., part reliability) decreases the sensitivity of the object against failures which are caused by faults. As a result the object will have a higher life expectancy. The sensitivity of an object is therefore the result of its draft, production and history.

The following factors are relevant in order to answer the question which stress factors have an effect on the system:

- Chemical corrosion
- Electrical voltage / current supply
- Mechanical vibration / shock
- Temperature
- Humidity

The load factors are considered during the development of a product, and are best neutralized by the selection of different design parameters and safety components. The safety factor is introduced in order to determine the effect of the load factor on the object. It estimates the probability that the failure occurs due to loads. The lower the influence of the load factor on a single component, the higher its safety factor is. The design is thus a key element when determining the sensitivity of a component. Robust objects with high safety factors are less sensitive. The reliability increase by decreasing the failure causes.

---

[57] [DINb90] DIN 40041, Zuverlässigkeit, Begriffe

Considering potential system failures one can conclude that failures can be divided into two categories.

- Physical failures, e.g., component failure
- Functional failures, e.g., when a system works but the planned function is not carried out as a program fails (e.g., due to a erroneous data) and thus the system fails.

All most all failures occur constantly. They affect usually some parts and/or modules that fail due to physical reasons. Two examples are a dried out battery or an open circuit due to corrosion. On the other hand the most functional failures are caused due to design failures. A simple failure occurs, e.g., when the number π is need but cannot be represented numerically. Functional failures can temporary as well as constant.

When stochastical methods are used physical failures are easier to find than functional failures, as it is easier to generate these in larger quantities. Functional failures as well as physical failures can be examined from three different view points, which are all necessary for a probability analysis.

- What is the cause of the failure?
- Which effect does the failure have on the monitoring function?
- What is with the time-dependent behavior? E.g., how many parts have failed?

4.3 Failure sources

Failure causes can be divided into internal and external failure causes. The reasons for internal failures are design failures or product failures. Design failures can occur in any possible development stages. They are classified as human failures and consists of, e.g., understanding problems of the developer during the design of functional units or components where several assumptions have not been considered. A general design failure can occur when different developers work on different sub projects and these sub projects cannot be integrated in the overall system.

Production failures consist of failures that can occur during manufacturing, e.g., bad soldering, wrong component sequence, missing part or a not considered crystal failure.

External failure sources occur due to environmental influences, maintenance failures or operational failures. Causes for external failures, based on environmental influences, could for example occur due to corrosion or diffusion by temperature, humidity, which influences the evaporation rate or mechanical vibrations.

One can conclude from this that all failures due to environmental stress are in practice design failures. One can say that the developer did not consider all possible stressors that can influence the product. The division between design failures and failures is subjective and depends on the weight of the evaluation. When during the development of a product all possible environmental influences were considered then failures due to the environment are considered common cause failures and not environmental failures.

In most cases maintenance failures can be considered human failures. In other words the human factor should be considered an external failure. Several parameters are relevant. To be mentioned are complexity, reparability, the knowledge level of the employee, the familiarity with the system as well as time pressure for creating the product. Estimating maintenance failures is difficult, as long as not enough information is available.

Also operation failures can be considered human failures. Some typical examples are "it can happen", overworked personnel, too difficult instructions, inexperienced personnel or time pressure.

The more clear failure sources of a complex system are defined the better it is to avoid failures and the control them due to periodic maintenance. When the developer knows about the failure source it is possible to consider it during the development of the product. Furthermore is an analysis of the probability of failure useful particularly than when the resulting awareness leads to a better failure model for potential and probable failure sources. The exact determination of the failure rate is possible via exact failure statistics. In general there is a relationship between the failure sources and the time dependency of the failure rate.

4.4 Failure tolerance

Failure tolerance is a special technique that allows a system, despite the existence of failures, to carry out its expected or at least its minimum operation. In order to achieve this a redundancy is applied on architectural level. Redundancy can be divided into four groups:

- Hardware redundancy;
- Software redundancy;
- Time redundancy;
- Information redundancy.

In case of hardware redundancy the system is carried out with more components than required. In case one component fails, a replacement is activated. In case of software redundancy the system can be equipped with different versions of tasks. Different and independent programmers write tasks so that in case one task fails other tasks can take over and the problem can be solved in a safe way. In case of time redundancy new tasks are started in case of a scheduler lockup, in order to keep the time limit. In case of information redundancy data is coded in such way that a determined number of bit failures can be detected and/or corrected.

A system with failure tolerance can only fail when several failure events occur. The minimum combination of occurring failures (e.g., connected via an AND gate) that can lead to a system failure is called "Minimum cut set" (MCS). A failure tolerant system has at least one "Cut Set" which consists of several commands. The rank of a minimum Cut Set is the number of failures that lead to occurrence. This also defines the rank of a fault tolerant system. The rank of a fault tolerant systems is the rank of the minimal Cut Set that can lead to a critical failure. A system cannot have a Cut Set with rank 1 in order to be fault tolerant.

4.5 Common cause failures

The introduction of redundancy make the work of safety engineers more difficult as redundancy brings with it a new class of failures, that is failure due to common cause. Failures due to common cause influence failure analysis is such way that the probability of the minimal Cut Set is higher as the product of the probabilities of the minimal Cut Set's individual components. Failures due to common cause make the increase of the number of channels useless at a certain sum. If engineers could design redundant systems with independent channels then it would be unnecessary to analyze common cause.

Engineers could achieve the required level of safety (and dependability) if they could increase the level of quality. Unfortunately, it is practically impossible to build independent channels. That is why it is necessary to rate the failures with common cause to assure that the safety and dependability requirements are met. The easiest way to analyze failures due to common cause is to work with Cut Sets. Failures in minimal Cut Sets can represent the same failure mode in different components or represent different failure modes. The can occur due to the same cause or due to different causes.

The bottom line of this thesis is that when all failure in a minimal Cut Set are due to the same root cause and occur simultaneously then the fault tolerant systems fails as if the failures of the minimal Cut Set occur randomly. The probability that a minimal Cut Set occurs due to a common cause failure is normally extremely low but still larger than the probability that a minimal Cut Set occurs randomly. The objective of a common cause failure analysis is to rate this probability so that the design can be improved. The probability of a critical minimal Cut Set of a fault tolerant system would be underrated when common cause failures would not be considered.

A detailed analysis of failures due to common cause is presented in Chapter 15.

5 Parameter of Risk- and Reliability Analysis

Colloquially there are no distinctions for the concepts of reliability and safety. When it comes to technical systems, especially in an automation process, they clearly vary:

Reliability describes the prevention of losses of a process automation system.

With *safety* the possible effects of a failure within a process automation system are being described.

However, for process automation systems other issues have to be considered. Additionally to the definitions of the concepts more aspects will be mentioned in Table 5.1.

The definition of the security in Table 5.1 refers to an overall automation system. A danger to a process automation system can only come from the process itself, while dangers and risks for persons only appear in technical processes.

The higher the reliability within process automation systems with definite measures, the bigger is also the thrift of the system. A proof of reliability can be established through definite reliability calculations. A proof of reliability can also be established through an assurance of a long time guarantee.

The safety of a system however directs itself not against the appearance of mistakes and losses, but against the dangerous effects on people and nature. Due to the danger of such a process automation system, a permits to operate of a local authority are usually required. To get such permission, a so called proof of safety is to be produced, in which the safety has been documented.

Table 5.1: Definition of security and reliability according to DIN of 40041 respectively DIN 31004[58]

	Reliability of a process automation system	Safety of a process automation system
Definition	Totality of the attributes with given conditions for a given time-interval which refer to the fulfillment of certain requirements	Circumstances when the risk is not higher than the limiting risk, contains the ability within set borders and a given time to not cause or allow any danger
Measures according to	Occurrence of errors and failures	Dangerous consequences of errors and failures
Reasons for measures	Efficiency	Permission of an admission authority
Verification Procedure	Reliability calculations, long time guarantee	Safety verification

---

[58] [DINb90] DIN 40041, 1990-12, *Zuverlässigkeit, Begriffe*
  [DINj] DIN 31004, *Zuverlässigkeit, Begriffe*

For technical processes that fail but do not pose a danger and their failure only poses an occurrence of costs, a very high reliability is being demanded to keep the expenses low. High safety is demanded of technical process, at which at failure not only expenses originate, but still a danger for people and the environment exists. These are called safety relevant processes. These usually avoid both, expenses and dangers. High safety is equally requisite as the reliability.

5.1 Reliability Parameters

In Table 5.1 the concept of reliability was defined as qualitative. The reliability can be described quantitatively with the help of quantitative variables.

The investigation of the failure behavior is being divided into so-called subject items. Subject items in process automation systems can be as follows:

- The entire process automation system
- Subsystems (like a process automation system affiliated software-hardware-system)
- Individual function blocks (like a process automation as an appliance)
- Modules (like a circuit board)
- Modules (like integrated circuits, plug contacts)

A bigger number of different subject items of a module are viewed as example. Under the assumption that all subject items connect at the same time and work under the same operating conditions, all subject items would, under the assumption that the failure mechanisms function randomly, fail at different times. They change from the condition "able to function" into the condition "not able to function". In Figure 5.1 the condition of the time response is represented in subject-items.

Reliability

The failures of the subject items are being described with a reliability function (probability of survival):

$$R(t) = W(T > t). \tag{5.1}$$

Definition 1

The reliability function $R(t)$ shows with which probability uptimes T appear, that are bigger than a preset period under consideration $(0...t)$ (distribution-function of the uptimes T until failure).

Definition 2

The reliability-function $R(t)$ is the probability of a contemplation unit to be functional in a period under consideration $(0...t)$. In Figure 5.2 the course of $R(t)$ is represented.

The probability that the uptimes T are bigger than the contemplation-time period $(0...t)$ is for little values of t virtually 1. For bigger values of t the probability declines more and more.

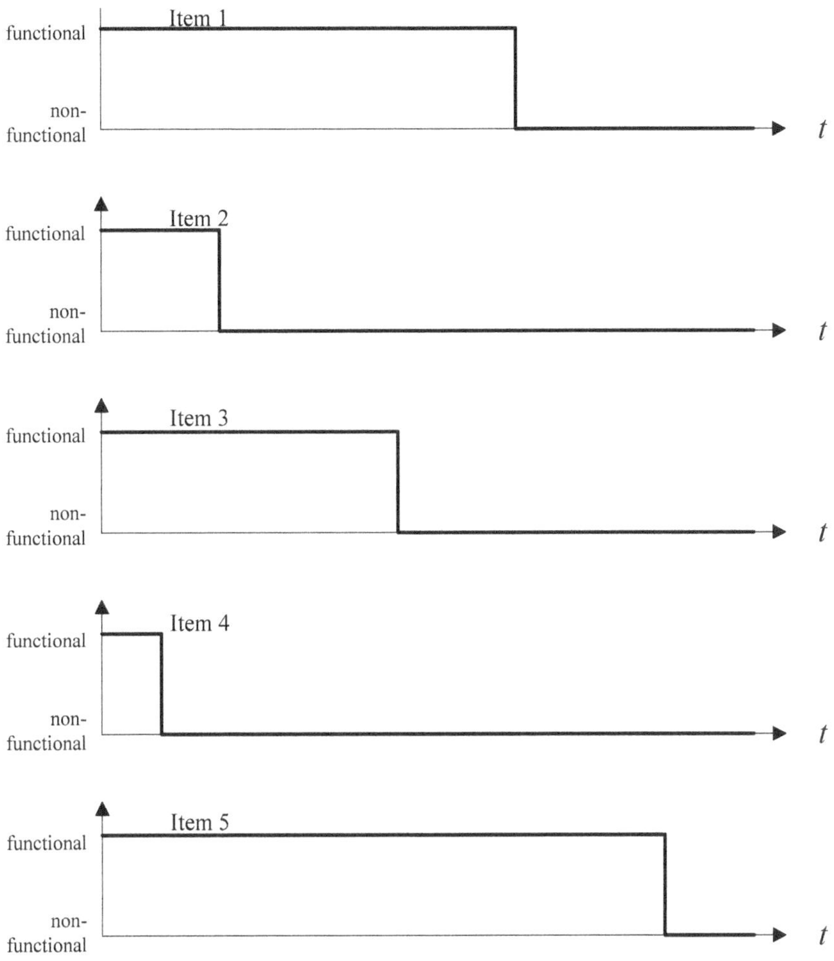

5.1 Reliability Parameters

Figure 5.1: States of Subject Items

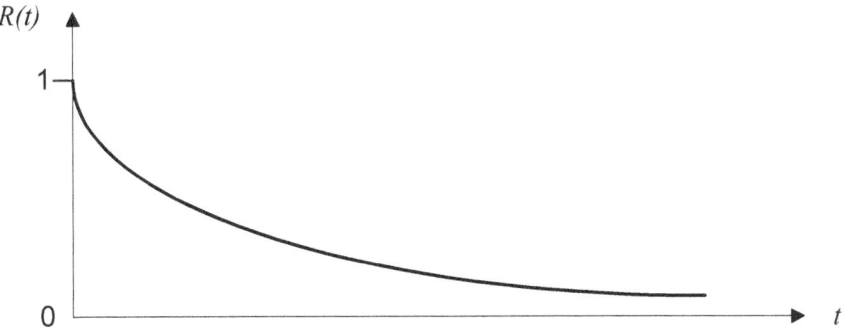

Figure 5.2: Chronological Sequences of the reliability-function $R(t)$

5.2 Probability of failure

If one forms the complement of the reliability function, one gets the probability of failure $F(t)$.

$$F(t) = 1 - R(t). \tag{5.2}$$

The probability of failure $F(t)$ is the probability that the uptimes T are not longer than a set time period t until failure.

5.3 Average Lifetime

As lifetime or failure-free operating time in the Figure 5.3 represents the uptime $T = 1$ to 3 of the subject items from operation to failure. The average of this lifespan of all subject items is being referred to as the average failure-free operation time or Mean Time To Failure *(MTTF)* and calculates itself by means of the reliability function itself $R(t)$.

For a constant random variable τ with the density $f(t)$ the expectancy-value (average value) calculates itself

$$E[\tau] = \int_{-\infty}^{\infty} t \cdot f(t) dt. \tag{5.3}$$

For a positive random variable the foregoing equation ($t > 0$) reduces itself to

$$E[\tau] = \int_{0}^{\infty} t f(t) dt. \tag{5.4}$$

Consequently the following is valid for a positive random variable:

$$E[\tau] = \int_{0}^{\infty} (1 - F(t)) \cdot dt = \int_{0}^{\infty} R(t) \cdot dt. \tag{5.5}$$

Normally $R(0) = 1$ is acceptable. The so calculated average is identical to the average of the failure-free operating time *[MTTF]*. Calculation is as follows:

$$MTTF = E[\tau] = \int_{0}^{\infty} R(t) dt. \tag{5.6}$$

For a single module as well as an entire system this equation is valid and can also be used for repairable subject items under the assumption that the system is like-new after the repair. With $R(0) = 1$ follows

$$R(t) = e^{-\int_{0}^{t} \lambda(x) dx}, \tag{5.7}$$

For a constant failure-rate $\lambda(t) = \lambda$ the following is valid:

$$R(t) = e^{-\lambda \cdot t}. \tag{5.8}$$

The mean time to failure in this event is

$$MTTF = \frac{1}{\lambda}. \tag{5.9}$$

The chronological sequence of the conditions for each subject item, as shown in Figure 5.3, represents itself when the subject items after failure through repair tasks are brought back into a working condition.

The probability in which both conditions itself are a subject item doesn't depend on time, if the subject items itself are stationary systems (systems that know only one condition if powered up). Under these conditions the average over many repair tasks calculates itself as "Average repair-time" and "Average duration of usefulness".

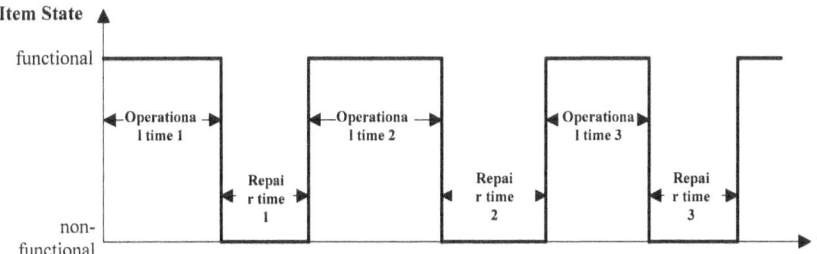

Figure 5.3: Operating and Repair Items of a Subject Item

5.4 Average Repair-Time

The average repair-times t of a subject item (MTTR, Mean Time To Repair) can be calculated with the following:

$$MTTR = \frac{t_1 + t_2 + \cdots + t_n}{n}. \tag{5.10}$$

With analytical calculation the *MTTR* of the distribution-function *G(t)* of the times of repair is:

$$MTTR = \int_0^\infty (1 - G(t))dt. \tag{5.11}$$

5.5 Average Duration of Usefulness

The average duration of usefulness (MTBF, Mean Time Between Failures) is the average time between two losses and calculates itself as follows:

$$MTBF = MTTF + MTTR. \tag{5.12}$$

If

$$MTTF \gg MTTR \tag{5.13}$$

is valid, then 5.12 can with very good approximation be written as follows:

$$MTBF = MTTF = \frac{1}{\lambda}. \tag{5.14}$$

5.6 Availability

The availability is the probability to find a repairable subject item at a set time t in the condition "able to function". The average availability calculates itself as follows:

$$PA = \frac{MTTF}{MTTF + MTTR}. \tag{5.15}$$

5.7 Failure Rate

The failure rate $\lambda(t)$, when it comes to safety systems, is the most important parameter. The single components, out of which the safety system is composed, are viewed specifically with their very own failure rates. Basically the failure rate for a given subject item is the same as the probability, referring to δt in Interval $(t, t + \delta t)$, under the condition that they have not failed up to the time t. The sequence consists of periods of early losses, the losses with a by approximation constant failure rate and the wear-out failures. In an industry virtually all systems that steer, regulate, or oversee security oriented processes are subjected to early losses, so that at the beginning of the actual utilization phase a constant or, in a strict scientific sense, an approximate constant failure rate can be determined:

$$\lambda(t) = \lambda. \tag{5.16}$$

The failure rate of a new module in general can only be determined experimentally. For known modules corresponding values can be found in failure rate catalogs.[59]

Basically the statement could be made that the failure rate $\lambda(t)$ is the negative value of the equation at the relevant time t differentiable, logarithmic reliability function:

$$\begin{aligned}\lambda(t) &= -\frac{d}{dt} \cdot \{\ln R(t)\} \\ &= -\frac{1}{R(t)} \cdot \frac{d R(t)}{dt}.\end{aligned} \tag{5.17}$$

Through 5.17 it is obvious, that the failure rate $\lambda(t)$ determines entirely the reliability function $R(t)$. With $R(0) = 1$ the following derivation emerges:

---

[59] [IECc97]   IEC 61709, *Electronic Components Reliability*
[CNET93]   CNET-Ausfallratenkatalog
[MIL]   MIL-HDBK-217
[SIEM91]   Siemens, SN 29500-2; *Ausfallraten Bauelemente*

5.7 Failure Rate

$$R(t) = e^{-\int_0^t \lambda(\tau)\,dt} \quad . \tag{5.18}$$

For a constant failure rate (approximate and time-independent) becomes $\lambda(t) = \lambda$. Consequently one gets the following equation

$$R(t) = e^{-\lambda t} \quad . \tag{5.19}$$

From 5.19 is evident, that for a constant failure rate the reliability function $R(t)$ is a falling exponential function, as shown in Figure 5.2.

Hereby λ is described as the average failure rate. It describes the expected average number of losses in a set time interval:

$$\lambda = \frac{1}{MTBF} \quad . \tag{5.20}$$

If the failure rate can be accepted as not constant, the function of the failure rate is as represented in Figure 5.4:

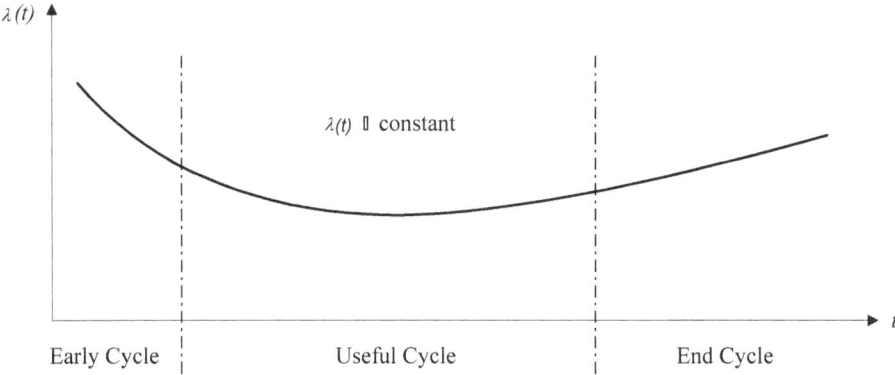

Figure 5.4: "Bathtub-curve" to describe the chronological sequence of the failure rate

The Weibull Distribution of the failure rate, also known as "bathtub-curve" is split into three sections:

- Early phase: Through early losses of modules (like bad modules, soldering mistakes, etc.) the failure rate is initially high.
- Utilization phase: In the utilization phase the failure rate $\lambda(t)$ reaches its minimum (the "bottom" of the bathtub-curve). The failure-rate can be assumed as constant in this phase.
- Wearout phase: Through the approach of the lifetime of the modules, the failure rate increases.

The failure rates of modules are only experimentally to be determined. So-called failure rate handbooks were established for standard modules. For all calculations in this work the failure rates from data sheets of the module manufacturer or the Siemens-Norm was used. The statements are so-called FIT-values, (Failure In Time), equal to the following unit:

$$[\lambda] = 1\,Fit = 1 \cdot 10^{-9} \frac{1}{h}. \quad (5.21)$$

5.8 SFF

The SFF parameter[60] defines the share of safety related failures. It is being formed through the quotient of the sum of all safe, dangerous, and detected failures and the sum of all safe and dangerous failures.

The formula for the SFF parameter is as follows:

$$SFF = \frac{\sum \lambda_S + \sum \lambda_{DD}}{\sum \lambda_S + \sum \lambda_{DD} + \sum \lambda_{DU}} \quad (5.22)$$

5.9 DC

The diagnostic coverage DC[61] is a measurement for the failure detection in safety related system. It calculates itself, in accordance with IEC 61508, as a ratio of known, dangerous failure rates to all dangerous failure rates of a system. In general a failure rate and effect analysis (FMEA) must be performed. The revision is pragmatic; typical DC-values are assigned to test measures. Here one refers to the IEC 61508 remitted possible measures to error detection. The for the overall system important average diagnostic coverage DC is being calculated according to a simple averaging formula in which all security related parts of a system are being added up:

$$DC_{avg} = \frac{\frac{DC_1}{MTTF_{d,1}} + \frac{DC_2}{MTTF_{d,2}} + \ldots + \frac{DC_N}{MTTF_{d,N}}}{\frac{1}{MTTF_{d,1}} + \frac{1}{MTTF_{d,2}} + \ldots + \frac{1}{MTTF_{d,N}}} \quad (5.23)$$

The *DC* parameter furthermore gives us the ratio between all through diagnosis detectable, dangerous failures and the overall number of dangerous failures.
The formula for the *DC* parameter is as follows:

$$DC = \frac{\sum \lambda_{DD}}{\sum \lambda_D} \quad (5.24)$$

---

[60] [IECb01] IEC 61508-2:2001, International Standard: 61508 Functional safety of electrical/electronic/programmable electronic safety-related systems, Part 2: Requirements for electrical/electronic/programmable electronic safety-related systems

[61] [IECb01] IEC 61508-2:2001, International Standard: 61508 Functional safety of electrical/electronic/programmable electronic safety-related systems, Part 2: Requirements for electrical/electronic/programmable electronic safety-related systems

Through reshaping of the foregoing equations for the DC factor and the SFF factor itself a connection between SFF and DC and the given S-factor is as follows:

$$DC = \frac{SFF - 1 + S}{S} \qquad (5.25)$$

$$SFF = 1 + S \cdot (DC - 1) \qquad (5.26)$$

5.9.1 Tests

Ein Test ist eine algorithmische Durchprüfung eines Systems in einem speziellen A test is an algorithmic examination of a system in a particular test-mode, in which a system is not or only partially available for other use. The reaction (test answer) of the subsystem depending on specific input signals (test pattern) is being observed and being compared with expected values and if a deviation exists an error will be reported. The totalities of the test patterns that are being used during the test are called test quantity. If the value reaches 100 %, the test is considered complete.

Usually a hierarchical sequence is being used. Beginning on the micro program level a primitive diagnostic routine (stored in redundant and read-only memory) determines the availability of other modules. According to result of the test a more extensive program starts that incorporates intact and functioning components. Eventually the complete system is included in the test. The test time complexity however must be restricted to obtain the minimal function of the system.

Failure detection distinguishes between the system being tested (test object) and the system doing the testing (test subject) which officiates as an official test executer. The subject-object-relation of all tests of a computer can be displayed graphically as a directed edge in a graph. There is a difference between a self test and an external test.

Within a self test the test subject and the test object are identical. So a system can perform a test for an array of failures F, there must be at least one test pattern for each failure that represents a valid input, so that the associated test answers does not show a valid combination of output signals.

If test subject and test object are not identical, we speak of an external test. The advantage of the external test as opposed to a self test is that the isolated system takes on the task of failure detection. Common mistakes of both systems are consequently improbable. On the other hand no additional system is required for a self test that has to be fault tolerant.

A differentiation in self and external test does not allow an opinion regarding the quality or quantity of the error detection.

The checkup necessary to detect a failure in the test-answer can be accomplished through comparison. Consequently there are two types of comparisons, the absolute and the relativity test.

- An absolute test examines if a conjectural conclusion of a system has been fulfilled. It is being used typically at dynamic redundancies, because results normally only exist from the primary but not the substitute system. Only the mistakes that harm a given conclusion over the results could be recognized through an absolute test. A conclusion, that includes all to be discovered mistakes, is mostly difficult to find and requires long term testing.

- A relativity test compares results that were generated while using different systems or through different calculations. It is being used in the case of static redundancies. Mistakes that exist on all systems are not found. The mistake of absolute and relativity tests differ depending on the kind of mistakes and the quantity of the mistakes that are to be tolerated, which or both test types are to be applied, or if both procedures should be applied simultaneously.

5.10 MTTF

MTTF (Mean Time To Failure) is a value for the average statistical duration of operation of a device up to the first failure. At a constant failure rate, the average of the failure-free working time MTTF equals $1/\lambda$, with λ being the failure rate of the subject item.

5.10.1 MTTF – Spurious Trip Rate

A safe failure of a component could conceivably cause a faulty operation (spurious trip). The average time up to a safe failure is known as average mean time to failure (MTTF-spurious). This is the estimated time between two safe failures of a component.

If through losses of the system components circuits of the safety system are affected, the predicted spurious trip rate (STR) can be calculated, to determine if further steps are justified in order to increase the reliability of the safety system.

The Norm ANSI/ ISA TR84.0.02[62] describes STR as the value with which a disturbance or a spurious operation in a safety system could occur.

5.11 PFD

PFD[63] is the probability that a system fails during a moment in which it is being required to function. The smaller this value, the better the system. Only dangerous failures are required for the calculation of the PFD-value. Because the system goes into a faulty condition if the failures are dangerous, detected or undetected. The safe failures can be neglected; they do not influence the system as a whole. For the calculation of the systems mainly the PFD_{avg} is being used. See Figure 5.5. The general formula for PFD_{avg} is

$$PFD_{avg}(T) = \frac{1}{T}\int_0^T P(t) \cdot dt \qquad (5.27)$$

$P(t)$ is the probability of a failure, frequently also described as probability of failure. The probability of failure is defined as follows:

$$P(t) = 1 - R(t). \qquad (5.28)$$

---

[62] [ISA-02] ISA-TR84.00.02-2002, Parts 1-5. Safety Instrumented Functions (SIF) – Safety Integrity Level (SIL) Evaluation Techniques
[63] *en.:* probability of failure, see [IECf00]

5.11 PFD

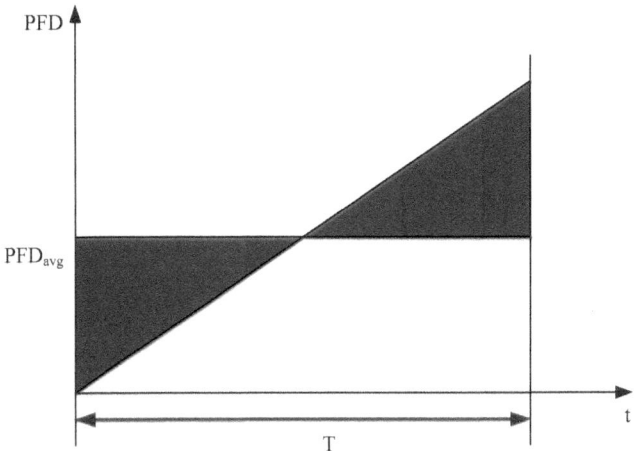

Figure 5.5: PFD$_{avg}$-value

$R(t)$ is the reliability of the system and calculates itself as follows:

$$R(t) = e^{-\lambda \cdot t}. \tag{5.29}$$

Because only the dangerous failures are relevant for the calculation, (5.28) and (5.29) with $\lambda = \lambda_D$ in (5.27) result in (5.30).

$$PFD_{avg}(t) = \frac{1}{T}\int_0^T P(t)\cdot dt = \frac{1}{T}\int (1 - e^{-\lambda_D \cdot t})\cdot dt$$

$$= \frac{1}{T}\left[\int_0^T 1\cdot dt - \int_0^T e^{-\lambda_D \cdot t}\cdot dt\right]$$

$$= \frac{1}{T}[t]_0^T - \frac{1}{T}\left[-\frac{e^{-\lambda_D \cdot t}}{\lambda_D}\right]_0^T$$

$$= \frac{1}{T}[T-0] + \frac{1}{T\cdot \lambda_D}\left[e^{-\lambda_D \cdot T} - e^{-\lambda_D \cdot 0}\right]$$

$$= 1 + \frac{1}{T\cdot \lambda_D}\left[e^{-\lambda_D \cdot T} - 1\right]$$

$$\Rightarrow PFD_{avg} = 1 + \frac{e^{-\lambda_D \cdot T} - 1}{T\cdot \lambda_D} \tag{5.30}$$

Equation (5.30) shows the calculation of the PFD_{avg} for a 1oo1 system für a 1oo1-system. In order to simplify this equation, MacLaurinsche row of the exponential function[64] is used. The general formula of this row is shown in (5.31).

$$e^x = \sum_{0}^{\infty} \frac{x^n}{n!} = 1 + x + \frac{x^2}{2!} + \frac{x^3}{3!} + \frac{x^4}{4!} \cdots \quad (5.31)$$

In this example one can use the exponential function

$$e^{-\lambda_D \cdot T}$$

for the MacLaurinsche row as follows:

$$e^{-\lambda_D \cdot T} = 1 - \lambda_D \cdot T + \frac{\lambda_D^2 \cdot T^2}{2!} - \frac{\lambda_D^3 \cdot T^3}{3!} + \cdots \quad (5.32)$$

For the calculation of a 1oo1-system the first three terms suffices for the equation (5.32). After linking the equation (5.32) into (5.30) one ends up with a PFD_{avg}-value of the 1oo1-system as in equation (5.33) following equation.

$$PFD_{avg} = 1 + \frac{(1 - \lambda_D \cdot T + \frac{\lambda_D^2 \cdot T^2}{2!}) - 1}{T \cdot \lambda_D} = 1 + \frac{2 - 2 \cdot \lambda_D \cdot T + \lambda_D^2 \cdot T^2 - 2}{2 \cdot T \cdot \lambda_D}$$

$$= \frac{2 \cdot T \cdot \lambda_D - 2 \cdot T \cdot \lambda_D + \lambda_D^2 \cdot T^2}{2 \cdot T \cdot \lambda_D}$$

$$\Rightarrow PFD_{avg} = \frac{\lambda_D \cdot T}{2} \quad (5.33)$$

With

$$\frac{T}{2} = t_{CE} = \frac{\lambda_{DU}}{\lambda_D}\left(\frac{T1}{2} + MTTR\right) + \frac{\lambda_{DD}}{\lambda_D} \cdot MTTR \quad (5.34)$$

The PFD_{avg}-value of the 1oo1-system is as follows

$$\Rightarrow PFD_{avg} = \lambda_D \cdot t_{CE} \quad (5.35)$$

t_{CE} stands for channel equivalent Mean Down Time.[65]

Mean Down Time, MDT, is an expression of the time in which a computer system is not available or not functioning.

---

[64] The derivation is described in [BÖRC04].
[65] [IECf00] IEC 61508-6:2000, International Standard: 61508 Functional safety of electrical/electronic/programmable electronic safety-related systems, Part 6: Guidelines on the application of IEC 61508-2 and IEC 61508-3

5.11 PFD

Corresponding one could calculate the PFD-values of different systems. The following equations show the results of some systems.

PFD_{avg}-Value of a 1oo2 system is presented in (5.36)

$$PFD_{avg} = 2 \cdot [(1-\beta_D) \cdot \lambda_{DD} + (1-\beta) \cdot \lambda_{DU}]^2 \cdot t_{CE} \cdot t_{GE}$$
$$+ \beta_D \cdot \lambda_{DD} \cdot MTTR + \beta \cdot \lambda_{DU} \cdot (\frac{T_1}{2} + MTTR) \quad (5.36)$$

with

$$t_{CE} = \frac{\lambda_{DU}}{\lambda_D} \left(\frac{T_1}{2} + MTTR \right) + \frac{\lambda_{DD}}{\lambda_D} \cdot MTTR \quad (5.37)$$

$$t_{GE} = \frac{\lambda_{DU}}{\lambda_D} \left(\frac{T_1}{3} + MTTR \right) + \frac{\lambda_{DD}}{\lambda_D} \cdot MTTR \quad (5.38)$$

t_{GE} stands for group equivalent mean down time.

The PFD_{avg}-value of a 2oo2-system is shown in (5.39).

$$PFD_{avg} = 2 \cdot \lambda_D \cdot t_{CE} = 2 \cdot PFD_{avg,1oo1} \quad (5.39)$$

The PFD_{avg}-value of a 2oo3-system in (5.40).

$$PFD_{avg} = 6 \cdot [(1-\beta_D) \cdot \lambda_{DD} + (1-\beta) \cdot \lambda_{DU}]^2 \cdot t_{CE} \cdot t_{GE}$$
$$+ \beta_D \cdot \lambda_{DD} \cdot MTTR + \beta \cdot \lambda_{DU} \cdot (\frac{T_1}{2} + MTTR) \quad (5.40)$$

t_{CE} = see equation (5.37)

t_{GE} = see equation (5.38)

For the calculation of failure probabilities some important associations have been made.

6 Measures for a Risk Analysis

6.1 Basic Concepts

A range of safety tasks in complex, process plants like production, supervision, danger identification, security analysis, procedure development and plant-planning necessitates an extensive understanding of the mode of operation of single components as well as the interaction of the components and the resulting system behavior. The kinds of processes that are being investigated range from microscopic (molecular) up to macroscopic dimensions of the entire plant. The analysis methods and procedures that are put in place are divided into quantitative and qualitative. Quantitative methods are mainly being used for limited operations like particle simulation, thermodynamics, and hydrodynamics. A quantitative description or simulation of an entire plant often fails because of the required calculating capacity or the difficulty to formulate mathematical equations for a description of a procedural technical plant. In consideration of the investigation and mastery of the system behavior, qualitative extensions in contemplation are therefore being considered. A qualitative and model-based approach points out advantages for the investigation of more technically complex systems:

- The human thinks qualitative. The expert knowledge of a process technical plant represents itself in a qualitative manner. A qualitative approach therefore allows an intuitive model. The use of qualitative words (more, fewer, …) on reference functions in the framework of PAAG[66] and HAZOP[67] studies corresponds more to the human approach as the wording of mathematical equations.

- Qualitative results of an analytic system study let themselves interpret more comprehensibly than quantitative results.

- Qualitative approaches offer promising possibilities in order to reduce the required extent of calculations.

Limiting risks for the employed, the environment, or the plant or product, necessitates three elementary steps. First step is the recognition of danger sources, second step is the evaluation of those sources and third the meeting of counteractive measures. In the following sections methods for the discovery of sources of interference and danger are described.

---

[66] PAAG: prognosis, find the cause, effect estimation, counteractive measures, in German: „**P**rognose, **A**uffinden der Ursachen, **A**bschätzen der **A**uswirkungen, **G**egenmaßnahmen".
[67] HAZOP: Hazard and Operability. Hazard describes a situation, in which a real or potential danger for human, material or environment can occur.

6.2 Methods of Danger Analysis

Under the concept of hazard analysis different technologies are combined, they check a system for potential hazards. One can divide the variety of these technologies in types like forward- and backward-search or top-down- and bottom-up-search. The names of these types give a clue as to the different approaches and procedures that are being used to analyze the dangers. One does not concentrate on a single aspect of the system and its development, but views the system in its entirety and attempts to recognize overall dangers and their causes and, where possible, to reduce or eliminate them. These technologies have usually been developed for a specific industry; others have been developed for a wider field of application. One is advised to combine several technologies because they shed light on different dangers and bring with it new findings, that could not be discovered with one technology by itself.

6.2.1 Forward- and Backward-Search

The **forward-search** follows the real time sequence, starts out with an incident and the resulting danger as a consequence. With a hazard analysis one looks for possible effects so they can be controlled or eliminated.

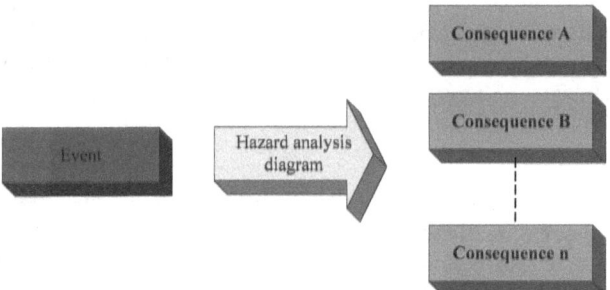

Figure 6.1: Forward Search

The **backwards-search** uses as a basis the effects that have already been caused by an incident (cause). The diagram attempts to locate different causes, the so called incidents that result in the same effect. One follows a reverse chronological sequence.

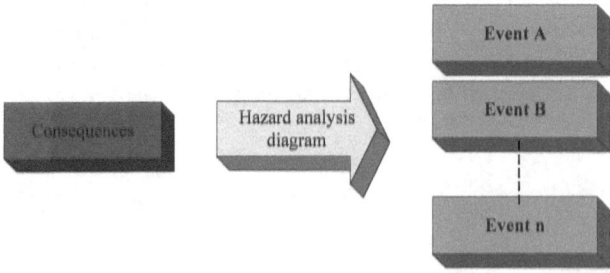

Figure 6.2: Backward Search

6.2.2 Top-Down and Bottom-Up Search

This search analysis puts an emphasis on the aspect of failure propagation. An example is a pressure transducer that encounters a disruption. From this incident different disruptions on a bigger scale result, like the failure of the heating-cooling-cycle etc. and eventually results in danger for people or machines like an explosion of the reactor.

The **Bottom-Up Search** follows the chronological method of failure propagation and first views the original source of the incident, here the pressure transducer, in order to investigate little by little the facilities that have been affected by the failure. The complexity can be very far reaching.

The **Top-Down Search,** like in Figure 6.3, takes the equipment as the starting point that generates the immediate dangerous situation for people and/ or the environment. From there one searches for possible sources of the mistake and ends up with the actual cause of the mistake.

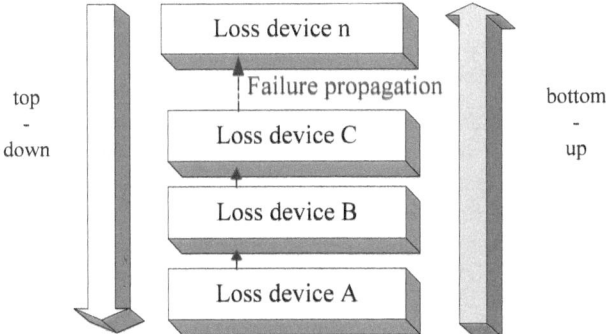

Figure 6.3: Top-Down and Bottom-Up Search Technique

6.3 Probability Analysis

In safety engineering probability analysis serves to better analyze the risk of a plant or a process, also called Likelihood Analysis. The knowledge over the magnitude of all consequences of an incident still does not give a complete overview over an existing risk. Only through the use of different methods of risk assessment can a risk be defined. The used methods should be qualitative as well as statistical and take into account several fault propagation dispersion models.

6.3.1 Statistical Analysis

In general failure rates are defined by the frequency of an incident by findings reached through past experience. In order to define the probability of a failure in a complex system, for example a failure in a procedural plant, there are no sufficient experience values for statistical use. Because of this reason and in order to quantify the failure of a complex system, it is being divided into units. This model is known as Fault Propagation Model. In contrast to a complex overall system, units are parts with different smaller units at least comparable to the statistical evaluation of failures. Thereby a failure probability or reliability for the partial units can be determined. The results of the partial unit's failure probability of the complex system can now be predicted.

General data about the failure frequency for a big number of well-known systems and subsystems can be found in industry reference books. Also, some companies that offer risk-management services or work for insurances have such data available.

Nevertheless, statistical failure analyses of subsystems cannot always directly determine the failure rate of an entire system. Two conditions must be met in order to reach a correct statistical conclusion for the failure rate of an entire system:

- A sufficient amount of data must be available for an analysis.
- All system data must originate from systems with a comparable structure. That ensures that all conclusions drawn can be inferred to other systems.

These two conditions are often contrary to each other and occur very seldom simultaneously when an analysis of the failure frequency is being done. If no data of such definite types of failure is available, estimations must be determined in order to decide on a failure probability.

For big processing plants the data set of like units is required. Through a combination of the data sets a sufficient risk judgment of the plant being investigated can be justified. One should take under consideration the fact that well-known data sets of the individual units are based on the foundation of different parameters. Each single parameter influences the failure probability. Therefore the findings and the results of like systems cannot simply be transferred one to one onto a different system since the conditions of framework and application differ strongly.

6.3.2 Fault Propagation Model

Provided the statistical analysis is not sufficient to determine the failure probability or the failure rate of a unit, Fault Propagation Models can be used alternatively. These models are based on a method that out of the likeliness of a sequence of incidents the failure probability of a unit can be determined. In order to determine the probability of failure, the logical connections between the initial incidents and the following incidents that led to a mistake or failure are being determined.

Different Fault Propagation Models are available for the calculation. In order to select a suitable model one must take into consideration the sophistication of the incident as well as the demanded accuracy of the result. Today the following models are being used:

- Fault tree
- Event tree
- Block diagram
- Markov

7 Risk Matrix

The increasing complexity of technical development always require better analysis methods. An overview can be seen in Figure 7.1.

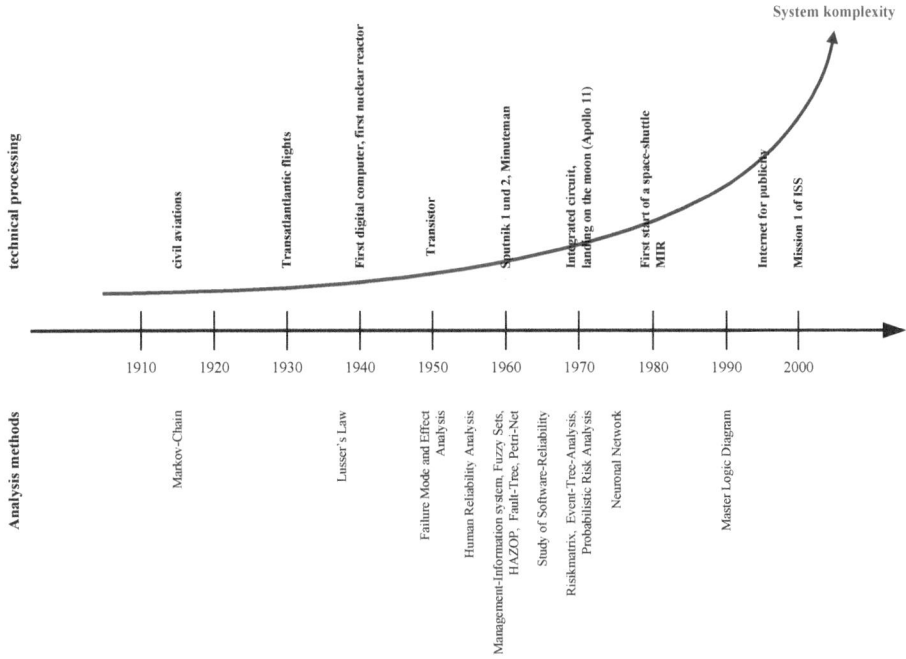

Figure 7.1: Risk Matrix with probability- and consequence Categories

Nowadays the Risk Matrix method, well-known in North America, will preferably be used to define the risk of a system. This is a simple qualitative method which does not require any special abilities.

The composition of a Risk Matrix starts with the probabilities classification for dangerous events in low medium and high levels. Their consequences will be classified with low, serious and grave.

An assignation of the probabilities and consequences in definite categories will follow as clarified in the following formula:

$R = H \cdot S$

(Risk R, Frequency H, Damage extent S)

This assignation can occur using quantitative tools such as LOPA (layer of protection analysis) or pure qualitative tools having an available technical discernment.[68]

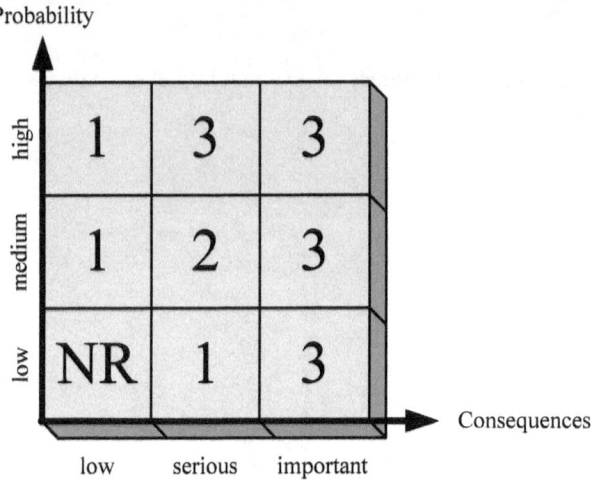

Figure 7.2: Risk Matrix with probability- and consequence Categories, NR: Not Relevant

In Figure 7.2 a Risk Matrix is represented with the two axes for the likelihood and the consequences of a dangerous event. Each Matrix cell is associated with a definite category, like the safety integrity level. If one now determines the frequency and the consequences for an incident, one could find the corresponding category in the Risk Matrix. For these categories norms or guidelines measures have been established, which reduce a risk into a residual risk.

The classical division of the damage with serious consequences has been discussed above. The gradation is sometimes divided in up to five categories. In table 7.1 the categories are being qualitatively described.

Table 7.1: Classical categories for effects with the example of a risk matrix

Consequence Category	Description
Low	Marginal injuries possible, low material damages
Serious	One or several considerable injuries possible, damage up to 1 million Euro
Massive	One or several fatalities, damage at least in the millions

This table elaborates on what signifies a "low", "serious" or "massive" incident. Not only material damage, but also human damages are taken into account. An incident becomes qualitative with a quantitative calculating tool or according to an expert opinion. A quan-

---

[68] [MARS02]   Marszal, E., Scharpf, E., Safety Integrity Level Selection, Systematic Methods Including Layer of Protection Analysis

titative determination for example could have the value of 0.05 which signifies that an incident with a probability of 5% will be accompanied by a fatality.

Just like the effects of an incident, one can divide the probabilities into three classical categories. It does not only include a qualitative term and a verbal description, but also a probability of occurrence -. See the division in table 7.2.

Table 7.2: Classical probability-categories at the example of a risk-matrix

Probability Category	Occurrence / year	Description
Low	$<10^{-4}$	The probability of a failure is very low and is not expected within a lifetime of a plant.
Medium	10^{-2} bis 10^{-4}	The probability of a failure is low is not expected within a lifetime of a plant.
High	$>10^{-2}$	The probability has increased, so that a failure within the lifetime is expected.

In this example in the categories "low" and "medium" no failure within the lifetime is expected. The probability of appearance at "medium" has increased however somewhat. Incidents with a "high" probability of an occurrence of a safety failure within the lifetime of a plant occurs. In the frequencies the protective layers of the processes must be taken into consideration. However, through a safety technical function (engl: Safety Instrumented Function, SIF) these probabilities are valid without the consideration of any preventive measures.

The determination of the frequency can be determined with the help of experts or with a method of probability analysis. The Risk Matrix is two-dimensional and determines a SIL-value. There are also variations with a third dimension regarding the protective layers. These protective layers reduce the probability of occurrence of a dangerous incident. These three-dimensional matrix types are rarely being used, because tools like LOPA take the attributes of the protective layers more and more under consideration.

The matrix of Figure 7.2 includes NR for an event. This means that a SIL-decrease through a SIF is necessary, because the risk is tolerated. In three events the SIL has the value of three. According to IEC 61511-3 further analysis is needed because a system with a SIL-value of 3 does not reduce the risk enough. A SIL of 4 does not appear in this event, although it exists. However, its development is often very difficult and expensive. A better possibility is the application of multiple dependent systems that possess a low SIL or carry improvements for the process within themselves.

Like in Figure 7.3 it is also possible to equip the cells of the Risk Matrix with SIL-values instead of defined risk classes. Thereby one refines the graduation. A table that among other things describes the risk classes and descriptions that include the degree of risk could show the possible measures to minimize risk.

A red line in the Risk Matrix separates the conditions with tolerable and intolerable risk. The blocks A to C are still tolerated; the critical steps D to F must be subjected to some measures in order to reduce the risk.[69]

---

[69] [BRAA00]   Braasch, W., *Workshop on Risk Assessment*
    [KUHN04]   Kuhn, Risk Matrix as a Tool for Risk Assessment in the Chemical Process Industry

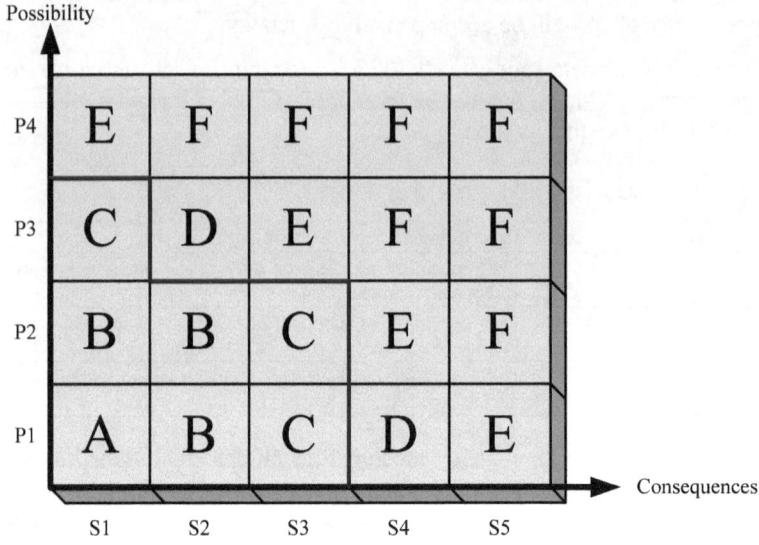

Figure 7.3: Refined Risk Matrix with Possibility and Consequence Categories

8 Risk Graph

Different methods can be applied to ensure the technical evaluation of different plants. One of these methods is the risk graph, a qualitative procedure that results in requirement classes. The procedure was initially developed for personal safety reasons but extends within the framework of the statutory order on hazardous incidents into the environmental safety.

8.1 Risk Graph according to DIN V 19250

An exact description of the method is shown in DIN V 19250 „Fundamental safety aspects to be considered for measurement and control equipment".[70] This norm however limits itself to the valuation of safety functions and can therefore not be applied to an entire system. The norm delivers these three examples:

1. Not the totality of a process plant is being viewed, but for example the safety equipment that protects the vessel from overpressure within this plant.

2. Not the entire traffic system is being viewed, but for example only the protectors that protect at a crossing point at faulty signals.

3. Not only the control of a complex industrial machine is being viewed, but only the part of the control (safety equipment) that is responsible for the supervision of the position of the safety grid.

The risk graph DIN V 19250 is being implemented for risks that originate at the failure of specific measurement and control safety equipment. It explicitly is not referring to risks that emanate from the system as a whole. The procedure is application and technology independent. It can therefore be used for many different applications like medical technology, chemical plants or mechanical engineering and is principally applicable for all measurement and control technologies like Electronics, Pneumatics or Electro Mechanics. The norm further limits itself to the discovery of requirement classes. The meaningful requirements of each class and resulting measures apply to DIN V 19251[71]. This norm is not considered at this point.

---

[70] [DINa98]　　DIN V 19250, *Grundlegende Sicherheitsbetrachtungen für MSR- Schutzeinrichtungen*
[71] [DINa95]　　DIN V 19251, *Leittechnik; MSR-Schutzeinrichtungen; Anforderungen und Maßnahmen zur gesicherten Funktion*

8.1.1 Correlation between Risk, Acceptable Risk, Residual Risk and the Risk Reduction

Figure 8.1[72], according to DIN V 19250, shows in a plausible manner the influence of specific safety measures on risk reduction. The existing risk must be reduced to at least to the acceptable risk. This acceptable risk is not always purely objective but is strongly influenced by subjective opinions. The standard here attempts a rather objective observation.

Figure 8.1 makes it obvious that the risk is reduced not only through MSR safety measures, but also through different measures like shut-off devices or training. These non-MSR safety measures influence the risk parameters and consequently lead to a decrease of the requirement class. The safety of a system could be reached therefore in different but equivalent ways. The conducted measures can be mutually supplemental or replace each other (see Figure 8.2).

The requirements arise out of a given safety goal and the partial risk that should be covered by the MSR-safety equipment. This partial risk is being described qualitatively through risk parameters. With their help the request classes become gradated. An exact gradual allocation of measures to the different requirements however is not possible, mainly because the possible measures are very diverse.

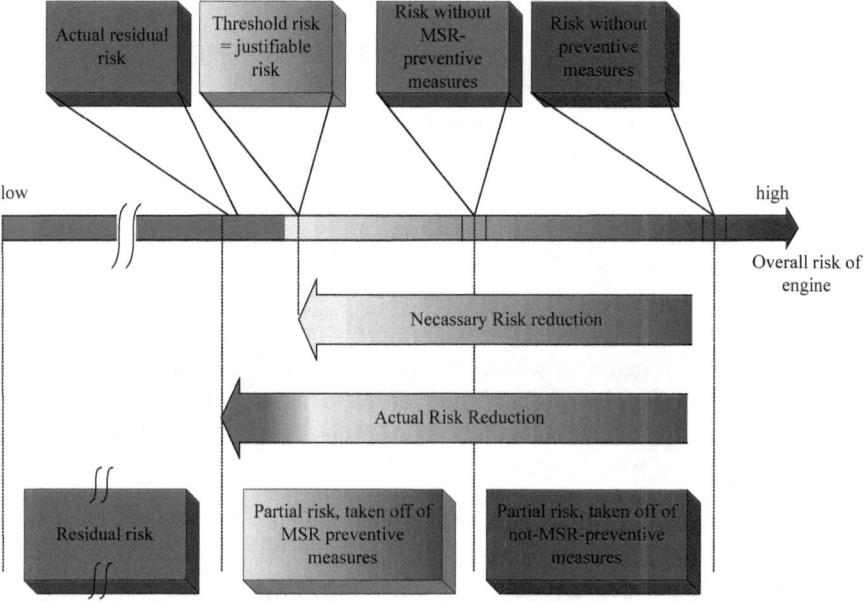

Figure 8.1: Risk Concept in accordance with DIN V 19250

---

[72] [DINa98]   DIN V 19250, *Grundlegende Sicherheitsbetrachtungen für MSR-Schutzeinrichtungen*

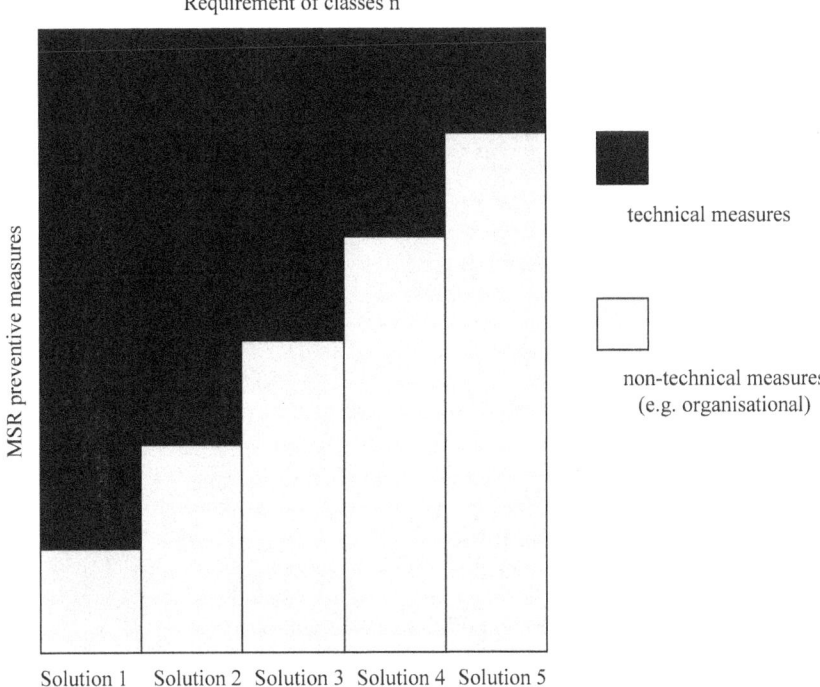

Figure 8.2: Shares of the to be covered partial risk (aus DIN V 19250)[73]

8.1.2 Risk Parameter

In order to determine the risk of a technical process or a condition, one gathers the combinations (X) out of the frequency (H) of an incident and the resulting extent of the damage (S)

$$R = H \times S \quad [74] \tag{8.1}$$

Because an exact quantification of risks is often not possible or at least very difficult, DIN V 19250 for the simplification of the quantification of the frequency speaks of three further influencing variables (parameters) and they are:

- Duration in danger-area (A)
- Possibility of the prevention of hazards (G)
- Probability of the result without any safety equipment (W)

---

[73] [DINa98]   DIN V 19250, *Grundlegende Sicherheitsbetrachtungen für MSR- Schutzeinrichtungen*
[74] [DINa87]   DIN VDE 31000 Teil 2:1987-12, *Allgemeine Leitsätze für das sicherheitsgerechte Gestalten technischer Erzeugnisse – Begriffe der Sicherheitstechnik – Grundbegriffe*

Extent of Damage

Under this parameter the following criterions exist:

1. Object of legal protection
   a. Persons
   b. Environment
   c. Material values or comparables
2. Extent of the damage (regarding persons)
   a. A person
   b. Several persons
   c. Very many persons (catastrophe)
3. Degree of injury
   a. Easy (usually reversible) injuries
   b. Heavy (usually irreversible) injuries
   c. Death

From there derives the parameter S "Extent of damage" as follows:

- S1: Light injury of a person; damaging environmental influences that do not fall under the statutory order on hazardous incidents.
- S2: Serious irreversible injury of one or several persons or death of a person; temporarily bigger damaging environmental influences, according to the statutory order on hazardous incidents.
- S3: Death of several persons; extensive and larger damaging environmental influences according to the statutory order on hazardous incidents.
- S4: Disastrous effect, extensive death toll.

Exposure Time

The stay in the danger area (temporal duration, frequency) is viewed under following criterions

- Seldom
- Frequent
- Very frequent/permanent

The parameter A "duration of stay" infers as follows:

- A1: Rare to frequent stay in the area of danger.
- A2: Frequent to permanent stay in the area of danger

Avoidance

Under this parameter the following criterions are viewed:

1. Handling of a Process
   a. Handling under supervision (by experts/laymen)
   b. Handling without supervision
2. Development/ Origination of danger (temporal)
   a. Sudden, quickly
   b. Slow
3. Recognition of danger
   a. Immediate observation
   b. With technical aids (instruments)
   c. Without technical aids
4. Prevention of danger (like a possibility of escape)
   a. Possible
   b. Conditionally possible
   c. Not possible
5. Practical safety experience, for example with a process/ plant/ appliance
   a. With same processes (processes well-known)
   b. With comparable processes
   c. None (no safety related experiences before).

The criterions "recognition of danger" and "practical safety experience" are not only objective and were reviewed only indirectly during the creation of the parameter.

From there derives the parameter G "prevention of danger" as follows:

- G1: Possible under specific conditions
- G2: Hardly possible

Probability of Occurrence of an Undesirable Event

Under this parameter following criterion are viewed:

Probability of occurrence of the undesirable incident without existence of the MSR safety equipment

- very low
- low
- relatively high

The parameter W "probability of occurrence" infers as follows:

- W1: Very low probability of the undesirable incident means that during the viewed processes or applicable processes (without existence of safety equipment) only very few undesirable incidents will have to be expected.

- W2: Low probability of the undesirable incident means that during the viewed processes or applicable processes (without existence of safety equipment) few undesirable incidents will have to be expected.

- W3: Relatively high probability of the undesirable incident means that during the viewed processes or applicable processes (without existence of safety equipment) undesirable incidents will have to be expected frequently.

8.1.3 Further Risk Parameters

In individual cases further risk parameters can be introduced. This could be conceivably under circumstances. At the employment of new technologies in safety equipment or in the process requirements could increase or decrease.

8.1.4 Risk Graph

If one generates all combinations of the risk parameters with each other, one gets 48 so-called "risk assessment packages". Figure 8.3 shows this in a graph.

The definition of the risk in accordance with equation (8.1) treats the influencing variables S and H (here represented through the risk parameters A, G, and W) equivalent. This shows that a small S and large H or large S and small H result in the same risk. In different words: Even at a large extent of the damage S, (like S4) and a corresponding low frequency H an optional low risk R and therefore a low request class will be reached.

8.1 Risk Graph according to DIN V 19250

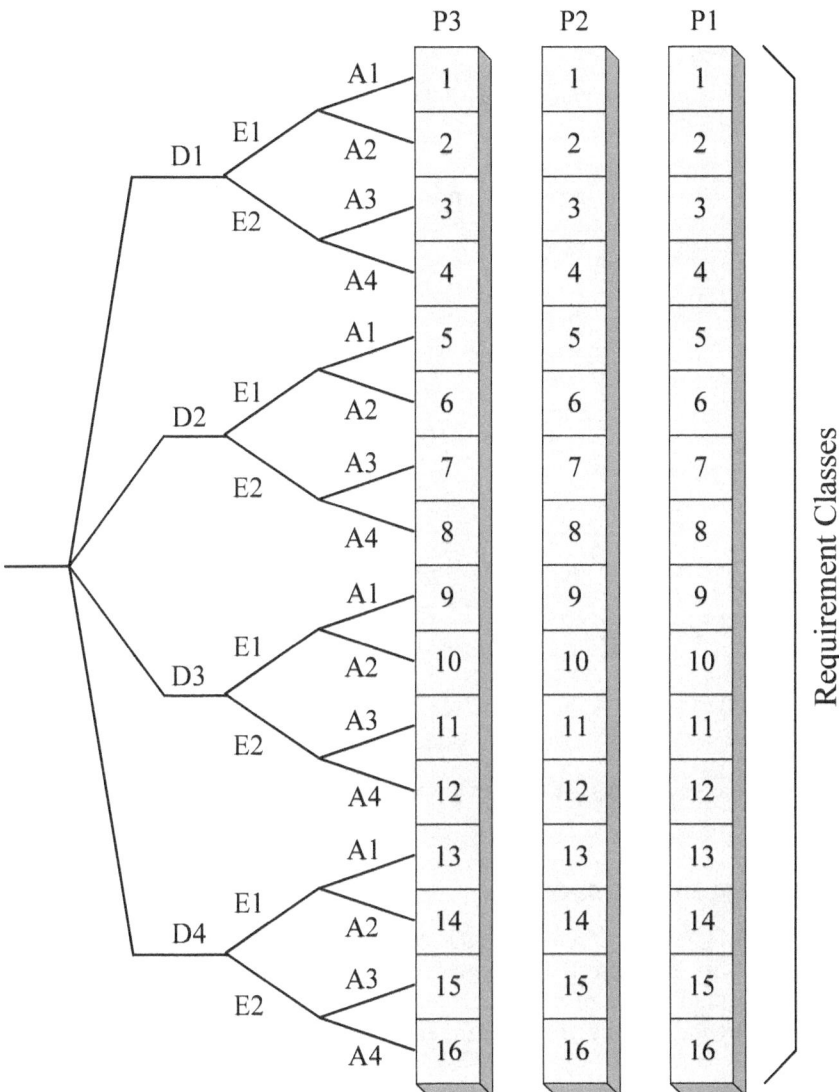

Figure 8.3: Entire Risk Graph

In reality the risk parameter S is dominant. This is founded to a large part on subjective points of view! The risk graph is supposed to deliver an objective assessment of risk, subjective aspects and practical actualities should however not be left unconsidered. Of the 48 theoretical combinations of the risk parameter therefore only few are meaningful. For the parameter S4 (Disastrous consequences, very many dead) a further gradation through the parameters A or G in the extended population is not applicable. Since persons are always in an area of danger, a differentiation through parameter A is not advised.

For the risk parameter S1 the requirements are so low that a further gradation is possible but not really necessary.

In reality there are only eight meaningful combinations for the parameters S, A and G, see the risk tree in Figure 8.4

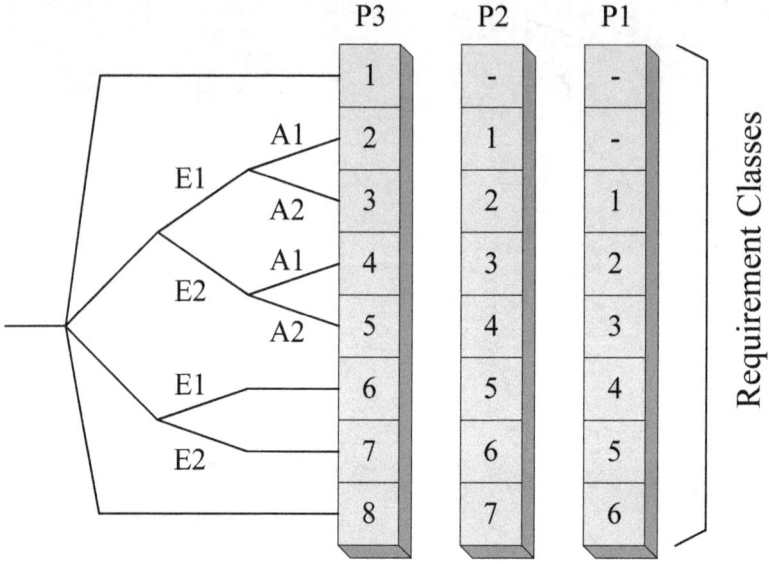

Figure 8.4: Risk Graph and Request Classes according to DIN V 19250

8.1.5 Requirement Classes

In order to attain the safety goal, requirements are being assigned to the risk parameters. Those are gathered in eight objective requirement classes (see Figure 8.4). The follow applies: The higher the number of a requirement class, the bigger the partial risk being kept under control by the safety equipment, the higher the requirements and the resulting measures.

The definition of the requirements of the eight requirements classes can be found in DIN V 19251 and for systems with safety functions additionally in DIN V VDE 0801.

Non-measurement and control safety measures are necessary for risk reduction in requirement class 8 because up to this day such incidents cannot be avoided with the level of technology by MSR safety facilities alone.

The measures can roughly be divided in two types:

- Failure avoidance (before operation)
- Failure control (during opeation)

A reciprocal substitution of both types of measure is partly possible.

8.2 Risk Graph according to IEC 61508-5 and IEC 61511-3

The standards IEC 61508-5 (appendix D) and IEC of 61511-3 (appendix E) are mainly derived from the DIN V 19250. The most important differences are special abbreviations for the risk parameters and especially the use of the level of safety integrity (SIL) instead of requirement classes. The differences are represented in the following chart.

Table 8.1: Comparison of the Denominations

	DIN V 19250	IEC 61508
Risk	R	R
Frequency	H	f
Extent of Damage	S	C
Extent of Stay	A	F
Avoidance of Danger	G	P
Probability of Occurence	W	W

Table 8.2: IEC 61508/61511, DIN V 19250 and VDI/VDE 2180

IEC 61508/61511	DIN V 19250	VDI/VDE 2180
	AK1	
SIL 1	AK2	Risk Region I
	AK3	(low risk)
SIL 2	AK4	
SIL 3	AK5	Risk Region II
	AK6	(high risk)
SIL 4	AK7	PLT Safety equipment
	AK8	not enough

Additionally to the comparison of the SIL- and AK-grade from Table 8.2 one can find a comparison like the one in Figure 8.5 in the appropriate literature. If one compares both more closely, their discrepancies attract attention. Like in the event of SIL 1 and SIL 2, respectively AK1 to AK4. The assignments of the level of safety integrity to the request classes differ. Because the origin of Figure 8.5 is not questionable, the comparison in Table 8.2 should be preferred. This corresponds precisely to IEC 61511-3, appendix E, section 4.

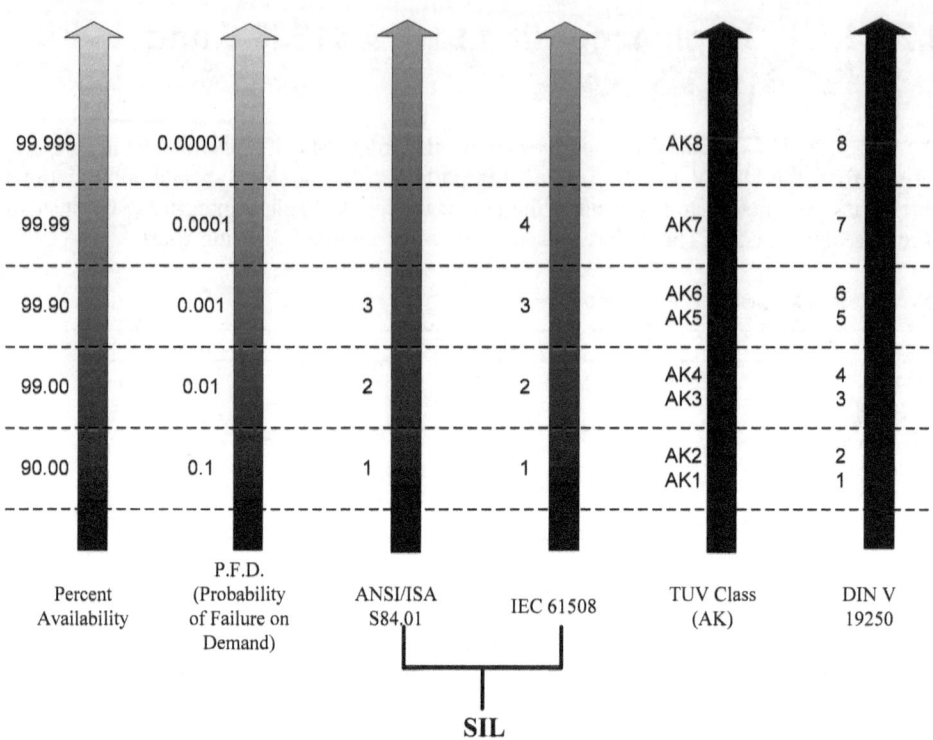

Figure 8.5: AK and SIL Classification

8.3 Risk Graph according to DIN EN 954-1

The basic framework of the risk graphs in accordance with DIN V 19250 was taken from the current European norm "Safety of machines, safety related parts of controls, general composition guidelines" (DIN EN 954 Part 1). The norm pursues the goal in connection with different risk standards according to suitable controls. Basically the DIN EN 954-1 risk graph possesses similarity with the already introduced graph DIN V 19250. The current risk graph shows only five different categories as possible results of the observation of risk. These categories are represented in Table 8.3.

As changes compared to DIN V 19250 besides the categories also different risk parameter show. The risk parameter S (here "extent of an injury") leaves only one differentiation in S1 (easy, usually reversible injuries) and S2 (heavy, usually irreversible injuries, including death). Since DIN EN 954-1 was created for the judgment of machines, S3 (death of several persons) or S4 (disastrous effect, very many dead) are not applicable. The risk parameter W (Probability of Occurrence of an Undesirable Incident) is not being used at all. Figure 8.6 shows the risk graph in accordance with DIN EN 954-1. The preferred category is marked here with a big red circle. In some applications the designer can branch out on a different category, one that is marked with either a little or big blank circle.

8.3 Risk Graph according to DIN EN 954-1

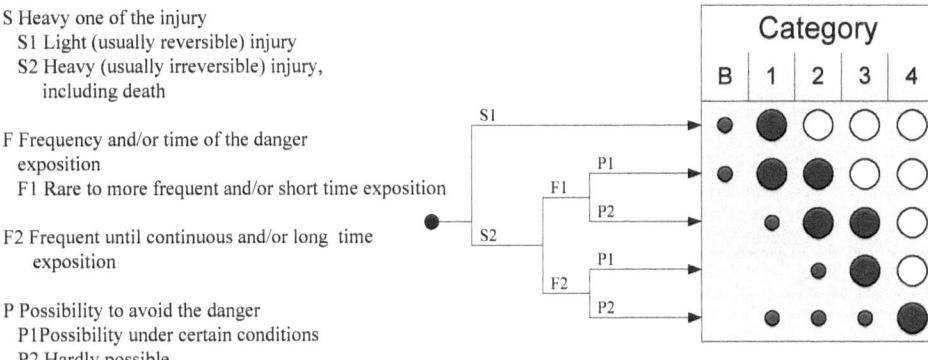

S Heavy one of the injury
  S1 Light (usually reversible) injury
  S2 Heavy (usually irreversible) injury, including death

F Frequency and/or time of the danger exposition
  F1 Rare to more frequent and/or short time exposition
  F2 Frequent until continuous and/or long time exposition

P Possibility to avoid the danger
  P1 Possibility under certain conditions
  P2 Hardly possible

Selection of the category

B, 1 to 4 categories for safety-referred parts of controls

● Preferred categories for points of reference

● Possible category, which require additional measures

○ Measures, which can be oversized regarding the applicable risk

Figure 8.6: Risk graph according to DIN EN 954-1[75]

---

[75] [DINa96]   DIN EN 954-1, *Sicherheit von Maschinen – Sicherheitsbezogene Teile von Steuerungen*

Table 8.3: Categories and Requirements according to DIN EN 954-1

Category[1]	Summary of Requirements	System Behavior[2]	Principles for safety goals
B	The safety related parts of controls and/or their protectors as their parts must be in accordance with applicable norms as to building, selecting, composing and combining in a way that they can withstand the expectant influences.	The occurrence of an error could lead to the damage of the safety function.	Predominantly characterized through selection of components
1	The requests of B must be fulfilled. Approved components and approved safety principles must be used.	The occurrence of an error could lead to the loss of the safety function, but the likeliness of the occurrence is lower than in category B.	
2	The requirements of B and the application of approved safety principles must be fulfilled. The safety function must be tested in timely intervals by mechanical means.	• The occurrence of an error can lead to the loss of the safety function. • The loss of the safety function will be recognized through examination.	Predominantly characterized through structure
3	The requirements of B and the application of approved safety principles must be fulfilled. Safety related parts must be constructed, so: • That a single mistake in each of these parts does not result in a loss of the safety function. • That a single mistake in each of these parts does not result in a loss of the safety function.	• If one error occurs, the safety function is still conserved. • Some but not all errors are recognized. • An accumulation of errors could lead to a loss of the safety function.	
4	The requirements of B and the application of approved safety principles must be fulfilled. Safety related parts must be designed in a way, so that: • A single mistake in each of these parts does not lead to the loss of the safety function, and • The single error at or before the next requirement according to the safety function is recognized, or that, if this is not possible, an accumulation of errors then do not lead to the loss of the safety function.	• If errors occur, the safety function always remains. • The errors are being recognized on time in order to prevent damage of the safety function.	

(1) The categories are not meant to be applied in any given definite sequence or hierarchy regarding the technical safety requirements.
(1) The assessment of the risk will determine if the acceptance of the entire or partial loss of the safety function due to errors is acceptable.

Others can also be used as preferred different categories, but in the event of an error the intended behavior of the system should remain. The reasons for the deviation should be stated clearly. One of the reasons for the different selections as preferred categories could be the application of different technologies, like approved hydraulic or electro-mechanical

8.3 Risk Graph according to DIN EN 954-1

parts in combination with electric or electronic systems. The selection of the categories, which are marked by a little circle in Figure 8.6, can require additional measures like:

- Over-dimensioning or the application of technologies that leads to the exclusion of errors;
- Application of dynamic supervision.

A risk assessment with parameter $S1$ (see Figure 8.6) for example has category 1 as the category for the safety related part of the control. At some applications the designer can choose different protective measures category B.

9 Fault tree analysis

Fault Tree Analysis (FTA) has been initially developed in 1961 by G. Watson together with A. Mearns in the "Bell Laboratories". Since that time it had been used by the US Air Force for evaluation of the "Minuteman" rocket launch systems. Already in 1963 Dave Haazl realized how useful this method was, and expanded its usage, having implemented this tool at Boeing. In the 60's the sphere of usage of the Fault Tree Analysis method had expanded greatly in the aerospace industry, and in the 70's they started to implement it into nuclear engineering. The American government decided in the beginning of the 80's to formalize this method and in 1981 the U.S. Nuclear Regulatory Commission published the „Fault Tree Handbook"[76], which served as a basis for development of various methods and tools for FTA support. In the 80's this method was widely implemented in the chemical industry, and in the 90's – in robotics and in software development.

The German Standards Institute (DIN) has described the FTA method in the DIN 25424 standard[77]. Its first part is devoted to the description of the method and the use of symbols. The second part tells about methods of FTA implementation. The „Fault Tree Handbook" and DIN 25424 served as the basis of this abstract.

Fault Tree Analysis should not be confused with Event Tree Analysis (ETA) (described in DIN 25419[78]). Event Tree Analysis uses an inductive approach. In other words, it is a search for a failure, which lead to undesirable consequences (Bottom-up or forwards analysis). The FTA method uses an opposite or deductive approach. The undesirable event is present and one searched all causes (Top-Down or backward analysis).

9.1 Field of application and purpose Fault Tree Analysis

Fault Tree Analysis is an effective tool for revealing logical relations between failing components or subsystem. Interesting are those combinations of failures which lead to undesirable event, as these are to be avoided or, at least, the probability of their occurrence to be minimized.

---

[76] [NURE81] Fault Tree Handbook, als Download unter http.//www.nrc.gov, Dokumentname: NUREG-0492
[77] [DINc90] DIN 25424, 1990-04, *Fehlerbaumanalyse; Handrechenverfahren zur Auswertung eines Fehlerbaumes*
[78] [DINa85] DIN 25419, *Ereignisablaufanalyse; Verfahren, graphische Symbole und Auswertung*

Fault Tree Analysis has the following objectives:

- Systematic identification of all possible failure combinations (causes) leading to a given undesirable event (qualitative analysis);
- Evaluation of the system reliability attributes (e.g., frequency of failure combinations, frequency of undesirable events occurrence or unavailability of the system on demand) by calculating reliability attributes of the units of the system (quantitative analysis);

9.2 Terms

Item

Is the subject matter of the reliability specification (definition from DIN 40042). Such an object can be represented by: a system, a part of a system, a component, a functional; element, etc. At that, one should differentiate between technical and functional objects.

System

Integration of technical and organizational means for independent processing of a complex of functions (definition from DIN 40042).

Functional unit

A technical system can consist of several functional units. Each function of the system is allocated to a specific functional unit.

Sub system

A system part is a combination of components, used for performing tasks in a functional unit.

Functional part of a system

A functional part of a system is a combination of functional elements, used for performing tasks in a functional unit.

Component

The smallest item of a system, which can be evaluated according to its reliability attributes. In this case, each component is subject to one or more functional elements.

Functional element

Functional element is the smallest object of a functional group. It can only be used for primitive tasks (for example, switch on, steer, lock, open, etc.)

Procedure

Work regulations for normal operation, emergencies, maintenance, repair works, transportation, etc.

9.2 Terms

System analysis

In this context a system analysis is an analysis of a technical system. It can address the following areas:

- System functions (especially working objectives and deviations from them),
- Environmental conditions (which the system cannot influence),
- Resources,
- System components,
- System structure.

Failure (defect)

A failure is a excessive deviation from the set working objectives of the technical unit. The following types of failures exist:

- Primary failure (by acceptable environmental use),
- Secondary failure or consecutive fault (in unacceptable environmental use),
- Forced failure (operational components, which are used for achieving wrong or missing goals or resource fault).

Failure mode (type of failure)

The different ways a component can fail.

Basic event

Only failed functional elements should be chosen as basic events of a fault tree.

Top event (or undesired event)

Failure of the functional unit in hand. Undesirable events can be caused by various combinations of failures.

Failure combinations

This notion refers to combinations of faults of functional elements, which caused the present undesirable event.

Fault tree

A fault tree is a graphical representation of relations between the basic events and the present undesirable top event.

9.3 Graphical representation

The fault tree consists of the graphical representation of the basic events and their relations. The relations show the logical interaction between the basic events and the top-event. The basic and top-events can take on the following values:

- „0", „false" ... functional
- „1", „true" ... failed

Next the graphical elements will be listed and explained in brief.

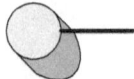

Basic event: This symbol represents a primary failure of a functional element, in case failure is possible for this element. The parameters for primary failure and the down time of the functional element are attributed to this symbol.

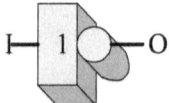

NO-Relation: This symbol demonstrates negation. When the input E is „0", the output A is „1" and vice versa.

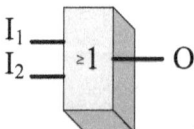

OR-relation: This symbol shows logical combination with the general known functional table. In this case, an unlimited number of inputs are possible.

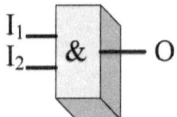

AND-relation: This symbol shows the logical cross-section in the general known functional table. Also in this case, an unlimited number of inputs are possible.

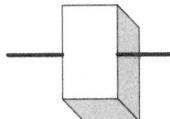

Comments: This symbol describes inputs and outputs of the relations. In general, output of one relation should not lead directly to the input of another relation. Instead, this symbol is used to describe an event in more detail

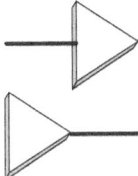

The fault tree can be cut and then continued from another point by using the transition output and the transition input.

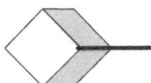

Secondary input: This symbol shows the input of a second relation. The symbol holds the parameters of the corresponding secondary input.

9.4 Analysis procedure

9.4.1 Analysis steps

The Fault Tree Analysis should be divided into 8 separate steps. Only this separation can guarantee accurate system evaluation. These steps are:

1. Precise structural analysis of the system
2. Determination of undesirable events and the failure parameters
3. Determination of the relevant probability parameters and the time intervals
4. Determination of the component failure modes
5. Creation fault tree
6. Determination of the basic events (such as: types of failures, time of fault occurrence and unavailability)
7. Analyze fault tree
8. Evaluation of the results

The following paragraphs will explain the steps in more detail.

9.4.2 System analysis

System functions

In general it is sufficient to analyze system functions by treating the technical system as a „black box" with different inputs and outputs. The output values of this box are defined by its functions.

The following parameters should be taken into account:

- Input values system,
- Output values system,
- Objectives system functions,
- Acceptable deviations from objectives.

It is possible that the system analysis is carried for different operational phases.

Working environment

Technical systems usually operate in an environment, which they are not able to influence. This environment and its influence on the system should be analyzed for each operational phase separately.

Resources

In normal operation technical systems use various resources. Such resources can be energy supply, informational support, or services. Deviation from the working objectives of these resources (up to a failure of one or more resources) can affect under certain conditions the system functions. This should be closely examined.

Only now look inside the "black box". Its composition should be examined and all the components should be identified.

For some components it is sometimes necessary to carry out an analysis as described above.

Organization and behavior

To complete the system analysis the technical system is analyzed according to the following three questions:

- How do the system components interact to carry out the system function?
- How does the system react to the environment?
- How does the system react in case of internal failures and the failure of support resources

At this point it is also analyzed how the human element influences the system (for example, unacceptable implementation of working regulations).

9.4.3 Undesirable event and failure criteria

Before it is possible to create the failure tree it is necessary to clearly define the top event to be examined. Here the question to solve is whether system safety or an operational function is examined.

When system safety is analyzed then the undesired top event represents also dangerous failure of the systems as derived during the system analysis.

9.4.4 Relevant reliability parameters and time intervals

In general two relevant reliability parameters with corresponding time intervals are distinguished:

- Frequency of failure in a specific time interval (for example, between two system tests),
- Unavailability on a given moment (or average unavailability in a given period).

9.4.5 Component failure modes

As soon as the to be examined undesired top event is clear the failure modes of the components relevant for the fault tree become clear.

It can prove useful to carry out separately a failure mode and effects analysis. This independent deductive method can simplify the creation of the fault tree, because it improves the understanding of the functionality of the given technical system (see DIN 25448 "Failure Mode and Effect Analysis" – FMEA).

9.4.6 Fault tree creation

Starting from the undesired event the fault tree is created from the top. Basic events and relations are presented in alphanumeric form, and commentaries on the failures of the examined functional units are situated in the corner.

The fault tree creation process is a simple algorithm, which is presented as a process chart in figure 9.1

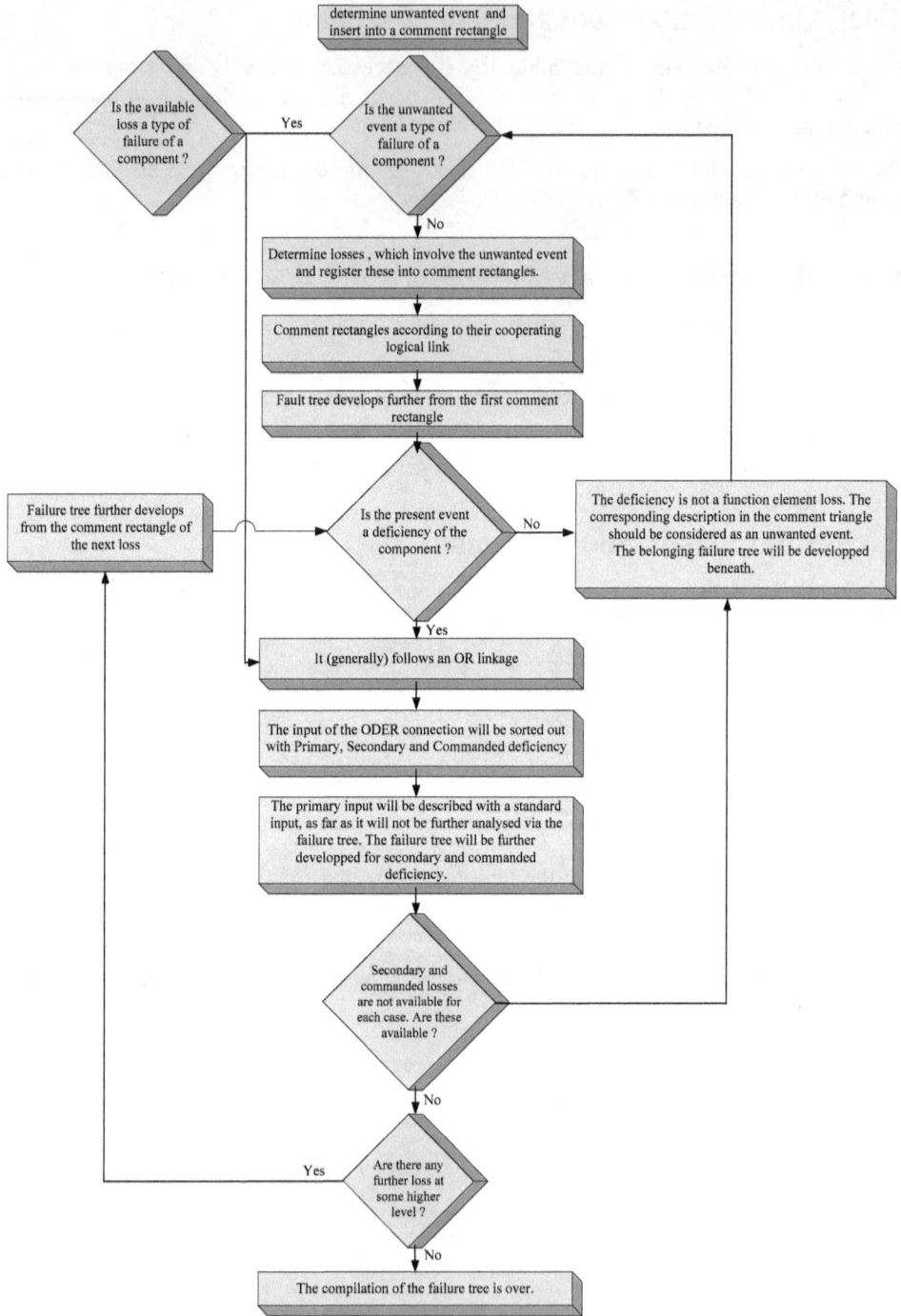

Figure 9.1: Process chart for fault tree creation

9.4 Analysis procedure

For better understanding the process chart is explained by a simple example.

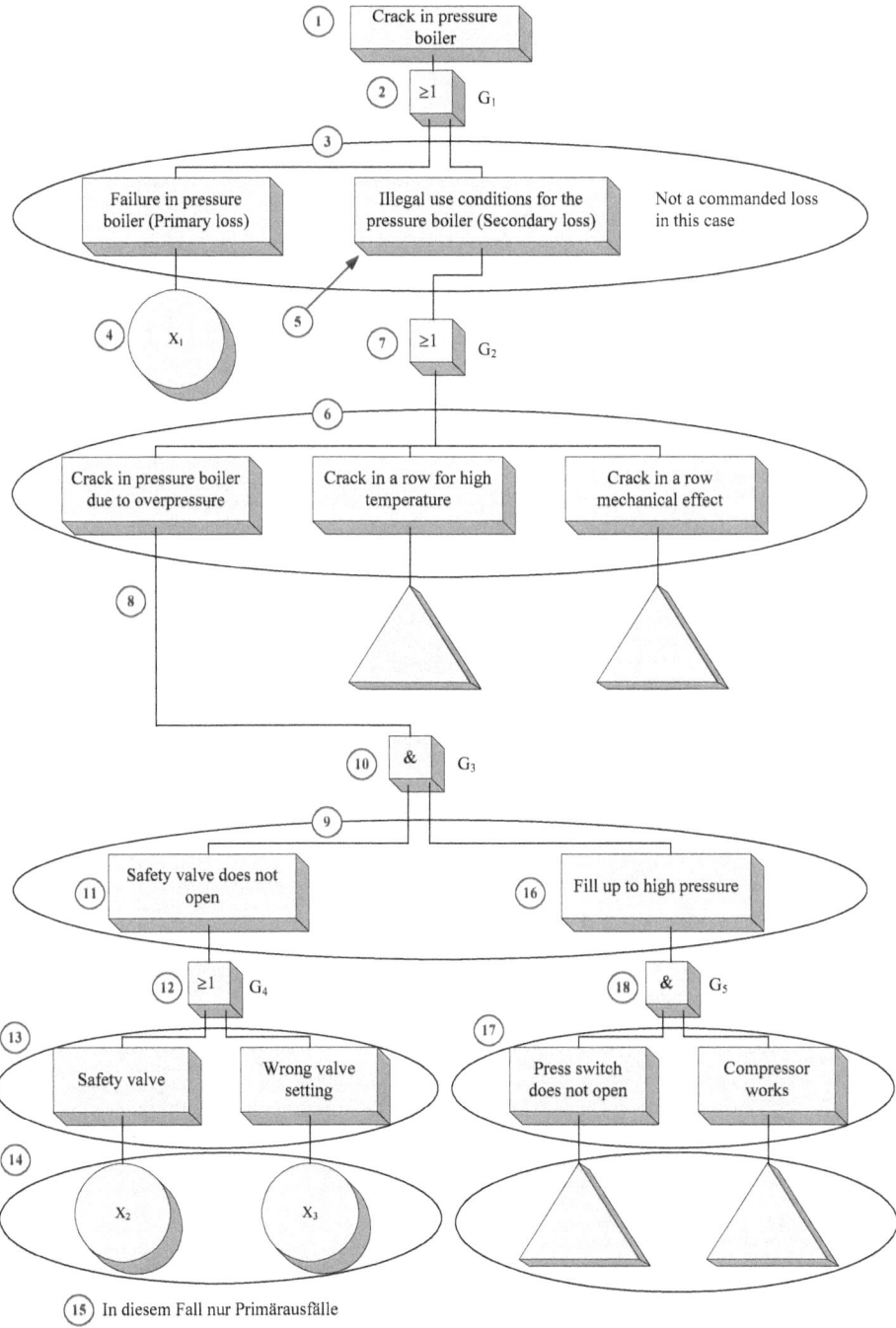

Figure 9.2: Fault tree for the event „Crack in pressure vessel"

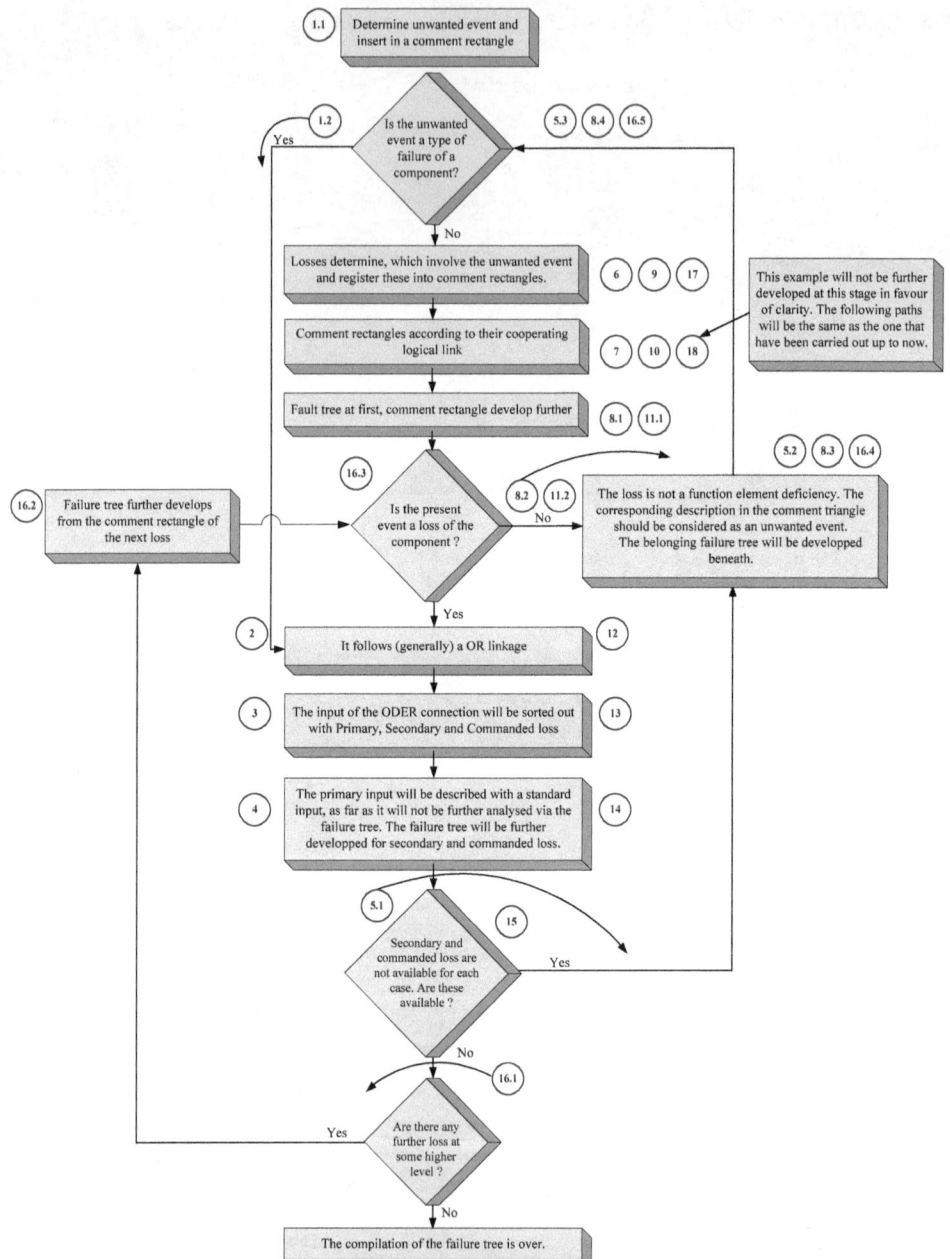

Figure 9.3: Process chart for the event „Crack in pressure chamber"

Subject to analysis is a system for the generation of pressure in a pressure vessel. The compressor delivers air or another gas into the pressure chamber. The compressor starts by pressing a switch or when the low pressure limit is reached. When the maximum pressure is reached the pressure switch is opened and the compressor is shutdown. A mechanical pressure relieve valve will vent air the atmosphere when the maximum pressure is exceeded, e.g., due to undesired further operation of the compressor or due to thermal pressure expansion.

The question for the analysis is: What is the probability of a crack in the pressure vessel within a year.

Figure 9.2 shows the Fault Tree, respectively its fragment, derived from the analysis of this event. The numbers in the circles show the time sequence of the Fault Tree creation. The scheme of sequence of operations for the Fault Tree creation is shown again on Figure 9.3. The index numbers on this figure relate to Figure 9.2, and differ only in the way that some steps are divided into several sub steps for better understanding.

9.4.7 Evaluation of the fault tree

Task description

Evaluation of the Fault Tree is based on both a qualitative and quantitative task description.

In case of the qualitative evaluation of the Fault Tree, it is necessary to consider all possible combinations of failures leading to the given TOP-event. These combinations of faults are sometimes called "Cut Sets". This is a term that is known from reliability block diagrams or similar graphs. Most interesting, in this regard, are "Minimal Cut Sets". These are combinations of faults, which do not contain other combinations of faults. For example, see Figure 9.4, which only shows the failure combinations A, B und D, the minimum cutsets. Failure combination A C contains A and is thus not a minimum cutest.

Minimal path sets are analogous to minimal cut sets. In English "Path Sets" are often called "Minimal path sets". Minimal cut sets are the smallest combinations of primary events, which can prevent the occurrence of the top-event, if they do not occur themselves. In other words, which components of the system should function for the top-event not to occur?

Quantitative analysis follows after qualitative analysis. Here, the probabilities of occurrence of the separate failure combinations, as well as top-events occurrence are examined to arrive at the system's availability and failure frequency.

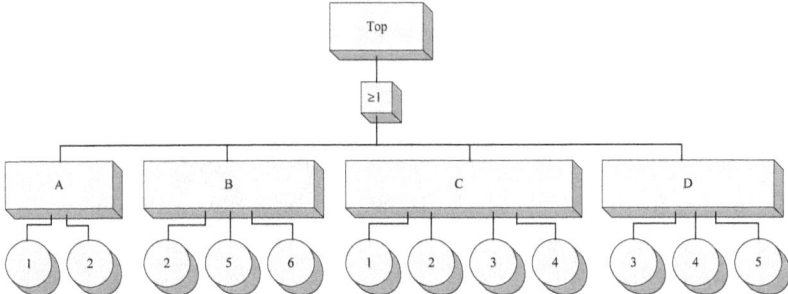

Figure 9.4: Example failure combinations and minimal cut sets

Minimal cut sets and -Paths in Reliability block diagrams

The meaning of minimal cut sets and paths is more clear in a reliability block diagram. A fault tree can easily be converted into a reliability block diagram, as shown here by the example of the pressure chamber in Figure 9.2.

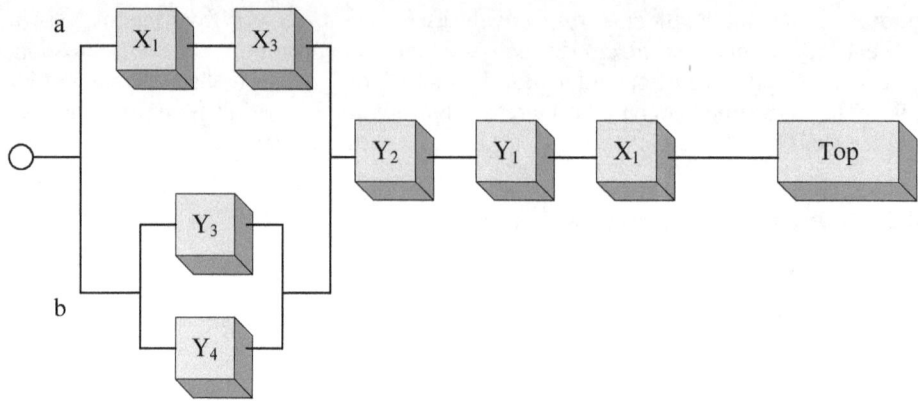

Figure 9.5: Reliability block diagram pressure vessel example

The following notations are used:

Primary event:

- X_1: failure in pressure vessel
- X_2: safety relieve valve fails
- X_3: Wrong valve configuration

Secondary events (without further development):

- $Y1$: crack caused by high temperature
- $Y2$: crack caused by mechanical effect
- $Y3$: pressure switch did not open
- $Y4$: compressor running

Path a represents in this case the event "safety relieve valve did not open" and path b – the event "Repletion until pressure to high". Parallel arrangement of both paths represents the event "a crack in a pressure vessel caused by over pressure".

A correlation between a fault tree and a reliability block diagram is easy to describe

OR-gates result into rows in a reliability block diagram, but an AND-gate into a parallel connection.

Minimal cut sets are the smallest unities of failed elements, which block the path from the input to the output. In this example, these are the cut sets $S1=\{X1\}$, $S2=\{Y1\}$, $S3=\{Y2\}$, $S4=\{X2;Y3;Y4\}$ and $S5=\{X3;Y3;Y4\}$.

Minimal path sets are similar to the smallest unities (functional) elements, which keep the path the "input" to the "output" open. In this example, the path sets are P1={X1;X2;X3;Y1;Y2}, P2={X1;Y1;Y2;Y3} and P2={X1;Y1;Y2;Y4}.

Minimal cut sets and minimal path sets can be easily derived from a reliability block diagram of this size. However, converting complex systems from a fault tree into a reliability block diagram is usually a difficult task. In addition, even if it succeeds, the search for correct cut sets and paths in the resulting block diagram will be impossible through visual inspection.

Deriving minimal cut sets and paths from a fault tree

As it already mentioned above, the search for minimal cut sets and minimal path sets is complicated for large systems when converting a fault tree into a reliability block diagram. If a fault tree is presented in the form as shown in Figure 9.4, then the minimal cut sets can directly be seen. Graphical conversion of fault tree into this form is unlikely and is never used for complex systems.

If a fault tree is presented as a Boolean function, then it can be converted into a disjunctive normal form and minimized. The resulting conjunctions are the searched the minimal cut sets and fully describe the fault tree. For this problem many algorithms exist which are easily implemented in software. One of the most famous algorithms is the Top-Down algorithm, which is presented here. This algorithm can be divided into 5 steps.

- A matrix is created, who's elementare are primary events or intermediate events.
- Starting from the TOP event, all events are successively replaced with inputs of the subordinated gate.
- An OR gate with n inputs is replaced with n new lines; in each of these lines, the gate is replaced with one of its inputs.
- For a AND gate with m inputs, the gate is replaced with its inputs on the appropriate line.
- The replacement is repeated from the TOP any number of times until the matrix will contain only primary events.

Elements in one line are now linked with AND operators, the lines locating one below the other are linked with OR operators. The matrix is virtually a Boolean function of the fault tree being in the disjunctive normal form, where each line represents a failure combination of the fault tree.

To be able to find minimal cut sets from the failure combinations, one applies, if possible, the Idempotence Law ($a \wedge a = a$) to each line of the matrix, what results in the elimination of multiple existing elements. After that, by using the Law of Absorption ($a \vee (a \wedge b) = a$), the lines containing other lines are deleted. In this way, only cut sets remain.

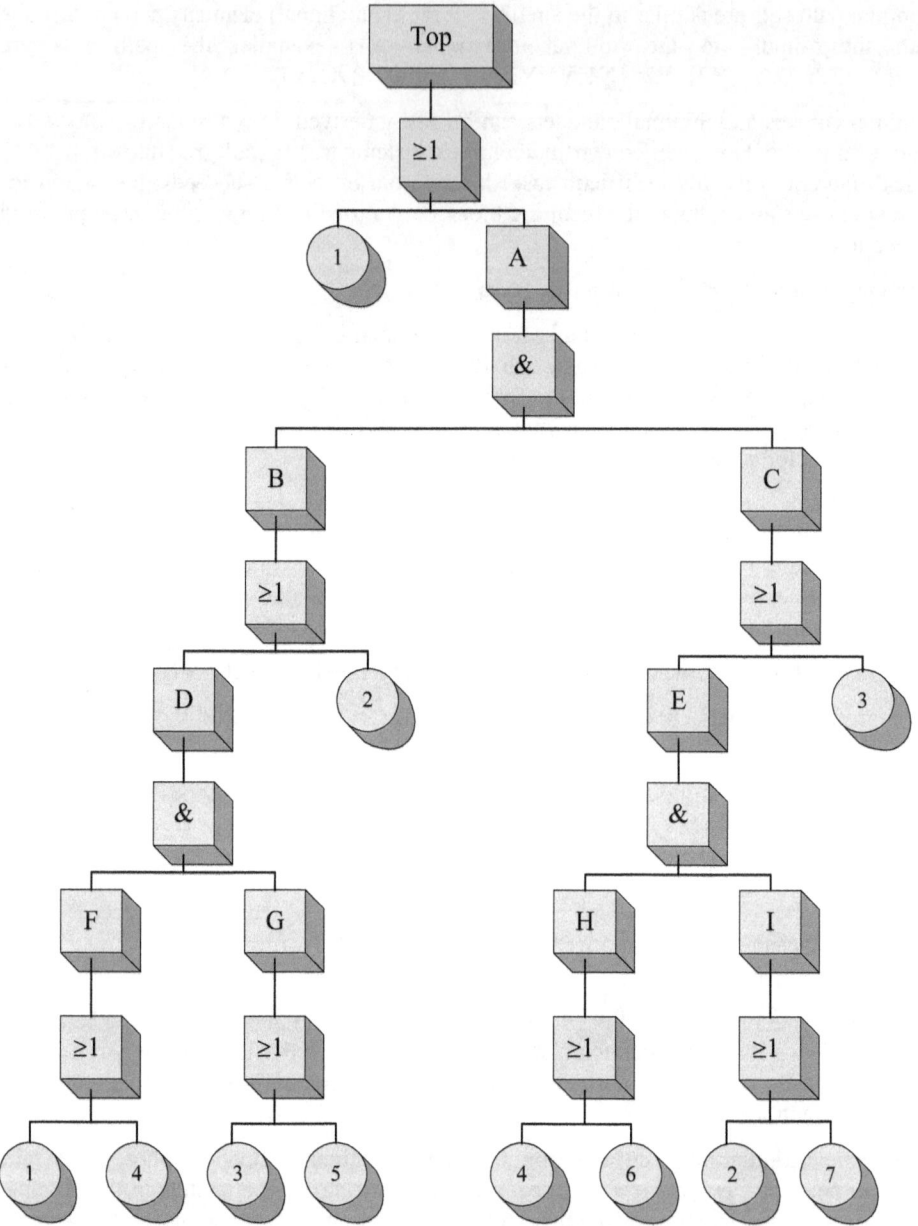

Figure 9.6: Example fault tree

The Top Down algorithm is shown in the fault tree of Figure 9.6. Table 9.1 contains the matrices for the Top Down algorithm, which results from the example according to Figure 9.6. The Top Event is the only entry in the M1 Matrix. Since an OR gate is located under the TOP event, the TOP is replaced in matrix M2 with two lines having the events 1 and A. A is the output of an AND gate, therefore this event in M3 is replaced with events B

9.4 Analysis procedure

and C in the same line. In M4, events B and C are replaced simultaneously, at the same time all mutual combinations of events D, 2 and E, 3 are generated. This procedure is repeated until matrix M6 contains only primary events.

After that, by applying the Idempotence Law, conjunctions marked with arrows are shortened. During the subsequent use of the Absorption Law, 1 located in the first line "absorbs" all conjunctions, which also contain a 1. After that, the conjunction of events 2 and 3 in the following line "absorbs" all other conjunctions containing these events, and so on. At the end, only six conjunctions remain. They are cut sets of the fault tree that can be created only in the form illustrated on Figure 9.4

Path sets can be created using the same approach. Since path sets are complements of cut sets, Morgan's theorem can be applied by denying all primary events and replacing all AND gates with OR gates and all OR gates with AND gates. Should the Top Down Algorithm be used again, you receive the fault tree as a conjunctive normal form. Individual disjunctions will be then path sets of the fault tree.

Table 9.1: Matrices for the Top-Down-Algorithm

M1		M2		M3			M4			M5			
TOP		1		1			1			1			
		A		B	C								
							2	3		2	3		
							2	E		2	H	I	
							3	D		3	F	G	
							D	E		F	G	H	I

M6					Idempotenz				Absorption			
1					1				1			
2	3				2	3			2	3		
2	4	2			2	4			2	4		
2	4	7			2	4	7					
2	6	2			2	6			2	6		
2	6	7			2	6	7					
3	1	3			1	3						
3	1	5			1	3	5					
3	4	3			3	4			3	4		
3	4	5			3	4	5					
1	3	4	2		1	2	3	4				
1	3	4	7		1	3	4	7				
1	3	6	2		1	2	3	6				
1	3	6	7		1	3	6	7				
1	5	4	2		1	2	4	5				
1	5	4	7		1	4	5	7				
1	5	6	2		1	2	5	6				
1	5	6	7		1	5	6	7				
4	3	4	2		2	3	4					
4	3	4	7		3	4	7					
4	3	6	2		2	3	4	6				
4	3	6	7		3	4	6	7				
4	5	4	2		2	4	5					
4	5	4	7		4	5	7			4	5	7
4	5	6	2		2	4	5	6				
4	5	6	7		4	5	6	7				

9.4 Analysis procedure

Calculation of the Unavailability and Failure density

In order to calculate from a fault tree the unavailability U and the failure density $\dot{H}$ of a system, it is first necessary to calculate the unavailability U_i and the failure density $\dot{H}_i$ the basic units. They can be calculated from the failure rates λ_i and repair rates μ_i, when the failure and repair rates of a basic unit are independent of the states of other basic units.

In the stationary case $(t \to \infty)$ the failure rates and repair rates result in:

$$U_i(\infty) = \frac{\lambda_i}{\lambda_i + \mu_i} \tag{9.1}$$

$$\dot{H}_i(\infty) = \lambda_i(1 - U_i(\infty)) = U_i(\infty) \cdot \mu_i = \frac{\lambda_i \cdot \mu_i}{\lambda_i + \mu_i} \tag{9.2}$$

After that, we calculate all gates available in the failure tree starting from the lowest level of primary events up to the TOP event. DIN 25424 Section 2 provides calculation procedures, which are specified in the table below. The availability is here a complement of the unavailability:

$$V_i = 1 - U_i \tag{9.3}$$

Table 9.2: Calculation procedures for gates in fault trees

Gates	Unavailability	Failure density
OR-Gate with n inputs	$U = 1 - \prod_{i=1}^{n} V_i \leq \sum_{i=1}^{n} U_i$	$\dot{H} = V \sum_{i=1}^{n} \frac{\dot{H}_i}{V_i} \leq \sum_{i=1}^{n} \dot{H}_i$
AND-Gate with n inputs	$U = \prod_{i=1}^{n} U_i$	$\dot{H} = U \sum_{i=1}^{n} \frac{\dot{H}_i}{U_i}$
Negation	$U = 1 - \overline{U}$	$\dot{H} = \dot{U} + \overline{\dot{H}}$ $\dot{H} = \overline{\dot{H}}$ (stationary)

However, this procedure provides only an exact result, if each primary event exists in the failure tree only once. They say: "Failure tree is unmeshed". If at least one of primary events exists in more than one copy, then the tree is not a real tree, they call it "meshed tree". In this case, calculation regulations provide only an upper (conservative) estimate.

They say: "Failure tree is unmeshed". If at least one of primary events exists in more than one copy, then the tree is not a real tree, they call it "meshed tree". In this case, calculation regulations provide only an upper (conservative) estimate.

9.5 Fault tree analysis

When using fault tree analysis, more complex logic types can be taken into consideration compared to event tree analysis or reliability block diagrams. However, fault tree analysis is limited to extend that failure rates of basis events must be calculable, because the fault tree can only be calculated if such results are available. Fault tree analysis is not applicable, if the system performance must be analyzed over time. In contrast to other methods, for this analysis this diagram is a construction consisting of logical relations between initiated, intermediated, and top events.

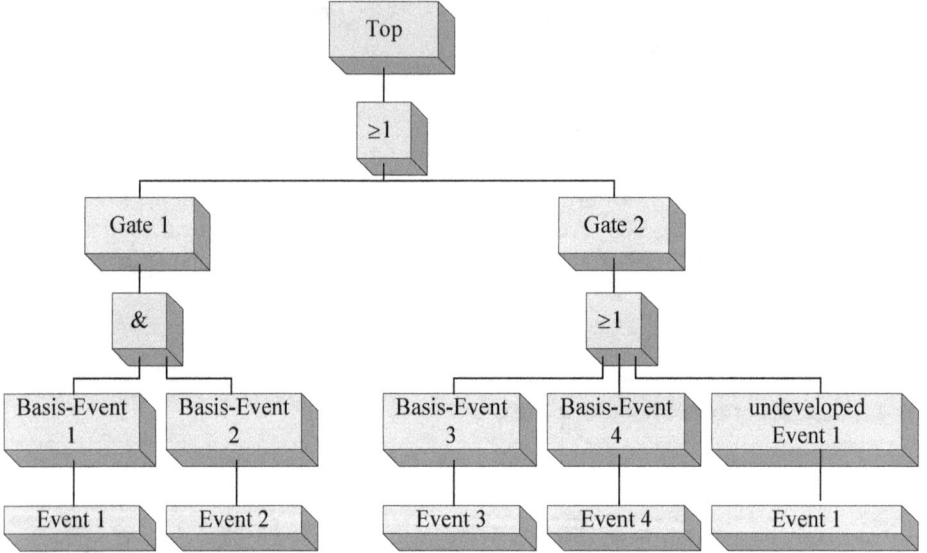

Figure 9.7: Fault tree analysis

The goal of creating a fault tree is to determine the probability of a top event at which a general system failure or a harmful state occur. The top event is based on the order of localized basic events at the bottom of the tree.

Basis events are linked to the top event by logical gates. For OR gates, probability values of inputs are added; for AND gates they are multiplied. As soon as the overall model is generated, the calculation of the top event probability is possible.

10 Event tree analysis

Event Tree Analysis[79] (ETA) belongs to the inductive type of analysis. According to this method, subsequent events are concluded from the initial event. According to this method, subsequent events are concluded from the initial event. Initial events – which are considered to be basic events of the Event Tree – can be, for instance, a condition of certain components, actions of the operators or other reliability parameters of the system to be modeled. Though binary notation of the events is usually sufficient, there is the option, with smart thinking to avoiding the binary way of thinking. The subsequent events are in simple cause-effect relation with the initial event. They take place after the initial event is finished.

In the past Event Tree Analysis was only of importance for reliability analysis of nuclear power plants. Today it is used for any systems, which behavior can be described by cause-effect chains.

Event Tree Analysis is a simple and easily to implement technique. The Event Tree begins with the initiating event, which is represented in the left part of the tree. From the initial event, the tree deviates into several branches, which represent subsequent events. Each branch leads to a situation with a different outcome. The Event Tree can lead to different outcome scenarios.

Despite the simplicity of the method, its field of application is quite limited, because it does not allow analyzing complex situations.

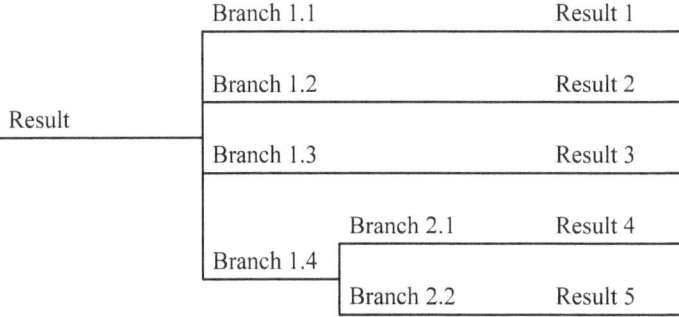

Figure 10.1: Event tree analysis

---

[79] [DINa85]   DIN 25419, *Ereignisablaufanalyse; Verfahren, graphische Symbole und Auswertung*

10.1 Components Event Tree Analysis

- *Initial Event*

  The initial event E_0 is the basis of any Event Tree. This event causes subsequent events through a chain of circumstances.

- *States and Events*

  Event tree analysis differentiates between an Event and a State.

  A state, for instance, can be represented by one of the component properties, module characteristics or human action. All states of a branch should be mutually exclusive. A state z_j in one of the branches becomes an event E_j. A functional binding line is connected to it, and in case of further division of the branch E_j new states leading to new events appear.

- *Functional binding lines*

  The connection between the event and its consequences is indicated by functional binding lines.

- *Division*

  If more than one state results from one event, then the event tree is represented in bifurcated form. In this case decision trees are differentiated on the basis of simple and complex bifurcations.

- *Sequences*

  A sequence begins with the initial event and consists of consequences, which lead to the final outcome. They are considered to be the end of the cause-effect chain.

- *Relations*

  Computational complexity can be lowered, using two or more identical residual sequences, which are connected according to „Exclusive-Or" principle.

Event E_i with $i \in \{0, 1, 2, \ldots, n\}$ as well as state z_j and its consequence E_j with $j \in \{0, 1, 2, \ldots, m\}$ act as random „events" in the analysis of the tree. The probability $\Pr(E_j) \geq 0$ is subordinate to the event E_j. In case of an event with bifurcation of the state $m + 1 \geq 2$, each conditional probability becomes a separate branch j, which will be described by $\Pr(z_j | E_j)$. A conditional probability shows that an event E_i acts as a consequence of a state z_j (which will then become an event E_j), at that j is not equivalent to i in its codomain. If an event E_j with $j \neq i$ comes as an outcome from a circumstance z_j, then the connection between E_i and z_j is specified. This connection can be represented in the following way:

10.1 Components Event Tree Analysis

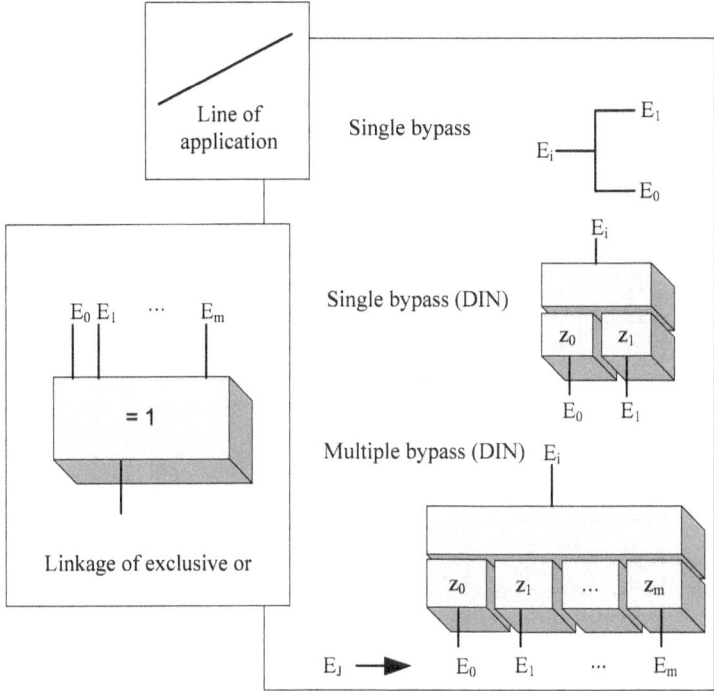

Figure 10.2: Event tree symbols according to DIN 25419/5.3-1/.

$$Pr(E_j) = Pr(E_i) \cdot Pr(z_j \mid E_j) \tag{10.1}$$

In practice probabilities Pr(E_i) and Pr($z_j \mid E_j$) are described, using the following residual probabilities for all states $m+1$ and their bifurcations.

$$\sum_{j=0}^{m} Pr(z_j \mid E_j) = 1 \tag{10.2}$$

Sum of the probabilities of the $m+1$ results calculates the probability of input events and can be described by the following formula:

$$Pr(E_i) = \sum_{j=0}^{m} Pr(E_j) \tag{10.3}$$

In the case, when all events of one sequence Seq_* are described as initial events (E_0), then the following sequence applies for the input probability:

$$Pr(Seq_*) = Pr(E_0) * \prod_{j=1}^{m} Pr(z_j * \mid E_j *). \tag{10.4}$$

A binary representation of the events is usually preferred, because in most cases it is sufficient. In implementation of this model, usage of event trees with simple bifurcations is often sufficient, as it is shown on Figure 10.3. Based on the differences between circumstances and consequences in this tree, the following applies:

$$Pr(E_7) = Pr(Seq_1) \tag{10.5}$$

$$Pr(E_8) = Pr(Seq_2) \text{ usw.} \tag{10.6}$$

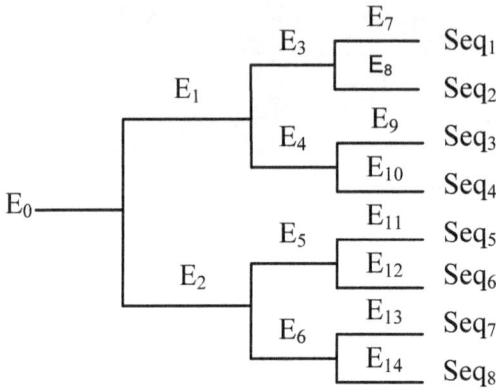

Figure 10.3: Event tree with single branches

The number of relations in one tree is indicated by the letter b (branches), and the number of sequences, in this case is $b+1$. Approximation is often used to reduce the number of sequences, thus reducing the number of computations. It is only possible if the input probability of the states z_j is very low in comparison with the probability of its consequences appearance (same as comparing to the present event). In this case, all relations, coming out of the present consequence E_j can be left unconsidered. A simplified event tree is the result with approximate residual sequences, as it is shown on Figure 10.4. This simplified event tree relates to the probabilities of Figure 10.3.

$$Pr(E_0) \gg Pr(z_2 \mid E_0). \tag{10.7}$$

If multiple states arise from the event E_j, then they are represented as multiple relations, as shown on Figure 10.5.

10.1 Components Event Tree Analysis

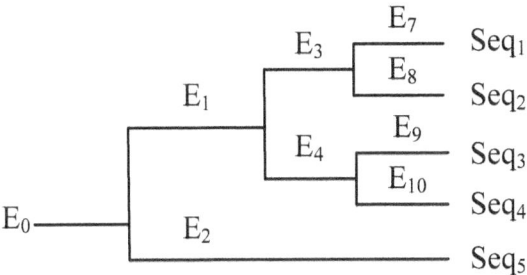

Figure 10.4: Reduced event tree with approximated residual sequences

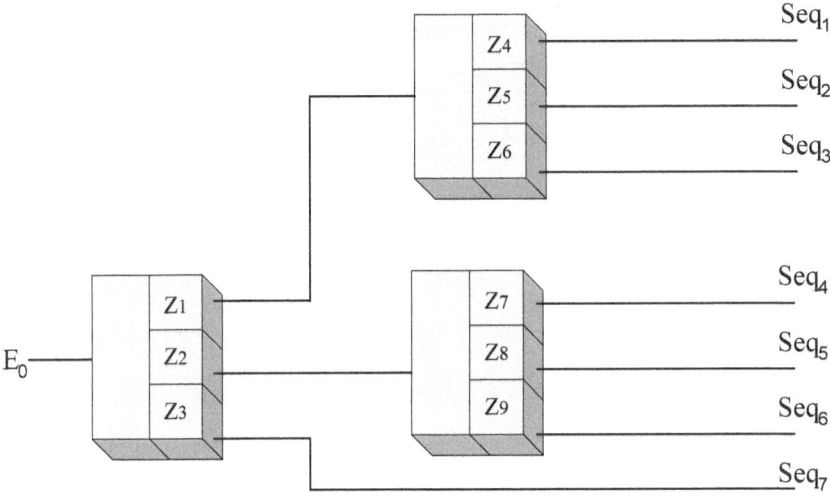

Figure 10.5: Example event tree with multiple branches

Simple relations in such event trees can be represented as Exclusive-Or relations, thus reducing the number of necessary computations. In this case, not states, but identical residual relations are analyzed.

In case the initial events from E_{1*} to E_{m-1*} have an Exclusive-Or relation, the probability of the consequence event E_{m*} is calculated as follows:

$$Pr(E_m*) = \sum_{j=1}^{m-1} Pr(E_j*). \qquad (10.8)$$

Figure 10.6 shows an example of a relation.

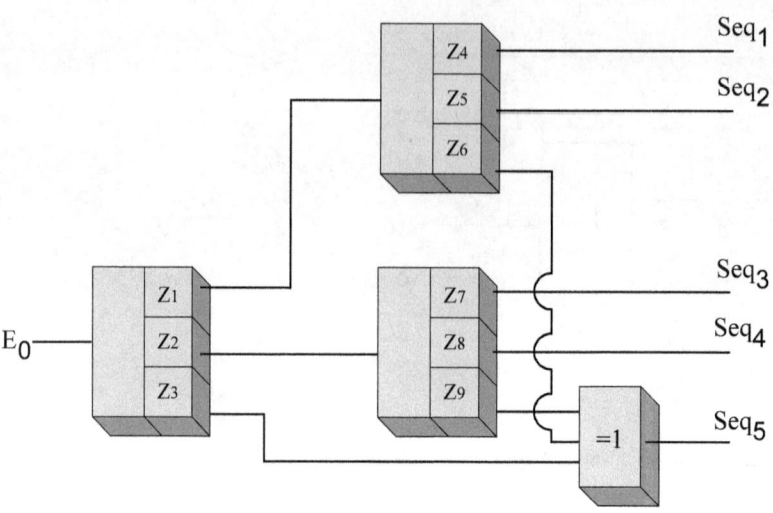

Figure 10.6: Event tree with combined identical residual sequences

11 LOPA

LOPA (Layer of Protection Analysis) is a modified event-tree-analysis. It is being used specifically for a risk analyses in the chemical industry and was developed in the 1990s in the USA. LOPA determines the Safety Integrity Level (SIL) for safety oriented processing plants. The principle of protection levels, their number and their valuation, was published for the first time by R. Growland[80] in 1993 by the Center for Chemical Process Safety (CCPS). The different protection levels within a plant fraught with risk is being described most descriptively through the onion-peel-model, see Figure 11.1. The single levels are independent and physically separate.

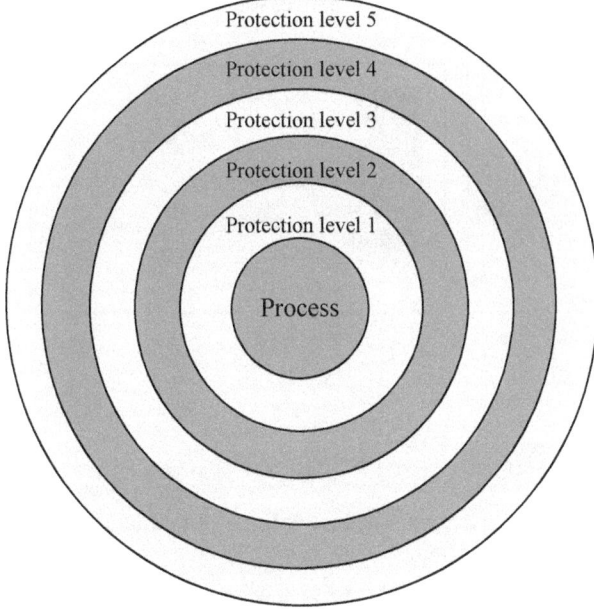

Figure 11.1: Onion Peel Model according to IEC 61511[81]

---

[80] [CCPS93] Center for Chemical Process Safety, *Guidelines for Safe Automation of Chemical Processes*

[81] [IECb00] IEC 61511, *Functional Safety: Safety Instrumented Systems for the Process Industry Sector*

Based on this layer of protection analysis model this method was adapted by different industry sectors according to their specific needs[82]. The following criterions are valid for all protection levels that are being developed in order to reduce a risk:[83]

- Specific potency: An independent protection level must be developed specifically for a precise requirement in order to prevent the consequences being observed.
- Independence: The protection level must function completely independent from all different protection levels, no equipment may be used simultaneously with different protection levels.
- Reliability: The protection level of the safety system should reliably protect from the occurrence of a consequence. Systematic as well as random failures must be taken into consideration at the development of the device.
- Verifiability: The protection layer/safety system must be tested and maintained. Such functional tests are necessary to ensure the reduction of the risk.

Instead of the Onion Peel Model the so-called LOPA-diagram according to the event-tree-analysis is being used. This shows with the help of two symbols (arrow and block) the direction of action of a failure/ a disturbance and the different independent protection layers that are supposed to neutralize or prevent the disturbance, see also Figure 11.2.

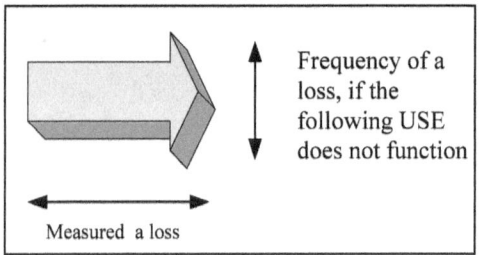

Figure 11.2: Structure of a LOPA diagram

11.1 Layers of Protection

Just like the general form of event tree analysis, one begins with an initiating accident, a disturbance and extends it into a chain of incidents.

These incidents are complementary incidents: failure of the protection level or safety via a IPL (Independent Protection Layer).

---

[82] [DOWa97]   Dowell, A. M., III, *Layer of Protection Analysis: A New PHA Tool, After HAZOP, Before Fault Tree Analysis*
    [CCPS01a]   Center for Chemical Process Safety, *Layer of Protection Analysis, Simplified Process Risk Assessment*
[83] [MARS02]   Marszal, E., Scharpf, E., *Safety Integrity Level Selection, Systematic Methods Including Layer of Protection Analysis*

11.1 Layers of Protection

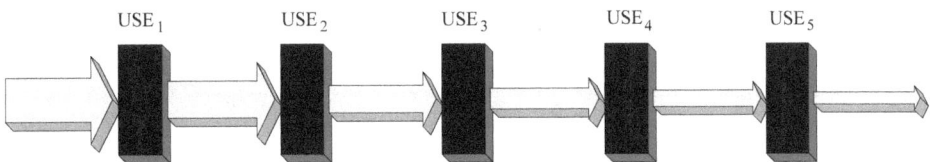

Figure 11.3: Protection Level Concept for five independent protection levels (USE), according to CCPS[84]

By conducting a LOPA, for each incident/disturbance and the developing consequence a separate risk analysis has to be conducted. This is especially necessary when it comes to common cause failures. If common cause failures occur, different occurrences of disturbances lead, according to the chosen protection levels, to different consequences frequencies. For example a fire braking through a protective fire wall as a protection level can be detained or be delayed at least for a certain time. For example, a firewall can protect against or at least delay a fire for a period of time. If on the other had a safety related system is used as protection layer then this will certainly result in a different failure rate as a firewall. The result of the risk assessment is too optimistic if the single consequence of a common cause failure is not being viewed in a LOPA.

The Layer of Protection Analysis (LOPA) is, like mentioned before, a special type of the event-tree-analysis. It optimized in deciding the frequency of an undesirable incident that can be prevented through one or more protection layers. One could determine the safety integrity level (SIL) by comparing the resulting frequency with the tolerable risk. LOPA uses actuating incidents similar to the event-tree-analysis. For the quantitative determination of the risks one must decide the frequency of these initiating events. The protection layers at a LOPA are analogies to the branches in an event-tree-analysis. In a LOPA each branch represents a set of complementary events, originating from the statement "protection layer functioning successfully" or "protection layer is failing". One calculates the probability of an undesirable event comparable to the event-tree-analysis, with the difference that one is only interested in one event; the undesirable event. Therefore only the frequency of the dangerous event is being calculated.

By conducting a LOPA, it is sometimes necessary to view several actuating events. This is done by analyzing each one incident with its own LOPA-diagram and then selecting the SIL (safety integrity level) on the basis of the incident with the highest requirements.

Figure 11.4 shows a typical diagram of a protection layer analysis. Similar to diagrams being used in an event-tree-analysis, the LOPA-diagrams are being viewed from left to right, beginning with the actuating event. There are five protection layers that follow the actuating event. Each layer consists of a set of complementary events: either protection-layer-failing or protection-layer-success. If the protection layer functions accordingly, there will be no further effects of the assumed original disturbance and the analysis of this branch is complete. If the protection layer fails, the analysis of the error branch to the next protection layer or to the relevant dangerous outcome will continue.

---

[84] [CCPS01a]   Center for Chemical Process Safety, *Layer of Protection Analysis, Simplified Process Risk Assessment*
     [DOWa02]   Dowell, A. M., III, Hendershot D. C., *Simplified Risk Analysis - Layer of Protection Analysis (LOPA)*
     [GOWL04]   Gowland, R., *Practical Experience of applying Layer of Protection Analysis For Safety Instrumented Systems (SIS) to comply with IEC 61511*]

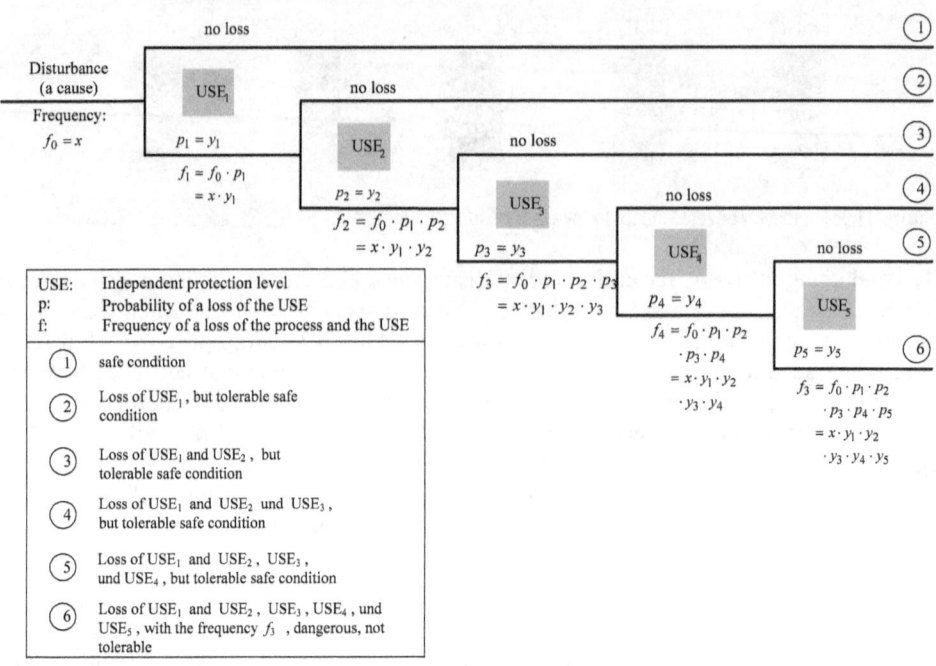

Figure 11.4: LOPA calculation for five independent protection levels (USE), according to CCPS[85]

In order to declare a protection layer an independent protection layer (IPL), the following criterions are useful and are known in the English language literature as "3 D's", "4 E's" and "Big I": [86]

- Detect: a IPL should be able to recognize a danger
- Decide: a IPL causes a corresponding action
- Deflect: a IPL reduces the consequence of a danger

The effect of a USE could be classified through the"4 E's":

- sufficient size ("big enough")
- sufficient speed ("fast enough")
- sufficient strength ("strong enough")
- intelligence ("smart enough")

---

[85] [CCPS01a] Center for Chemical Process Safety, *Layer of Protection Analysis, Simplified Process Risk Assessment*

[DOWa02] Dowell, A. M., III, Hendershot D. C., *Simplified Risk Analysis - Layer of Protection Analysis (LOPA)*

[GOWL04] Gowland, R., *Practical Experience of applying Layer of Protection Analysis For Safety Instrumented Systems (SIS) to comply with IEC 61511]*

[86] [DOWa02] Dowell, A. M., III, Hendershot D. C., *Simplified Risk Analysis - Layer of Protection Analysis (LOPA)*

The most important criterion is however the independence ("Big I"): A IPL must under all circumstances be independent of the actuating incident and of the different protection levels!

The concept of "independent protection layer" is being applied when used for an existing dangerous event. Thereby the risk of a dangerous event is being reduced to a residual risk. Usually a protection layer can be used against more than just a dangerous event. Therefore the risk is not being reduced to a residual risk, but the protection layer is reducing the risk per se. Therefore two categories of safety levels can be used to limit the risk for a system fraught with risk; the "independent protection layer" and the "risk reducing protection layer".

An "independent protection layer" reduces the risk of the appearance of a specifically dangerous event to a minimal residual risk, while a "risk reducing protection layer" reduces the risk only by a certain amount. The "risk reducing protection level" additionally incorporates safety measures into a system fraught with risk. Each independent protection layer and each risk reducing protection layer can only be taken into consideration once. LOPA shows by means of a probability of failure of the respective protection layers referring to the dangerous event how high the influence of a protection layer on the risk reduction is.[87]

11.2 LOPA Valuation

LOPA is a semi-quantitative risk analysis method.[88] Ideally it will be used after a qualitative risk analysis like a HAZOP (Hazard and Operability). It is a semi-quantitative method because the dangerous events can be assigned generic failure rates by assigning a probability of failure according to a certain timeframe and different layers of protection with a specific dangerous incident. These frequencies and probabilities of failure are determined and displayed in dimensions only. These statements are conservative in comparison to exact, quantitative risk calculations. Quantitative risk calculations are possible through a fault tree/event tree analysis or a reliability block analysis.[89]

A LOPA always begins with the definition of a dangerous event in form of a disturbance of the normal business process, see Figure 11.4. Could the disturbance result in different grave consequences, called scenarios, one must determine the frequency for each single scenario of the occurrence f_c. A LOPA views only the frequency of the occurrence of a dangerous event in connection with the probability of failure of the planned protection levels. The frequency f_I of an actuating incident will be multiplied with the specific probabilities of failure of the protection levels at request p_l. If risk reducing protection levels are part of the overall safety concept, failure probabilities will be regarded in the same type and manner as the protection levels. The frequency of occurrence of a single scenario calculates itself consequently with equation 11.1, see also Figure 11.4.

---

[87] [LEES92] Lees, F. P., *Loss Prevention for the Process Industries*
[88] [DOWa02] Dowell, A. M., III, Hendershot D. C., *Simplified Risk Analysis - Layer of Protection Analysis (LOPA)*
[89] [SFKa04] Störfallkomission, SFK-GS-41, *Risikomanagement im Rahmen der Störfall-Verordnung*

$$f_C = f_I \cdot \prod_{j=1}^{J} p_j \qquad (11.1)$$
$$= f_I \cdot p_1 \cdot p_2 \cdot \ldots \cdot p_J$$

By determining the probability of a failure of protection one assumes that no other protection layer exists. Meaning all protection layers before the protection layer being observed and meant to protect from the dangerous event are being regarded as either failures or none existent. Taking this condition under consideration, one avoids low values of frequency appearances of a scenario and consequently the pretense of safety.

11.3 Typical Protection Levels

According to IEC 61511 a protection layer reduces the risk by controlling and regulating, applying safety and damage prevention measures[90]. Protection layers can be procedural measures like building of receptacles, technical facilities like safety related systems or organizational measures like a response plan. Three typical protection layers are being described, which are deployed in the process industry quite frequently.[91]

- Basic Process Control System (BPCS)
- Physical facilities
- External measures to reduce risks

These protection layers are independent from each other and fulfill the above mentioned criterions for a IPL. A LOPA ensures that each protection level is being considered exactly one time. Figure 11.5 shows a suggestion for a protection level according to IEC 61511.

---

[90] [IECb00]   IEC 61511, *Functional Safety: Safety Instrumented Systems for the Process Industry Sector*
[91] [DOWa02]   Dowell, A. M., III, Hendershot D. C., *Simplified Risk Analysis - Layer of Protection Analysis (LOPA)*

11.3 Typical Protection Levels

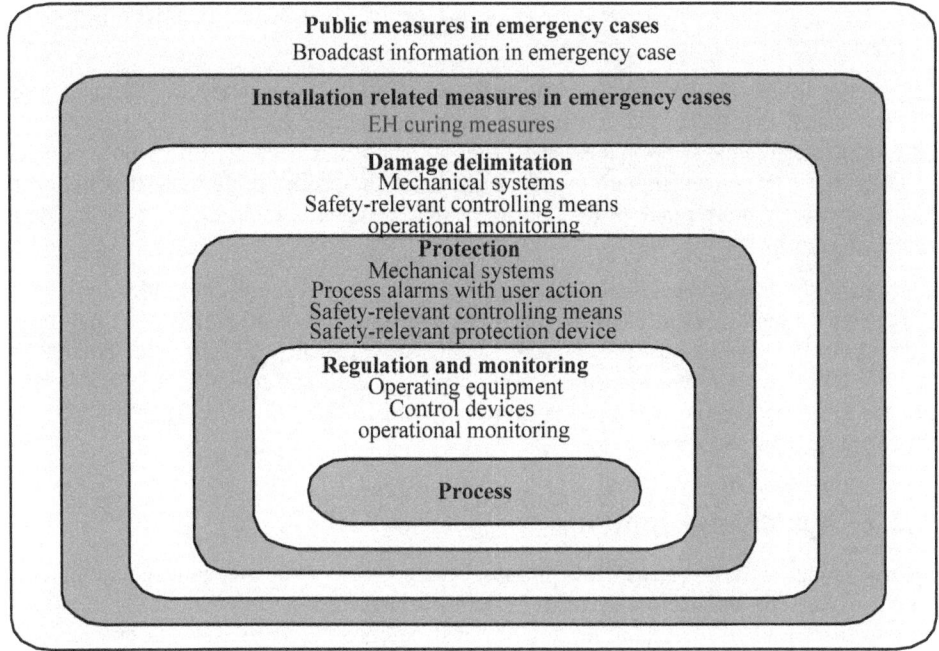

Figure 11.5: Suggestion for a Protection Level according to IEC 61511[92]

11.3.1 Basic Process Control System

The Basic Process Control System is a system that receives input signals, evaluates them and generates output signals accordingly. The BPCS can, under specific circumstances, be a protection layer, if it can interfere in a process and steer that process in a subsequent manner. A BPCS can be configured so it reproduces the actions of a safety instrumented system (SIS). If the BPCS is being used as a protection layer, the BPCS must precede the SIS, i.e., the BPCS responds before the SIS. A BPCS can be considered a protection layer of the following three conditions apply:[93]

- The BPCS and the SIS are separate physical units, including the sensors, the logical unit and the actors.
- A failure of the BPCS is not responsible for the triggering of an involuntary event.
- The BPCS has suitable sensors and actors in order to produce a function similar to the IS.

If a BPCS is set up as a protection layer, its failure probability will be given when requiring the PFD value[94]. The PFD- value of a system is defined through an availability analy-

---

[92] [IECb00]   IEC 61511, *Functional Safety: Safety Instrumented Systems for the Process Industry Sector*
[93] [DOWa02]   Dowell, A. M., III, Hendershot D. C., *Simplified Risk Analysis - Layer of Protection Analysis (LOPA)*
[94] PFD: Probability of failure on demand

sis. As an option there are many databanks' protection level, which receive information about overall control loop's failure rate.

Moreover it is important to explain the reaction of a BPCS on user interferences. One should consider that the BPCS- Protection layer is connected to an automatic action and that a user interference is not necessary. When an alarm has been set off, which requires the interference of a user, though this first generates the proper protection, it is better to consider the user interference protection layer. Therewith it is important not to valuate a protection layer twice.

In any case the effectiveness of a BPCS will be achieved through the under limit of a safety related function with SIL 1. When one carries out a protection analysis then make sure that the PFD- value of a BPCS is not smaller as 0,1. When a PFD- value smaller as 0,1 for a BPCS is required, then one should create, install, and maintain a BPCS as one does for a SIS according to IEC 61511 and/or ANSI/ISA-84.01-1996. Thereby a BPCS will be put on a par with a SIS.

11.3.2 Physical equipment

Physical equipments to release the pressure are not technical, mechanical equipments which by exeeding the normal pressure or temperature work areas of an action lead to a pressure discharge itself. Examples for physical equipments to release the pressure are pressure relief valve, temperature relief valve, bursting disks, breaking bolt or melting fuse plugs.

A pressure relief valve opens when the pressure during the process is high and leaves any process material leak in a secondary container or in the atmosphere. There are many different sorts of valves and technologies but the main are direct or indirect acting. Direct acting valves open under the pressure of the process liquid, which works against the spring in the valve. When the pressure created through the process liquid is stronger than the one of the spring, the valve opens and releases consequently the pressure. As soon as the pressure regularise itself, the valve will close. Indirect acting valves work similar to the direct valves, but they use a process liquid mechanism to open and close .

Bursting disks are weak points in the integrity of the tank. They will be so dimensioned that they will break under a pressure which is lower than the one laid out for the tank. This allows to leak safe the content of the tank in the atmosphere or in a secondary container. If they are accurately chosen and installed the bursting disks have an extremely small probability of a dangerous failure. This can only happen through a plugging of the way to the disc or a too high back pressure. Bursting disks are certainly very efficient, but their installation is limited because of a too long process down time when changing a disc.

Breaking bolts are also weak points. A deformable pin will be twisted under the pressure of the process liquid and lead to a ventilation or to a secondary container. As far as their function is similar to the one of a disc, pins also have a low dangerous failure rate. Moreover the down time is generally shorter as the one of a disc, though it also depends on both installations.

Melting fuse plugs work similar to bursting disks. However they react to over temperature instead of over pressure. Melting fuse plugs are often used in transport container, where fire could be the only cause of a high pressure. Melting fuse plugs consist of a solder which melting point has been chosen so that it melts before an overpressure causes an external fire ignition

The probability for the failure of a physical equipment to release the pressure varies with the sort of the equipment. The failure rate for all these different sorts of pressure relief valves can be found in many databanks or reference books. The failure probability of a bursting disk is principally a probability function that a plugging in the way to the disc and or a high back pressure. Human factors play a part depending on the sort of construction that has been used. Failure rate of melting fuse plugs are also very low about the size of the bursting disks.

It is important to fix the size of the here described equipments to release pressure and its corresponding leading ways. Make sure that you can reduce the pressure faster than the potential expansion rate. Otherwise the pressure will grow further in another system, though the equipments to reduce the pressure are opened. Hereby the corresponding protection layer failed as if the equipment had never been opened.

11.3.3 External systems to reduce the risk

External risk reduction facilities – ERRF are active protection layer which first act when chemical products have been released in the environment. On that account they are active systems because they use energy. The opposite of active systems are passive systems such as dike for example. One suppose that passive systems never fail, and that they only diminish the consequences gravity of a release.

An ERRF will be considered as a protection layer if it fully protects against any consequences. If the ERRF only diminish the gravity of the consequences it will not be taken into account. In this case the risk reducing effect of the ERRF should be considered into a consequences analysis. The failure probability of a ERRF will be determined and its components will be analysed. In some case this requires a pattern to predict failures such as a failure tree.

As well as the equipments to release the pressure, ERRF systems must be created and sized very carefully. A well-known example in which this has not been made was a system for a water quench in 1984 in Bhopal (India). The methylisocyanat-water quench system was far too under dimensioned for the released amount of methylisocyanat. The retention basin of the water quench was also too small for the released amount so that the catastrophe aggravated. Indeed, the methylisocyanat did not mix to the water so that it dispersed outside the retention basin.

11.4 Several actuating events

The LOPA procedure, as presently described, considers that only a triggering event can be the cause of an incident. Such incidents generally have but more than one potential trigger.

LOPA can efficiently be inserted with several triggering events in certain cases. If it is the case for a potential danger, this dangerous event will be splitted into single danger points and solve each triggering event in a separate LOPA. Once all analysis have been carried out, a frequency of the resulting incident will be given for each triggering event. One define the incident overall frequency by adding the frequency of each event which lead to danger.

12 Reliability Block Diagram Analysis

The reliability block diagram (RBD) is a stable probability model that is quite easy to use for reliability and failure probability calculations. Each block in a diagram represents a component of the system. The configuration of a block represents the logical relation between the potential losses of the components. Through the vertical configuration of the blocks parallel functional ways are being represented, which are OR-functions. If horizontal blocks are being displayed, two specified conditions are being described resulting in an AND-function. In the graph one can also depict complex component groups like 2oo3 Systems, see Figure 12.1. The failure rate of these blocks is determined by particular mathematical calculations. The RBD is strictly mathematical and therefore easy to apply. In reality the application is restricted because only mathematical computable events lead to a result.

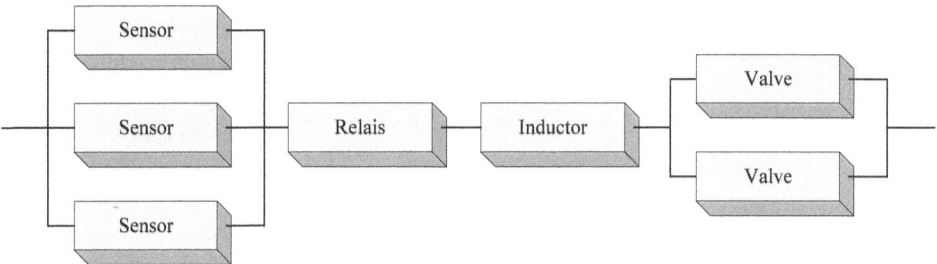

Figure 12.1: Reliability Block Diagram

The Reliability Block Diagram (RBD) shows which elements of a system fulfill the demanded function and which might fall out. The RBD is one of the most widely used methods to represent systems in a graphical form. The system is dismantled into elements in order to prepare a RBD; those elements fulfill a specific task. Each block relays a reliability characteristic of a component. If a component has several types of failure each type must be represented by way of a block.

There is an essential difference between a RBD and a function diagram. In a RBD elements can occur several times even though they exist only once as hardware.

Serial and parallel structures are the simplest kinds to link components. Figure 12.2 shows the linking of n components into a serial structure.

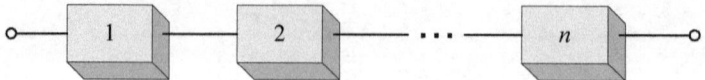

Figure 12.2: Linkage of n components into a serial structure

With the assumption that these n components are independent from each other, one can describe the probability of failure $F(q)$ of a series system with equation 12.2. Here, q_i is the probability of failure of component i. The description of the reliability $R(p)$ of a series system is possible in way of equation 12.3. Here p_i describes the function probability with the component i. The connection between p and q are shown in equation 12.1.

$$p = 1 - q \qquad (12.1)$$

Because the system attributes, probability of failure and reliability

$$F(q) = 1 - \prod_{i=1}^{n}(1 - q_i) \qquad (12.2)$$

and

$$R(p) = \prod_{i=1}^{n} p_i \qquad (12.3)$$

can be decided directly in this case, the formation of the Boolean equation y with

$$y = \bigvee_{i=1}^{n} x_i \qquad (12.4)$$

and the structure function Φ with

$$\Phi(x) = 1 - \prod_{i=1}^{n}(1 - x_i) \qquad (12.5)$$

is not necessary. This is also valid for the parallel structure.

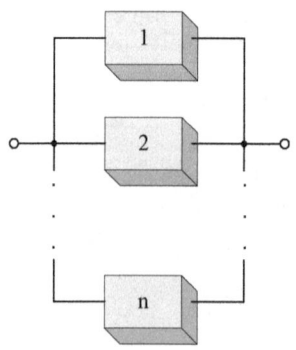

Figure 12.3: Linkage of n components into a parallel structure

For parallel structure the following is valid:

$$y = \bigwedge_{i=1}^{n} x_i \tag{12.6}$$

$$\Phi(x) = \prod_{i=1}^{n} x_i \tag{12.7}$$

The parameters, the probability of failure and the reliability are consequently:

$$F(q) = \prod_{i=1}^{n} q_i \tag{12.8}$$

$$R(p) = 1 - \prod_{i=1}^{n} (1 - p_i) \tag{12.9}$$

The shaping of the cross linking (like meshing) is also possible with block diagrams. Shannon's decomposition theorem is being used with smaller systems in order to appoint the parameters of probability.

$$\Phi(x) = x_i \cdot \Phi(x_1, x_2, ..., x_{i-1}, 1, x_{i+1}, ..., x_n) + (1 - x_i) \cdot \Phi(x_1, x_2, ..., x_{i-1}, 0, x_{i+1}, ..., x_n) \tag{12.10}$$

Shannon's decomposition theorem is one of many algorithms that allow the methodical development of Boolean functions.

With the general function

$$F(q) = q_i \cdot F(q_1, q_2, ..., q_{i-1}, 1, q_{i+1}, ..., q_n) + (1 - q_i) \cdot F(q_1, q_2, ..., q_{i-1}, 0, q_{i+1}, ..., q_n) \tag{12.11}$$

one gets the probability of failure for any system.

The following example represents the use of the theorem.

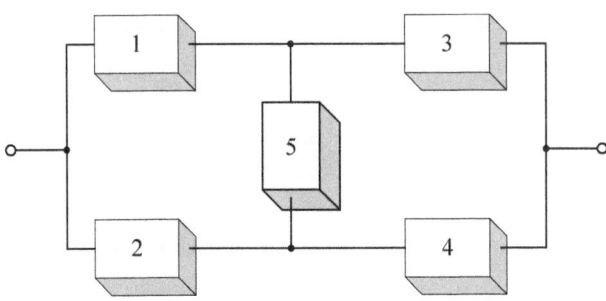

Figure 12.4: Bridge structure with five components

Figure 12.4 shows a bridge structure which fifth component is a meshing which is not presentable through serial or parallel structure.

The first event to be viewed is a failure of component 5, $x_5 = 0$.

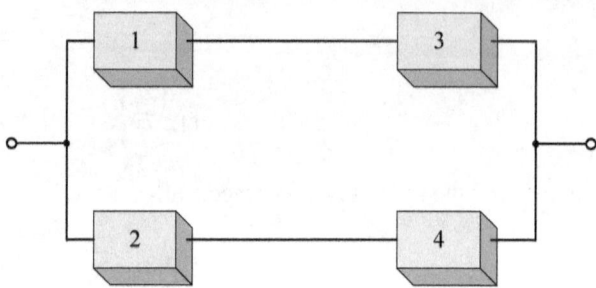

Figure 12.5: $x_5 = 0$ (failure)

With $x_5 = 0$ follows:

$$\Phi(x_1,..., x_4, 0) = (1-(1-x_1)\cdot(1-x_3))\cdot(1-(1-x_2)\cdot(1-x_4)) \qquad (12.12)$$

The second event being viewed is $x_5 = 1$.

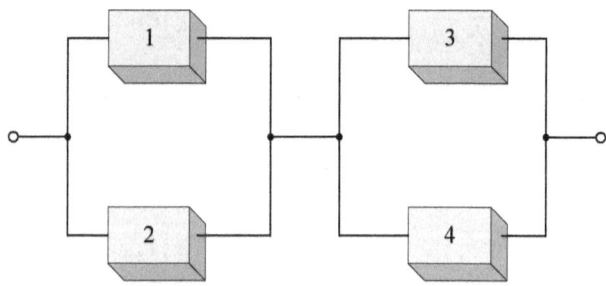

Figure 12.6: $x_5 = 1$ (able to function)

With $x_5 = 1$ follows:

$$\Phi(x_1,..., x_4, 1) = 1-(1-x_1 x_2)\cdot(1-x_3 x_4) \qquad (12.13)$$

For the entire bridge structure the function Φ is

$$\Phi(x) = x_5 \cdot \Phi(x_1,..., x_4, 1) + (1-x_5)\cdot \Phi(x_1,..., x_4, 0) \qquad (12.14)$$

Comparable to Φ the probability of failure F with

$$F(q_1,..., q_4, 0) = (1-(1-q_1)\cdot(1-q_3))\cdot(1-(1-q_2)\cdot(1-q_4)) \text{ und} \qquad (12.15)$$

$$F(q_1,..., q_4, 1) = 1-(1-q_1 q_2)\cdot(1-q_3 q_4) \text{ zu} \qquad (12.16)$$

$$F(q) = q_5 \cdot F(q_1,..., q_4, 1) + (1-q_5)\cdot F(q_1,..., q_4, 0) \qquad (12.17)$$

can be determined.

12 Reliability Block Diagram Analysis

With bigger systems the structure function can be developed by finding the minimum cuts or minimum paths. In the past years numerous calculator supported procedures have been developed that are based on this method. The assumption here is that the analyzed systems fulfill the monotonous characteristics.

Minimum cuts are the subset of all components $C_{Sch} \subseteq C$, for which

$$x_i = 1 \quad \forall c_i \in C_{Sch} \text{ UND } \bar{x}_i = 0 \quad \forall \bar{c}_i \notin C_{Sch} \Rightarrow \Phi(\underline{x}) = 1 \qquad (12.18)$$

The system has failed if all components of a minimum cut have failed and all residual components of the system, which do not belong to this cut, are able to function.

Minimal paths are the subset of all components $C_{Pf} \subseteq C$, for which

$$x_i = 0 \quad \forall c_i \in C_{Pf} \text{ UND } \bar{x}_i = 1 \quad \forall \bar{c}_i \notin C_{Pf} \Rightarrow \Phi(\underline{x}) = 0 \qquad (12.19)$$

The system is still able to function if all components of a minimum path are able to function and all residual components of the system, which do not belong to this path, have failed.

If a system is represented through minimum cuts or paths, the structure function Φ follows

$$\Phi(\underline{x}) = 1 - \prod_{C_{Sch}} \left(1 - \prod_{c_i \in C_{Sch_{ii}}} x_i \right) = \prod_{C_{Pf}} \left(1 - \prod_{c_i \in C_{Pf_{ii}}} (1 - x_i) \right) \qquad (12.20)$$

The ii is the continuous index of all cuts or paths. The cut for the bridge structure in Figure 12.4 is being determined as follows:

$$C_{Sch1} = \{c_1, c_2\} \qquad (12.21)$$

$$C_{Sch2} = \{c_3, c_4\} \qquad (12.22)$$

$$C_{Sch3} = \{c_1, c_4, c_5\} \qquad (12.23)$$

$$C_{Sch4} = \{c_2, c_3, c_5\} \qquad (12.24)$$

The structure-function Φ is

$$\Phi(x) = 1 - (1 - x_1 x_2) \cdot (1 - x_3 x_4) \cdot (1 - x_1 x_4 x_5) \cdot (1 - x_2 x_3 x_5) \qquad (12.25)$$

The paths of the bridge structure are

$$C_{Pf1} = \{c_1, c_3\} \qquad (12.26)$$

$$C_{Pf2} = \{c_2, c_4\} \qquad (12.27)$$

$$C_{Pf3} = \{c_1, c_4, c_5\} \qquad (12.28)$$

$$C_{Pf4} = \{c_2, c_3, c_5\} \qquad (12.29)$$

Consequently the structure function is

$$\Phi(x) = \left[1-(1-x_1)\cdot(1-x_3)\right]\cdot\left[1-(1-x_2)\cdot(1-x_4)\right]\cdot \\ \left[1-(1-x_1)\cdot(1-x_4)\cdot(1-x_5)\right]\cdot\left[1-(1-x_2)\cdot(1-x_3)\cdot(1-x_5)\right] \quad (12.30)$$

12.1 Reliability models

Except for few areas of application[95] a failure-free system, and therefore a reliable construction, is impossible to implement due to expenses. Mechanical constructions and its safety and reliability can within reason be influenced through over dimensioning.

The design of electronic circuits - apart from industrial, military and aviation technical purposes - have undergone few or no considerations regarding this goal. In the past years opinions have shifted. Less than ten years ago the components of electronic systems went hand in hand with a nominal value load without thinking of the safety factor twice, today, if not regarding the safety, a strong modification of design and coding standards has occurred regarding the operational availability. Especially today with its complex and fast circuits a temperature control for example is a central part of the design. The approach of the manufacturer's temperature limits results normally in an alteration of the design, wanting to avoid destruction through excess temperature.

With an exponential distribution of failure of a system with N components follows with a non-redundant structure the failure rate

$$\lambda_B = \sum_{i=1}^{N} \lambda_{Bi} . \quad (12.31)$$

Hereby λ_{Bi} is the failure rate of a single i-th element.

12.1.1 Systems without Redundancy

Safety related systems exist in general of a number of components, which must basically fulfill all their functions so that the safety related systems are capable of fulfilling its requirements. A failure of only one component leads to the failure of the entire system. When it comes to safety all components of the system are "switched into series". A series architecture of elements is shown in Figure 12.7. In order to achieve a preferably reliable system is it imperative to plan the over dimensioning of the subsystems and partial circuit components.

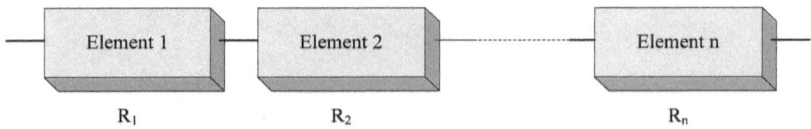

Figure 12.7: Series Architecture of Elements

---

[95] For example the wing of a air plane. This needs to be „Fail-Safe".

12 Reliability Block Diagram Analysis

The reliability function in Figure 12.7 of a given series architecture R_s[96] calculates itself out of the product of the reliability functions $R_i(t)$ of the individual elements. One assumes that the failure free working time of $\tau_1, \tau_2, ..., \tau_n$ of the elements $E_1, E_2, ..., E_n$ are independent random variables. With the reliability function of an individual component being

$$R_i(t) = e^{-\lambda_i t} \tag{12.32}$$

and under the condition, that the failure rate $\lambda(t)$ is constant ($\lambda(t) = \lambda$), the reliability function of the entire system calculates itself to:

$$R_s(t) = R_1(t) \cdot R_2(t) \cdot ... \cdot R_n(t)$$
$$= \prod_i^n R_i(t). \tag{12.33}$$

Contemplating the probability, one can proceed as follows:

$$R_s = P(E_1) \cdot P(E_2) \cdot ... \cdot P(E_n)$$
$$= \prod_i^n P(E_i). \tag{12.34}$$

It is further presumed that all blocks of the series are identical and $P(E_i) = p = 1 - q$ so this consideration leads to the following equation:

$$R_s = p^n = (1-q)^n. \tag{12.35}$$

Under the condition, that $q \ll 1$, the equation can be replaced with the following approximation:

$$R_s = 1 - n \cdot q. \tag{12.36}$$

Hereby the well-known rule is confirmed that the entire reliability of a series system can not exceed the reliability of the single element. Viewing the failure probability and the reliability under the following conditions

$$R + F = 1 \tag{12.37}$$

and

$$p + q = 1, \tag{12.38}$$

so for a series system, consisting of two elements, $n = 2$, the following equation for the failure probability F can be reached:

$$F_s = F(E_1) + F(E_2) - F(E_1)F(E_2). \tag{12.39}$$

With $F(E_i) = q$ the failure probability for this series system is

$$F_s = 2q - q^2. \tag{12.40}$$

---

[96] The index s describes a serial structure

With equation 12.37 one gets the reliability of this series system as the equation

$$R_s = 1 - F_s = 1 - (2q - q^2) = (1-q)^2. \qquad (12.41)$$

Consequently by predicting the failure probability the reliability of a system can also be predicted.

The *MTTF*-value for a series system calculated itself with the equation:

$$\begin{aligned} MTTF_s &= \int_0^\infty R_s(t)dt \\ &= \int_0^\infty R_1(t) \cdot R_2(t) \cdot \ldots \cdot R_n(t)dt. \end{aligned} \qquad (12.42)$$

If all elements have a constant failure rate $\lambda(t) = \lambda$, meaning the equation 12.32 is valid, then the mean time to failure is:

$$\begin{aligned} MTTF_s &= \int_0^\infty e^{-\left(\sum_{i=1}^n \lambda_i\right)t} dt \\ &= \frac{1}{\sum_{i=1}^n \lambda_i} \\ &= \frac{1}{\lambda_1 + \lambda_2 + \ldots + \lambda_n}. \end{aligned} \qquad (12.43)$$

12.1.2 Systems with Redundancy

It was said in the previous section that in simple structured systems, meaning systems without redundancy[97], the reliability[98] and the safety is limited. Especially in safety related systems model constructions are necessary that guarantee the functioning of the system (the entire system operation) or the plant in case of a failure. In connection with system reliability and ssystem safety the redundancy as a possibility is being defined in the following chapter and more than one way shown to comply with the requirements.

This definition includes the assumption that all paths have failed in order to speak of a system failure. A broad application can be derived in order to describe a safe and reliable system operation. The first step is the determination of a reliability block diagram[99] that describes the successful system operation. Upon completion of this step, meaning the exact analysis of the system regarding its correct function and possible failures, the relia-

---

[97] The real meaning of the word Redundancy is abundance or diffuseness. Since many years this word has been relevant in the information theory as far as it describes the difference between the maximum possible information content of a source and really describes the delivered information.
[98] [IECa]   IEC61508, *International Standard 61508 Functional Safety: Safety Related System*
[99] [HENL92]   Henley, E. J.; Hiromitsu, K.; Probabilistic risk assessment and management for engineers and scientists

12 Reliability Block Diagram Analysis

bility block diagram only differs in comparison with the circuit block diagram since not the signals but the concrete function is being observed.

The second step is the determination of the mathematical model for the reliability. Like mentioned before, in a redundant structure of a safety related system the elements can replace each other mutually in case of a failure. Basically there are three kinds of redundancies:

- **Hot redundancy** (active or parallel redundancy). The redundant element is being subjected to the same stress from the start. This procedure is being applied today almost exclusively in the safety related process and automation industry.
- **Warm redundancy** (low loaded redundancy). The redundant element is subjected to minimal stress up to the failure. This procedure is not applicable for the safety related industry and the automation industry.
- **Cold redundancy** (unstressed redundancy). The redundant element has not been subjected to stress up to the point of failure. This procedure is not applicable for the safety related and automation industry.
- **Redundancy through switching** to reserve facilities (a variation of the unstressed redundancy, in case of a failure a reserve unit takes over the function of the main unit in order to maintain the entire system operation[100]. This variation's applicability is not common in the automation technology any more.

This listing of the different redundancy procedures shows that for safety related systems only the hot redundancy control is applicable and the other procedures will not be discussed any further. In the discussion of the different types of redundancy arrangements[101] one assumes that the probability of failure is time-independent and that single blocks can fail independently. A redundant architecture of elements is shown in Figure 12.8. Basically these contemplations state that at all resulting observations, systems with "cold reserve" are not applicable.

It must be mentioned that observations on the basis of the probability block diagram and consequently the observations, apply to the redundancy of components and the redundancy of systems as well.

The reliability function of a redundant architecture $R_r(t)$[102] calculates itself - just like the series structure - under the assumption, that the failure free working time of $\tau_1, \tau_2, ..., \tau_n$ the elements of $E_1, E_2, ..., E_n$ are independent parameters. With the reliability function of the single element

$$R_i = e^{-\int_{-\infty}^{\infty} \lambda_i t \, dt} \qquad (12.44)$$

and under the condition that the failure-rate $\lambda(t)$ is constant ($\lambda(t) = \lambda$), the reliability function R_r for a system with two redundant system components with the equation is

---

[100] As an exemple diesel generators to in airports and hospitals are power supply sources.
[101] In general a distinction is made between serial, parallel and quadruple structures or a mixture of these. E.F. Moore and C.E. Shannon recommended the quadruple structure as a general redundancy principle. Another interesting structure is the parallel function blocks with voting, e.g., the 2oo3 Architecture.
[102] The index stands for redundant systems

$$R_r(t) = R_1(t) + R_2(t) - R_1(t)R_2(t) \tag{12.45}$$

and for the fault probability

$$F_r(t) = F_1(t)F_2(t). \tag{12.46}$$

The connection between the reliability function and the error probability is as follows

$$R_r(t) = 1 - F_r(t) \tag{12.47}$$

If one observes – similar to the observation of the serial system – and assumes the condition, that both blocks are independent from each other and should fully execute several system functions, the results are as follows:

$$R_r(t) = P(E_1) + P(E_2) - P(E_1) \cdot P(E_2) \tag{12.48}$$

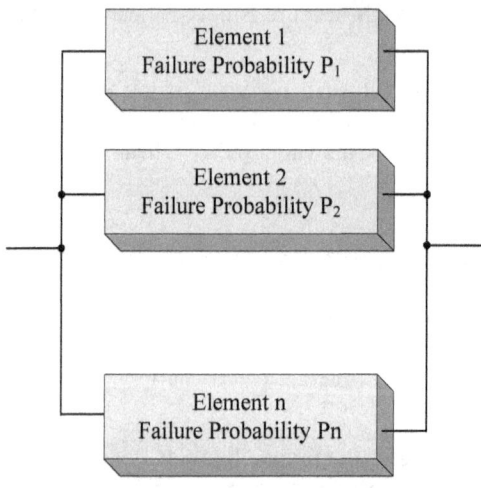

Figure 12.8: Basic Redundant Architecture of Elements

If both redundant blocks are functionally the same, and $P(E_i) = p = 1 - q$, then:

$$R_r(t) = 2p - p^2 = 1 - q^2 \tag{12.49}$$

Observing – just like the serial system structure – next to the reliability the failure probability with following conditions,

$$R + F = 1 \tag{12.50}$$

and

$$p + q = 1 \tag{12.51}$$

Therefore the following equation is valid:

$$F_r = F_1(t) \cdot F_2(t) = q^2. \tag{12.52}$$

12 Reliability Block Diagram Analysis

With equation 12.50 the following is valid

$$R_r = 1 - F_r = 1 - q^2 = 1 - (1-p)^2 = 2p - p^2. \tag{12.53}$$

The n redundant component is as follows:

$$R_r(t) = \sum_{i=1}^{n} \binom{n}{i} \cdot R^i(t) \cdot (1 - R(t))^{n-i}. \tag{12.54}$$

For the error probability F_r for n redundant components the following is valid

$$F_r(t) = \prod_{i=1}^{n} F_i(t). \tag{12.55}$$

The reliability function calculates consequently to

$$R_r(t) = 1 - \prod_{i=1}^{n} F_i(t). \tag{12.56}$$

The reliability of n redundant components could consequently be described with the following equation:

$$R_r = 1 - F_r = 1 - q^n = 1 - (1-p)^n. \tag{12.57}$$

This equation obviously shows that the system reliability is clearly higher then in a serial system. The *MTTF*-value for a redundant system consists of two elements and calculated as follows:

$$MTTF_r = \int_0^\infty R_r(t) dt$$
$$= \int_0^\infty R_1(t) + R_2(t) - R_1(t) \cdot R_2(t) dt \tag{12.58}$$

$$MTTF_r = \int_0^\infty (e^{-\int_{-\infty}^\infty \lambda_1 \cdot t\, dt} + e^{-\int_{-\infty}^\infty \lambda_2 \cdot t\, dt} - e^{-\int_{-\infty}^\infty (\lambda_1 + \lambda_2) \cdot t\, dt}) dt. \tag{12.59}$$

If all elements have, as presumed, a constant failure-rate $\lambda(t) = \lambda$, an exponential distribution of the parameters τ_n is evident and the average error-free working time calculates as:

$$MTTF_r = \frac{1}{\lambda_1} + \frac{1}{\lambda_2} - \frac{1}{\lambda_1 + \lambda_2}. \tag{12.60}$$

12.1.3 Mixed systems

True safety related systems are based on these simple combinations; however, these simplified systems consist of several blocks. The previous observations can however apply to systems with mixed parallel series combinations[103]. The simplest of these combinations are shown in Figure 12.9 and Figure 12.10.

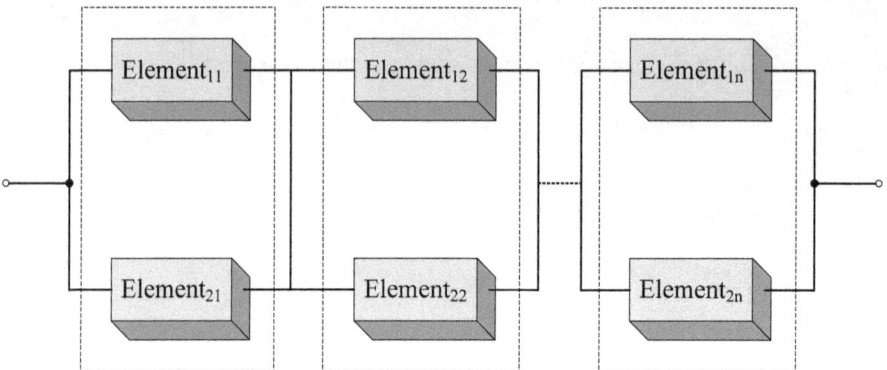

Figure 12.9: Combined Series Parallel Systems

For the series parallel system (Figure 12.9), viewed as a series system, the following is valid:

$$R_{sp}(t) = R_1(t) \cdot R_2(t) \cdot \ldots \cdot R_n(t)$$
$$= \prod_i^n R_i(t). \tag{12.61}$$

The similarity of the parallel blocks is as follows:

$$R_{sp} = \left(1 - (1-p)^m\right)^n. \tag{12.62}$$

This general description takes the number of the serial structure elements into account and expresses through the exponent n and the number of the redundant single block structures expressed through the exponent m. Under the conditions that the elements and the failure probability are the same, the equation is as follows:

$$R_{sp} = \prod_{i=1}^{n}[1 - \prod_{j=1}^{m} F(A_{ij})]. \tag{12.63}$$

Basically one begins the calculation of parallel series systems with the calculation of the series system. This method assumes that the single parallel branches are identical and the single elements show identical failure rates. Like in Figure 12.10 the configuration for a string of parallel series systems are as follows:

---

[103] [BIRO97]   Birolini, A.; Reliability of devices and systems

$$R_j = \prod_{i=1}^{n} P(A_{ij}) = p^n . \tag{12.64}$$

From this equates for the entire system of the parallel system:

$$R_{ps} = 1 - F = 1 - \prod_{i=1}^{m} Fi = 1 - \prod_{i=1}^{m} [1 - \prod_{j=1}^{n} P(A_{ij})] = 1 - (1 - q^m)^n . \tag{12.65}$$

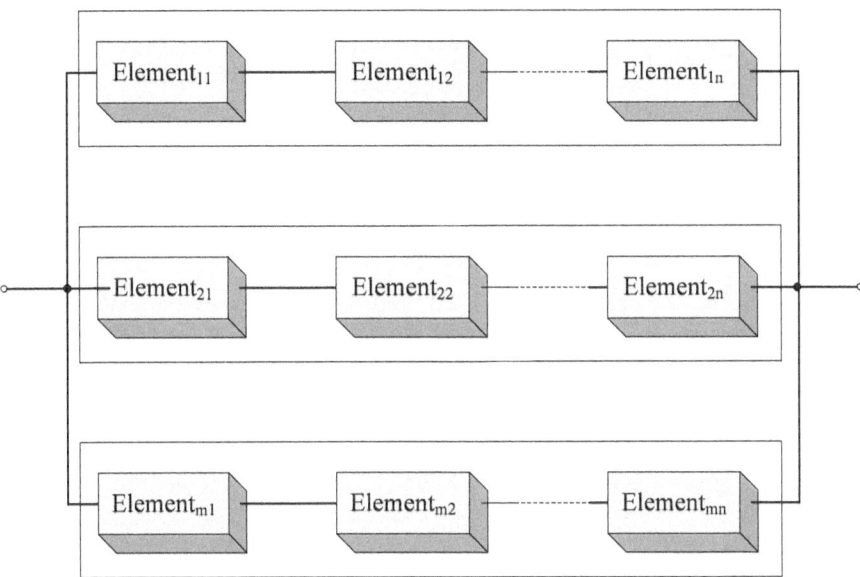

Figure 12.10: Combined Parallel Series Systems

This method of observation already allowed the transition to different system architectures, as described in DIN EN 801 and IEC 61508[104]. By observing the existing redundancy circuits, one assumes that the system function exists as long as a functioning string exists in the entire system (Figure 12.10). Using the system configurations for industrial use, different kinds of redundant elements can be observed. Depending on the application, input units, processing units or output units are deemed redundant. The following models show especially the systems of 1oo1, 1oo2 and 2oo3. With these systems, in order to warrant a sure function, a certain number of redundant branches are necessary. The combination is basically a combination without repetition. In this context the question is concerning the most favorable events, in order to find the functionality of the system[105]. Equation is as follows:

$$R_k = \sum_{i=k}^{n} \binom{n}{i} q^i p^{n-i} . \tag{12.66}$$

---

[104] 1oo1, 1oo2, and 2oo3 systems are involved.
[105] Literature designates this as partial redundancy

It is taken into consideration that the possible n redundant subsystems must be k functioning in configuration for guarantee the entire function.

In order to get a quantitative estimation, the following conditions

$$R + F = 1 \qquad (12.67)$$

as well as

$$p + q = 1 \qquad (12.68)$$

apply. Further, under the condition that λ is constant, the following is true:

$$R = p = 1 - q = e^{-\lambda t}. \qquad (12.69)$$

With $t = 8760$ hours (this corresponds to a year) the values calculate as shown in Table 12.1 The used corresponding real industrial values can consequently be deemed authentic. These values shall be the basis in all consequent calculations. Furthermore they allow for ensuing observation purposes quantitative comparison possibilities for the different architectural concepts.

A frequent structure is the 1oo1 architecture with redundant entrances and exits. For the calculation of the system in Figure 12.11, under the condition that at least one entrance/exit system is functioning, and the redundant connected subsystems are identical, the following approach is possible:

$$R_{sys} = \left[\sum_{i=k}^{n} \binom{n}{i} q_E^{\ i} \cdot p_E^{\ n-i} \right] \cdot q_C \cdot \left[\sum_{i=k}^{n} \binom{n}{i} q_A^{\ i} \cdot p_A^{\ n-i} \right]. \qquad (12.70)$$

Table 12.1: Failure rates, and reliability, respectively, failure probability values

Element	Failure Rate λ [1/h]	Reliability $R = p$	Error Probability $F = q$
Input	$1{,}60 \cdot 10^{-06}$	0,9861	0,0139
Processor	$2{,}50 \cdot 10^{-06}$	0,9783	0,0217
Output	$4{,}20 \cdot 10^{-06}$	0,9639	0,0361

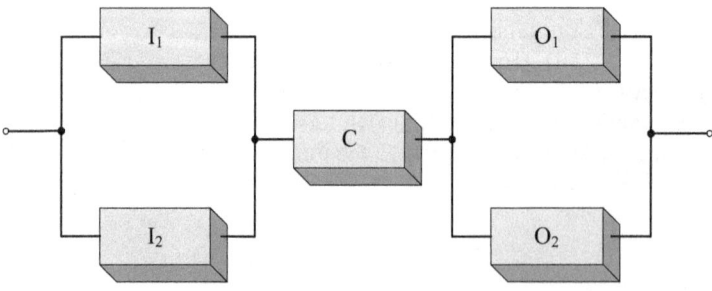

Figure 12.11: 1oo1-Architectur with Redundant Inputs and Outputs

12 Reliability Block Diagram Analysis

For the following calculation the conditions $q = q_E$, $q = q_A$ respectively $p = p_E$, respectively $p = p_A$ are valid.

$$R_{EA} = \binom{2}{1} p^1 \cdot q^{2-1} + \binom{2}{2} p^2 \cdot q^{2-2} \qquad (12.71)$$
$$= 2pq + p^2$$

$$R_{EA} = 2p(1-p) + p^2 \text{ with } q = 1-p \qquad (12.72)$$

$$R_{EA} = 2p - p^2. \qquad (12.73)$$

Therefore

$$R_{sys} = \left(2p_E - p_E^2\right) \cdot p_C \left(2p_A - p_A^2\right) \qquad (12.74)$$

The following value displays Figure 12.11 and the values in Table 12.1:

$$R_{sys} = \left(2 \cdot 0{,}9861 - 0{,}9861^2\right) \cdot 0{,}9783 \cdot \left(2 \cdot 0{,}96390 - 0{,}96390^2\right) \qquad (12.75)$$
$$= 0{,}97684.$$

The in Figure 12.12 represented systems show a standard architecture (1oo2) of a safety related automation technology. Through a redundant safe processing unit a high availability of the actual computing element is being reached.

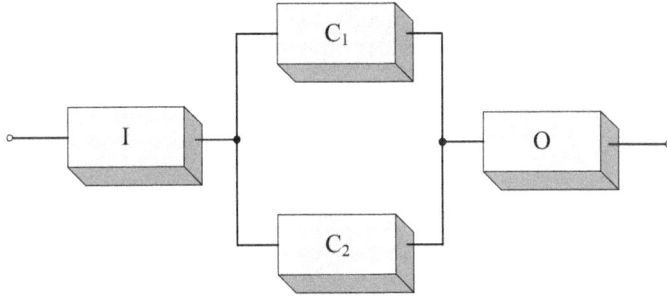

Figure 12.12: 1oo2-Architectur with simple Entrances and Exits[106]

The calculation of the reliability is computed by the following functions.

$$R_{sys} = \left[\sum_{i=k}^{n} \binom{n}{i} p_C^i \cdot q_C^{n-i}\right] \cdot p_E \cdot p_A \qquad (12.76)$$

---

[106] The shown architecture is the most used architecture in the safety industry. Considering the fact that a minimum of diagnostics is required in safety systems, 60% is generally accepted, especially those systems that require certification.

$$R_C = \binom{2}{1} p^1 \cdot q^{2-1} + \binom{2}{2} p^2 \cdot q^{2-2} \tag{12.77}$$

$$R_C = 2pq + p^2$$

$$R_C = 2p(1-p) + p^2 \text{ with } q = 1-p \tag{12.78}$$

$$R_C = 2p - p^2. \tag{12.79}$$

Above system the simplification $q = q_E$, respectively $q = q_C$, respectively. $q = q_A$ is valid. The reliability for the entire system is as follows:

$$R_{sys} = (2p_C - p_C^2) \cdot p_E \cdot p_A. \tag{12.80}$$

If one adds the values from Table 12.1 to this equation, the following value for the reliability of the system is:

$$R_{sys} = (2 \cdot 0{,}9783 - 0{,}9783^2) \cdot 0{,}9861 \cdot 0{,}9639 \tag{12.81}$$
$$= 0{,}95005.$$

A system with multiple redundancies is represented in Figure 12.13. In this configuration both the input and the output circuits, therefore the safe part of the computer, are implemented redundantly. For the safe functioning at least one functional branch of the redundant assembly is necessary.

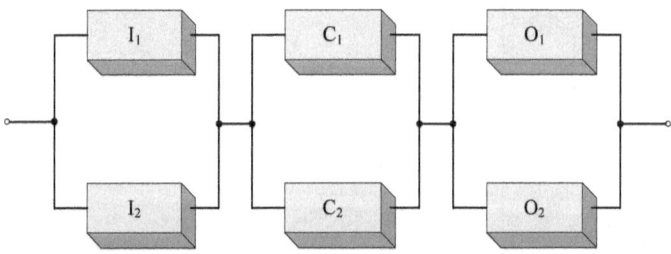

Figure 12.13: 1oo2-Architectur with redundant Inputs and Outputs

This architecture variation is a change of the structure in Figure 12.12 and is found frequently in areas of technical processes. This approach represents a standard architecture for safety related systems.

$$R_{sys} = \left[\sum_{i=k}^{n} \binom{n}{i} p_E^i \cdot q_E^{n-i}\right] \cdot \left[\sum_{i=k}^{n} \binom{n}{i} p_C^i \cdot q_C^{n-i}\right]_c \cdot \left[\sum_{i=k}^{n} \binom{n}{i} p_A^i \cdot q_A^{n-i}\right] \tag{12.82}$$

$$R_{C/E/A} = \binom{2}{1} p^1 \cdot q^{2-1} + \binom{2}{2} p^2 \cdot q^{2-2} \tag{12.83}$$

$$R_{C/E/A} = 2qp + p^2$$

$$R_{C/E/A} = 2p(1-p) + p^2 \text{ with } q = 1-p \qquad (12.84)$$

$$R_{C/E/A} = 2p - p^2. \qquad (12.85)$$

For the system in Figure 12.12 with $p = p_E$, respectively $p = p_C$, respectively $p = p_A$ follows

$$R_{sys} = \left(2p_E - p_E^2\right)\cdot\left(2p_C - p_C^2\right)\cdot\left(2p_A - p_A^2\right). \qquad (12.86)$$

With the values of Table 12.1 for the single blocks one can determine the following value for the reliability:

$$R_{sys} = \left(2\cdot 0{,}9861 - 0{,}9861^2\right)\cdot\left(2\cdot 0{,}9783 - 0{,}9783^2\right)\cdot\left(2\cdot 0{,}9639 - 0{,}9639^2\right) \qquad (12.87)$$
$$= 0{,}99803.$$

Figure 12.14 shows a system architecture that is used for small and smallest applications in safety related industrial automation. Here, through a redundant safe processing unit, a high availability of the actual element is being reached. Goal and purpose of the high availability is the possibility to reach a long period of communication and control of the associated processes in case of a failure.

For the safe function are at least two functioning branches of redundant configurations necessary.

$$R_{sys} = \left[\sum_{i=k}^{n} \binom{n}{i} p_C^{\,i} \cdot q_C^{\,n-i}\right] \cdot p_E \cdot p_A \qquad (12.88)$$

$$R_C = \binom{3}{2} p^2 \cdot q^{3-2} + \binom{3}{3} p^3 \cdot q^{3-3} \qquad (12.89)$$
$$= 3p^2 q + p^3.$$

With

$$q = 1 - p \qquad (12.90)$$

resulting

$$R_C = 3p^2(1-p) + p^3 \qquad (12.91)$$

$$R_C = 3p^2 - 2p^3. \qquad (12.92)$$

For the represented system in Figure 12.14 resulting under the condition

$$p = p_C \qquad (12.93)$$

$$R_{sys} = \left(3\cdot p_C^{\,2} - 2\cdot p_C^{\,3}\right)\cdot p_E \cdot p_A. \qquad (12.94)$$

With the values of Table 12.1 for the single blocks one can determine the following value for the reliability:

$$R_{sys} = (3 \cdot 0{,}9783^2 - 2 \cdot 0{,}9783^3) \cdot 0{,}9861 \cdot 0{,}96390$$
$$= 0{,}94917. \qquad (12.95)$$

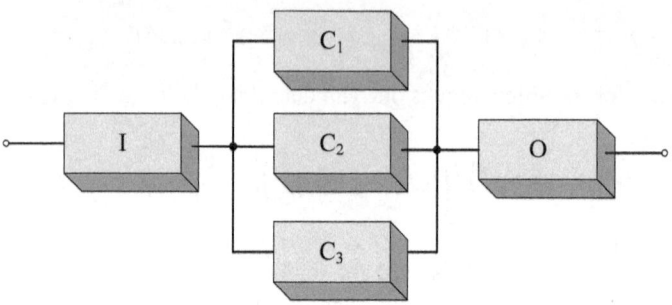

Figure 12.14: 2oo3-Architectur with simple Input and Output

The in Figure 12.15 shown circuit variations represent a change of the circuit of Figure 12.14. This precise structure is hardly put in use in the industry. 2oo3-systems or TMR (Triple Module Redundancy) usually consist of three identical organized strings with a voting-unit. It should be mentioned that this circuit is being applied in a different form; the safe computer unit involves only two instead of the three safe computer units. This variation then is however part of the 1oo2-Architecture.

The method of calculation corresponds to the one for different structures. The assumption is that at least two of the three redundant subsystems are able to function.

$$R_{sys} = \left[\sum_{i=k}^{n}\binom{n}{i} p_E^i \cdot q_E^{n-i}\right] \cdot \left[\sum_{i=k}^{n}\binom{n}{i} p_C^i \cdot q_C^{n-i}\right]_c \cdot \left[\sum_{i=k}^{n}\binom{n}{i} p_A^i \cdot q_A^{n-i}\right] \qquad (12.96)$$

$$R_{C/E/A} = \binom{3}{2} p^2 \cdot q^{3-2} + \binom{3}{3} p^3 \cdot q^{3-3} \qquad (12.97)$$

$$R_{C/E/A} = 3p^2 q + p^3$$

$$R_{C/E/A} = 3p^2(1-p) + p^3. \qquad (12.98)$$

Under the condition, that

$$q = 1 - p \qquad (12.99)$$

follows

$$R_{C/E/A} = 3p^2(1-p) + p^3 \qquad (12.100)$$

$$R_{C/E/A} = 3p^2 - 2p^3 \qquad (12.101)$$

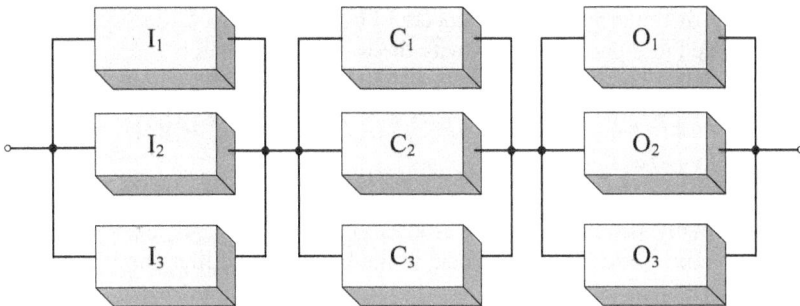

Figure 12.15: 2oo3-Architectur with triple Inputs and Outputs

Just like in the previously described variations, the following conditions must be valid for the system in Figure 12.15; $p = p_E$, respectively. $p = p_C$, respectively $p = p_A$. From this follows

$$R_{sys} = \left(3p_E^2 - 2p_E^3\right)\cdot\left(3p_C^2 - 2p_C^3\right)\cdot\left(3p_A^2 - 2p_A^3\right). \qquad (12.102)$$

With the values of Table 12.1 the following value calculates:

$$\begin{aligned}R_{sys} &= \left(3\cdot 0{,}9861^2 - 2\cdot 0{,}9861^3\right)\cdot\left(3\cdot 0{,}9783^2 - 2\cdot 0{,}9783^3\right)\cdot\\&\quad\left(3\cdot 0{,}9639^2 - 2\cdot 0{,}9639^3\right)\\&= 0{,}990741.\end{aligned} \qquad (12.103)$$

Specifically by observing the circuit variations in picture of Figure 12.11, Figure 12.12 and Figure 12.13, an important aspect must be considered. The former observations imply no influence of the blocks among each other, and they always fail as a part of the redundant exit blocks. Observing the previous mentioned concepts it is concluded that the safe function of a energy less condition is no longer attainable. The literature[107] refers to the *Rubrum element* with more than one kind of failure. Basically, in order to observe the reliability, the failure mechanism which influences the arrangement in their reliability strongest must always be taken into consideration. The redundant circuitry of two elements shows in equation 12.92 under the condition

$$q = q_u + q_k \qquad (12.104)$$

Is to be presented in the following form

$$R_p = 1 - F = 1 - q^2 = 1 - (q_u + q_k)^2 \qquad (12.105)$$

Multiplying and summarizing shows the following and

$$R_p = 1 - (q_u^2 + 2q_k q_u + q_k^2) \neq (1 - q_k)^2 - q_u^2 \qquad (12.106)$$

---

[107] [BIRO97]   Birolini, A.; Zuverlässigkeit von Geräten und Systemen
[LEWI96]   Lewis, E. E.; Introduction to reliability engineering
[SIEW82]   Siewiorek, D. P.; et al., The theory and practice of reliable system design
[STOR96]   Storey, N.; Safety Critical Computers Systems

one gets the reliability of the redundantly connected blocks with two failure types. This equation can be expanded to n parallel blocks with the following equation:

$$R_p = 1 - q^n = 1 - (q_u + q_k)^n = 1 - \sum_{i=0}^{n} \binom{n}{i} q_u^i \cdot q_k^{n-i} \qquad (12.107)$$

Blocks switched in series can be describes like redundantly connected blocks. The application of 1oo2, respectively 1oo2D architectures as well as the different variations of the 2oo3 and 2oo4-architectures, are being described in the below stated literature[108, 109]. The equation can now be written as follows:

$$R_s = (1-q)^2 = 1 - 2q + q^2 \qquad (12.108)$$

$$R_s = 1 - 2 \cdot (q_u + q_k - q_u q_k) + q_u^2 + q_k^2. \qquad (12.109)$$

This equation can also be expanded to any number n of serial blocks:

$$R_s = (1-q)^n = (1-(q_u + q_k))^n = (1 - q_u - q_k)^n \neq (1-q_u)^n - q_k^n \qquad (12.110)$$

This method of observation applies to output circuits in safety related systems of significance. Specifically with systems like the 1oo2-, 2oo3- and 2oo4-Architectur, in which a big number of blocks are connected in serial-parallel combination due of safety and availability reasons. A difficulty, also described in the literature[110] described, is the following event; $\frac{q}{2} = q_k = q_u$.

It is obvious in this case that no reliability increase could be reached, but is being aided by a so-called quartet configuration[111] of the observed blocks.

It is clear that at an interruption of blocks in a branch does not create a critical situation, under the condition that the energy less condition is safe. In the event that a failure in two blocks was caused by a shortage in a branch, the entire configuration is in danger.

---

[108] [BIRO97]   Birolini, A., *Zuverlässigkeit von Geräten und Systemen*
    [LEWI96]   Lewis, E. E., *Introduction to Reliability Engineering*
    [SIEW82]   Siewiorek, D. P., et al.; *The theory and practice of reliable system design*
    [SIEW90]   Siewiorek, D. P., *Fault Tolerance in Commercial Computers*
    [STOR96]   Storey, N., *Safety Critical Computer Systems*
[109] In the output blocks will be implemented redundant switched serial systems.
[110] [BIRO97]   Birolini, A., *Zuverlässigkeit von Geräten und Systemen*
    [SIEW82]   Siewiorek, D. P., et al., *The theory and practice of reliable system design*
[111] This has already been shown by Moore and Shannon 1956.

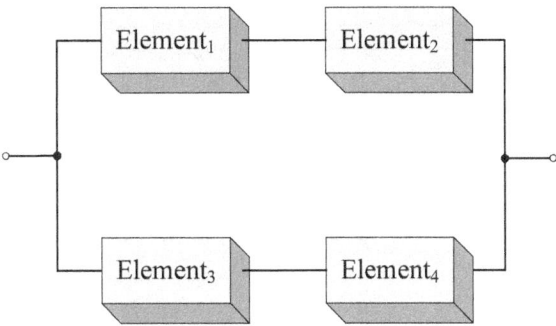

Figure 12.16: Parallel-Serial Connection [112]

First of all an analysis of the failure behavior must be conducted. The event in Figure 12.16 can determine the following result for the failure of the configuration:

$$(A_{1k} \wedge A_{2k}) \vee (A_{3k} \wedge A_{4k}) \tag{12.111}$$

or

$$(A_{1u} \vee A_{2u}) \wedge (A_{3u} \vee A_{4u}). \tag{12.112}$$

A failure occurs if an interruption in the redundant branches - a sure failure - or two failures in one branch – a dangerous error - occur. The second failure mechanism is consequently the one that causes the failure of this configuration. In order to get a meaningful statement regarding the reliability, the failure as well as the short circuit of blocks must be investigated: It is valid

$$R_{ps} = 1 - F = 1 - (F_k + F_u). \tag{12.113}$$

The following equation is valid for the failureshort circuit failure:

$$F_k \neq 2q_k^2 - q_k^4 \tag{12.114}$$

and for the interruption

$$F_u \neq 2p_u^2 - p_u^4 \tag{12.115}$$

After insertion and combination as follows:

$$R_{ps} = 1 - F = 1 - \prod_1^2 \left(1 - \prod_1^2 (1-q)\right) = 1 - \prod_1^2 \left(1 - \prod_1^2 (1-(q_u + q_k))\right)$$

$$R_{ps} = 1 - (F_k + F_u) \neq 1 - [(2p_k^2 - p_k^4) + (2p_u^2 - p_u^4)] \tag{12.116}$$

---

[112] The failure behavior of the individual blocks A_n can be due to interruption or short circuit.

Or in short

$$R_{ps} \neq (1-p_k^2)^2 - [1-(1-p_u)^2]^2. \tag{12.117}$$

Equally for the event in Figure 12.17, the analysis is the first step for the determination of reliability of the configuration. For this variation a failure occurs, if a short circuit in the redundant branches occurs (a dangerous fault), or two interruptions occur in the redundant branch (sure fault). The first failure mechanism is consequently the one that causes the failure in the configuration.

$$(A_{1u} \wedge A_{3u}) \vee (A_{2u} \wedge A_{4u}) \tag{12.118}$$

or

$$(A_{1k} \vee A_{3k}) \wedge (A_{2k} \vee A_{4k}). \tag{12.119}$$

In order to make a sensible statement about the reliability of the quartet configuration, the failure as well as the short circuit of the blocks must be investigated:

The procedure is similar to the one in Figure 12.16. The assumption is, that two failures of the redundantly connected blocks or a short circuit in every block in the series exist. From this follows the equation for the probability of error of the quartet configuration as follows:[113]

$$F = 2(q_u + q_k)^2 - (q_u + q_k)^4. \tag{12.120}$$

Through insertion and multiplication the equation for the reliability is as follows:

$$R_{Quartett} = 1 - F = 1 - 2(q_u + q_k)^2 + (q_u + q_k)^4 = \left(1 - (q_u + q_k)^2\right)^2 \tag{12.121}$$

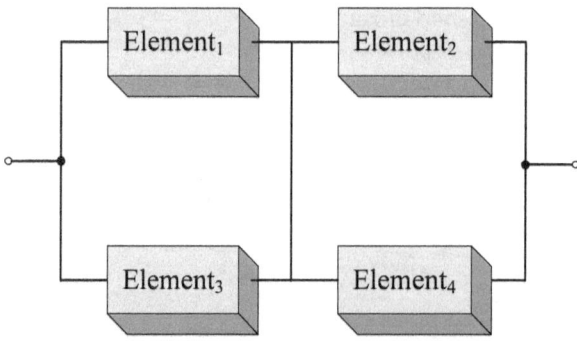

Figure 12.17: Quartet Configuration of Four Blocks[114]

Which circuit variation proves meaningful and applicable depends on the failure behavior of the block and must be decided case by case. Furthermore the marginal cases must be considered and can be described as follows:

---

[113] [BIRO97]   Birolini, A.; Zuverlässigkeit von Geräten und Systemen
[114] The failure behavior of the individual blocks A_n can be due to interruption or short circuit.

- $q_u = 0$ and $q_k = p$
- $q_u = q$ and $q_k = 0$
- $q_u = q_k = q/2$.

Considering the above presented conditions, the investigations shows that it appears to be meaningful to choose the purely redundant configuration in Figure 12.16. This is the case in most industrially used output circuits. If predominantly interruption disturbances occur, the quartet circuit is more advantageous. For the equality of short circuits and interruptions both variations are equivalent.

12.2 Redundant Systems with Different Failure Rates

The previous observations concerning the reliability were applied under the assumption that definite probabilities of failure for the single elements were accepted. However, their time-dependency must be considered so that the reliability of failure of the redundant circuits is also time-dependent. For a timely constant failure rate λ one gets the failure probability density function

$$f(t) = \lambda e^{-\lambda t} \tag{12.122}$$

The probability of failure of a single element calculates with the equation

$$p(t) = \int_0^t f(\tau) d\tau = \lambda \int_0^t e^{-\lambda \tau} d\tau = [-e^{-\lambda \tau}]_0^t = 1 - e^{-\lambda t}. \tag{12.123}$$

According to the observations for the description of the redundant system configuration as follows:

$$R_r(t) = R_1(t) + R_2(t) - R_1(t) R_2(t) \tag{12.124}$$

Two possibilities exist to predict the *MTTF* of this configuration. One possibility is to use the reliability equation. One gets the following equation:

$$\text{MTTF} = \int_0^\infty R(t)\, dt = R_1(t) + R_2(t) - R_1(t) R_2(t)$$

$$= -\frac{1}{\lambda_1}\left[e^{-\lambda_1 t}\right]_0^\infty + \frac{1}{\lambda_2}\left[e^{-\lambda_2 t}\right]_0^\infty - \frac{1}{\lambda_1 + \lambda_2}\left[e^{-(\lambda_1+\lambda_2)t}\right]_0^\infty \tag{12.125}$$

From this calculates

$$\text{MTTF} = \frac{1}{\lambda_1} + \frac{1}{\lambda_2} - \frac{1}{\lambda_1 + \lambda_2} \tag{12.126}$$

The second possibility to calculate the MTTF-value for a redundant circuit of two blocks, the following applies:

$$R = 1 - q^2(t) = 1 - (1 - e^{-\lambda \tau})^2 = 2e^{-\lambda \tau} - e^{-2\lambda \tau}. \tag{12.127}$$

If different failure rates are being used, the following applies:

$$R = 1 - q'(t)^2 = e^{-\lambda_1 t} + e^{-\lambda_2 t} - e^{-(\lambda_1+\lambda_2)t} \tag{12.128}$$

The mean time to the failure can be determined by the following equation:

$$MTTF = \int_0^\infty R(t)\, dt = -\left[\frac{1}{\lambda_1} e^{-\lambda_1 t}\right]_0^\infty - \left[\frac{1}{\lambda_2} e^{-\lambda_2 t}\right]_0^\infty + \left[\frac{e^{-(\lambda_1+\lambda_2)t}}{\lambda_1+\lambda_2}\right]_0^\infty \tag{12.129}$$

The MTTF-value results as follows:

$$MTTF = \frac{1}{\lambda_1} + \frac{1}{\lambda_2} - \frac{1}{\lambda_1+\lambda_2} \tag{12.130}$$

It is obvious that both possibilities lead to a solution of the problem. For the sake of completeness the architecture variations (Figure 12.12 to Figure 12.14) should be calculated with different failure rates of single blocks. The following MTTF-value according to the architecture in Figure 12.12 calculates as follows:)

$$R_g(t) = R_1(t) \cdot (R_2(t) + R_3(t) - R_2(t) \cdot R_3(t)) R_4(t) \tag{12.131}$$

$$MTTF = \int_0^\infty R_g(t) \cdot dt$$
$$= \int_0^\infty R_1(t) \cdot (R_2(t) + R_3(t) - R_2(t) \cdot R_3(t)) R_4(t) \cdot dt \tag{12.132}$$

$$MTTF = \int_0^\infty e^{-\lambda_1 \cdot t}\left(e^{-\lambda_2 \cdot t} + e^{-\lambda_3 \cdot t} - e^{-\lambda_2 \cdot t} \cdot e^{-\lambda_3 \cdot t}\right) e^{-\lambda_4 \cdot t} \cdot dt \tag{12.133}$$

$$MTTF = \int_0^\infty e^{-(\lambda_1+\lambda_2+\lambda_4)\cdot t} + e^{-(\lambda_1+\lambda_3+\lambda_4)\cdot t} - e^{-(\lambda_1+\lambda_2+\lambda_3+\lambda_4)\cdot t} \cdot dt \tag{12.134}$$

$$MTTF = \frac{1}{\lambda_1+\lambda_2+\lambda_3+\lambda_4} - \frac{1}{\lambda_1+\lambda_2+\lambda_4} - \frac{1}{\lambda_1+\lambda_3+\lambda_4} \tag{12.135}$$

12 Reliability Block Diagram Analysis

For the configuration in Figure 12.13 the MTTF-value are as follows:

$$R_g(t) = (R_1(t) + R_2(t) - R_1(t) \cdot R_2(t)) \cdot (R_3(t) + R_4(t) - R_3(t) \cdot R_4(t)) \cdot$$
$$\cdot (R_5(t) + R_6(t) - R_5(t) \cdot R_6(t)) \tag{12.136}$$

$$MTTF = \int_0^\infty R_g(t) \cdot dt$$

$$= \int_0^\infty (R_1(t) + R_2(t) - R_1(t) \cdot R_2(t)) \cdot (R_3(t) + R_4(t) - R_3(t) \cdot R_4(t)) \cdot \tag{12.137}$$
$$\cdot (R_5(t) + R_6(t) - R_5(t) \cdot R_6(t)) \cdot dt$$

$$MTTF = \int_0^\infty \left(e^{-\lambda_1 t} + e^{-\lambda_2 t} - e^{-(\lambda_1 + \lambda_2)t}\right) \cdot \left(e^{-\lambda_3 t} + e^{-\lambda_4 t} - e^{-(\lambda_3 + \lambda_4)t}\right) \cdot$$
$$\cdot \left(e^{-\lambda_5 t} + e^{-\lambda_6 t} - e^{-(\lambda_5 + \lambda_6)t}\right) \cdot dt$$

$$MTTF = \int_0^\infty \left[e^{-(\lambda_1+\lambda_3)t} + e^{-(\lambda_1+\lambda_4)t} - e^{-(\lambda_1+\lambda_3+\lambda_4)t} + e^{-(\lambda_2+\lambda_3)t} + e^{-(\lambda_2+\lambda_4)t} + \right.$$
$$\left. - e^{-(\lambda_2+\lambda_3+\lambda_4)t} - e^{-(\lambda_1+\lambda_2+\lambda_3)t} - e^{-(\lambda_1+\lambda_2+\lambda_4)t} + e^{-(\lambda_1+\lambda_2+\lambda_3+\lambda_4)t} \right] \cdot$$
$$\cdot \left(e^{-\lambda_5 t} + e^{-\lambda_6 t} - e^{-(\lambda_5+\lambda_6)t}\right) \cdot dt$$

$$MTTF = \int_0^\infty \left[e^{-(\lambda_1+\lambda_3+\lambda_5)t} + e^{-(\lambda_1+\lambda_3+\lambda_6)t} - e^{-(\lambda_1+\lambda_3+\lambda_5+\lambda_6)t} + e^{-(\lambda_1+\lambda_4+\lambda_5)t} \right.$$
$$+ e^{-(\lambda_1+\lambda_4+\lambda_6)t} - e^{-(\lambda_1+\lambda_4+\lambda_5+\lambda_6)t} + e^{-(\lambda_1+\lambda_3+\lambda_4+\lambda_5+\lambda_6)t} - e^{-(\lambda_1+\lambda_3+\lambda_4+\lambda_5)t}$$
$$- e^{-(\lambda_1+\lambda_3+\lambda_4+\lambda_6)t} + e^{-(\lambda_2+\lambda_3+\lambda_5)t} + e^{-(\lambda_2+\lambda_3+\lambda_6)t} - e^{-(\lambda_2+\lambda_3+\lambda_5+\lambda_6)t}$$
$$+ e^{-(\lambda_2+\lambda_4+\lambda_5)t} + e^{-(\lambda_2+\lambda_4+\lambda_6)t} - e^{-(\lambda_2+\lambda_4+\lambda_5+\lambda_6)t} + e^{-(\lambda_2+\lambda_3+\lambda_4+\lambda_5+\lambda_6)t}$$
$$- e^{-(\lambda_2+\lambda_3+\lambda_4+\lambda_5)t} - e^{-(\lambda_2+\lambda_3+\lambda_4+\lambda_6)t} - e^{-(\lambda_1+\lambda_2+\lambda_3+\lambda_5)t} - e^{-(\lambda_1+\lambda_2+\lambda_3+\lambda_6)t}$$
$$+ e^{-(\lambda_1+\lambda_2+\lambda_3+\lambda_5+\lambda_6)t} - e^{-(\lambda_1+\lambda_2+\lambda_4+\lambda_5)t} - e^{-(\lambda_1+\lambda_2+\lambda_4+\lambda_6)t}$$
$$+ e^{-(\lambda_1+\lambda_2+\lambda_4+\lambda_5+\lambda_6)t} - e^{-(\lambda_1+\lambda_2+\lambda_3+\lambda_4+\lambda_5+\lambda_6)t} + e^{-(\lambda_1+\lambda_2+\lambda_3+\lambda_4+\lambda_5)t}$$
$$\left. + e^{-(\lambda_1+\lambda_2+\lambda_3+\lambda_4+\lambda_6)t} \right] \cdot dt$$

$$MTTF = -\frac{1}{\lambda_1 + \lambda_3 + \lambda_5} - \frac{1}{\lambda_1 + \lambda_3 + \lambda_6} + \frac{1}{\lambda_1 + \lambda_3 + \lambda_5 + \lambda_6} - \frac{1}{\lambda_1 + \lambda_4 + \lambda_5}$$

$$-\frac{1}{\lambda_1 + \lambda_4 + \lambda_6} + \frac{1}{\lambda_1 + \lambda_4 + \lambda_5 + \lambda_6} + \frac{1}{\lambda_1 + \lambda_3 + \lambda_4 + \lambda_5}$$

$$+\frac{1}{\lambda_1 + \lambda_3 + \lambda_4 + \lambda_6} - \frac{1}{\lambda_1 + \lambda_3 + \lambda_4 + \lambda_5 + \lambda_6} - \frac{1}{\lambda_2 + \lambda_3 + \lambda_5}$$

$$-\frac{1}{\lambda_2 + \lambda_3 + \lambda_6} + \frac{1}{\lambda_2 + \lambda_3 + \lambda_5 + \lambda_6} - \frac{1}{\lambda_2 + \lambda_4 + \lambda_5}$$

$$-\frac{1}{\lambda_2 + \lambda_4 + \lambda_6} + \frac{1}{\lambda_2 + \lambda_4 + \lambda_5 + \lambda_6} + \frac{1}{\lambda_2 + \lambda_3 + \lambda_4 + \lambda_5}$$

$$+\frac{1}{\lambda_2 + \lambda_3 + \lambda_4 + \lambda_6} - \frac{1}{\lambda_2 + \lambda_3 + \lambda_4 + \lambda_5 + \lambda_6}$$

$$+\frac{1}{\lambda_1 + \lambda_2 + \lambda_3 + \lambda_5} + \frac{1}{\lambda_1 + \lambda_2 + \lambda_3 + \lambda_6}$$

$$-\frac{1}{\lambda_1 + \lambda_2 + \lambda_3 + \lambda_5 + \lambda_6} + \frac{1}{\lambda_1 + \lambda_2 + \lambda_4 + \lambda_5}$$

$$+\frac{1}{\lambda_1 + \lambda_2 + \lambda_4 + \lambda_6} - \frac{1}{\lambda_1 + \lambda_2 + \lambda_4 + \lambda_5 + \lambda_6}$$

$$-\frac{1}{\lambda_1 + \lambda_2 + \lambda_3 + \lambda_4 + \lambda_5} - \frac{1}{\lambda_1 + \lambda_2 + \lambda_3 + \lambda_4 + \lambda_6} \quad (12.138)$$

$$+\frac{1}{\lambda_1 + \lambda_2 + \lambda_3 + \lambda_4 + \lambda_5 + \lambda_6}.$$

For the architecture variation in Figure 12.14 the MTTF-value can be calculated as follows:

$$R_g(t) = R_1(t) \cdot [(R_2(t) + R_3(t) - R_2(t) \cdot R_3(t)) + \\ + R_4(t) - (R_2(t) + R_3(t) - R_2(t) \cdot R_3(t)) \cdot R_4(t)] \cdot R_5(t) \quad (12.139)$$

$$MTTF = \int_0^\infty R_g(t) \cdot dt \quad (12.140)$$

$$MTTF = \int_0^\infty \bigl[(R_1(t) \cdot R_2(t) \cdot R_5(t)) + (R_1(t) \cdot R_3(t) \cdot R_5(t)) \\
- (R_1(t) \cdot R_2(t) \cdot R_3(t) \cdot R_5(t)) \\
+ (R_1(t) \cdot R_4(t) \cdot R_5(t)) - (R_1(t) \cdot R_2(t) \cdot R_4(t) \cdot R_5(t)) + \\
- (R_1(t) \cdot R_3(t) \cdot R_4(t) \cdot R_5(t)) \\
+ (R_1(t) \cdot R_2(t) \cdot R_3(t) \cdot R_4(t) \cdot R_5(t)) \bigr] \cdot dt \quad (12.141)$$

$$MTTF = \int_0^\infty [e^{-(\lambda_1+\lambda_2+\lambda_5)t} + e^{-(\lambda_1+\lambda_3+\lambda_5)t} - e^{-(\lambda_1+\lambda_2+\lambda_3+\lambda_5)t}$$
$$+ e^{-(\lambda_1+\lambda_4+\lambda_5)t} - e^{-(\lambda_1+\lambda_2+\lambda_4+\lambda_5)t} - e^{-(\lambda_1+\lambda_3+\lambda_4+\lambda_5)t} \quad (12.142)$$
$$+ e^{-(\lambda_1+\lambda_2+\lambda_3+\lambda_4+\lambda_5)t}] \cdot dt$$

$$MTTF = -\frac{1}{\lambda_1+\lambda_2+\lambda_5} - \frac{1}{\lambda_1+\lambda_3+\lambda_5} + \frac{1}{\lambda_1+\lambda_2+\lambda_3+\lambda_5} - \frac{1}{\lambda_1+\lambda_4+\lambda_5}$$
$$+ \frac{1}{\lambda_1+\lambda_2+\lambda_4+\lambda_5} + \frac{1}{\lambda_1+\lambda_3+\lambda_4+\lambda_5} - \frac{1}{\lambda_1+\lambda_2+\lambda_3+\lambda_4+\lambda_5}. \quad (12.143)$$

Observing the special case for a redundant system of two blocks with $\lambda_1 = \lambda_2 = \lambda$, the MTTF-value can be determined with the following:

$$MTTF = \frac{3}{2} \cdot \frac{1}{\lambda} \quad (12.144)$$

With n parallel elements of the same failure rate

$$R(t) = 1 - (1 - e^{-\lambda t})^n = n \cdot e^{-\lambda t} - \binom{n}{2} e^{-2\lambda t} + - \ldots e^{-n\lambda t} \quad (12.145)$$

with the average mean time to failure

$$MTTF = \frac{n}{\lambda} - \frac{n(n-1)}{2 \cdot 2\lambda} + \frac{n(n-1)(n-2)}{2 \cdot 3 \cdot 3\lambda} - + \ldots \frac{1}{n\lambda} \quad (12.146)$$

Redrafted

$$MTTF = \frac{1}{\lambda} \left[\binom{n}{1} - \frac{\binom{n}{2}}{2} + \frac{\binom{n}{3}}{3} - + \ldots \frac{1}{n} \right] \quad (12.147)$$

$$MTTF = \frac{1}{\lambda} \sum_{i=1}^{n} \frac{1}{i} \quad (12.148)$$

It is obvious from the previous executions that redundancy circuits always convey better values for the reliability than a single element.

12.3 Substitution of Redundant System Components through Single System Components

In the previous chapters the improvement of the system reliability through redundancy concepts was discussed and it was determined, that a redundancy circuit always delivers better values than the single element. Therefore an observation regarding a single configuration as opposed to redundancy configurations is in order while the mean time to failure of both components is equal. In this case the redundant order of two system blocks with

$$\lambda = \frac{2}{3}\lambda \qquad (12.149)$$

This aspect is important when it comes to lowering the expenses of material. Exemplary observations are to be made if single components or redundancy configurations are to be preferred[115]. The following condition applies (R_r redundancy configuration system with two blocks and the comparison of a single element R_e):

$$R_r = 2e^{-\lambda t} - e^{-2\lambda t} \qquad (12.150)$$

$$R_e = e^{-\lambda''} = e^{-\frac{2}{3}\lambda t} \qquad (12.151)$$

Figure 12.18 and Figure 12.19 respectively illustrate the distribution of the reliability and the distribution of probability of failure over time. The following applies; $e^{-\lambda t}$.

Broadly speaking the distribution also applies to systems with a redundancy > 2.

Broadly speaking the tolerance of failure as a criterion for the reliability of a safety related system is being considered. It can be applied, that a safety critical architecture, in which any element might fail, normally shows an advanced reliability compared to any other equivalent non-redundancy system configuration, in which certain elements may not fail.

---

[115] [BIRO97]   Birolini, A.; Zuverlässigkeit von Geräten und Systemen
  [GÖRK68]   Görke, W.; Probleme der Zuverlässigkeit elektronischer Schaltungen
  [NIED94]   Niedermeier, A.; Fehlertoleranz durch Kombination statischer und dynamischer Redundanz
  [O'CO90]   O'Conner, P.; Zuverlässigkeitstechnik

12 Reliability Block Diagram Analysis

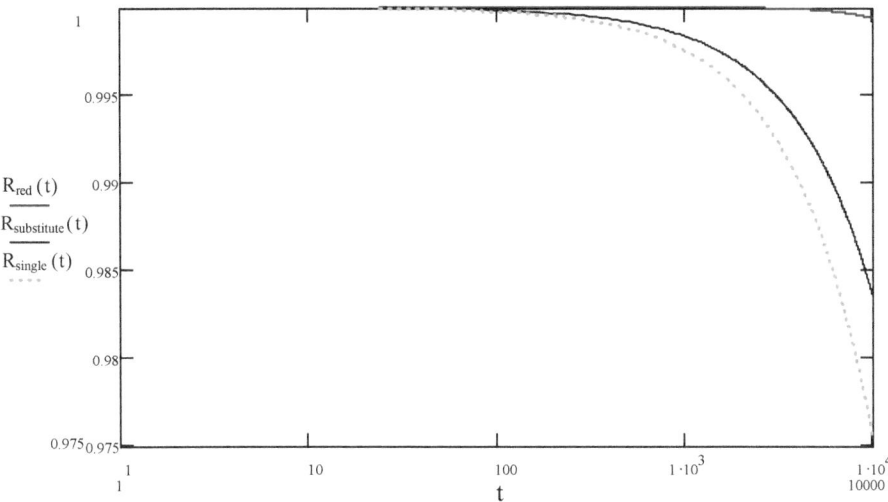

Figure 12.18: Distribution of Reliability [116]

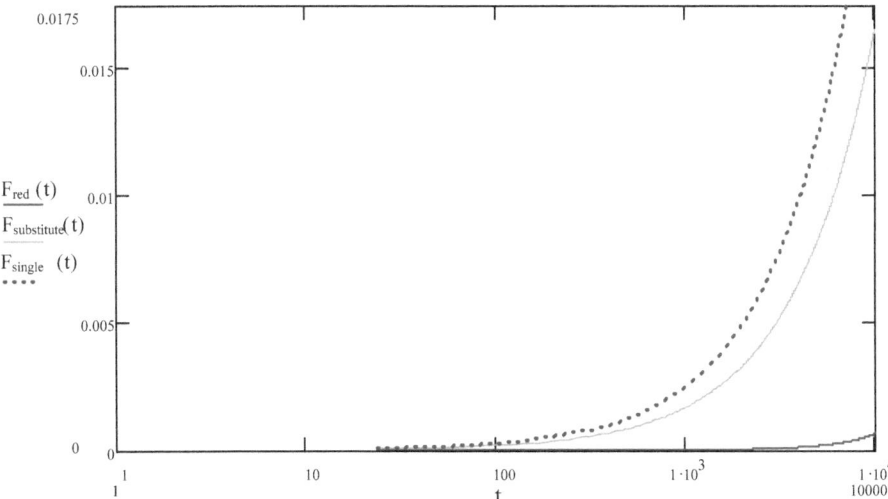

Figure 12.19: Distribution of Probability of Failure

---

[116] For both diagrams the top line represents reliability of the redundant architecture in Figure 12.18 (the lower in Figure 12.19), the middle line in Bild 12.18 the reliability/failure probability of the redundancy and the lower dotted line in Bild 12.19 the reliability/failure probability of the individual components.

13 Markov Model

13.1 Introduction

In safety related processes in the process and automation industry repairable systems are being used almost exclusively because they reach a high system availability and -safety. Reliability models of simple systems however do not take into account that the exchange of faulty modules takes time. In order to take the repair time into consideration, probability methods and models come into play. They must consider errors in entire and partial repairable systems, failure rates and repair times. Basically the methods must be able to determine how realistic repair times, system configurations, system specifications including diagnostic examinations are.

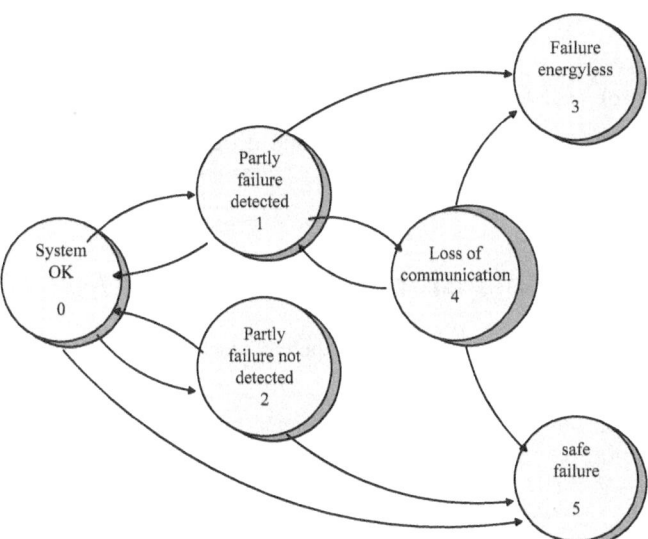

Figure 13.1: Example of a Markov Model

The Markov Model belongs to the reliability and safety procedures that work in conjunction with state diagrams and fulfill these demands. The Markov Model works with two symbols: the circle and the transition curve. The states of the components are displayed in circles; the transition between two states is displayed by a transition curve with a directional arrow. The Markov Model can consequently represent in a graph the entire configuration of a fault tolerant control system.

After the complete design of a Markov Model the depiction of states likes "system error free" or "system in operation" is possible. Further, in case of a failure, the system can be deemed in operation but susceptible for further failures.

Figure 13.1 shows an example of possible failure conditions.

13.2 Possibilities with Markov Models

The following reliability values, safety dimensions and different parameters can be derived from the Markov Model:

- **Point-Availability**: Probability, that a system executes its demanded function under set working conditions at a specific time
- **Availability**: Probability, that a system executes its demanded function under set working conditions in a specific time frame
- **MTTF (Mean Time To Failures):** average value of the failure free working time
- **Reliability:** Probability, that a system executes its demanded function in a specific time frame without failure.

With entirely repairable systems the availability can be time-dependent or constant.

Present day safety considerations put more and more emphasis on the probability conditions. Therefore the Probability of Failure on Demand ((*PFD*; probability of a failure by demand of a safety function) is being predicted with the addition of the probabilities of the conditions where a request of a safety function is not being answered by the system.

The Markov Model is capable of predicting *MTTF*-values and different average failure times very exactly. This model is the most flexible when it comes to calculating fault time conditions.

13.3 Theoretical Principals of the Markov Models

Following is a summary of the theoretical principals[117] of the calculation by means of the Markov Model.

The Markov Models[118] were constantly subject to enhancement. They apply the failure/ repair process because the failure combinations are followed by distinct system conditions. The fault/ repair process is the result of the current state and the current error and transitions only between distinct states.

---

[117] [BIRO97]   Birolini, A.; *Zuverlässigkeit von Geräten und Systemen*
      [HENL92]   Henley, E.; et al.; *Probabilistic risk assessment-reliability engineering, design and analysis*
      [KREY67]   Kreyszig, E.; *Advanced engineering mathematics*
      [MAKI73]   Maki, D. P.; et al.; *Mathematical models and applications*
      [PAPO84]   Papoulis, A.; *Probability, random variables and stochastic processes*

[118] Andrei Andreyevich Markov (1856-1922) defined this Markov process, where explicitly the current variable determines the future variable. The future variable is thus independent from the previous variable. Here the sequences are important where the variable values are discrete. This sequence is today knowns and Markov chain.

13.3 Theoretical Principals of the Markov Models

The correct calculation of the reliability functions for more complex redundancy systems frequently turns out to be difficult. Not all function possibilities of the circuit are being assessed because the reliability block diagram for example is hard to design, if at all. Reasons are the sporadic alterations of failure rates or the consideration for the repair of single blocks for example. Here the condition analysis is applied which interprets the distribution of the condition as homogeneous Markov chains.

In probability teachings a homogeneous Markov chain is a consequence of an experiment by which the conditional probability distribution of the state in the future, given the state of the process currently and in the past, depends only on its current state and not on its state in the past. This process does not have any memory, meaning the states have no effect on the future. Such chains can describe the different redundant systems. The event conditions are the possible system conditions that are defined through the functional condition or the fail state of all system elements. With this procedure all failure rates including their appearance, changing failure rates and reversible occurrences are regarded. Basis for this procedure is the defined condition (able to function, or failing[119]) of the system at a specific time. Furthermore the system must, within the *MTTR* and the operation respectively, be able to function because the system is safety related and can define the probabilities for the conditional transitions. Bases are constant failure rates and an exponential distribution. The probability that a condition remains or transitions can be defined for each condition.

The probability, that the system in a timed interval of the length dt transitions from the condition j into the condition k can be mathematically expressed as follows.

$$p(Z_j \to Z_k) = p_{jk} = \lambda_{jk} \cdot dt \ . \tag{13.1}$$

The coefficients $\lambda_{jk} \geq 0$ are the transition rates. If $\lambda_{jk} = 0$, then a transition of state Z_j into state Z_k is not possible. Since within the time interval dt only a change in state can occur[120]. A condition change is however not necessary[121]. A formula with n conditions is as follows:

$$p(Z_j \to Z_j) = p_{jj} = 1 - \sum_{k=1}^{n} \lambda_{jk} \, dt \tag{13.2}$$

with $(j = 1, 2, \ldots, n; \ k = 1, 2, \ldots n; \ k \neq j)$.

The sum in 13.2 can be simplified,

$$\sum_{k=1}^{n} \lambda_{jk} = \lambda_{jj} \tag{13.3}$$

with $(j = 1, 2, \ldots, n; \ k = 1, 2, \ldots n; \ k \neq j)$,

---

[119] The system goes into the safe state when the transition leads to a so called absorbing state.
[120] Achieved by choosing the corresponding interval length
[121] [GOBL95] Goble, W. M.; *Safety of programmable electronic systems*
  [GÖRK68] Görke, W.; *Probleme der Zuverlässigkeit elektronischer Schaltungen*
  [HENL92] Henley, E.; et al.; *Probabilistic risk assessment-reliability engineering, design and analysis*
  [MAKI73] Maki, D. P.; et al.; *Mathematical models and applications*
  [PAPO84] Papoulis, A.; *Probability, random variables and stochastic processes*

so that

$$p(Z_j \rightarrow Z_j) = p_{jj} = 1 - \lambda_{jj}\, dt \qquad (13.4)$$

with $(j = 1, 2, \ldots, n;\ k = 1, 2, \ldots n;\ k \neq j)$

can be displayed. For all states one acquires a transition matrix, known through the Markov Processes.

$$\mathbf{P} = \begin{bmatrix} p_{11} & p_{12} & \cdots & p_{1n} \\ p_{21} & p_{22} & \cdots & p_{2n} \\ \vdots & \vdots & \vdots & \vdots \\ p_{n1} & p_{n2} & \cdots & p_{nn} \end{bmatrix}$$

$$= \begin{bmatrix} 1 - \lambda_{11}\, dt & \lambda_{12}\, dt & \cdots & \lambda_{1n}\, dt \\ \lambda_{21}\, dt & 1 - \lambda_{22}\, dt & \cdots & \lambda_{2n}\, dt \\ \vdots & \vdots & \vdots & \vdots \\ \lambda_{n1}\, dt & \lambda_{n2}\, dt & \cdots & 1 - \lambda_{nn}\, dt \end{bmatrix}. \qquad (13.5)$$

The matrix includes the conditional transitional probabilities for the single states. Each sum must result in 1. If no conditional transitions are possible, jk[122] becomes 0.

Interesting are here the probability functions for the single states. [123] $p_i(t)$ is the probability, that the system at the time t is in the state Z_j. The probability functions $p_i(t + dt)$ at a time $(t + dt)$ can be calculated with help of the column j of the matrix $\mathbf{P}$ as follows:

$$p_1(t + dt) = p_1(t) \cdot (1 - \lambda_{11} dt) + \cdot p_2(t) \cdot \lambda_{21} dt + p_3(t) \cdot \lambda_{31} dt + \cdots$$
$$\ldots + p_n(t) \cdot \lambda_{n1} dt$$

$$p_2(t + dt) = p_1(t) \cdot \lambda_{12} dt + p_2(t) \cdot (1 - \lambda_{22} dt) + p_3(t) \cdot \lambda_{32} dt + \cdots$$
$$\ldots + p_n(t) \cdot \lambda_{n2} dt \qquad (13.6)$$

$$p_3(t + dt) = p_1(t) \cdot \lambda_{13} dt + p_2(t) \cdot \lambda_{23} dt + p_3(t) \cdot (1 - \lambda_{33} dt) + \ldots$$
$$\ldots + p_n(t) \cdot \lambda_{n3} dt$$
$$\vdots$$

---

[122] Transitions do not necessarily exist. A transition from state Z_1 to state Z_0 is e.g. denoted with μ. This variable is calculated with $\mu = 1/\tau_{LT}$, where τ_{LT} represents the lifetime of the system, for which counts

$$\lim_{\tau_{LT} \to \infty} \frac{1}{\tau_{LT}} = 0.$$

The system will first transfer ad infinitum in the next state.

[123] [GOBL95] Goble, W. M.; *Safety of programmable electronic systems*
[GÖRK68] Görke, W.; *Probleme der Zuverlässigkeit elektronischer Schaltungen*
[HENL92] Henley, E.; et al.; *Probabilistic risk assessment-reliability engineering, design and analysis*
[MAKI73] Maki, D. P.; et al.; *Mathematical models and applications*
[PAPO84] Papoulis, A.; *Probability, random variables and stochastic processes*

13.3 Theoretical Principals of the Markov Models

The equations for the different states are similar and can be expressed in vector way of writing.

$$\underline{p}(t + dt) = \underline{p}(t) \cdot P \qquad (13.7)$$

$\underline{p}(t + dt)$ and $\underline{p}(t)$ are the line vectors whose components occur in all states. The equation can be reshaped via the matrix calculation

$$d\,\underline{p}(t) = \underline{p}(t + dt) - \underline{p}(t) = \underline{p}(t) \cdot (P - I) \qquad (13.8)$$

by which I is the unit matrix. Subtracting the unit matrix from P, one gets the matrix $M \cdot dt$

$$P - I = M \cdot dt = \begin{bmatrix} -\lambda_{11} & \lambda_{12} & \cdots & \lambda_{1n} \\ \lambda_{21} & -\lambda_{22} & \cdots & \lambda_{2n} \\ \vdots & \vdots & \vdots & \vdots \\ \lambda_{n1} & \lambda_{n2} & \cdots & -\lambda_{nn} \end{bmatrix} dt \qquad (13.9)$$

It includes all transition rates as elements. The sum of the lines must be zero. One gets

$$d\,\underline{p}(t) = \underline{p}(t) \cdot M\,dt \qquad (13.10)$$

and

$$\underline{\dot{p}}(t) = \underline{p}(t) \cdot M. \qquad (13.11)$$

Equation 13.11 shows a system with coupled linear differential equations of the 1st order. Out of the root x_i the characteristic polynomial one gets the exponential time functions of the transformed coupled system of the differential equations. Out of the inverse transformation one receives the general solutions. The initial state of the system at the beginning of the observation must be taken into consideration. [124]

The Markov Model emphasizes mutually exclusive system states, for example "in operation" or "error status". A state is displayed via a circle with a condition description. The system state changes at the occurrence of an error. The transition curves represent the conditional transitions. They are labeled with a transition probability rate, which assigns a probability density for the state transition (often only the error rates/ /repair rates are given).

This depiction of the Markov Model describes the behavior of a system over time. If the time of a model can be displayed in specific increments (once per hour), simulations with probabilities can be conducted and therefore probability calculation for the conditional changes of a system are possible. By summarizing the probabilities of a functioning system, the system (reliability or the system (availability can be predicted.

To demonstrate the Markov Model on a single component, two examples follow. If a system is not repairable, the Markov Model displays itself as in Figure 13.2. The circles display the system state Z_0 for 'able to function' and Z_1 for 'system failure'. The only

---

[124] [MAKI73]   Maki, D. P.; et al.; *Mathematical models and applications*
   [PAPO84]   Papoulis, A.; *Probability, random variables and stochastic processes*

possible state change is represented as a transition curve of state Z_0 into the state Z_1 with a constant failure rate. An expansion shows Figure 13.3 in which the system is repairable, showing that there is a second transition curve; from state Z_1 to state Z_0 with the repair rate μ_R.

Beside the partial or entirely repairable systems one can capture through all mistakes and data into the Markov Model simultaneously occurring errors. A differentiation of errors that are discovered with help of the online diagnostics is possible through equivalent description.

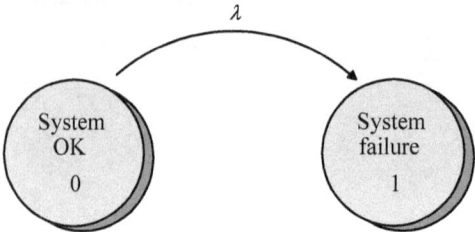

Figure 13.2: Markov Model of a single, not repairable component

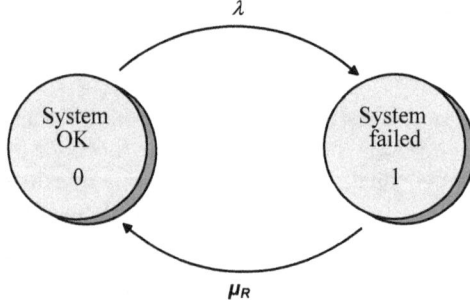

Figure 13.3: Example for a single, repairable component

Common cause failures can also be represented in the Markov Model[125]. Generally one could granulate Markov Models as long as necessary.

---

[125] One differentiates between detected, through computer diagnostics, and undetected failures.

13.4 Time dependent Markov Models

All processes, which transition probabilities depend only on current states, can be displayed in a Markov Model. They are also described as „ "memory-less" systems[126]. That depiction of the Markov Model occurs via a definition of the states and changes between the transitions.

If it is likely, that a state changes within a specific time interval, this condition transition is transferred into the Markov Model. If the probability is time independent, like it is assumed in these systems, the model itself can be solved with linear algebra or through approximate solutions of average condition probabilities or condition probabilities as a function of time between two states[127]. Are the probabilities time dependent, one can predict via computer simulations the state probabilities as a function of time.

Using Markov Models one can generally solve states that are time dependent. These time dependent Markov models change within a defined time increment[128]. If the states are time independent, one can proceed equally, however, one calculates lower limits and the time increment is approaching zero (the lower limit Δt approaches zero) [129].

13.5 Implementation of a Markov Calculation for a Safety Related System

First the system states have to be analyzed and described. In a simple example in Figure 13.4 seven different system states can be defined.

The state Z_0 stands for an failure free state. Six further states can originate from this condition.

In state Z_1, also called state S^{130}, a safe failure has occurred that is discovered through the diagnostic system. The system is in state Z_1, in a safe state. This can occur after the appearance of a safe, detectable or not detectable error. The transition probability rate is displayed via the failure rate $2 \cdot \lambda_S = 2 \cdot (\lambda_{SD} + \lambda_{SU})$ because a safe error is possible in both

---

[126] [GOBL95]   Goble, W. M.; *Safety of programmable electronic systems*
       [HENL92]   Henley, E.; et al.; *Probabilistic risk assessment-reliability engineering, design and analysis*
       [MAKI73]   Maki, D. P.; et al.; *Mathematical models and applications*
       [PAPO84]   Papoulis, A.; *Probability, random variables and stochastic processes*
[127] [GOBL95]   Goble, W. M.; *Safety of programmable electronic systems*
       [HENL92]   Henley, E.; et al.; *Probabilistic risk assessment-reliability engineering, design and analysis*
       [MAKI73]   Maki, D. P.; et al.; *Mathematical models and applications*
       [PAPO84]   Papoulis, A.; *Probability, random variables and stochastic processes*
       [STOR96]   Storey, N.; *Safety critical computer systems*
[128] Exemple for time increment are: once per hour, 10 times per hour, once per day, once a week.
[129] [GOBL95]   Goble, W. M.; *Safety of programmable electronic systems*
       [HENL92]   Henley, E.; et al.; *Probabilistic risk assessment-reliability engineering, design and analysis*
       [MAKI73]   Maki, D. P.; et al.; *Mathematical models and applications*
       [PAPO84]   Papoulis, A.; *Probability, random variables and stochastic processes*
[130] Safe Failure, safe state, the failure is anknowledged.

channels. With help of the diagnostic system this mistake can be detected. A repair returns the system back into the error free condition.

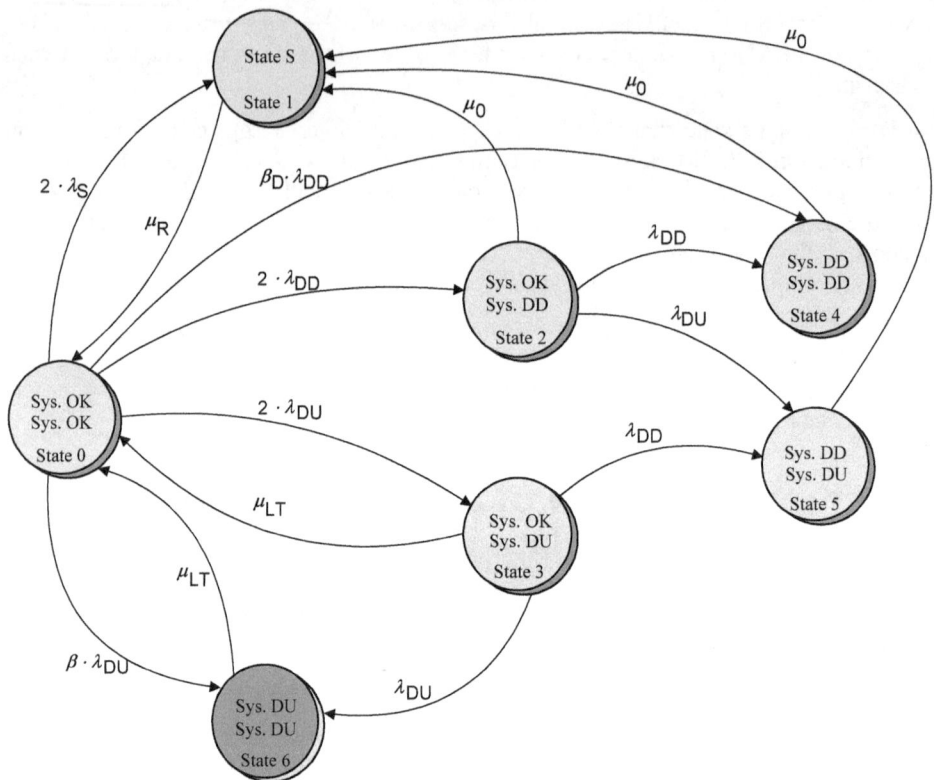

Figure 13.4: Markov Model of a simple, redundant system model

When a dangerous failure in one of the channels occurs and can be detected via the diagnostic system, the system is in state Z_2. The transition probability of state Z_0 to state Z_2 is $2\lambda_{DD}$. A transition into the safe state Z1 is possible if the diagnostic system finds the mistake. After repair, the system can be moved back into an failure free condition. If the failure is discovered within the test interval τ_{Test}, the transition rate then amounts to $\mu_0 = 1/\tau_{Test}$. If, within the test interval τ_{Test}, one or more dangerous, detectable or undetectable failures occur in state Z_2 and the first failure has not yet been discovered, the system changes from state Z_2 into state Z_4 or Z_5. Transition rate is λ_{DD} or λ_{DU} respectively.

If the system is in state Z_3 a failure occurred in a channel that is dangerous and cannot be detected through failure diagnostics. Because there can be a dangerous, non-detectable failure in both channels, the transition rate between the states is Z_0 and Z_3 $2 \cdot \lambda_{DU}$.

The system cannot, within the test interval τ_{Test}, transition into the safe state Z_1, because it lacks an failure correction procedure. If in state Z_3 appears a failure in the still faultless channel, the system transitions into the state Z_5 or Z_6 respectively. The system switches into the faultless state Z_0 if, during the overall lifetime of the system, no more failures occur in state Z_3.

13.5 Implementation of a Markov Calculation for a Safety Related System

A system in state Z_4 consists of one dangerous, detectable failure per channel[131]. The difference of state Z_5 is that in contrast to state Z_4, one or both failures are not detectable. Both states Z_4 and Z_5 can possibly transition through a failure correction procedure within the test interval τ_{Test} into the safe state Z_1.

In state Z_6 each of the two channels includes a dangerous, undetectable failure. A diagnosis of the failure is here not possible. The transition probability is described via λ_{DU}. A repair in order to achieve a working state is here no longer possible.

Looking at the common cause failure, two events can be distinguished in a redundant system model:

- A common cause failure leads to dangerous detectable mistakes. The system transitions from state Z_0 directly into state Z_4. The transition rate is $\beta_D \cdot \lambda_{DD}$.

- A common cause failure leads to dangerous, undetectable failures. The system transitions from state Z_0 directly into state Z_6. The transition rate is $\beta \cdot \lambda_{DU}$.

Recapitulating can be determined for this system model:

- If failures occur, that through the redundant system models transition into the states Z_2, Z_4 or Z_5, then the system transitions, after some time smaller than $2 \cdot \tau_{Test}$, into the safe state Z_1. The transition rate of these states after state Z_1 always amount to $\mu_0 = 1/\tau_{Test}$.

- The states Z_1, Z_4, Z_5 and Z_6 are absorbing states [132]

- In the states Z_0, Z_2 und Z_3 the system is in operation and must be taken into consideration with the MTTF-calculation of the redundant system model.

There are several possibilities to transition the model system in Figure 13.4 from the states of Z_1 to Z_6 into the faultless state Z_0. The transition probability of state Z_1 to state Z_0 is marked with the repair rate μ_R and is being calculated with $\mu_R = 1/\tau_{repair}$. It includes the repair and the restarting time of the system. The repair rate μ_0 displays the transition probability of state Z_2, Z_4 and Z_5 after state Z_1 and is determined by $\mu_0 = 1/\tau_{Test}$. Is the system in condition Z_3 and no further failure occurs, the transition into the state Z_0 is only possible at the end of its lifetime and subsequent exchange or repair. The transition rate is consequently $\mu_{LT} = 1/\tau_{LT}$. Is the system in state Z_6, the transition rate is the same. The failure rates are defined as losses per hour, so that the transition rates μ_0, μ_R and μ_{LT} are $1/h$.

---

[131] DU = Dangerous Undetected, failures
[132] An absorbing state is given when explicitly transitions into the safe state Z_1 or in the state „overall system operational, State Z_0" exist.

13.5.1 Transition Matrix *P* for System Model

The probabilities can be displayed in form of a matrix. If a system model has n states, a $n \times n$ matrix can be developed. This matrix is called transition matrix P. and includes the transition probabilities of the distinct states:[133]

$$P = \begin{bmatrix} P_1 & P_2 \\ P_3 & P_4 \end{bmatrix} \tag{13.12}$$

with

$$P_1 = \begin{bmatrix} 1 - A_1 \cdot dt & 2 \cdot \lambda_S \cdot dt & 2 \cdot \lambda_{DD} \cdot dt \\ \mu_R \cdot dt & 1 - \mu_R \cdot dt & 0 \\ 0 & \mu_0 \cdot dt & 1 - A_2 \cdot dt \end{bmatrix}$$

$$P_2 = \begin{bmatrix} 2 \cdot \lambda_{DU} \cdot dt & \beta_D \cdot \lambda_{DD} \cdot dt & 0 & \beta \cdot \lambda_{DU} \cdot dt \\ 0 & 0 & 0 & 0 \\ 0 & \lambda_{DD} \cdot dt & \lambda_{DU} \cdot dt & 0 \end{bmatrix}$$

$$P_3 = \begin{bmatrix} \mu_{LT} \cdot dt & 0 & 0 \\ 0 & \mu_0 \cdot dt & 0 \\ 0 & \mu_0 \cdot dt & 0 \\ \mu_{LT} \cdot dt & 0 & 0 \end{bmatrix}$$

$$P_4 = \begin{bmatrix} 1 - A_3 \cdot dt & 0 & \lambda_{DD} \cdot dt & \lambda_{DU} \cdot dt \\ 0 & 1 - \mu_0 \cdot dt & 0 & 0 \\ 0 & 0 & 1 - \mu_0 \cdot dt & 0 \\ 0 & 0 & 0 & 1 - \mu_{LT} \cdot dt \end{bmatrix}.$$

A characteristic of the transition-matrix P is that the sum of the elements of a line must be 1, because the sum of all state probabilities is 1, meaning the following is valid:

$$A_1 = 2 \cdot \lambda_S + \beta_D \cdot \lambda_{DD} + 2 \cdot \lambda_{DD} + 2 \cdot \lambda_{DU} + \beta \cdot \lambda_{DU}$$
$$A_2 = \mu_0 + \lambda_{DD} + \lambda_{DU}$$
$$A_3 = \mu_{LT} + \lambda_{DD} + \lambda_{DU}.$$

The matrix elements are defined as $p_{i,j}$. The states in picture 13.4 are $i, j \in [0...6]$. The value of a matrix element $p_{i,j}$ described with the row 1 index and column index i and the probability j that the state Z_i transitions into the state Z_j. If row 1 and column 2 have the value $2 \cdot \lambda_S \cdot dt$, then that is the probability of a transition of state Z_0 into state Z_1 at the next interval. If row 1 and column 1 have the value $1 - A_1 \cdot dt$, then that is the probability

---
[133] This matrix, which in literature is designated with P is known as transition matrix or „stochastical transition matrix".

13.5 Implementation of a Markov Calculation for a Safety Related System

of a transition of state Z_0 into state Z_0 at the next interval. Meaning the system remains in the state Z_0. All mandatory statements of a Markov Model are included in a transition matrix.

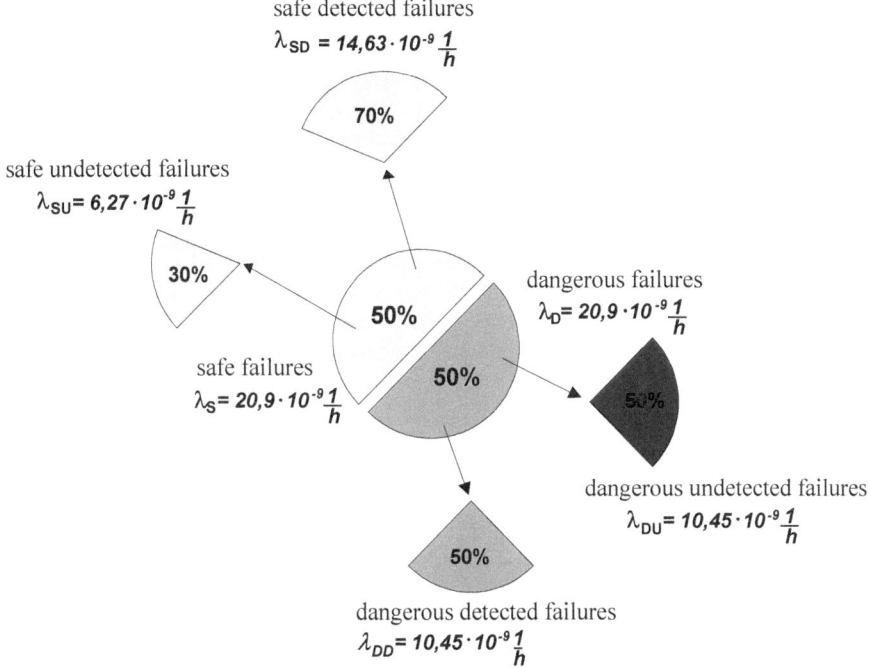

Figure 13.5: Failure Rates of the System Model

Further calculations follow from of this matrix and will be demonstrated in an example. The example assumes that the overall failure rate is $\lambda = \lambda_D + \lambda_S = 41,8 \cdot 10^{-9}$ 1/h and that 70% of the sure losses can be recognized[134]. Figure 13.5 shows the failure rates.

Following the factor 0.03 is assumed for the β-factors and the average time to repair (τ_{Repair}) for the system is eight hours according to the norm. From this the repair rate of the transition of state Z_1 into state Z_0, calculates as follows:

$$\mu_R = \frac{1}{\tau_{Repair}} = \frac{1}{8\,h} = 0{,}125\frac{1}{h}.$$

The following is valid

$$\mu_0 = \frac{1}{\tau_{Test}} = \frac{1}{24\,h} = 0{,}0416\frac{1}{h}$$

---

[134] For highly safe systems the diagnostic coverage should at least be 99%. The value of 70% is used here as an example for an industrial safety related system classified as "weak diagnostic coverage".

and for

$$A_1 = (2 \cdot 20{,}9 \cdot 10^{-9} + 0{,}03 \cdot 10{,}45 \cdot 10^{-9} + 2 \cdot 10{,}45 \cdot 10^{-9} \cdots$$
$$+ 2 \cdot 10{,}45 \cdot 10^{-9} + 0{,}03 \cdot 10{,}45 \cdot 10^{-9}) \frac{1}{h} = 8{,}4227 \cdot 10^{-8} \frac{1}{h}$$

$$A_2 = (0{,}0416 + 10{,}45 \cdot 10^{-9} + 10{,}45 \cdot 10^{-9}) \frac{1}{h} = 0{,}04160002 \frac{1}{h}$$

$$A_3 = (0 + 10{,}45 \cdot 10^{-9} + 10{,}45 \cdot 10^{-9}) \frac{1}{h} = 20{,}9 \cdot 10^{-9} \frac{1}{h}.$$

Now add the corresponding values into the transition matrix. The **P**-matrix was divided due to clarity reasons:

$$\mathbf{P} = \begin{bmatrix} \mathbf{P}_1 & \mathbf{P}_2 \\ \mathbf{P}_3 & \mathbf{P}_4 \end{bmatrix} \tag{13.13}$$

with

$$\mathbf{P}_1 = \begin{bmatrix} 1 - 8{,}4227 \cdot 10^{-8} \frac{1}{h} \cdot dt & 2 \cdot 20{,}9 \cdot 10^{-9} \frac{1}{h} \cdot dt & 2 \cdot 10{,}45 \cdot 10^{-9} \frac{1}{h} \cdot dt \\ 0{,}125 \frac{1}{h} \cdot dt & 1 - 0{,}125 \frac{1}{h} \cdot dt & 0 \\ 0 & 0{,}0416 \frac{1}{h} \cdot dt & 1 - 0{,}04160002 \frac{1}{h} \cdot dt \end{bmatrix}$$

$$\mathbf{P}_2 = \begin{bmatrix} \mathbf{P}_{21} & \mathbf{P}_{22} \end{bmatrix}$$

with

$$\mathbf{P}_{21} = \begin{bmatrix} 2 \cdot 10{,}45 \cdot 10^{-9} \frac{1}{h} \cdot dt & 0{,}03 \cdot 10{,}45 \cdot 10^{-9} \frac{1}{h} \cdot dt \\ 0 & 0 \\ 0 & 10{,}45 \cdot 10^{-9} \frac{1}{h} \cdot dt \end{bmatrix}$$

and

$$\mathbf{P}_{22} = \begin{bmatrix} 0 & 0{,}03 \cdot 10{,}45 \cdot 10^{-9} \frac{1}{h} \cdot dt \\ 0 & 0 \\ 10{,}45 \cdot 10^{-9} \frac{1}{h} \cdot dt & 0 \end{bmatrix}$$

13.5 Implementation of a Markov Calculation for a Safety Related System

$$P_3 = \begin{bmatrix} 0 & 0 & 0 \\ 0 & 0{,}0416\frac{1}{h} \cdot dt & 0 \\ 0 & 0{,}0416\frac{1}{h} \cdot dt & 0 \\ 0 & 0 & 0 \end{bmatrix}$$

$$P_4 = \begin{bmatrix} 1 - 20{,}9 \cdot 10^{-9} \frac{1}{h} \cdot dt & 0 & 10{,}45 \cdot 10^{-9} \frac{1}{h} \cdot dt & 10{,}45 \cdot 10^{-9} \frac{1}{h} \cdot dt \\ 0 & 1 - 0{,}0416 \frac{1}{h} \cdot dt & 0 & 0 \\ 0 & 0 & 1 - 0{,}0416 \frac{1}{h} \cdot dt & 0 \\ 0 & 0 & 0 & 1 \end{bmatrix}$$

Out of the transition matrix **P** the **Q**-matrix follows. It is considered to be a reliability matrix and its elements evolve out of the probabilities for which states the following is valid:

- System in operation
- No absorbing condition.[135]

Following states can therefore not occur:

- Safe state S, in the chosen example Z_1, with safe failures in the system.
- Condition *DU*, in the chosen example Z_6, where each channel includes a dangerous undetected failure.

The discussed redundant system models can therefore be only in the states Z_0, Z_2 and Z_3. The Q-matrix consists of

- the 1st row with the elements of the 1st, 3rd and 4th column,
- the 3rd row with the elements of the 1st, 3rd and 4th column and
- the 4th row with the elements of the 1st, 3rd and 4th column.

The conditions Z_1, Z_4, Z_5 and Z_6 are absorbent and can be disregarded at the MTTF-calculation[136]. The appearance of the Q-matrix looks therefore as follows:

$$Q = \begin{bmatrix} 1 - A_1 \cdot dt & 2 \cdot \lambda_{DD} \cdot dt & 2 \cdot \lambda_{DU} \cdot dt \\ 0 & 1 - A_2 \cdot dt & 0 \\ \mu_{LT} \cdot dt & 0 & 1 - A_3 \cdot dt \end{bmatrix} \quad (13.14)$$

---

[135] If a state has a failure transition to another state then this state is considered as an absorbing state.
[136] [GOBL95]   Goble, W. M.; *Safety of programmable electronic systems*
[HENL92]   Henley, E.; et al.; *Probabilistic risk assessment-reliability engineering, design and analysis*
[MAKI73]   Maki, D. P.; et al.; *Mathematical models and applications*
[PAPO84]   Papoulis, A.; *Probability, random variables and stochastic processes*

For the Markov Model used, $\tau_{LT} = \infty$ is valid and therefore

$$\mu_{LT} = \frac{1}{\tau_{LT}} = 0. \qquad (13.15)$$

μ_{LT} is equated with zero in the Q-matrix.

$$\mathbf{Q} = \begin{bmatrix} 1-A_1 \cdot dt & 2 \cdot \lambda_{DD} \cdot dt & 2 \cdot \lambda_{DU} \cdot dt \\ 0 & 1-A_2 \cdot dt & 0 \\ 0 & 0 & 1-A_3 \cdot dt \end{bmatrix} \qquad (13.16)$$

In order to determine the MTTF-value of the system, first the **(I- Q)** matrix is calculated. The **(I- Q)** matrix of the example looks then as follows

$$\mathbf{I-Q} = \begin{bmatrix} 1 & 0 & 0 \\ 0 & 1 & 0 \\ 0 & 0 & 1 \end{bmatrix} - \begin{bmatrix} 1-A_1 \cdot dt & 2 \cdot \lambda_{DD} \cdot dt & 2 \cdot \lambda_{DU} \cdot dt \\ 0 & 1-A_2 \cdot dt & 0 \\ 0 & 0 & 1-A_3 \cdot dt \end{bmatrix}$$

$$= \begin{bmatrix} A_1 \cdot dt & -2 \cdot \lambda_{DD} \cdot dt & -2 \cdot \lambda_{DU} \cdot dt \\ 0 & A_2 \cdot dt & 0 \\ 0 & 0 & A_3 \cdot dt \end{bmatrix}$$

$$\mathbf{I-Q} = \begin{bmatrix} A_1 & -2 \cdot \lambda_{DD} & -2 \cdot \lambda_{DU} \\ 0 & A_2 & 0 \\ 0 & 0 & A_3 \end{bmatrix} \cdot dt \qquad (13.17)$$

$$= \mathbf{M} \cdot dt.$$

From this the **M**-matrix can be determined:

$$\mathbf{M} = \begin{bmatrix} A_1 & -2 \cdot \lambda_{DD} & -2 \cdot \lambda_{DU} \\ 0 & A_2 & 0 \\ 0 & 0 & A_3 \end{bmatrix}. \qquad (13.18)$$

The **N**-matrix is the inverse matrix of the **M**-matrix.[137]

$$\mathbf{N} = [\mathbf{M}]^{-1} \qquad (13.19)$$

---

[137] [MAKI73]   Maki, D. P.; et al.; *Mathematical models and applications*
     [PAPO84]   Papoulis, A.; *Probability, random variables and stochastic processes*

13.5 Implementation of a Markov Calculation for a Safety Related System

For the **N**-matrix one gets:

$$\mathbf{N} = \begin{bmatrix} \dfrac{1}{A_1} & \dfrac{2\cdot \lambda_{DD}}{A_1 \cdot A_2} & \dfrac{2\cdot \lambda_{DU}}{A_1 \cdot A_3} \\ 0 & \dfrac{1}{A_2} & 0 \\ 0 & 0 & \dfrac{1}{A_3} \end{bmatrix} \qquad (13.20)$$

The MTTF-value is an average time between the appearances of two failures. The starting condition is the state 0 in which the system works failure free. After the inversion of the matrix all elements have the unit 'time'. If all elements in the first row of the **N**-matrix are added up, one gets the MTTF-value for the system. The appearance of the MTTF-Term looks therefore as follows:

$$\text{MTTF} = \frac{1}{A_1} + \frac{2\cdot \lambda_{DD}}{A_1 A_2} + \frac{2\cdot \lambda_{DU}}{A_1 A_3} \qquad (13.21)$$

With those calculated values one gets

$$\begin{aligned}\text{MTTF} = (&\frac{1}{8{,}4227\cdot 10^{-8}} + \frac{2\cdot 10{,}45\cdot 10^{-9}}{8{,}4227\cdot 10^{-8}\cdot 0{,}04160002} + \\ &+ \frac{2\cdot 10{,}45\cdot 10^{-9}}{8{,}4227\cdot 10^{-8}\cdot 20{,}9\cdot 10^{-9}})\cdot h \\ =\, & 23745360{,}78\,h \\ \approx\, & 3094{,}6\ \text{years}\end{aligned}$$

The average lifetime of the redundant system model amounts to consequently 3094.6 years.

14 Lifecycle Analysis of a Safety System

14.1 Hazard and Risk Analysis

Safety-instrumented systems are required for processes/ systems that could create a potential danger. Therefore it is a basic principle of a safety norm to perform risk analyses for the to be supervised processes/ systems, generally described as facilities, and also for safety-instrumented systems in order to develop corresponding specifications for safety conditions. Facilities that must be supervised are described as "Equipment Under Control" (EUC). In the process industry sector an entire analysis can be defined in three steps:

- Risk Assessment
- Danger Analysis
- Risk Assessment

A danger is defined in the norm as a "potential source of danger". A process and its control system can contain many sources of danger.

The norm does not specify how to apply the danger analysis. A good method being used in the industry is the HAZOP-method. The HAZOP concept stands for "Hazards and Operability", an analysis to reduce the possible risks or to avoid them entirely. This method is based on the methodical extension of a process in order to identify any and all possible dangers. This identification and calculation is being done in a team. The team should consist of members like developers, safety-, mechanical- and electrical engineers. The HAZOP-method offers an opportunity to combine the ingenuity of the team to contemplate all possibilities that can cause dangers. The HAZOP-analysis must list all potential sources of danger.

14.2 Execution of a Risk Evaluation Analysis

At the beginning of each risk decrease a HAZOP study is performed in which the potential danger sources have been displayed. The next step is the determination of the safety instrumented function (SIF) including its description and its consequences. Then the necessary risk reduction will be transferred to each SIF.

A risk reduction can be accomplished in three steps:

- External risk reduction of the plant
- Mechanical Safety
- Employment of a "Safety Instrumented System" (SIS) in case the resulting risk is still to high

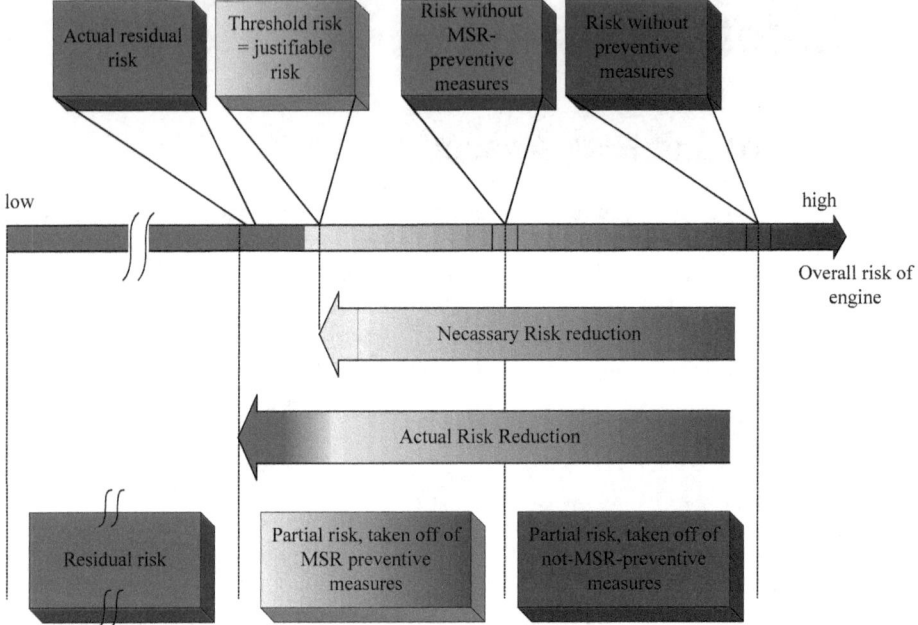

Figure 14.1: Riskreduction[138]

In the IEC/ EN 61508 several ways are mentioned to evaluate the SIL and the quantitative determination of the risk reduction requirements:

- ALARP (As Low As Reasonably Practicable)
- Risk Matrix
- Risk Graph
- Fault Tree Analysis
- Protective Layer Analysis

All methods are based on the expectant demand mode of a safe function of a EUC. If no danger reducing measures are being deployed, each system/ each process is a potential danger. The expected consequences of a danger without protection are analyzed and the maximum acceptable danger is defined.

---

[138] [IECe99]   IEC61508-5, Figure A.1

According to the norm, a risk evaluation analysis results in a certain number of SIFs, by which for each SIF a SIL is defined. Much depends on if the safety devices have already been installed or if the necessity exists to install different technologies, for example mechanical safety valves or break disks. Through already existing safety devices further SIFs might posses smaller SILs, under the assumption they have been designed correctly.

The implementation of SIFs by means of a SIS consisting of a selection of sensors, safety valves or several logic solvers and the calculation of the PFD-value of the SIS, in order to prove a required SIL of a SIF quantitatively, will be described in the following chapter.

14.3 Life Cycle Phases

All demands for each safety function must be documented. It is important that the function demands and the requisite SIL be specified for each single safety function. It is not enough to order a "PLC SIL3" and to think it will suffice according to IEC 61508. The following list shows some items that must be covered in order to certify a safety instrumented control according to IEC 61508. IEC 61511 shows a similar list of demands. Following items should be specified for a safety function:

- Description of the Safety Functions
- SIL Goal
- Safe Condition (open/ close, on/ off switches)
- Process Safety Time (and the derivative System Response Time)
- Excess Pressure
- Application Interface
- Relevant Mode: Warm-up, Durable Operation, Deactivation
- Foreseeable Exceptional Conditions
- Demand for Warm-up and Deactivation etc.

If one develops a safety-instrumented system, the following points must be considered:

- Safety Integrity Function (SIF)
- System Architecture
- Possibility of Errors and the Calculation of the probability of failure

These details are discussed in the following chapter.

14.3.1 Development of a safety-instrumented function

A safety-instrumented function, short SIF, is implemented through a specific number of facilities in order to revert the system back into a safe condition in the event of a danger. SIF also serves to minimize a process risk. Therefore the required Safety Integrity Level (SIL) for the risk decrease must already be documented in the safety requirement specification (SRS). The SIF necessary facilities, hardware and/ or software form a safety-instrumented system (safety integrity system (SIS). The definition of a safety-instrumented function according to IEC 61508 and IEC 61511 is:

"A function, through SIS, that achieves or retains a safe condition for the process, in reference to the specific dangerous event, is being implemented."

A SIF can be one of the following appliances, functions and energy sources:

1. Sensors (pressure, temperature or level sensors)
2. Field interfaces (Transmitter)
3. Logic processing units (programmable or not programmable controls with corresponding entrance, exit units)
4. Actuators (spools, safety valves, activation switches)
5. Safety communication
6. Wiring
7. Electric energy
8. Air supply
9. Maintenance of the overlap circuit
10. Start-up of the overlap circuit
11. Start-up of the logic and time functions
12. Alarm and indicator circuits
13. Status and incident sequence, monitoring circuit, connections
14. Logic solver loss alarm

The items 1 through 6 are important parts of an IF, items 7 through 8 are frequently used energy sources of a SIF and items 9 through 14 are auxiliary functions that have both the same integrity level and the same probability of failure as the SIF, or are not impaired by any disturbances in their function. Special attention must be given to item 14. A logic solver loss alarm is a warning signal that alerts the operator that the logic solver is no longer in the position to conduct its intended safety functions. The operator must follow specific instructions, mostly in form of manual operation, in order to return the process into a safe condition. This warning signal must at least possess the same probability of failure, defined by the SIL (safety integrity level), as the actual SIF; meaning a simple correlation between safety-instrumented control and a warning signal on a display is not advised!

A SIF can consist of several sensors and a single safety valve or one sensor and several safety valves. Each potential combination is possible, see Figure 14.2.

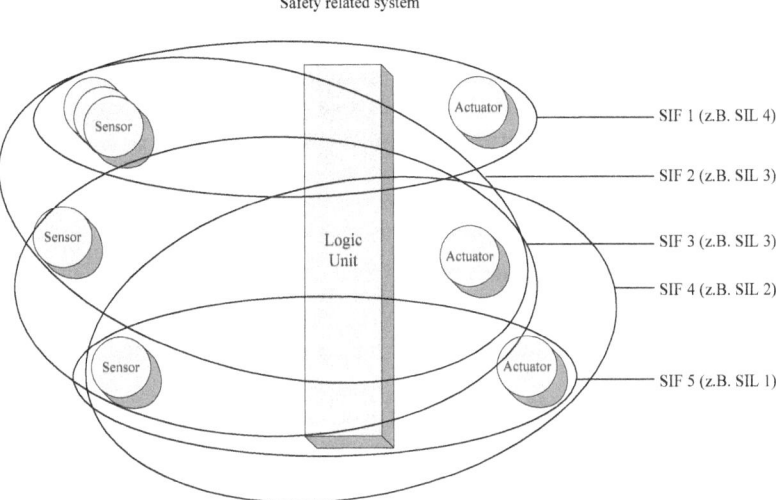

Figure 14.2: Multiple SIFs within a SIS

14.3.2 Failure models and PFD calculation

The SIL of a SIF is expressed by a *probability of failure on demand* (PFD). Column 1 in Table 14.1 specifies the four SILs according to IEC 61508, followed by the associated probabilities of failure expressed in ten orders of magnitude. A demand exists if a process parameter lies no longer within its tolerance level and consequently the process risk of an endangering event increases. The safety system is prompted to shut down the process into a safe condition. If a request appears and the safety system at this time cannot execute the request, then a dangerous incident might occur, meaning the process is in a dangerous condition. How high the probability of a failure of the safety system is, is mathematically expressed through the PFD-value. The PFD-value expresses the average probability of the occurrence of a dangerous, undetected failure of a SIF and the associated SIS respectively.

The norm defines the SIL for a specific safety function. This safety function can be implemented through a single subsystem or through an entire SIS, consisting of a sensor logic unit actuator chain. Table 14.1 gives an overview of the four safety standards for processes in the so called "low demand mode". "Low demand mode" signifies that a demand occurs fewer than once a year, a life-like situation in the process industry.

Table 14.1: Safety Integrity Level: Target failure measures for a SIF

Safety Integrity Level	Low demand mode of operation (Average probability of failure to perform its designfunction on demand)
4	$\geq 10^{-5}$ bis $< 10^{-4}$
3	$\geq 10^{-4}$ bis $< 10^{-3}$
2	$\geq 10^{-3}$ bis $< 10^{-2}$
1	$\geq 10^{-2}$ bis $< 10^{-1}$

Basic Rule of Development

The recommended rule of development for the realization of SIFs is the *de-energized to safe principle*, (DTS). With this basic rule it is possible to create SIL 4 applications conforming to standard equipment with a specific architecture.

Failures

In the safety technology failures can be divided into two types:

- Safe failures, meaning losses that do not influence the safety but consequently hinder activities.
- Dangerous failures, meaning failures that prevent the execution of a SIF.

With built-in diagnostic functions it is possible to determine parts of the dangerous losses and the corresponding suitable steps (like reactions regarding the redundant system, alarming of the operator, shut off). Through these measures a dangerous failure of a safety function is reduced to a safe failure.

The residual dangerous, undetectable losses are those that decide the residual risk of a SIF and can be determined by means of a PFD-calculation. It must be mentioned that the losses are random losses. Methodical losses that occur through technical and design errors or production errors or losses through persons (like operator errors) are not included in these calculations. The IEC 61508 requires here that specific measures are taken in advance in order to avoid these kinds of losses.

Necessary Parameters

In order to perform a reliability calculation of each SIF, one must know the following parameters of a safety-instrumented system:

1. Dangerous error rate
2. Diagnostic coverage factor for dangerous losses
3. Coverage factor of the validation and the checkup
4. Desired checkup-interval T_1
5. Safe failure rate
6. Diagnostic coverage factor for definite losses
7. Mean Time To Repair (MTTR)

Most of the end users are interested in parameters in order to determine the risk of a process and to take further safety precautions, if warranted. Additionally to external protection measures frequently double process measures are applied, meaning a process is being secured via a redundant process. Thereby one wants to avoid user errors that occur through an erroneous human reaction following an action of a SIS.

General proof test

In order to achieve required SILs, a frequent proof test and repairs are necessary to reduce or entirely avoid undetectable, dangerous losses. During this test all relevant devices are examined via a safety function. A sufficient manual checkup with a high coverage factor can be conducted if these relevant devices are off-line. This frequently means, especially in the case of a single process unit, that production will be halted. Redundant installations, specified redundant sensors and valve configurations are a method to avoid the shut down during the manual test. An alternative is the application of a production with a low dangerous failure rate.

Failures due to a common cause

In a redundant plant structure, losses through failures due to a common cause, so-called common cause failure, must be observed. A stress factor (for example electromagnetic disturbance) can cause multiple failures in redundant parts. If this is the case, some or all identical modules can be impaired simultaneously in a redundant structure. The common cause failure factor is expressed as a percentage of the overall number of losses. This overall number of failures is relatively small and lies between 0.5% and 10% (according to IEC 61508 examples).

Based on the probability of failure of a redundant system, the influence of common cause errors is higher than the influence of a simple error.

Reliability details and calculation methods

Reliability details of a safety-instrumented system can result from the manufacturer of such a system or from different data bases of particular testing institutes world-wide. It is important to check and compare that data with different existing data. It is a requirement of IEC 61508 that each step, starting with the specifications to used data up to the facilities, is documented. It later allows retracing the reasons for the employment of specific equipment, certain hardware or software. Furthermore through detailed documentation possible mistake could be discovered before the initial startup.

IEC 61508 presents some examples on how to predict the probability of failure of a unit. In this norm, as well as in IEC 61511 and in the ISA TR 84, different methods are mentioned. The probability of failure of a system can be determined with the help of

- a Reliability Block Diagram
- a Fault Tree Analysis
- or with a Markov Model

Each method has its advantages. For an average safety function all methods led to similar results.

14.3.3 System Architecture

The SIL-determination for a system does not only depend on the probability of failure. Equivalent to the probability of failure the system architecture must be observed. Only if both components are taken into consideration can a SIL according to IEC 61508 be determined. In order to assess the system architecture three parameters must be analyzed. First must be determined if the system consists of type A or type B components. Second the hardware fault\ tolerance must be decided and finally the safe failure fraction (SFF) out of different failure rates must be predicted. According to IEC 61508 a system is, in order to implement the safety function, viewed as type A if the components meet the following conditions:

- The failure behavior of all components is well-known and
- the failure behavior is completely predictable and
- Reliable failure data is available through experience in the field.

According to IEC 61508 a system can in order to implement the safety function is viewed as type B if the components meet the following conditions:

- The failure behavior of at least one included component is not sufficiently known or
- The failure behavior of the subsystem is not completely predictable or
- Reliable failure data is not available through experience in the field.

If at least one component of a system meets the conditions for a type B-system, then this system must be viewed as type B. Type A-components are components like transistors or resistors that have proven an extended lifetime and 100% verifiability. Type B-components are components without any extended experience values regarding their lifetime and are not 100% verifiable, like microprocessors for example.

The hardware-fault-tolerance *(HFT)* determines how many dangerous failures are caused by the architecture without endangering the execution of the safety function. In case of a 1oo1-architecture the system consists of a subsystem that executes the safety function, then this architecture has a HFT of zero because in case of a dangerous failure in the subsystem the safety function can no longer be executed. A 1oo2-architecture consists of two subsystems and one subsystem must work correctly in order to execute the safety function. This architecture possesses a HFT of One. A HFT of One can also be found in a 2oo3-architecture – in contrast to a 1oo2-architekture – two of three subsystems must operate in order to guarantee the safety function. Finally there are also systems with an HFT of Two like systems with a 1oo3- or 2oo4-architecture. Here two systems can fail and the safety function is still executed correctly. The industry speaks of systems that are 2-failure-tolerant.

The third parameter is the "safe failure fraction", (*"Sicherer Fehleranteil"*, SFF). The SFF is the correlation of the addition of the safe and the dangerous, but detectable failure rates and the overall failure rate of a defined unit. Table 14.2 and Table 14.3 show the architectural limitations of the HFT and the SFFs in correlation of the components to the attainable SIL.

Table 14.2: Hardware safety integrity: architectural limitations set up with type A components

Percentage of not dangerous failures	Hardware Fault Tolerance		
	0	1	2
< 60 %	SIL 1	SIL 2	SIL 3
60 % - 90 %	SIL 2	SIL 3	SIL 4
90 % - 99 %	SIL 3	SIL 4	SIL 4
> 99 %	SIL 3	SIL 4	SIL 4

14.3 Life Cycle Phases

Table 14.3: Hardware safety integrity: architectural limitations set up with type B components

Percentage of not dangerous failures	Hardware FaultTolerance		
	0	1	2
< 60 %	not allowed	SIL 1	SIL 2
60 % - 90 %	SIL 1	SIL 2	SIL 3
90 % - 99 %	SIL 2	SIL 3	SIL 4
> 99 %	SIL 3	SIL 4	SIL 4

Table 14.3, which can be applied to a system with type B components, indicates that in case of a HFT of zero, the system can reach a SIL of 3, if a SFF larger than 99% is available. If this system however consists of a 1oo2-architecture, for a SIL of 3, a SFF between 90% and 99% will suffice.

Following the subsystem sensors, actuators and Logic Solvers will be discussed.

Sensors

The appropriate sensors, depending on their use, must be selected, for example in dependence:

- Of the type of process and the process conditions
- Of the preferences respecting the guidelines of a company, of the empirical value and the valid and acceptable reliability data.

Analogs sensors have a higher reliability than digital sensors (for example pressure switches). A continuous transmitter test through comparison of the dynamic behavior in analog signals is easier than in digital signals. Both arguments are the reason why currently mainly analog sensors are in use. The in Figure 14.3 displayed redundant sensor constellations result in a smaller failure probability than the non-redundant 1oo1-sensor constellation.

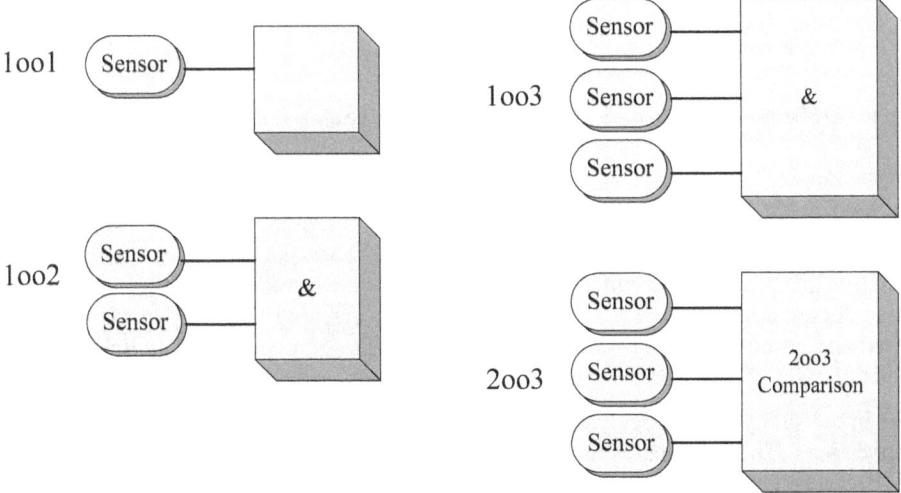

Figure 14.3: Simple sensor constellations

Actuators

The most frequently used actuator component is the valve. Frequently so-called "starter" components are used to start and stop engines. Most of the safety-instrumented valves are propelled by air pressure which requires the use of spools. In order to open the valves to the source of the air pressure, relays are being used. These relays have mostly trilateral, electric spools. By operating the relays, the valve of the pneumatic source opens and the air pressure reaches the safety-instrumented valve and opens it. In order to close the safety directional valve again, the valve must first be closed so that no air pressure reaches the safety-instrumented valve and second the air escape valve and the still existing air pressure must escape between the valve of the pneumatic source and the safety-instrumented valve. Only then the safety-instrumented valve will close. The failure rates of the relays and the valve of the pneumatic source must be added to the failure rate of the safety-instrumented valve. Different architectures, like valves of the pneumatic source and the safety-instrumented valves, could be interconnected and are displayed in Figure 14.4.

Most of the valves have no automatic diagnostic ability and the number of the undetectable, dangerous failures is high. There are special valve manufacturers which manufacture highly reliable valves for large gas and oil plants. These valves can be applied in 1oo1-architectures for SIL 2 uses.

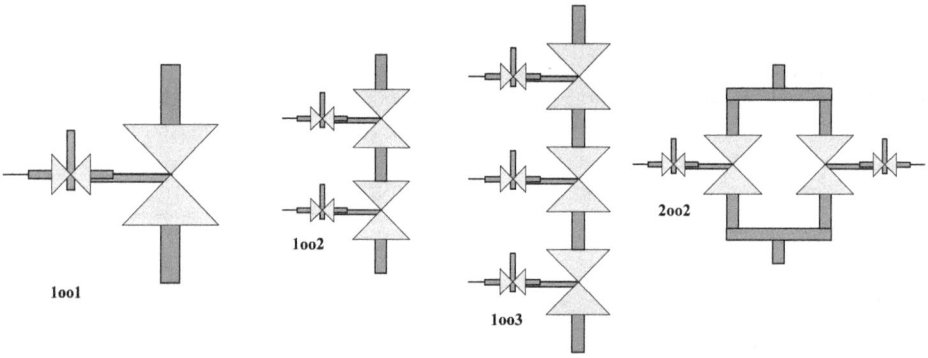

Figure 14.4: Simple actuator constellations with air pressure and safety-instrumented valve

Logic Solver

The Logic Solver is being mentioned last because in the field, compared with valves and sensors, it possesses a smaller probability of failure. There are several recognized manufacturers that produce accredited safety systems. The most frequent architectures in safety technologies possess either a natural failure safety (meaning the components are solidly wired and do not require any software) or they consist of a simple or multiple redundant structure (1oo2-, 2oo3- or - like with many a processor systems – a 2oo4-architecture).

The in the safety technology used programmable systems are suitable for SIL 2 to SIL 3 applications. The certificates are issued by well-known accreditation organizations like TÜV, Exida, Factory Mutual and Baseefa.

14.4 Overall Planning

Regardless if provider or end user, a safety-related plant must be planned according to IEC of 61508 or IEC 61511 respectively. This consists of several parts:

- Entire procedure and maintenance planning
- Entire safety validating planning
- Entire installation and initial operation planning

The operational stipulations of the concept and design phases and the consideration of the initial operation are defined in IEC 61508. Here planning phases of the entire plant and the entire maintenance are specified. These plans must, if necessary, include the standard routines for the maintenance of the functional safety. Likewise the measures of the maintenance for the safety during the maintenance must be planned. Among other things the entire safety validation is examined. A plan for the substantiation of the system is set up and all different potential operational conditions are contemplated. Criterions regarding the strategy of the substantiation as well as criterions for the existence or nonexistence of the plant must be laid out.

Further parts of the overall planning represent the planning of the entire installation and the entire initial operation. The planning must include the examined installation and initial operation. Here especially the schedule, the responsibilities and the procedure during the installation must be examined. Furthermore criterions must be established that consider the installation of the system or the plant to be completed.

Naturally one of the most important parts of the overall planning is the realization. In this phase the actual transposition of the plan's safety related systems must be planned and lead to actual developments. In this connection also the parts of 2 and 3 the IEC 61508must be viewed that regard the development and the requests of the hardware and software itself.

After the concept and realization phase the overall planning phase occurs; the entire installation, entire initial operation and overall substantiation phase in accordance with the guidelines. The IEC 61508 dictates the procedure for the entire modification and the overall back fitting. Here the functional safety must be maintained while allowing a modification. Important is that all activities concerning the modification are carefully planed and the effects are observed. Also the shutdown and disposal must be regulated. Here the effects of a shutdown on the functional safety with the system in connection with other standing systems are analyzed and taken into consideration. Only after this analysis the authorization for the initial operation can be granted.

14.5 Realization of a SIS

A specialized safety outfitter arranges the majority of the realization of the SIS. The IEC 61508 displays the requirements for the development of hardware and software in separated chapters.

Safety-instrumented systems that are developed according to IEC/ EN 61508 require that the norm and the maximum accepted probability in case of a failure of the safety-function are suitable for the specified operational requirements. Therefore a designated formative process and a procedure model are necessary so that each single step is documented, the calculations can be retraced and the system attributes can be checked. A suitable model is the V-model that was developed between 1987 and 1988 by request of the department of Defense, see Figure 14.5. This model adheres to the required documentation and takes functional, as well as safety related requirements, into consideration.

During the development of critical safety systems the verification and validation are a big part of the entire complexity. They are based on test procedures. Consequently the test planning represents an essential part of the developmental process. A part of this effort refers to the planning of modules and system integration tests. The actual validation of the complete system is normally tested in the very last steps of the development of a system.

The planning of the validation should begin in the very early phase of the development phase. The classical V-shaped model shows a secondary information flow of the early formative phases into the later phases. Figure 14.5 further clarifies that for each phase on the left side a test plan must be generated that describes how the implementation of this transformation must be verified. This test planning of each step must identify the traits that must be investigated in order to design meaningful and sufficient test. The latter depends on the safety integrity level of the test object. It is clear that before the system is implemented, the validation, including all factors, that can influence the safety, must be considered.

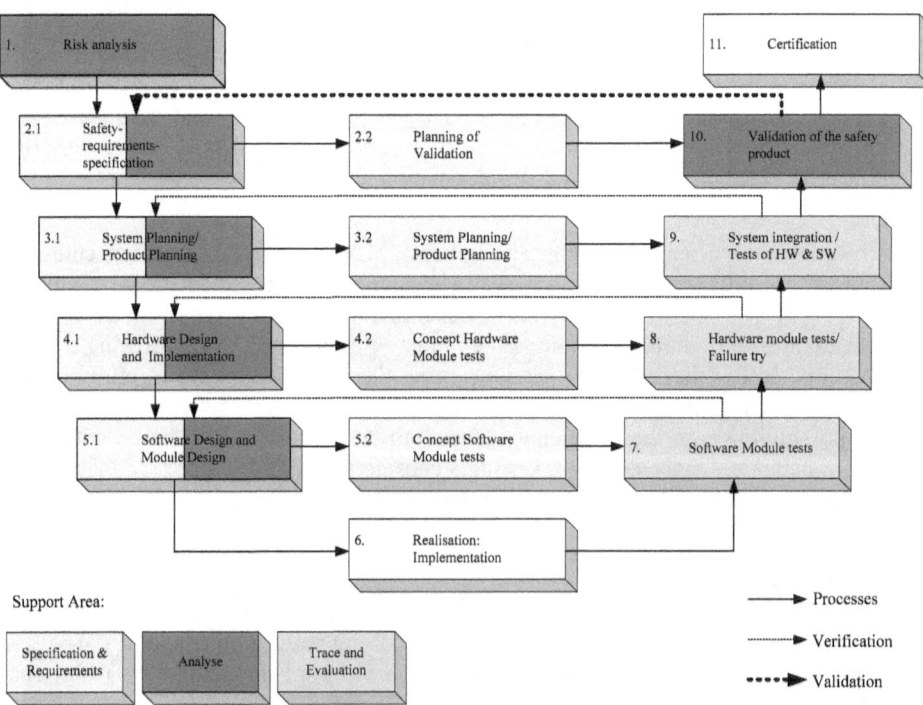

Figure 14.5: V-diagram for the system validation

The overall design of a SIS must follow the safety requirement specifications (SRS). The design and the implementation of the safety-instrumented system must adhere to the appointed requirements of the safety functions and the safety integrity. The design of the system includes the overall hardware- and software-architecture, sensors, actuators, programmable electronics, embedded software, application software etc.

The following general requirements for the design of a safety-instrumented system must be met:

- the requirements of the safety integrity of the hardware, consisting of:
  - the restrictions of the safety integrity of the hardware based on the architecture and
  - the requirements of the probability of random endangering hardware losses
- the requirements of the methodical safety integrity, consisting of:
  - the requirements of the avoidance of losses and the requirements of mastery of methodical losses, or
- the proof that the operating methods are "operating proven":
  - the requirements of the system behavior at detecting an failure.

The architecture requirements for hard- and software must be in accordance with the necessary SIL. The possible SIL is restricted through the failure tolerance of the hardware and its share of harmless failures. The highest SIL that can be taken into account is specified in both above Tables 14.2 and 14.3.

14.6 Installation, Startup and Validation

The installation and the initial operation of a SIS must be in accordance with the appropriate planning. All activities must be documented; including the adjustment of failures.

After the initial operation, an overall validation, before the plant starts, must occur. The validation must be in accordance with the appropriate planning. A detailed documentation is required.

14.7 Operation, Maintenance and Repair

During operation the maintenance data must be collected regarding failures, test results, requirements and losses. That data can be used to determine if the assumptions from the HAZOP-analysis were correct. Further can be determined if the failure rate, that was used for the probability of failure calculation and the SFF, is correct. If failures within the calculations are recognized, an adaptation must follow and parts of the HAZOP- and the SIL-calculations must be repeated, if warranted. This can lead to changes in the extent of lifetimes or the checkup-intervals of the safety system.

Mechanical Test

After the mounting of appliances it can be necessary to run a test in order to identify the still existing dangerous failures that can reinforce itself over time. The time interval between such checkups depends on the calculated reliability of the plant. The scope in such an examination does not always include the entire plant, but is frequently somewhere between 75% to 100%. One must consider that a repair is never 100% correct. The result will show that the appliance over a certain time will no longer be suitable to execute its intended SIF and to respectively fulfill the required SIL. The time interval between the tests, for example a partial stroke test, and the lifetime of a unit before a general overhaul must occur, should be part of the reliability calculations. The test must be conducted according to specified requirements. The test results must be documented and made available for further examinations.

14.8 Modification and Retrofit

Is the system in operation, the likelihood of a modification increases. These modifications must follow under the same strict rules as the first realization of the SIS. Figure 14.6 shows how such modifications are to be conducted. While a modification of a subsystem of the existing plant can be examined and be tested, the documentation should also be adjusted.

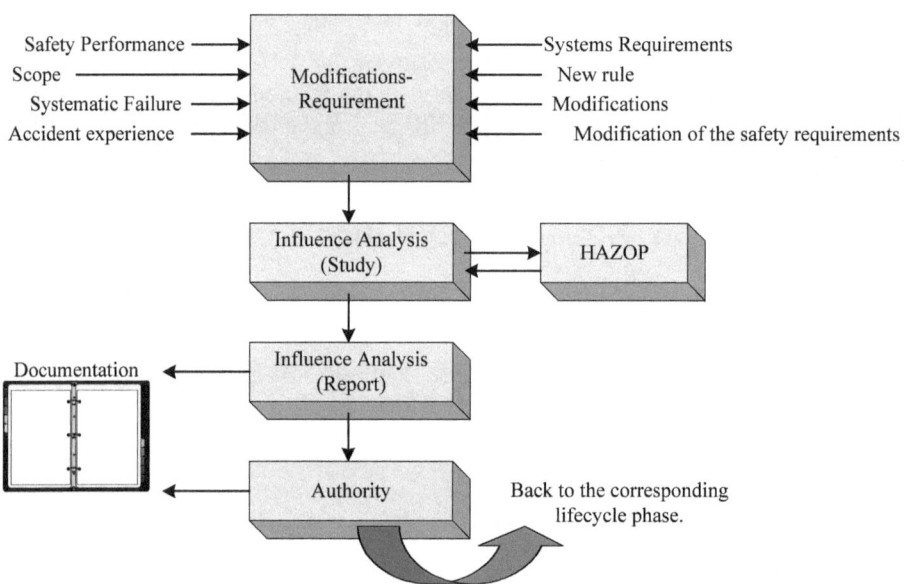

Figure 14.6: Modification process

14.9 Summary

The order of the design, the development and the installation of a safety system of a new or already established project will result in an alteration of the work cycle in order to adhere to IEC 61508.

The rules regarding the structure of a SIS according to IEC 61508 must already be considered in the early phase of the construction process. The selection of the appropriate

14.9 Summary

types of sensors, the number of sensors as well as the number of valves has an impact on the construction and erection phase. An untimely consideration in the formative phase can be too late and results in large alterations at a later point in the erection phase.

The implementation of the in the safety cycles of specific guidelines at the right time and their observation in the remaining course of the project helps to save money and avoid discrepancies with the sponsors. The experience and range of a well-known, certified supplier of safety systems enables the end user and contractual partner to fulfill the safety goals according to IEC 61508.

15 Common Cause Failure

15.1 General

The observation of failures in general took place in chapter 4. Especially in the safety-technology additional sources of failures, that result from a common cause, must be discussed. There are different types of failures. Next to common cause failures (or CCF for short) there are also

- Design Failures
- Production Failures
- Maintenance Failures
- Operation Failures

It is important to know the sources of failures and their meaning. The clearer the meaning of sources of failures in a complex system is, the better failures can be controlled and avoided. Is the source known to the developer, the failure can be taken into consideration during the development of the product. Furthermore, an analysis of the probability of failure is meaningful because the understanding of potential and probable sources of failure leads to the development of a better failure model.

Fault tolerance is a special technique that allows a system to pick up the expected (or the minimal) operation, despite the existence of failures. In order to achieve this behavior, redundancy is being used. Redundancy can be divided in four groups:

- Hardware-Redundancy,
- Software-Redundancy,
- Time-Redundancy and
- Information-Redundancy.

In case of a hardware-redundancy the system is equipped with more components than required. If a component fails, a substitute is in place. In case of a software-redundancy, the system can be equipped with different versions of tasks. Independent programmers write tasks so that, if a task based on specific inputs fails, a different task can take over and solve the problem safely. In case of a time-redundancy, if the scheduler fails, some tasks are restarted, so the problem can be solved. In the event of information-redundancy the data is coded in a way, so that a definite number of bit errors is discovered and corrected.

A fault tolerant system only fails if several errors occur. The smallest combination of errors occurring together (connected through an AND gate) leading to the failure of a system is called a "Minimal Cut Set" (MCS). A fault tolerant system has minimal "cut sets" which include specific commands. The rank of a minimal "cut sets" is the number of errors leading to its occurrence. Consequently the rank of a fault tolerant system is defined. The rank of a fault tolerant system is the rank of the minimal "cut sets" that lead to the critical error. In order to be fault tolerant, a system may not include any "cut sets" with a rank of 1.

15.2 Common cause failures

The frequency of common cause failures is intricate in its assessment. Modeling methods and existing failure rates make the predicative calculation of these losses difficult and sometimes the results are even debatable. A CCF occurs, if a single error leads to the failure of several components. CCFs can therefore lead to a failure of the SIS-function because they can influence several channels of the system.

It is important to recognize, that during the developmental process CCFs and their possible effects on the SIS-functionality are understood. Unfortunately, there are many different expert opinions on how a CCF ought to be defined and which specific events constitute a CCF. Following frequently used examples for CCFs and its failures are described:

- Faulty calibrating of sensors
- Improper servicing
- Incorrect maintenance
- Inappropriate by-passing
- Environmental impact of the device

Regardless of the type of applied SIS-designs, each of these five examples can bring each single I/O-system as well as the redundant I/O-systems to failure. Unfortunately, many of the proposed methods ignore this fact. The implementation of redundancy makes the tasks of safety engineers more difficult because an additional type of failure must be considered; namely the "common cause failures". Common cause failures influences the error analysis so that the probability of the minimal "cut set" is higher than the sum of probability of the minimal "cut sets" of single components alone. They render the increase of the number of channels at a specific number meaningless. If engineers could construct redundant systems with independent redundant channels, a consideration of failures would not be necessary. Engineers could reach the demanded amount at safety and reliability by increasing the level of the redundancy. Unfortunately, it is virtually impossible to construct independent redundant channels. Therefore the contribution of common cause failures must be assessed in order to comply with the safety and reliability demands. The simplest way to view common cause failures is to work with minimal "cut sets". Incidents of a minimal "cut sets" can represent the same failure-type in different components, like the same type, or different failure types. They can be caused through the same cause (like a common cause) or through different causes.

The core of this thesis is that, if all incidents, in a minimal "cut set" caused by the same root cause, occur simultaneously, the fault tolerant system fails as if the incidents of the minimal "cut sets" occurred coincidentally. The probability, that a minimal "cut set" appears to have failed due to a common cause, is extremely low but still higher than the probability, that a minimal "cut set" occurs coincidentally. The task of the analysis of common cause failures is to assess this probability and to facilitate the improvement of the design. Without any regard for the incidents of a common cause, the probability of a critical minimal "cut set" for error tolerant systems would be undervalued.

15.2.1 Analysis of Common Cause Failures

These failure incidents are usually not viewed as independent, in the system occurring incidents, but as influences on the system from a source which shares redundant components and leads to abnormal output conditions. If one deals with common cause failures, the first problem is the classification of an obvious terminology. The terminology of common cause failures has changed over the years. In the beginning, only common mode failures were taken into consideration. Later the definition of common cause failures was introduced, covering a bigger group of failures and replaced the definition of the common mode failures. There was, however, a wide spread opinion, that common cause failures and common mode failures were synonymous.[139]

The difference between cause and mode was clarified in 1985 when the concept of dependent failures was introduced in order to replace the definitions of

- common mode failures
- common cause failures as well as
- cascade failures

and in order to combine these concepts.

Common mode failures are a subgroup of the common cause failures. The cascading failures include all dependent failures that are not failures with a common cause. These are for example multiple failures due to the failure of a component (chain reaction or domino effect). If several components are subjected to the same load, the failure of one component can lead to an increased load for the other components and so lead to an elevated probability of failure.

Common cause failures

Common cause failures are multiple failures that can be directly attributed to a common or shared cause.

Examples for root causes are:

- Extreme environmental impact (fire, flood, earthquakes, lightning strikes, etc.),
- Failure of an external hardware part of the system
- Human error.

The root cause is not the failure of a different system component.

---

[139] [MAUR00]   Mauri, G., *Integrating Safety Analysis Techniques, Supporting Identification of Common Cause Failures*

Table 15.1: Dependent failures

Dependent Failure (DF)			The probability of a group of events which probabilities cannot be expressed as a simple product of unconditional probability of failure of single components.
	Common Cause Failure (CCF)		This is a kind of dependent failure which occur in redundant components in which a common cause – simultaneously or near simultaneously – leads to failures in different channels.
		Common Mode Failure (CMF)	This definition applies to failures of common causes in which multiple elements fail similarly.
	Cascade Failure (CF)		These are all dependent failures that do not share a common cause, meaning they do not affect redundant components.

Additionally:
The definition of „dependent failures" includes all definitions of failures that are not independent.
This definition of dependent failures clearly implies that an independent failure in a group of events can be expressed as a simple product of conditional probabilities of failures of a single event.

It has been reported frequently that operation and maintenance errors are the root cause of failures (carelessness, faulty calibration, and faulty procedures). If a higher number of components fail due to a common cause, these failures are so called "multiple failures".

There are two reasons for system failures:

- Coincidental hardware failures and
- Systemic failures.

It is accepted that coincidental hardware failures can result anytime and in any component and due to a failure in a channel of the system in which the component is part of. Consequently the probability that such a failure applies to parallel channels simultaneously is low compared to the probability of failure of a single channel. This probability can be predicted with the help of current technologies.

Failures with common causes, due to a single failure function, involve more than one channel. This can either result from a systemic failure or through an external load which led to a premature coincidental hardware failure. Because failures with common causes are prone to involve more than one channel in a multiple channel environment, they are the dominant factor for the determination of the entire probability of failure in a multiple channel environment. Failures with common causes are less than 25% of all failures in any applicable system.

The graph of the relationship between common cause failures and failures of single channels, as well as dependent failures and common cause failures, is depicted in Figure 15.1 (on the left and right).

15.2 Common cause failures

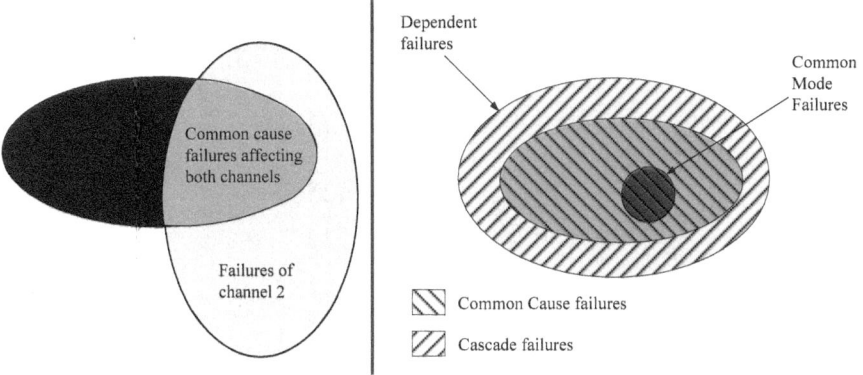

Figure 15.1: Dependent Failures

The reference document for the analysis of the common cause failures is NUREG 4780. In order to understand and model the mechanisms that lead to the dependent events, the following questions must be answered:

- Why do components fail or why are they not available?
- What can lead to several losses?
- Has a specific piece of equipment an attribute that can avoid the occurrence of such failures?

The *root cause, the coefficient of coupling* and the existence or non-existence of *technical or operational protection* against unexpected plant failures is the answer to such questions.

The transition of an operational to a failed system is due to a common cause, also called root cause.

If, for example, two components are employed in the same area and they are sensitive to humidity, an external incident outside the area, which causes extensive moisture in the area, can cause a failure with a common cause. In this event, extensive moisture is the root cause for the failure of the two components.

If a root cause exists, *the coefficient of coupling* indicates the force of the cause on multiple components. It creates the coupling requirements that lead to the failure of the components. The deployment in a common area for example is the coefficient of coupling for components that are highly sensitive to moisture.

Figure 15.2 displays the connection of root cause, coefficient of coupling and the final failure of the components. This signifies that, as soon as a coefficient of coupling exists, (like deployment in a common area) and a causing event occurs (like the failure of an air-conditioning unit), a failure of several components can occur.

Technical protections are all measures that counteract the formation of root causes and coefficient of coupling. In order to avoid the formation of root causes, the sensitivity of the components (like sensitivity to severe moisture) must be reduced. In order to avoid a coefficient of coupling, the diversity must be increased. This can be achieved through technologies like design and quality control. This can be supported by the separation of the appliances and the production of a high-qualitative construction.

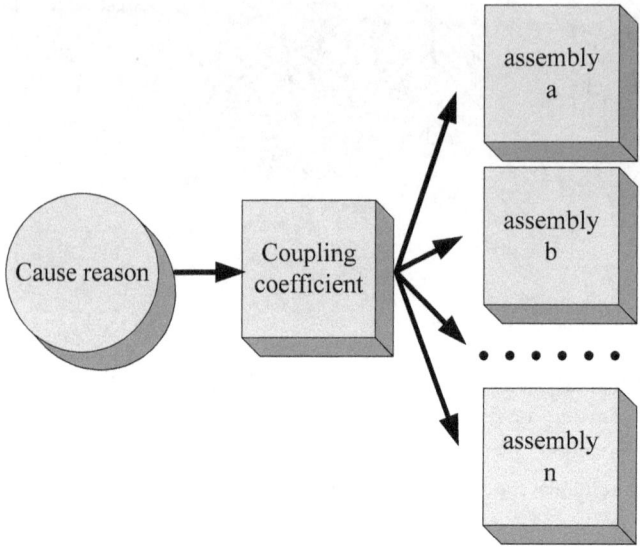

Figure 15.2: The root cause influences several components

15.3 Common Mode Failure

Redundancy Systems, as well as fault tolerant systems in general, can be operational despite the failure of a specific number of hardware or software components. This occurs when single components fail independently, but the systems are endangered by common mode failures. These failures can involve safety critical systems which qualifies them for a safety analysis. Four different types of common mode failures are generally accepted and they are:

- The coincidental failure of two or more similar components in separate channels of a redundant system due to a common cause (most likely the failures are due to a similar common failure type).
- The coincidental failure of two or more different components in different channels of a redundant system due to a common cause (the failures will also likely have a common failure type).
- The failure of one or several components (as cause of a causing incident), which leads to a failure of one or several different components which not necessarily are of the same type.

A consideration: In the above mentioned events, the failure can occur simultaneously or at different times, but ultimately the failure condition exists at the same time.

- The failure of a single component or service, which has all channels in a redundant system in common, (like common maintenance, testing). Here only components / services are considered, that are an essential and part of the system and the correct functioning of the system depends on them.

On the basis of these failures, Edward and Watson defined common mode failures as follows:

"A common-mode failure (CMF) is the result of an event(s) which, because of dependencies, causes a coincidence of failure states of components in two or more separate channels of a redundancy system, leading to the defined system failing to perform its intended function".

In order to investigate common mode failures, the concept *system* must be defined explicitly, meaning the inclusion and exclusion of a system. The inclusion therefore must be treated with technologies of the safety analysis and the exclusion is the object of the common mode failure analysis.

A system is a *"connection of components which together form a specified functional relationship of Inputs and Outputs"*, according to Edward and Watson. Everything a system does not require, in order to achieve an input-output-relationship while under normal operation, is not a component of the system and therefore a possible cause for common impact. These impacts can cause failures like fires, explosion, rocket impact, contamination etc. Indirect sources that can have an important effect on the system are weather, earthquakes, flooding etc.

15.4 Examples for Failures through a Common Cause

A simple example is the failure of data for a processor complex that consists of multiple processors. A failure of the data itself affects all channels of a redundant computer and their results. Figure 15.3 depicts a "processor complex" consisting of two processors with a common data failure.

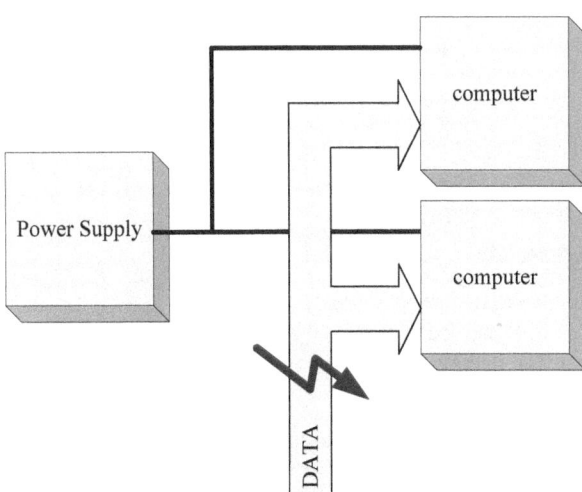

Figure 15.3: Failure through common cause

One of the most well known fault functions of a common cause is a fire in the nuclear power plant Brown Ferry in Alabama, USA in 1975. Two technician used a candle to locate ducts between the cable area and one of the reactor blocks, in which an under pressure prevailed. The flame of the candle was sucked into the ducts and the methane sealing, which surrounded the led lines, caught fires. The fire damaged about 2000 cables including the one for the automatic reactor shut down system, and all cables leading to the manually operational valves excluding the ones that led to the four relief valves. With these four valves it was possible to shut down the reactor and avoid a meltdown. As a consequence of this accident the cables for the Electronic Shut Down (ESD) were placed into separated ducts and combustible fillings (like Urethane) are no longer being used.

15.5 Technologies for the Evaluation of SIS Designs for CCF

The selection of the evaluation technology depends typically on the experiences of the users with particular SIS design and on the knowledge of different design methods. The aspect of the installation detail must also be taken into consideration. Finally a sufficient experience with the specific application environment is necessary, because a component that works during an application, might not work without an expansion at a later time.

15.5.1 Industrial Standards

The binding part of ANSI /ISA S84.01-1996 regulates detailed demands that the SIS-design must fulfill. The not binding part and the informative appendix respectively offer a guideline. Additionally IEC 61508 specifies precise design requests for safety-related systems as well as special measures and methods.

Planed or installed SIS-designs with these specific requests can be developed with regards of an admission/certification.

Even though these guidelines represent an important part of the process industry, no general, extensive industry standard can capture all potential shortcomings of a certain use. The comparison between the different guidelines is important in order to avoid many potential errors in a SIS-design.

15.5.2 Technical organization-specific Guidelines and Standards

In order to support the design engineer, many users develop engineering guidelines and standards (EGS) for the SIS-design. How precise a technical guideline or norm is formulated, depends on the application and the users which take part in the different processes within the organization.

Technical guidelines and norms can include approved architectures, appliance types, suppliers, frequency of tests and installation details. They shall include whatever is necessary for sound engineering and effective design practice within the business.

In general business internal guidelines are an excellent opportunity to deal with the SIS-design because the user can test and incorporate his specific use area and risk tolerance directly. Many users have difficulties within their business to come to a unification of what an acceptable design is. In most cases it is common knowledge how an initial design could be improved. There must be a strong advocate for the technical guidelines and standards within the company so they can be developed further. This advocate must have a direct line to someone in upper management, someone who is a strong advocate for the revision process which is required to formulate new technical guidelines and standards, which will actually be implemented in the end.

15.5.3 Qualitative hazard identification methods

Qualitative danger identification methods have been deployed for several years in order to recognize potential danger sources within the process units. These methods necessitate experts with an extensive knowledge of processes as well as experts with an extensive practical experience regarding the transaction of the different analytical methods. Typical methods for danger identification are the what-if-analysis, computer aided danger identification (HAZOP), as well as failure and effect analysis (FMEA).

Each one of the methods can be changed in order to evaluate the SIS. Qualitative evaluation procedure represents the check list method, which can easily be adapted for a SIS-evaluation. In reality they are a part of many international norms, like the API 14C for the design of safety-related systems on oil platforms.

15.5.4 Qualitative Valuation

For a qualitative evaluation a review of general attributes and mechanisms of failures, which could lead to potential common cause failures, is implemented. A useful check list on unit dependencies in order to determine common key attributes could be as follows:

- Type of component
- Component deployment
- Component manufacturer
- Internal component conditions (pressure, temperature, chemical composition)
- Construction unit dimension and system binding
- Component positioning
- Area attributes (moisture, temperature, pressure)
- Component requirements (standby, active)
- Component test routines
- Component servicing

Current guidelines for the design of subsystems are as follows:

- Identical components, which should generate redundancy, should always be combined in a common cause group.

- If several redundant components possess single parts that are also redundant in nature, the components should not be classified as independent.

The sensitivity of a component group does not only depend on the level of similarity, but also on the counteractive measures against common cause failures.

15.5.5 Checklists

Check lists consist of a simple list of questions, that can be answered with "yes", "no" or "not applicable". A check list analysis serves the identification of

- specific dangers,
- deviations of norms or guidelines,
- fault of the design and
- potential accidents.

Historically check lists are used to improve the human reliability regarding the design and the observance of the different rules and technical regulations. While normally the quantitative analysis is executed after the PIDs (piping & instrument diagrams) is fully completed, the check list analysis can be deployed in any design phase like the phase of

- concept design,
- detail designs and the
- field construction.

Check lists can be deployed for the SIS evaluation in general or for specific uses. It is the simplest method for the identification of design faults.

15.6 Quantitative Evaluation of Common Cause Failures

Sometimes it can be necessary to consider the effects of potential common cause failures on the SIS-performance. In such events the common cause failures at a quantitative performance interpretation of the systems must be considered. There are two procedures[140] that are applicable for common cause failures:

- the explicit model and
- the method of approximation.

Multiple failure incidents, for which no root cause can be found, are displayed by way of application of implicit parameters models.

---

[140] [ZIO-02]   Zio, E., Common Cause Failures, *An Analysis Methodology and examples*
    [SUMM99]   Summers, A. E.; Ford K. A.; Raney G., Estimation and Evaluation of Common Cause Failures in SIS

15.6.1 Explicit Methods

The explicit model is used for specific and well understood sources of common cause failures. A fault analysis tree with such specific sources of common cause failures, are represented as explicit basis events. The failure rates for these events are determined with help of business internal data, published data, if existing, or conservative estimations of the failure rates.

The following illustration displays the use of the explicit models for the interpretation of common cause failures in connection with a group of redundant transmitters.

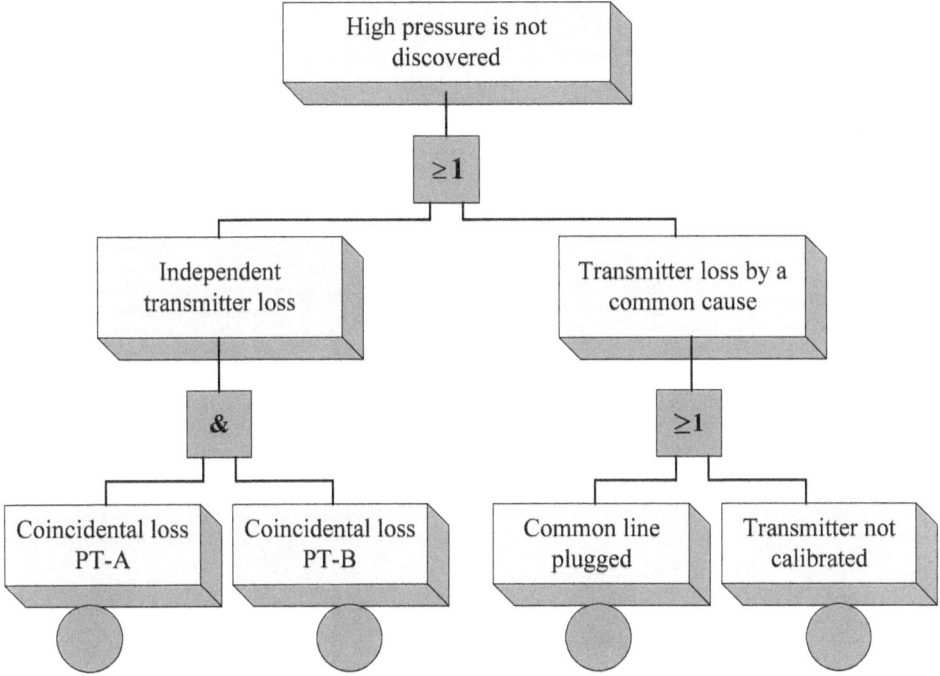

Figure 15.4: Example for an explicit model for the interpretation of common cause failures

In order to implement a quantitative examination of a system on failures with a common cause, a very simple evaluation model is used for all components of the system, see Figure 15.5. For each component an individual fault tree analysis is set up. It is assumed, that all components could fail due to a common cause, displayed in Figure 15.5 with the symbol C_{ABC}. If a fault tree has been implemented, then the minimal "cut sets" can be determined. For each "cut set" $\{A_I\}$, $\{B_I\}$ and $\{C_I\}$, there is a corresponding C_{ABC}. Numerical values for the "common cause failures" can be estimated with the beta-factor-model:

$$P(C_{ABC}) = \beta \cdot P(A) \quad \text{mit} \quad \beta = 0{,}1 \tag{15.1}$$

$P(A)$ is the absolute coincidental failure frequency for failure A without a common cause. Losses with a common cause, the few that contribute to the entire system failure frequency, are neglected.

If inclined to implement a more elaborate common cause analysis, a fault tree on the component level for every single component must be developed, see right part of Figure 15.6. Additionally to the common cause failures, which involve all components, the paired common cause failures can be taken into consideration. Figure 15.6 shows the component A with the common cause failure C_{AB} and C_{AC}.

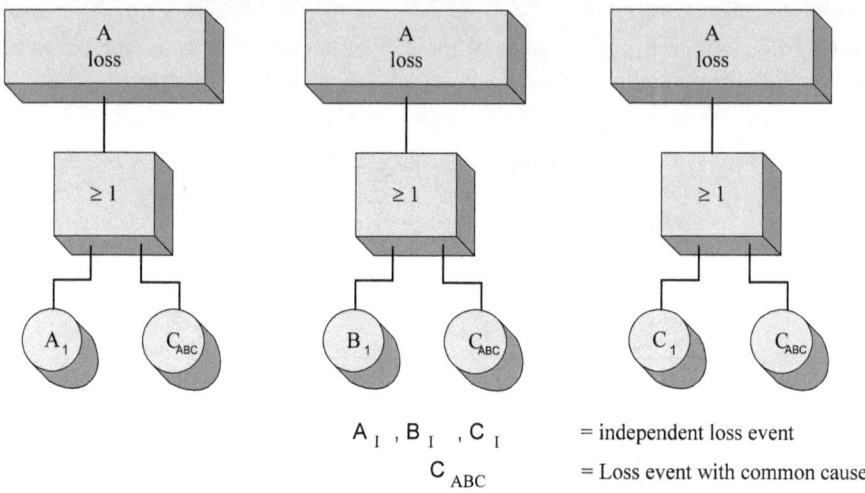

A_I, B_I, C_I = independent loss event
C_{ABC} = Loss event with common cause

Figure 15.5: Fault Tree Analysis for three components (A, B, C) for a 3oo3-system

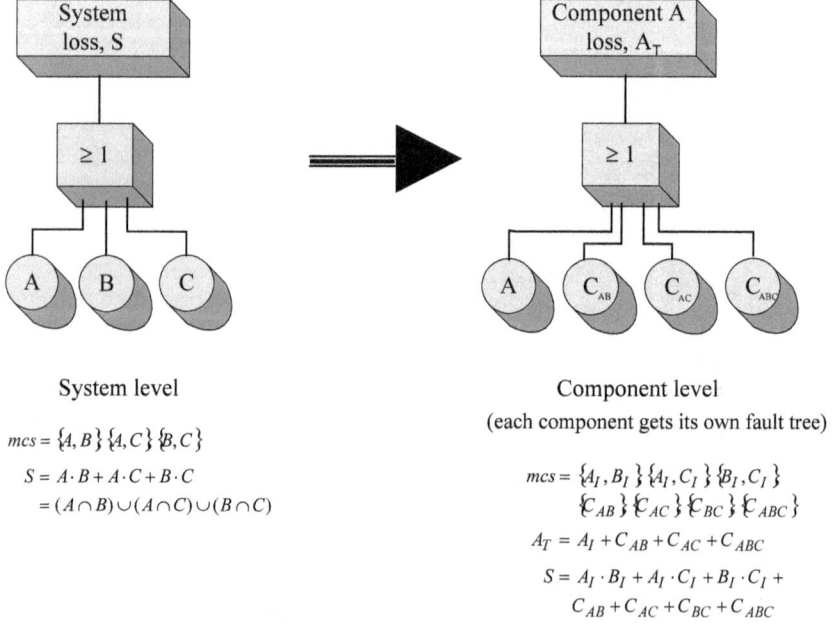

System level

$mcs = \{A,B\}\{A,C\}\{B,C\}$
$S = A \cdot B + A \cdot C + B \cdot C$
$= (A \cap B) \cup (A \cap C) \cup (B \cap C)$

Component level
(each component gets its own fault tree)

$mcs = \{A_I, B_I\}\{A_I, C_I\}\{B_I, C_I\}$
$\{C_{AB}\}\{C_{AC}\}\{C_{BC}\}\{C_{ABC}\}$
$A_T = A_I + C_{AB} + C_{AC} + C_{ABC}$
$S = A_I \cdot B_I + A_I \cdot C_I + B_I \cdot C_I +$
$C_{AB} + C_{AC} + C_{BC} + C_{ABC}$

Figure 15.6: Comparison system fault tree and component fault tree for a 3oo3-system

15.6 Quantitative Evaluation of Common Cause Failures

The probability of failure $P(A_T)$ for the component A is:

$$P(A_T) = P(A_I) + P(C_{AB}) + P(C_{AC}) + P(C_{ABC}). \qquad (15.2)$$

For the entire system, under the assumption that a 3oo3-architecture is available, the probability of failure is as follows

$$\begin{aligned} P(S) &= P(A_I) + P(B_I) + P(C_I) \\ &+ P(C_{AB}) + P(C_{AC}) + P(C_{BC}) + P(C_{ABC}). \end{aligned} \qquad (15.3)$$

Function Dependency

If a system consists of two series connected systems (S1 and S2), the system of S2 (NN) depends on the functional capability of the system S1. If system S1 fails, system S2 is no longer able to function. The single possibilities of failure are incident sequences and are marked with Greek lowercase letters. The fault tree for this system consisting of the subsystems S1 and S2 with the described dependencies are represented in Figure 15.7. At each knot are two decision possibilities: Yes, the subsystem is able to function, or No, the subsystem is not able to function.

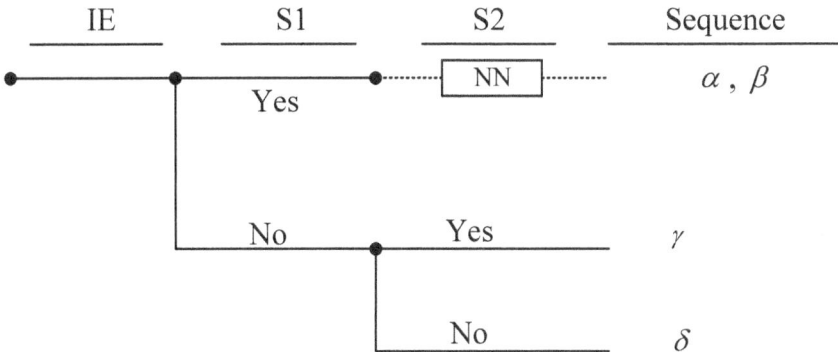

Figure 15.7: Event Tree for a system consisting of two function dependent subsystems

Shared-equipment

Event trees can be used to analyze the dependencies of commonly used components, called "shared-equipment". If one wants to develop these event trees out of fault trees, the fault trees for the above described systems S1 and S2 must experience the same component failure. For example, a fault analysis has determined, that the components A to F of the subsystems, can fail due to primary incidents, meaning incidents that are the cause of the failure. Both fault trees of the subsystems S1 and S2 can appear as displayed in Figure 15.8. If the components A and F are common components of both subsystems (shared-equipment), then the corresponding fault tree can appear as displayed in Figure 15.9.

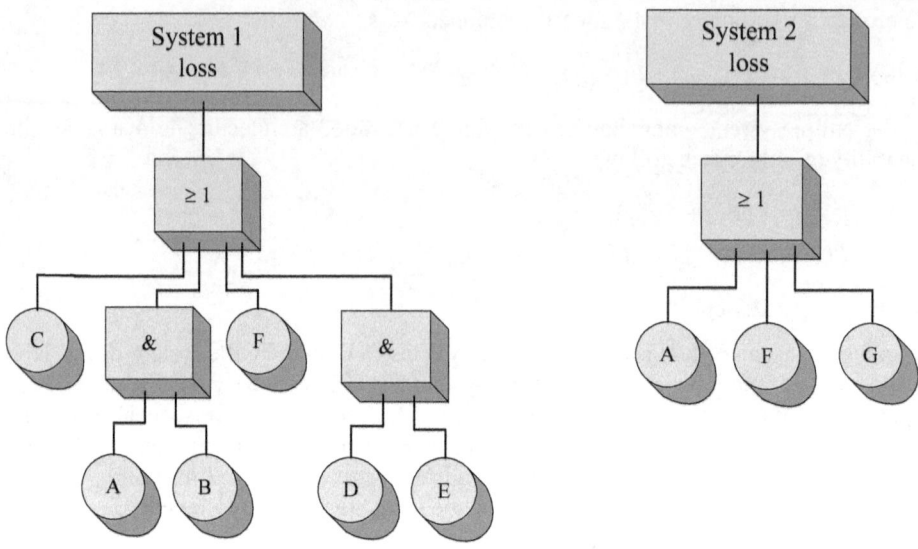

Figure 15.8: Fault Tree of two subsystems S1 and S2 with the same components A and F

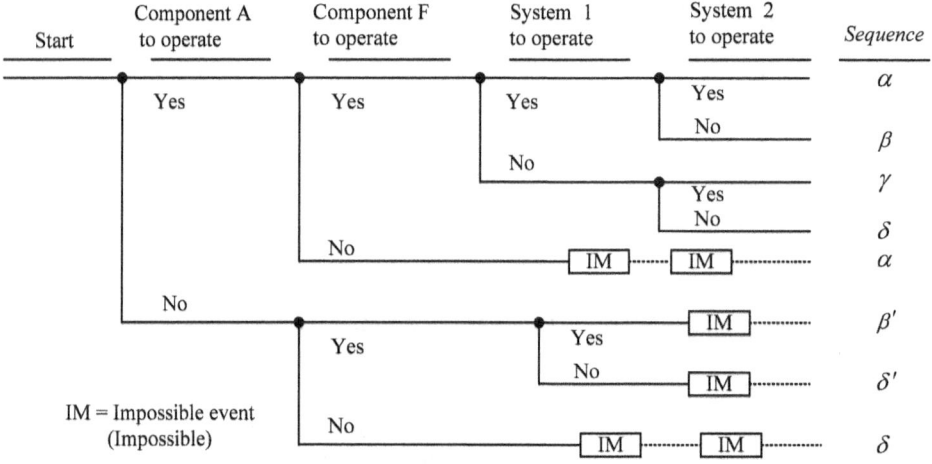

Figure 15.9: Event Tree

Method of the event tree linkage

The fault trees of the subsystems 1 and 2 are linked together so that for each incident sequence one big fault tree develops. Meaning that for sequence γ two events must be considered:

- S1 fails
- S2 functions

15.6 Quantitative Evaluation of Common Cause Failures

Mathematically this is represented by an AND-relation,

$$FT = FT1 \cap FT2, \tag{15.4}$$

by which *FT1* the fault tree of the subsystem S1, describes the failure of the subsystem in form of a failure logic, and *FT2* the fault tree of the subsystem S2 describes the condition "subsystem in operation" in form of a success logic, see Figure 15.10.

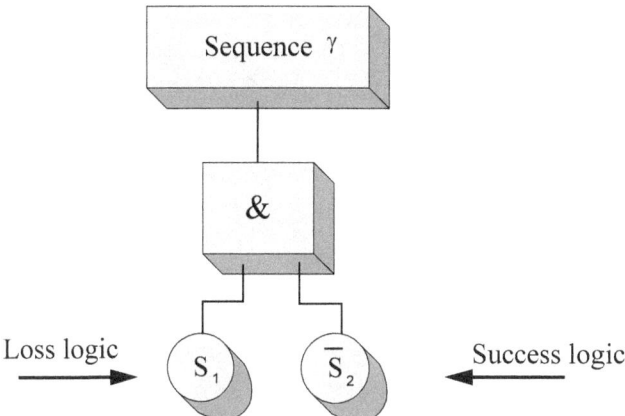

Figure 15.10: Sequence γ

$\overline{X}$ is the successful function of a component. X signifies the failure of a component. The sequence γ develops the fault tree in Figure 15.11. The fault tree and the minimal "cut sets" can then be displayed as follows:

$$\begin{aligned}\gamma = S_1 \cap \overline{S_2} &= [(A \cap B) \cup C \cup (D \cap E) \cup F] \cap \overline{[A \cup F \cup G]} \\ &= [C \cup (D \cap E)] \cap (\overline{A} \cap \overline{F} \cap \overline{G})\end{aligned} \tag{15.5}$$

$$mcs = \{\overline{AFGC}; \overline{AFGDE}\} \tag{15.6}$$

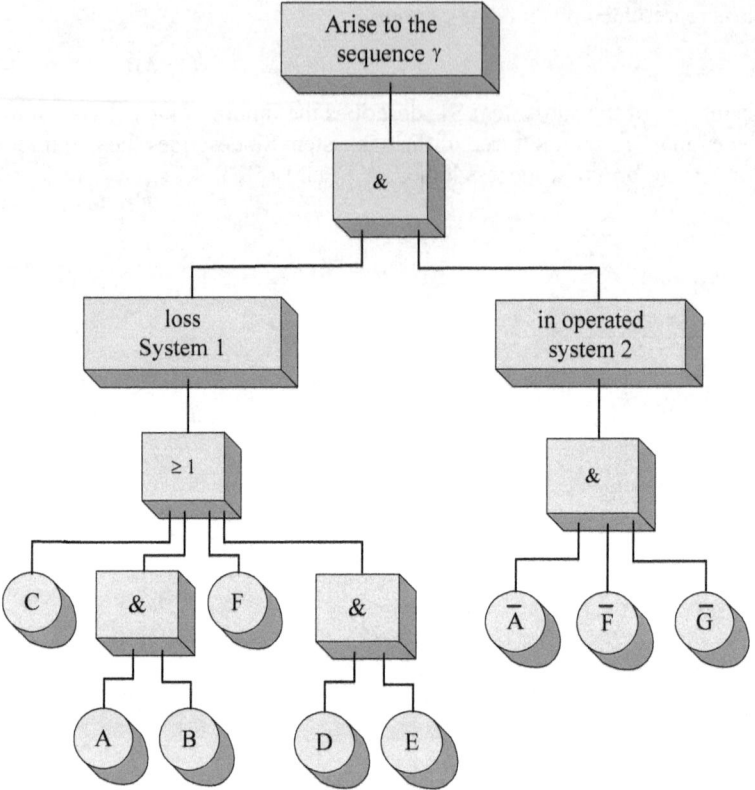

Figure 15.11: Fault Tree for the Sequence γ

Also component overlapping dependencies, like in a three component system, see Figure 15.12, can be represented as an event tree.

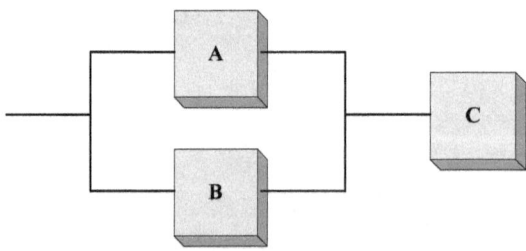

Figure 15.12: Three-component System

The corresponding system unavailability Q for the three component system can be expressed as follows:

$$Q = P(A \text{ und } B) + P(C) - P(A \text{ und } B \text{ und } C). \tag{15.7}$$

15.6 Quantitative Evaluation of Common Cause Failures

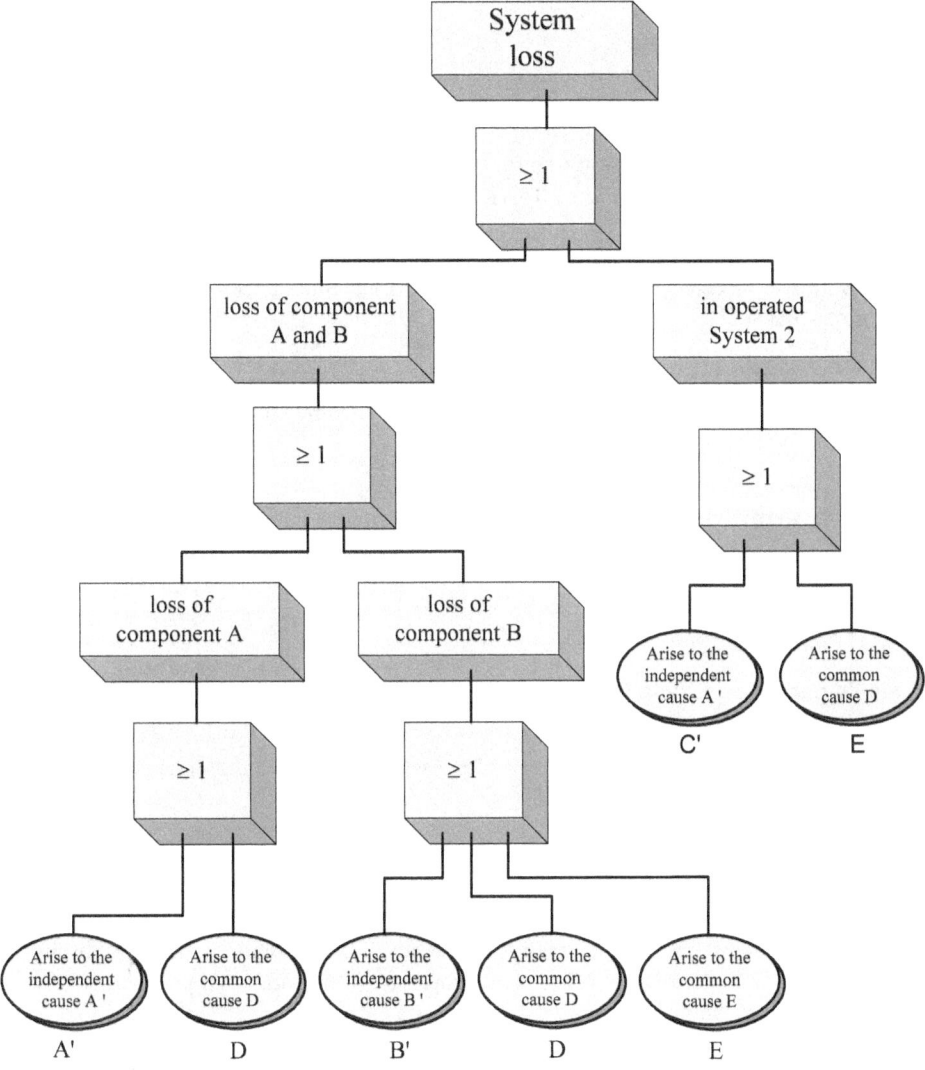

Figure 15.13: Fault tree for a three component system with independent and common causes

Explanations of parameters in Figure 15.13:
D: Common cause failure regarding the redundant components A and B, must be considered for the set-up of the minimal cut set.
E: Common cause, regarding the serial wired components, can be disregarded for the set-up of the minimal cut set.
A', B', C': Independent cause of failure of the components A, B, C.
D, E: Common cause of failure of the components A and B, B and C respectively.

Each cause of failure appearing simultaneously in one of the pairs of all three components, has a negative effect on the availability of the system. The components A, B and C can fail through independent causes A', B' or C', or through failures with a common cause. Failures with common causes can be divided into two categories; failures of the components A and B (primary event is displayed with the letter D) and failures of the components B and C (primary event displayed with the letter E). While primary events influence the redundant component, and must be considered for the set-up of the minimal cut sets, primary events that cover serial wired components, no influence on the minimal "cut sets" can be observed. A fault tree for the three-component system with the described primary-events is displayed in Figure 15.13.

It is a common belief that the neglect of common causes of losses can lead to

- an optimistic forecast of the system reliability of components which are part of the minimal "cut set"

and to a

- conservative forecast of the system reliability of components which are part of the minimal "cut set".

15.6.2 Implicit Methods of Common Cause Failures

Different failure models can be divided according to their number of parameters or different categories according to the type of occurrence:

Parameter models are divided into:

- **Single parameters**: Beta-factor-model, delivering conservative results for redundancy systems.
- **Multiple parameters**: delivers a better assessment of the common cause failure frequencies for a redundancy level > 2.

The different categories take the occurrence of multiple failures into account. These are the models:

- **Shock model:** the binomial failure rate model assumes that the system is part of a "common cause shock" with a specific frequency. The frequency of failures through a common cause is the result of the "shock rate" and the conditional probability of failure to provide a "shock".
- **Non shock model:**
    - **Directly:** according to the probabilities of the common cause events.
    - **Indirectly**: is being estimated indirectly from the probabilities of the different parameters.

15.6.2.1 Basic-Parameter-Model

This model is based, like all existing parameter models, on the "symmetry hypothesis". It uses the "rare incident approach"

$$P(S) = P(A_I) \cdot P(B_I) + P(A_I) \cdot P(C_I) + P(B_I) \cdot P(C_I) + P(C_{AB}) + P(C_{AC}) + P(C_{BC}) + P(C_{ABC}) \quad (15.8)$$

with the following assumptions:

- The probability, that similar events involve similar components, is equally high.
- The probability of failure of the basic events of a common cause component group depends only on the number and not on the components.

The absolute probability for the failure of a system, which components are wired in a 2oo3-Logic, amounts to:

$$\left. \begin{array}{l} P(A_I) = P(B_I) = P(C_I) = Q_I \\ P(C_{AB}) = P(C_{AC}) = P(C_{BC}) = Q_2 \\ P(C_{ABC}) = Q_3 \end{array} \right\} \begin{array}{l} Q_t = Q_1 + 2 \cdot Q_2 + Q_3 \\ Q_S = 3 \cdot Q_1^2 + 3 \cdot Q_2 + Q_3 \end{array} \quad (15.9)$$

$$Q_t = \sum_{k=1}^{m} \binom{m-1}{k-1} \cdot Q_k \quad (15.10)$$

15.6.2.2 Beta-Factor-Model

The beta factor model considers two types of dependent failures:

- Component related physical interactions
- Human interactions

The model assumes, that Q_t, the absolute probability of failure of each component, can be divided into an independent share Q_l and a dependent share Q_m, by which m is the number of the components in the common cause group. All Q_k's are 0, except Q_l and Q_m:

$$Q_t = Q_l + Q_m \quad (15.11)$$

The parameter β is defined as a quotient, consisting of the probability of the dependent failures and the total probability of failure:

$$\beta = \frac{Q_m}{Q_t} = \frac{Q_m}{Q_l + Q_m} \Rightarrow \left. \begin{array}{l} Q_m = \beta \cdot Q_t \\ Q_l = (1-\beta) \cdot Q_t \end{array} \right\} Q_s = 3(1-\beta)^2 Q_t^2 + \beta Q_t \quad (15.12)$$

In systems with more than two units, the beta factor model does not differentiate between the different numbers of multiple failures. This simplification can lead to conservative forecasts, if all units fail and if a CCF appears. The strength of the beta factor model lies in its direct use of experience values and its flexibility. The overall construction part probability of failure and β must be estimated.

15.6.2.3 Multy Greek Letter Model

The following equation allows the calculation of the probability of CCFs of the order k with (m- 1) parameters:

$$Q_k = \frac{1}{\binom{m-1}{k-1}} \cdot \left(\prod_{i=1}^{k} \rho_i\right) \cdot (1-\rho_{k+1}) \cdot Q_t \qquad (15.13)$$

$\rho_1 = 1$

$\rho_2 = \beta$: Conditional probability of failure of at least one additional component if one component has failed.

$\rho_3 = \gamma$: Conditional probability of failure of at least one additional component, if two components have failed.

$\rho_4 = \delta$: Conditional probability of failure of at least one additional component, if three components have failed.

15.6.2.4 α-Factor Model

The α-factor model consists of m parameters. With help of the following equations, the probability of failure through a common cause can be determined.

$$\alpha_k^{(m)} = \frac{\binom{m}{k} \cdot Q_k^{(m)}}{\sum_{k=1}^{m} \binom{m}{k} \cdot Q_k^{(m)}} \qquad (15.14)$$

Normalization:

$$\sum_{k=1}^{m} \alpha_k^{(m)} = 1 \qquad (15.15)$$

with

$k = 1, 2, \ldots, m$

and

$$Q_t = \sum_{k=1}^{m} k \cdot \alpha_k \qquad (15.16)$$

result in

$$Q_k^{(m)} = \frac{k}{\binom{m-1}{k-1}} \cdot \frac{\alpha_k^{(m)}}{\alpha_t} \cdot Q_t \qquad (15.17)$$

15.6.2.5 Binomial Failure Rate Model (BFR)

The examination of a system consisting of m identical components. Each component can fail due to a coincidental moment in time, independent from each other, with a failure rate λ. Additionally, a common cause shock can hit the system with an appearance rate μ. Whenever this shock occurs, each of the single components m can fail with the probability p, regardless of the condition of the different components. The designation "binomial disturbance rate" is being used, because the main consequence of a single component failure is considered a consequence of the shock, so that a binomial distribution with the parameters m and p follows:

$$p[I = i] = \binom{m}{i} \cdot p^i \cdot (1-p)^{m-i} \qquad i = 0, 1, \ldots, m \qquad (15.18)$$

Two further assumptions:

- Shocks and individual failures appear independently from each other.
- All failures are discovered immediately and the repair time is minimal.

The assumption that the components fail independently from each other due to a shock, is not realistic. The problem can be remedied by defining parts of the shocks as "lethal shocks", meaning as shocks that lead to an automatic failure of the component (p= 2). If all shocks are lethal, the β-factor model must be applied. It must be mentioned that the event p= 1 is a situation, in which no built-in protection against these shocks exists.

The BFR-model differs from the β-factor in the numbers of multiple failures in a system with more than two units.

$$\lambda_1 = m\lambda + \mu\left[\binom{m}{1} p(1-p)^{m-1}\right] \qquad \text{Failure rate of a Unit}$$

$$(15.19)$$

$$\lambda_i = m\lambda + \mu\left[\binom{m}{1} p(1-p)^{m-i}\right] \qquad \text{Failure rate of } i \text{ Units}$$

15.7 β-Factor

The sensitivity of a system for common load factors is usually being displayed by the so called "beta-factor" (β-factor)[141]. The value range of the β-factor is 0 to 0.25 (0 = no failure with common cause) and is a percentile of all failures in a string of a multiple channel system which lead to the failure of one or more strings due to a common load.

A dual system shows only the β-factor. Triple- or multiple redundancy systems show several sub factors, see Figure 15.14. A triple redundant PLC-system has four different factors with common cause: Three string combinations which show β1, β2 and β3. 0 represents the only load factor that leads to the failure of all strings.

The load factors that express β1, β2 and β3, are possibly caused by methodical (design-) errors, while β0 results due to environmental causes. For practical safety integrity calculations, all β-factors are combined into one value. The formulas in the next chapters are also based on a β-factor, assuming that all strings of a redundant system will fail due to only one common cause.[142]

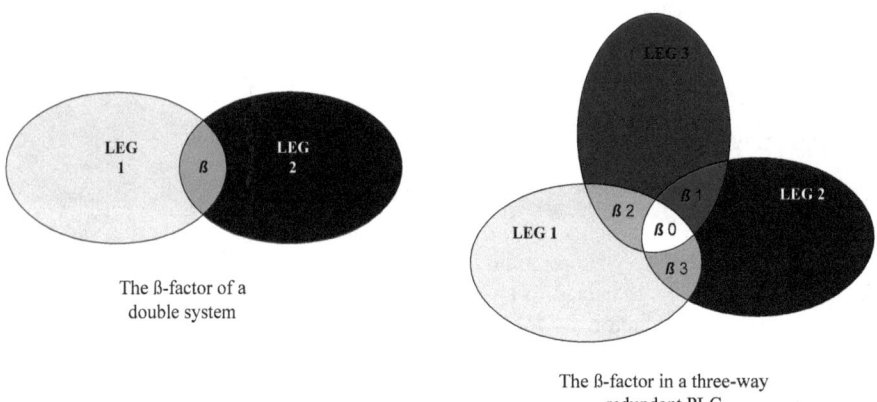

The ß-factor of a double system

The ß-factor in a three-way redundant PLC

Figure 15.14: β-Factor for redundant systems[143]

The investigation of the β-factor is difficult and is mostly based on an expert analysis as well as on several debates and assumptions. That could be the reason for the countless publications on the topic. Regardless of the split opinions of the experts, all publications point out a common opinion:

With a decrease of the safety integrity by a certain magnitude, the common cause plays an important role.

The method of approximation is the most common procedure for the interpretation of CCFs. In the application of this method, commonly called the "β-factor method", the probability of a CCF correlates with the rate of coincidental failures of the plant. This

---

[141] [IECa00]   IEC 61508, International Standard 61508: Functional safety of electrical/electronic/ programmable electronic safety-related systems
[142] [ISA-96]   ISA-84.01-1996. Application of Safety Instrumented Systems for the Process Industries
[143] [TIEZ98]   Tiezema, R., Common Cause effects on safety rated PLCs

method makes it possible to appraise CCFs without identifying the specific sources of dependent mistakes and their corresponding probability.

The β-factor can be estimated as follows:

- Identification of the overall failure rate of the plant according to published or internal data;
- Revision of the failure modes in order to determine which part can, as expected, be influenced by a common cause;
- Calculation/ assessment of the percentile share of the failure rate that correlates with the CCFs (β-factor);
- Application of the β-factor in order to calculate the failure rate for dependent and independent failures of the plant.

The β-factor can reach from almost 0 up to 25%, according to plant and specific considerations regarding a common cause. The estimation of the β-factor can be done by either quantitative or qualitative methods. The experience with systems can be used for the calculation of the β-factor for a specific plant, if service and inspection reports are available. In these events the following equation can be used:

$$\beta = \frac{m}{n+m} \quad (15.20)$$

n = number of problems or events in which only a single component has failed,

m = number of facilities that have failed due to certain problems or events at which similar parallel components have failed.

In the event that no sufficient internal data is available, qualitative methods can be used to determine the β-factor. A number of public sources provide limited instructions based on competent judgment in order to choose a β-factor.

There are methods for the qualitative determination of the β- factor which are founded on the existence of different common cause events. The IEC-commission has developed test aspects which are founded on definite measures that reduce the potential of common cause failures or increase the quantitative factors X and Y. These factors X and Y are used to determine the overall β-factor for each component. The β-factor is used in connection with the coincidental hardware failure rate in order to determine the CCF-rate for redundant systems.

The comparison list of the dimension unit is captured in the design. ISA TR84.0.02 Part 1 Appendix A regarding IEC 61508. The proposed dimension unit is still in development and numerous alterations are expected until the final issue.

15.7.1 The Effect of the β-factor on safety

There are different quantitative models that predict the effect of common cause failures on a safety system. The simplest is the "β-model". This model is based on the same formulas as in the previous chapter, but the β-factor is based on a common stress factor that impairs all areas of the redundant PLC at the same time. Differentials of the ISA-TR84.02 and the β-factor can be displayed as in Table 15.2: The table shows the probability of an error at the request of a safety function of the commonly used PLC-architectures (1oo2, 2oo3, 2oo4). The table shows clearly that the formulas consist of two parts. The left part describes the effect of the redundant architecture. The right part of the formula is identical to the formula for a single channel PLC and indicates the effect of the β-factors on PFD_{avg}.

The higher the β-factor, the more important the effect. Expert opinions put the β-factor right about 0.1% to 10% regarding the hardware failures. Especially parts for safety circuits are prone to environmental influences due to their sensitivity and high β-factors. Typical values for those elements lie around 10%.

Table 15.2: Formulas with β-factors

Safety System	PFD_{avg}
1oo2D	$\frac{1}{3} \cdot ((1-\beta) \cdot \lambda_{DU} \cdot T)^2 + \frac{1}{2} \cdot \beta \cdot \lambda_{DU} \cdot T$
2oo3	$((1-\beta) \cdot \lambda_{DU} \cdot T)^2 + \frac{1}{2} \cdot \beta \cdot \lambda_{DU} \cdot T$
2oo4	$((1-\beta) \cdot \lambda_{DU} \cdot T)^3 + \frac{1}{2} \cdot \beta \cdot \lambda_{DU} \cdot T$

The graphics in Figure 15.15 show how PFD_{avg} with increasing β-factor decline. This is logical if the β-factor and the resulting effects on the outcomes are looked at. The comparison of a 1oo2D-architecture calculation (= 2.56E-5) with β = 0% and a calculation with β = 10% (PFD_{avg} = 3.0E-4) results in a deterioration of PFD_{avg} to a factor 10!

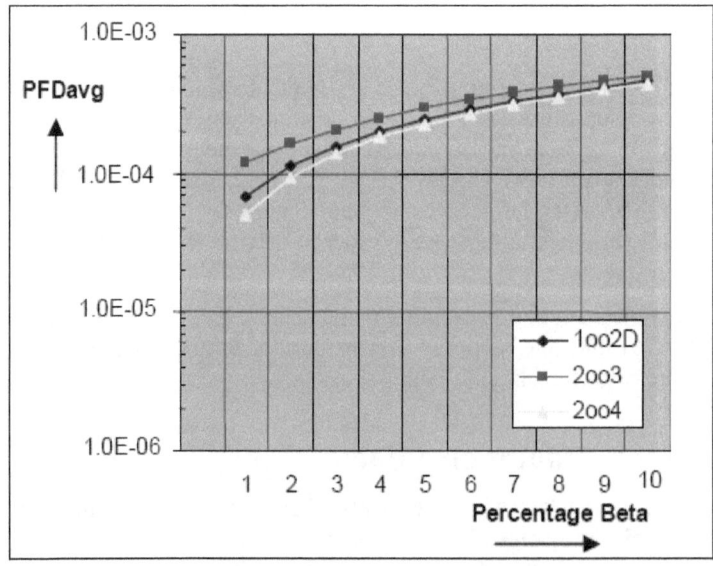

Figure 15.15: Relationship between PFD- and beta-values

15.7.2 Assessment of the β-factor

As previously mentioned, many publications exists regarding a realistic value for the β-factor. One of the most interesting documents regarding the evaluation was developed in the early 70ies; data of the power plant industry was gathered and published as UPM 3.1 of AEA. Goal of this document was the development of a practical method to develop a reliable β-factor for reliability assessment. The document has a clear structure and serves as a reference for the IEC 61508.

Table 15.3: Application and weight factors

Application	Sub application	Weight Factor
Design	Redundanz (& Diversität)	6
	Trennung	8
	Verstehen	6
	Analyse	6
Betrieb	MMI	10
	Sicherheitskultur	5
Umwelt	Kontrolle	6
	Tests	4

Recapitulating the report presents the following approach: First the areas of the dependent failures are defined and divided into subfields. Each subfield is assigned a weighting factor. The weighting factor results from the "debates" of experienced engineers in the area of reliability.

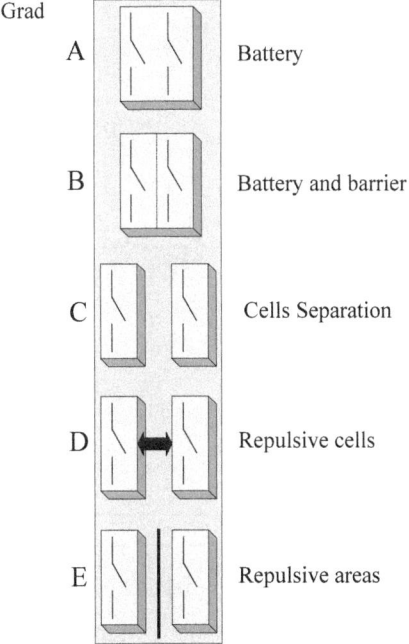

Figure 15.16: Separation Levels

The higher the weighting factor, the higher the final β-factor. Following the degree is defined, up to which value the redundant strings of a system of dependent failures can be involved (A-E). As example Figure 15.16 shows the division into subfields, namely the partition.

The alteration of the degree of A to E corresponds to a low β-factor. From this results the "rough separation", in which E is the best attainable value for the β-factor and A the value for the lowest attainable β-factor for this sub field. The numbers in the columns B, C and D are based on an "even distance".

Table 15.4: Beta-Factor estimation

Sub a	Degree and sub factor				
	A	B	C	D	E
Redundancy and Diversity	600	190	60	19	6
Partitioning	800	250	80	25	8
Understanding	600	190	60	19	6
Analysis	600	190	60	19	6
MMI	1000	320	100	32	10
Safety culture	500	160	50	16	5
Control	600	190	60	19	6
Tests	400	130	40	13	4

In order to complete the table, the same procedure is applied to the other defined sub fields.

Then the β-factor is calculated by

- Selecting the value of the corresponding sub factor for each subfield;
- Adding up these values;
- Dividing the sum by a so called nominator;

The nominator results from the sum of all values of column A.

Table 15.5: Chosen sub factors

Sub aplications	Nominator value	Sub factors
Redundancy und Diversity	600	19
Partitioning	800	8
Understanding	600	60
Analysis	600	19
MMI	1000	320
Safety culture	500	16
Control	600	190
Tests	400	40
	Total: 5100	Total: 672

In this case the common nominator is 5100. The sum of the chosen sub factors is 672.

In this case the MMI and control fields show a potential for improvement. Many experts in the area of reliability do not consider this method for the evaluation of the β-factor to be sound. It is missing the scientific reasoning for issues like weighting factors and the selection of the sub factors, which are based on debates of engineers and the assumption of even distances. This method does not claim to be correct, is however viewed as a meaningful assessment. A similar method can be found in part 6, appendix D of the IEC 61508. The report "Simulation through load factors for the depiction of the common cause" of the "University of Eindhoven" includes a more scientific approach.

The report is a written article for lecture purposes regarding the simulation of load factors. It shows a simulation program for different kinds of load parameters of different redundant systems. The simulation proves, that the β-factor for physically separate systems is smaller, than the β-factor for physically not separate systems. The results show furthermore, that the β-factor for physical not separate systems is relatively stable and high despite an alteration of the load capacity. The value of the β-factor for separate systems reaches from 1E-01 to 1E-05. The load parameters have been selected randomly. Safety experts consider the report as limited, because it does not take the effects of different causes of failure into consideration, like the human factor.

15.8 1oo2 System

The 1oo2-system consists of two independent channels. In order to execute the safety function correctly, both channels have to be connected in a way, that one of the channels will suffice for the safety function to fulfill its task. A 1oo2-system will fail, if both channels are simultaneously faulty. The probability of failure of a 1oo2-system can be determined with help of the fault tree. The equation for the probability of failure is:

$$P(t) = P_1(t) \cdot P_2(t) + P_{DUC}(t) + P_{DDC}(t) \tag{15.21}$$

with

$$P_1(t) = 1 - e^{-\lambda_{D1} \cdot t} \tag{15.22}$$

$$P_1(t) = 1 - e^{-\lambda_{D2} \cdot t} \tag{15.23}$$

$$P_{DDC}(t) = 1 - e^{-\beta_D \cdot \lambda_{DD} \cdot t} \tag{15.24}$$

15.8.1 Probability of Failure with Common Cause Failures

First of all the probabilities of failure for dangerous and unrecognizable and dangerous and recognizable common cause failures P_{DUC} and P_{DDC} are calculated. Common cause failures are failures that appear simultaneously in all system channels and have a common cause. This failure type is weighted at the PFD_{avg}-determination of a multiple-channel system via the β-factor.

$$P_\beta(t) = P_{DUC}(t) + P_{DDC}(t) \tag{15.25}$$

These error probabilities can be determined analogously in a 1oo1-system with

$$\lambda_{D,1oo1} = \beta \cdot \lambda_{DU} \qquad \text{bzw.} \qquad \lambda_{D,1oo1} = \beta_D \cdot \lambda_{DD}$$

respectively.

A common cause failure represents a 1oo1 functional block. It is possible, for the calculation of the probability of failure through common cause failures, to use the derivation of the PFD_{avg}-equation. From the fault tree of the 1oo2-systems for common cause failures the following equation can be derived:

$$\begin{aligned} PFD_{avg,\,\beta\,faulttree} &= \beta \cdot \lambda_{DU}(T_1 + MTTR') + \beta_D \cdot \lambda_{DD} \cdot MTTR' \\ &= \lambda_D \cdot \tilde{t}_{CE,\beta} \end{aligned} \qquad (15.26)$$

A PFD_{avg}-equation contains the probability, that a dangerous, undetectable failure of a common cause appears within the $T_1 + MTTR$ t,

$$PFD_{avg,\,\lambda_{DU},\,\beta} = \beta \cdot \lambda_{DU}(T_1 + MTTR') \qquad (15.27)$$

And according to the probability that a more dangerous, undetectable common cause failure within the repair time MTTR appears:

$$PFD_{avg,\,\lambda_{DD},\,\beta_D} = \beta_D \cdot \lambda_{DD} \cdot MTTR' \qquad (15.28)$$

From this follows:

$$\tilde{t}_{CE,\beta} = \beta \cdot \frac{\lambda_{DU}}{\lambda_D}(T_1 + MTTR') + \beta_D \cdot \frac{\lambda_{DD}}{\lambda_D} MTTR' \qquad (15.29)$$

With $t_{ce,\beta} = T$ follows:

$$\begin{aligned} PFD_{avg} &= \frac{\lambda_D}{2} \cdot \tilde{t}_{CE,\beta} \\ &= \frac{\lambda_D}{2} \cdot \left[\beta \cdot \frac{\lambda_{DU}}{\lambda_D}(T_1 + MTTR') + \beta_D \cdot \frac{\lambda_{DD}}{\lambda_D} MTTR' \right] \\ &= \left[\beta \cdot \lambda_{DU} \left(\frac{T_1}{2} + \frac{MTTR'}{2} \right) + \beta_D \cdot \frac{\lambda_{DD}}{\lambda_D} \frac{MTTR'}{2} \right]. \end{aligned} \qquad (15.30)$$

With the assumption, that

$$\frac{T_1}{2} >> \frac{MTTR'}{2} \approx MTTR' = MTTR \qquad (15.31)$$

The PFD_{avg}-value for common cause failures can be calculated with

$$\begin{aligned} PFD_{avg,\,\beta} &= \left[\beta \cdot \lambda_{DU} \cdot \left(\frac{T_1}{2} + MTTR \right) + \beta_D \cdot \lambda_{DD} \cdot MTTR \right] \\ &= \lambda_D \left[\frac{\beta \cdot \lambda_{DU}}{\lambda_D} \left(\frac{T_1}{2} + MTTR \right) + \frac{\beta_D \cdot \lambda_{DU}}{\lambda_D} \cdot MTTR \right] \\ &= \lambda_D \cdot t_{CE,\beta} \end{aligned} \qquad (15.32)$$

with

$$t_{CE,\beta} = \beta \cdot \frac{\lambda_{DU}}{\lambda_D}\left(\frac{T_1}{2} + MTTR\right) + \beta_D \frac{\lambda_{DD}}{\lambda_D} MTTR \qquad (15.33)$$

Adding the share of common cause failures to the probability of failure, the PFD_{avg}-formula for a 1oo2-system calculates as follows:

$$\begin{aligned}PFD_{avg,\,1oo2} &= 2 \cdot \left((1-\beta_D)\cdot\lambda_{DD} + (1-\beta)\cdot\lambda_{DU}\right)^2 \cdot t'_{CE} \cdot t'_{GE} \\ &\quad + \beta_D \cdot \lambda_{DD} \cdot MTTR + \beta \cdot \lambda_{DU} \cdot \left(\frac{T_1}{2} + MTTR\right)\end{aligned} \qquad (15.34)$$

This equation for the PFD_{avg}-value is identical to the corresponding formula of IEC 61508.

15.9 Measures against Failures through Common Cause

The best method to prevent common cause failures is to pay attention to those failures at the beginning the life cycle. This is predominantly the task of engineers who are consciously aware of the sources of the common cause failures and consequently can take measurements against their formation. Through early recognition and prevention of common cause failures in the design phase, expenses can be prevented at a later point. If however the decrease of common cause failures is not feasible, measures can be applied that are founded on the specific requirements of the respective company.

Possible measures for common cause failures are:

- Careful project administration,
- Sound planning,
- Diverse assembly of the modules,
- Diversity of the equipment,
- Protection and separation of the equipment,
- Barriers,
- Performance reduction and simplicity of the equipment,
- Quality control,
- Prophylactic maintenance and
- Supervision.

In order to verify that certain measures for each potential cause of common cause failures have been considered, specific technologies have been deployed. The development of such a technology, that includes measures against common cause failures, was supported by the US nuclear regulatory committee and presented under the name "Cause Defense Matrix".[144]

---

[144] [PAPA90]   Paula H. M., Parry G. W., *A Cause Defence Approach to the Understanding and Analysis of Common Cause Failures*

[MOSL93]   Mosleh A., Fleming K., Parry G., Paula H., Warledge D., Rasmusson D., *Procedures for Analysis of Common-Cause Failures in Probabilistic Safety Analysis*

16 Proof Test

The first proof tests were implemented early on in airplane maintenance. They served mainly to test highly stressed components. Nowadays, proof tests are a standard used in a variety of industrial sectors, especially for the purpose of systematically examining safety-critical systems for malfunctions. Die individual components of a system are subjected to stepwise tests in order to determine the critical phase transitions. Each proof test must be adequately documented. By doing so it becomes possible to prove that components can be reinserted into the system "like new" after completion of the test. In the process industry, the following points should be examined in a proof test and registered in the test documentation.

- System interfaces
- Diagnostic units
- Trigger values
- Tolerance
- Communication paths (system release after test)
- Shunting units, if required

16.1 Monitoring and Conducting of Proof Tests

Proof tests serve to detect dangerous failures which otherwise remain undetected until demand and can then become hazardous to a process. A well-planned and executed test program can minimize test time and time until recommissioning during the off-line execution of the proof test. Reliability programs make it possible to precisely determine repeat intervals and thus plan off-line tests in an economic manner.

On-line tests entail the danger that additional failures can be triggered during the tests. On the other hand, unless adequate planning is performed, a test of the demands on system integrity is not always possible if a proof test is conducted off-line. Suitable monitoring and execution of proof tests must therefore ensure that a complete scrutiny of all components and of the system itself take place, and with adequate effort.

The following three important aspects of a safety-critical component are tested in a proof test:

- Functionality
- Is the component suitable for the predominant operating conditions
- Are the interfaces with additional components defined and in order

All critical components must be completely, that is 100%, tested with the proof test. Non-safety-critical components, on the other hand, need only be spot-tested.

What are the Advantages of Proof Tests?

Proof tests prove that functionality meets requirements even in the appropriate operating conditions, and that intersections with other components meet requirements. Furthermore, proof tests confirm that the component can be implemented without endangering its environment, and thus they ascertain that these components comply with ISO 9001. In addition, proof tests can be a cost-efficient test alternative to a very expensive and time-consuming prototype, which may possibly be subjected to a demolition test.

16.2 Types of Proof Tests

Simple Proof Test

In this test, the component to be examined is powered up to the desired operating range, this state is then maintained for a defined period, and subsequently the component is brought back to its original state. Only general data are recorded.

Stepwise Proof Test

The stepwise proof test is conducted analogously to the simple proof test, although here the data are recorded in a stepwise manner.

Heat Proof Test

This test is also conducted analogously to the simple proof test, yet here the behavior of components in high ambient temperature is examined.

Cold Proof Test

In this test, as in the previous procedures, the component is also powered up into the desired state and tested at low temperatures.

Comparison Proof Test

In this test, the heat and cold proof test are combined. The component is powered up into a specified operating range and then exposed to rapidly changing extreme temperatures.

Differential Proof Test

This test examines the behavior of the component if differing temperatures exist in its individual operating areas. This test provides an improved means of modeling reality.

Fatigue Test

In this test, the component is examined for a specific period of time outside of its operational limits, which are exceeded by a specific factor. The functionality of the component before and after the test is compared.

16.3 Reliability Function and MTTF

The reliability function $R(f)$ is the probability that a unit under observation is functional in a period under observation $(0...t]$. The reliability function $R(f)$ is determined absolutely by the failure rate $\lambda(t)$.

$$R(t) = e^{-\int_0^\infty \lambda(t)dt} \tag{16.1}$$

If the distribution is exponential, that is, if the failure rate $\lambda(t) = \lambda$ is constant with regard to t, then the reliability function is simplified to

$$R(t) = e^{-\lambda \cdot t}. \tag{16.2}$$

One of the most important parameters of reliability is the expected value, better known as MTTF (Mean Time To Failure):

$$MTTF = \int_0^\infty R(t)dt = \frac{1}{\lambda} \tag{16.3}$$

If you chart R over t, you get the progression shown in Figure 16.1.

16.3.1 Failure Probability

Failure probability $P(f)$ is calculated by means of the reliability function $R(f)$. While $R(f)$ specifies the probability that no failure will occur in the interval $(0, f)$, the failure function specifies the probability that an error occurs which leads to the failure of a component. Failure probability is described in equation 16.4.

$$\begin{aligned} P(t) &= 1 - R(t) \\ &= 1 - e^{-\lambda \cdot t} \end{aligned} \tag{16.4}$$

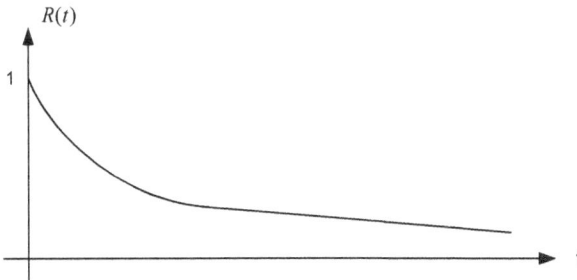

Figure 16.1: Reliability function for constant failure rate

For very small failure rates and a very big time interval t, that is, if the following inequality

$$\lambda \cdot t \ll 1 \tag{16.5}$$

can be set, the approximation for failure probability is shown as

$$P(t) = \lambda \cdot t \quad \text{für} \quad \lambda \cdot t \ll 1. \tag{16.6}$$

The correlation between failure probability and reliability is shown in Figure 16.2.

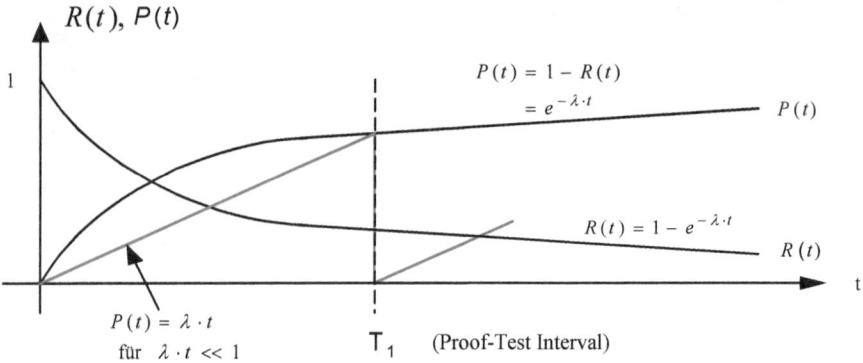

Figure 16.2: Correlation between failure probability function and reliability function for constant failure rate

16.3.2 Probability of Failure on Demand

In safety technical applications, the special term "Probability of Failure on Demand", in short PFD, is more frequently used than the general term "failure probability". This term stands for the probability of an error on demand on a safety function and indicates the quality of a system with respect to accuracy. The smaller the PFD-value, the better the system. It is assumed that not all failures can be detected, for 100% safety can not be achieved in technical systems.

16.3.3 Proof Test Interval T_1

The proof test interval T_1 defines a time interval between two periodic tests, respectively, maintenance, performed on a safety system. These tests are conducted to identify undetected, dangerous errors. The correlation between T_1 and $MTTF$ is deduced from equation 16.6. If one substitutes the condition on the right side of the equation for $t = T_1$, the proof test interval, and for λ the reciprocal of the mean failure-free time, $MTTF$, then one gets:

$$T_1 \ll MTTF. \tag{16.7}$$

This stipulation should be observed when calculating the *PFD*-value of a safety function.

16.4 Definition of the Proof Test according to IEC/EN 61508

The following definition is taken from IEC/EN 61508:

A proof test is a recurring test for uncovering failures in a safety-related system, so that, if necessary, the system can be brought into an "as new" state, or brought as close as possible to this state, from a practical standpoint.

NOTE: The effectiveness of the proof test depends on how close the system is brought to the "as new" state. In order for the proof test to be completely effective, it is necessary to detect all dangerous failures. Although this can not be achieved for other, simple E/E/PE-safety related systems, at the very least all completed safety functions must be tested according to E/E/PES specification of safety requirements. If separate conduits are used, these tests are performed separately for each conduit.

Part 6, supplement B1, of the standard shows that the proof test interval must be at least ten times higher than the diagnostic test interval.

16.5 Consequences of an Insufficient Proof Test

One example from IEC/EN 61508, part 6, appendix B 2.5, shows the consequences of an insufficient proof test.

Failures in the safety system that remained undetected by the diagnostic test, and by the proof test, can only be found in the event of demand on a safety function affected by the failure. Therefore, the expected demand rate on the safety system and the actual downtime for these completely unknown failures are determined.

A 1oo2-architecture is shown as an example. The known PFD formula for a 1oo2 system has been supplemented by introducing the additional time T_2. T_2 is the time between two demands on the system. Equations 16.8 and 16.9 indicate that T_2 only has to be considered for dangerous, non-identifiable error rates.

$$t_{CE} = \frac{\lambda_{DU}}{2 \cdot \lambda_D} \cdot \left(\frac{T_1}{2} + MTTR\right) + \frac{\lambda_{DU}}{2 \cdot \lambda_D} \cdot \left(\frac{T_2}{2} + MTTR\right) + \frac{\lambda_{DD}}{\lambda_D} \cdot MTTR$$

$$t_{CE} = \frac{\lambda_{DU}}{2 \cdot \lambda_D} \cdot \left(\frac{T_1}{3} + MTTR\right) + \frac{\lambda_{DU}}{2 \cdot \lambda_D} \cdot \left(\frac{T_2}{3} + MTTR\right) + \frac{\lambda_{DD}}{\lambda_D} \cdot MTTR$$

(16.8)

$$PFD_G = 2 \cdot ((1-\beta_D) \cdot \lambda_{DD} + (1-\beta) \cdot \lambda_{DU})^2 \cdot t_{CE} \cdot t_{GE}$$

$$+ \beta_D \cdot \lambda_{DD} \cdot MTTR + \beta \cdot \frac{\lambda_{DU}}{2} \cdot \left(\frac{T_1}{2} + MTTR\right)$$

$$+ \beta \cdot \frac{\lambda_{DU}}{2} \cdot \left(\frac{T_2}{2} + MTTR\right)$$

(16.9)

Table 16.1: Example of an insufficient proof test[145]

Architecture	DC	$\lambda = 1{,}0 \cdot 10^{-0{,}5}$	
		100 % Proof test $\beta = 10\ \%,\ \beta_D = 5\ \%$	50 % Proof test ($T_2 = 10$ years) $\beta = 10\ \%,\ \beta_D = 5\ \%$
1oo2	0 %	$2.7 \cdot 10^{-03}$	$6.6 \cdot 10^{-02}$
	60 %	$9.7 \cdot 10^{-04}$	$2.6 \cdot 10^{-02}$
	90 %	$2.3 \cdot 10^{-04}$	$6.6 \cdot 10^{-03}$
	99 %	$2.4 \cdot 10^{-05}$	$7.0 \cdot 10^{-04}$

Table 16.1 shows the numerical results of a 1oo2-system, taken from IEC 61508, part 6, with a 100% annual proof test compared to a 50% proof test. The demand time interval T_2 between two demands is defined as 10 years, the basic failure rate is $1 \cdot 10^{-5}$/h, and the values for β and β_D are 10% and 5%, respectively.

16.6 Differences between Diagnostic Test and Proof Test

Safety systems perform safety functions and consist, as a rule, of three groups: sensors, such as sensing elements; logic units, usually in the form of storage-programmable units, and actuators, such as alarm units, pumps, or valves. The implementation and operating conditions of a safety function are determined by the benchmark data of the individual components. Benchmark data include architecture, the probability of the occurrence of hardware and/or software failures, system failures, or common-cause failures, online diagnosis, and periodic test intervals, that is, the so-called proof tests.

The following segments discuss the consequences of online diagnosis and periodic proof tests on the implementation of the safety function with regard to the PFD-value, which is a measure of the failure probability of a safety function on demand.[146]

16.6.1 Definition of Diagnostic and Proof Test

Diagnostic tests are used to identify errors in safety systems. These tests are performed during the operation of the safety system. Thereby it is not sufficient to detect errors, but decisions must be made on how to deal with an error. One of the typical decisions is the activation of the emergency shutdown, or the powering down into a safe state.

The proof test is defined both in IEC 61508 and IEC 61511. Proof tests are performed when the safety system is not operational. If, according to a proof test, failures are detected in the safety system, these failures are corrected so that the original state can be recreated completely or approximately. According to IEC 61508, a test is a diagnostic test if it proceeds automatically and in a lower magnitude than the demand rate. All other cases can be defined as proof tests.

---

[145] [IECf00]   IEC/EN 61508, part 6, appendix B2.5
[146] [VELT06]   Velten-Philipp, W.; Houtermans, M. J. M., *The Effect of Diagnostic and Periodic Testing on the Reliability of Safety Systems*

16.6 Differences between Diagnostic Test and Proof Test

According to certification authorities[147], a test is defined as a diagnostic test if it fulfills the following criteria:

- The test is performed automatically, without human intervention and in specific intervals.
- The purpose of the test is to detect errors which could disrupt the safety function.
- The system reacts automatically to the test results.

One typical example of a diagnostic test is a CPU-test.

A proof test, on the other hand, is performed manually, and external hardware/software is frequently required to determine the result. When compared with the requirements of a diagnostic test, the interval between two proof tests is considerably greater than the safety. period of the process. A typical example is the valve cap test which is rarely performed without human intervention and usually requires additional equipment.

The advantage of diagnostic tests is that errors can be identified and fixed quickly. With a proof test, on the other hand, it is possible that the system will run with an error that remains undetected until the next text. Then again, a good diagnosis also requires additional hardware/software which can be complicated and expensive.

A further aspect that distinguishes the two tests is that the IEC 61508 requires a calculation of the percentage of harmless errors for classifying a safety system. This parameter, *SFF*, or Safe Failure Fraction, is defined as follows:

$$SFF = \frac{\lambda_S + \lambda_{DD}}{\lambda_S + \lambda_{DD} + \lambda_{DU}} \tag{16.10}$$

A high *SFF* can only be guaranteed if many safe failures(λ_S) and many of the dangerous detectable failures(λ_{DD}) can be identified by a diagnostic test. Failures which are ferreted out during proof tests are not considered in the *SFF*.

16.6.2 Performance Indicators

So as to demonstrate the effects that altering the diagnostic and proof tests have on the PFD-value of a system, performance indicators are introduced. To determine the effect that a change in parameters has on the PFD-value, one refers to a value of 50%. In the case of diagnostic coverage, this value means that 50% of errors can be detected through diagnosis.

To get an overview of the consequences which different values have, the calculations are performed for 25% and 75%, respectively, followed by comparison of the values. It can thereby be determined whether a change in parameters will have a small or a large effect on the PFD-value.

---

[147] z. B. TÜV

16.6.3 Results of Calculations with or without Diagnosis

In this segment, the effects of diagnosis on the PFD-value of the different architectures are examined. For this purpose, the performance indicators of diagnostic coverage are regarded more closely, in particular its influence of the PFD-value. Diagnostic coverage varies from 0 %, 25 %, 50 %, 75 %, to 99 %. Figure 16.5 shows the influence of diagnostic coverage on the PFD-value for three different architectures. It is readily apparent that the influence of diagnostic coverage is greater for a 1oo2-system than for a 2oo3 or a 1oo1-system.

Figure 16.6 shows the change of the PFD-value for 25 % diagnostic coverage compared to 50 % diagnostic coverage, and the change for 75 % diagnostic coverage compared to 50 %.

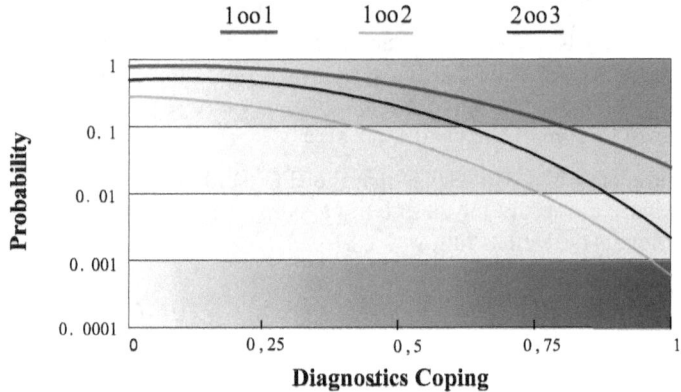

Figure 16.3: Test detection for different DC-factors

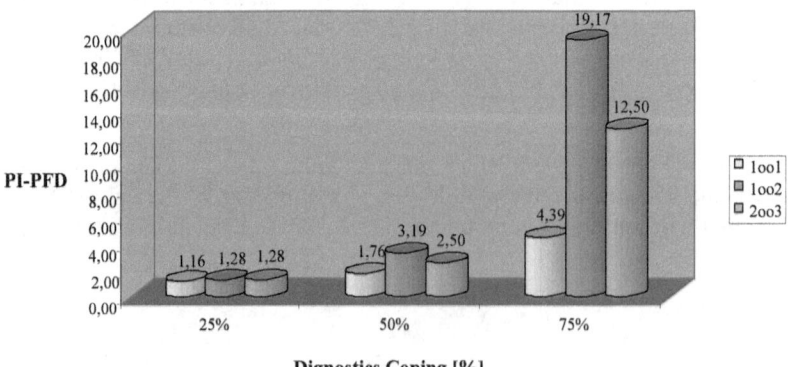

Figure 16.4: PFD-values for different DC-factors (25 %, 50 %, and 75 %)

16.6 Differences between Diagnostic Test and Proof Test

The following conclusions can be drawn from this result: diagnostic coverage factor has a significant influence on the PFD-value. The 1oo1-system is the least sensitive to modifications of diagnostic coverage factors.

The 1oo2-system, on the other hand, is very sensitive to modifications of the diagnostic coverage factors. The 1oo2- and 2oo3-systems are both systems that are tolerant of one error only. Both systems require two errors in two different strands in order to lose their safety function.

It pertains to all three architectures that better diagnostic procedures produce the greatest ramifications. Therefore all three architectures are highly sensitive with regard to diagnostic coverage factors. Thereby, redundant structures are albeit more sensitive than individual structures.

16.6.4 PFD-Calculation with Variable Proof Test Coverage

The influence of proof test coverage on the PFD_{avg} values of different architectures is shown below.

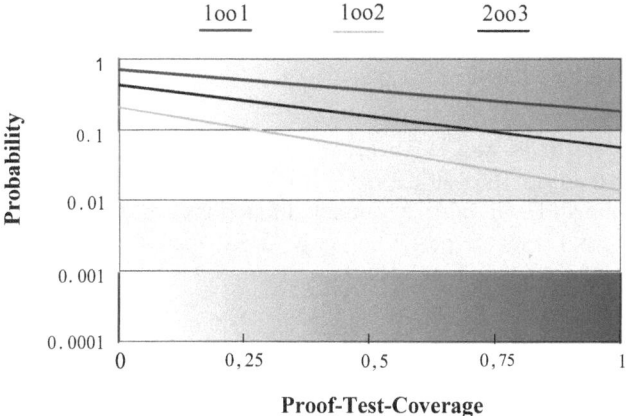

Figure 16.5: Proof test coverage for different architecture

The change in the PFD-value for 25 %, 50 % and 75 % proof test coverage, compared to 0 % proof test coverage is compared by means of the performance indicator.

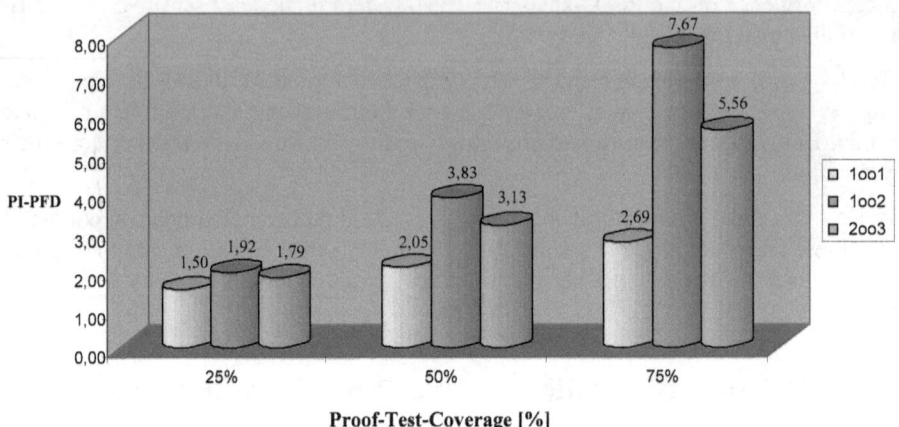

Figure 16.6: Proof test coverage for different proof test factors (25 %, 50 %, and 75 %)

A higher proof test coverage improves the PFD_{avg}-value, although not as significantly as an increase in the DC-value. If proof test coverage is increased from 0% to 75%, the improvement of the PFD-value is facilitated – depending on the architecture selected – by one order of magnitude.

The following is valid both for DC as well as for proof test coverage: the effects on redundant systems are greater than on 1oo1-structures. Error detection through diagnosis has a greater influence on the PFDavg-value because malfunctions can be recognized within seconds. In contrast, proof tests are performed only rarely, therefore system malfunctions can remain undetected for a long time.

16.7 Influence of Proof Test Interval on PFD_{avg}-Value

In the following segment, changes of the PFD_{avg}-value for a proof test interval of 5 years, respectively, 10 years, are compared with the PFD-value for a proof test interval of 1 year, by means of a performance indicator. As can be seen in Figure 16.9, the effect of a proof test coverage on the failure probability of a 1oo2-architecture is greater than for a 2oo3- and 1oo1-architecture. Furthermore, the decreasing slope of the curves shows that, the smaller the proof test interval, the higher the influence on lower PFD-valules. This is also apparent in Figure 16.10. In this figure, the change of the PFD_{avg}-value between the proof test interval of 1 and 5 years, in relation to the PFD-value of 5 years, respectively, between the proof test intervals of 5 and 10 years compared to the PFD-value at 10 years, is shown by means of the performance indicator. It is obvious that, for a 1oo1-system, a longer proof test interval has a greater influence on the PFD-value than for a redundant structure, while it must still be taken into account that the PFD-value of a 1oo2- architecture is better than for a 2oo3- or even a 1oo1-architecture.

16.7 Influence of Proof Test Interval on PFDavg-Value

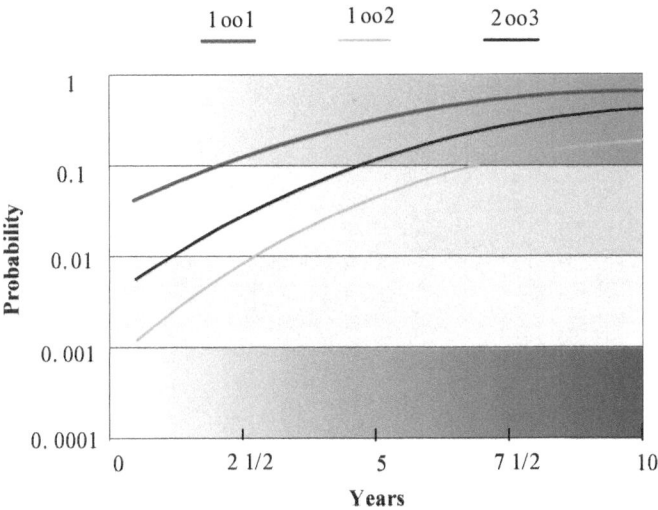

Figure 16.7: PFD-values for three different architectures (1oo1, 1oo2 und 2oo3)

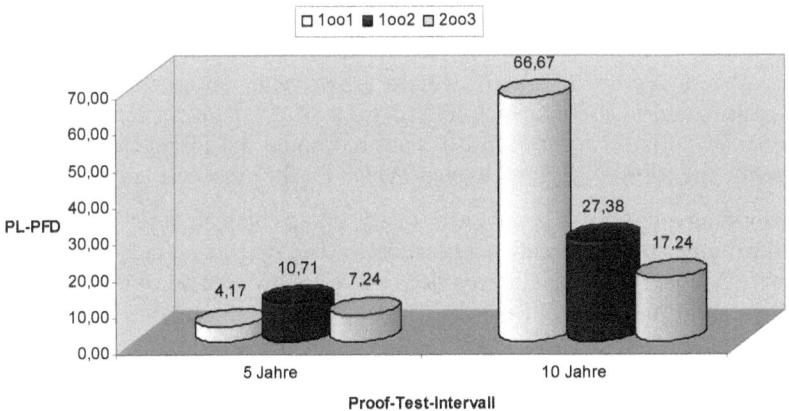

Figure 16.8: Changes of the PFD$_{avg}$-value for a proof test interval of 5 years, resp., 10 years compared with the PFD-value for a proof test interval of 1 year

In analyzing Figure 16.8, the strong influence of intervals on the PFD$_{avg}$-value can be seen, by a factor of 10 - 100 (for coverage of over 99%: factor 1000). Redundant architectures are influenced more strongly. A 1oo2-architecture derives the greatest use from frequent proof tests.

As for failure probability, there is only a minor difference between a 5- and a 10-year proof test interval. For this reason it usually makes more sense to perform proof tests more frequently than to try and have a proof test with high coverage available. A simple but frequently performed proof test decreases the PFD$_{avg}$-value very strongly.

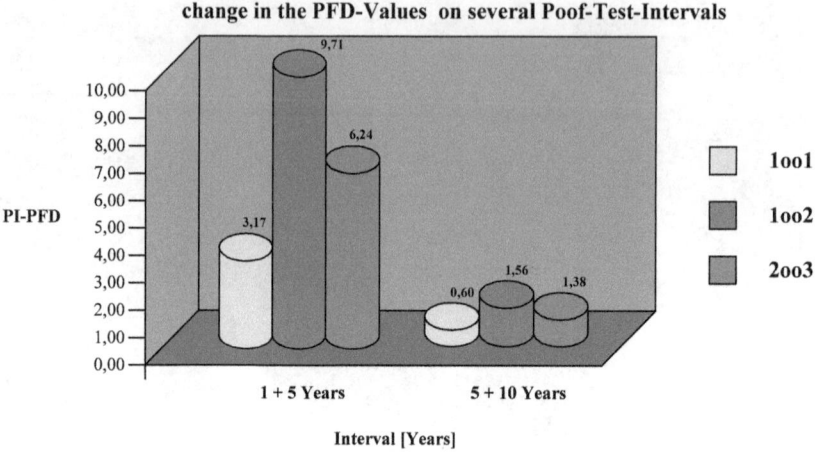

Figure 16.9: Changes of the PFD$_{avg}$-value for a proof test interval of 1 and 5 years, referring to the PFD-value at 5 years, respectively, between the proof test intervals of 5 and 10 years, relative to the PFD-value at 10 years.

16.8 Risk Reduction

A periodic (manual) test, a proof test, can decrease the demands on a safety technical function (SIF) with regard to the maximal desired value of dangerous errors.[148] Proof tests can lead to lower costs for a required SIF but one should not forget that the cost advantage is offset by the costs which need to be provided for the proof test itself.

Furthermore, performing a proof test is not always easy, and a complete repair following the occurrence of a malfunction is usually not practicable. The statement made by a number of experts, whereupon a system can be regarded as new following a proof test, can not be verified, and it is just this statement which is usually used as the foundation for the definition of SILs for a proof-tested SIF. Apart from performing the tests and the obvious repairs, there is always the possibility of additional dangerous (human) malfunctions. According to a publication by the HSE (Great Britain)[149] of a variety of processes, approximately 30% of all accidents are caused by faulty maintenance and faulty implementation of modifications. Once the decision is made to perform a proof test, the test frequency must be specified.

There are many solutions which facilitate the required risk reduction. One such solution is the use of a safety instrumented system or SIS. Generally, an SIS consists of a number of safety technical functions (SIF) each of which is implemented by an appropriate safety loop and its associated risk parameters. If we assume that a SIF is used for risk reduction, then the combination of demand and the occurrence of a dangerous malfunction of the SIF must be considered dangerous, that is, the probability of real danger.

---

[148] [TIEZ03] Tiezema, R., *Risk Reduction in the Process Industry, Proof testing*
[149] [HSEb87] UK Health and Safety Executive, *Programmable Electronic Systems in Safety-Related Applications*

16.8 Risk Reduction

That is, the average risk rate (H) depends on the average dangerous failure rate of the SIS (λ_D) and the average demand rate (d). Because they have independent sources, the relationship between these average parameters can be described as follows:

$$\frac{1}{H} = \frac{1}{\lambda_D} + \frac{1}{d} = \frac{\lambda_D + d}{\lambda_D \cdot d} \tag{16.11}$$

The formula shows that, in the absence of any safety system, the expression $1/\lambda_D$ does not exist, which means that $H = d$. In other words, if a SIS does not exist, then there is a latent danger whenever an abnormal operation-dependent state occurs in the process. The risk reduction factor *(RRF)* can be calculated by comparing the risk rate of the installation with a SIF (H) and without a SIS (H_1). It follows that:

$$RRF = \frac{H_1}{H} = \frac{d}{H} = \frac{\lambda_D + d}{\lambda_D} \tag{16.12}$$

16.8.1 Risk Rate and Average Failure Probability

A danger exists if a demand occurs during a period in which the SIF is not able to react safely. This is the average safety unavailability *(SU)*, respectively, "failure probability on low demand" *(PFD)*, according to IEC 61508. The average risk rate can then be determined as follows:

$$H = d \cdot SU = d \cdot PFD \tag{16.13}$$

In practice, this period of unavailability is longer than the maximum required life time. Assuming that failure probability after start is as big as failure probability just before the end of the lifetime, then the average safety unavailability (SU_L), respectively, failure probability PFD_L during a lifetime is as follows:

$$PFD_L = SU_L = \frac{1}{2} \cdot \lambda_D \cdot T_L \tag{16.14}$$

Next, the following definitions can be made:

$$H = d \cdot SU_L = d \cdot PFD_L = \frac{1}{2} d \cdot \lambda_d \cdot T_L \tag{16.15}$$

The risk reduction faktor:

$$RRF = \frac{d}{H} = \frac{1}{SU_L} = \frac{1}{PFD_L} \tag{16.16}$$

These two formulas can be applied to define the safety integrity level (SIL) for the applied SIF. Due to the known limitations

$$\lambda \cdot T \ll 1 \tag{16.17}$$

and

$$d \cdot T \ll 1 \tag{16.18}$$

the practical implementation is restricted to processes with very low demand rates. This can be explained in a simple example.

$T_L = 25$ years means: $d \ll 0{,}04$/year = 1 time in 25 years, that is, one demand on the safety function in 25 years. Figure 16.14 shows the characteristic line of a perfect proof test with $PFD_{avg} = SU_{L,\ avg.}$, according to equation 16.14:

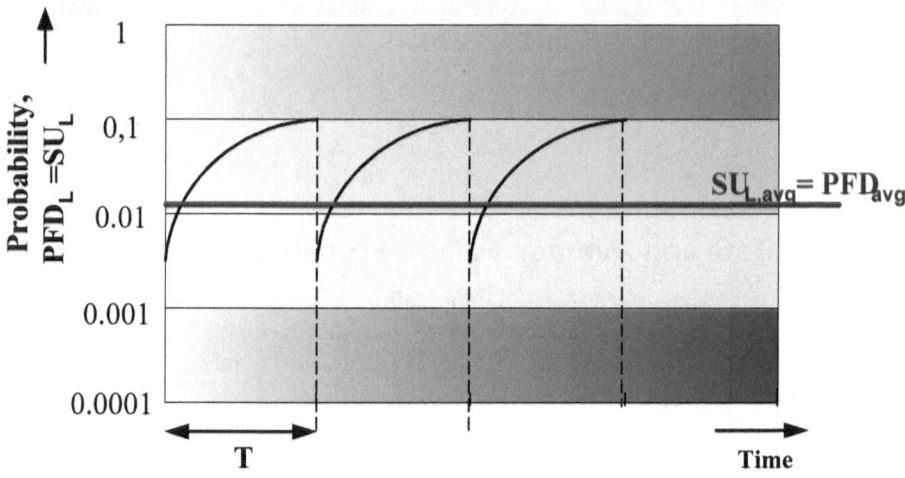

Figure 16.10: Average $PFD_L = SU_L$ for perfect proof testing

In industry, there are not many processes with such low demand rates. Therefore a different approximation must be found in order to determine a SIL for a SIF. The solution is in observing the average failure probabilities over a shorter period of time. This approach is used for the so-called "proof tested" SIF. The *RRF* for a proof-tested SIF is defined as $1/PFD$.

16.8.2 Proof Test Frequency

A proof test only makes sense if an improvement in risk reduction can be proven. In other words, when

$$\frac{H_t}{H} < 1 \tag{16.19}$$

is, where H is the risk rate that is specified by the application of an untested SIF, and H_t is the rate of a SIF which is periodically tested at a certain time point T. The definition of H_t makes it necessary to define the average *PFD* of the SIF over the test interval T. This can be calculated in the same way and with the same restrictions as for the calculation of PFD_L

16.8 Risk Reduction

$$PFD = \frac{1}{2}\lambda_D \cdot T \qquad (16.20)$$

$$H_t = d \cdot PFD = \frac{1}{2} d \cdot \lambda_D \cdot T_L \qquad (16.21)$$

The risk rate of a non-proof-tested SIF can be described as follows:

$$H = \frac{d \cdot \lambda_D}{d + \lambda_D} \qquad (16.22)$$

Let's return to the statement that testing only makes sense if

$$\frac{H_t}{H} < 1 \qquad (16.23)$$

By using this formula for H_t und H, the statement can be rewritten as follows

$$\frac{1/2 \cdot d \cdot \lambda_D \cdot T}{d \cdot \lambda_D / (d + \lambda_D)} < 1. \qquad (16.24)$$

From this follows that

$$\frac{1}{2}(d + \lambda_D) \cdot T < 1 \quad \text{oder} \quad T < \frac{2}{(d + \lambda_D)}. \qquad (16.25)$$

In practice, one can assume that $\lambda_D \ll d$.

Practical examples are λ_D = 1.4E-06 1/h, d = 1 time in 5 years = 2.3E-05 1/h. Consequently the formula can be simplified as follows

$$T < \frac{2}{d} \qquad (16.26)$$

The formula $T < 2/d$ provides the maximal acceptable test interval to ascertain an improved risk reduction by means of a proof test, with regard to the demand rate. For a safety management in practice values are obtained that lie at around $T <=$ 10 years.

What does this mean in practice? To answer this question, let's assume a process that has a demand rate of 2 times per year. This would mean that T would need to be a little smaller than 1 year, that is, a test interval of just a few months. If one examines the costs involved, the test frequency would not be an acceptable situation, especially because most proof tests have to be performed while the process is off-line.

In this respect it was assumed that proof tests were introduced for the purpose of reducing dangerous errors and of achieving the required PFD_{avg}. Which alternative is offered if the goal is not achieved? By using the original formula

$$PFD_{avg} = \frac{1}{2} \cdot \lambda_D \cdot T \qquad (16.27)$$

it appears as though the answer to the question on how to reduce PFD_{avg} can only be found by reducing T, that is, a higher test frequency.

16.8.3 Proof Test Expansion Factor

In general, undiscovered hazardous failures harbor a potential risk. The last step of the proof test is the reduction of the rate of dangerous failures (λ_D) of a SIF. This is based on the assumption that already identified hazardous failures are safe. The reduction is defined through the Coverage Factor, CF. The result of this formula is $\lambda_{DU,\,PT}$, the remaining, undetectable, dangerous error rate after such a test.

$$\lambda_{DU,PT} = (1-CF)\lambda_D \qquad (16.28)$$

It is assumed that the remaining undetected failures are based on the fact that not all parts of a component are tested for everything, but that the remaining parts were tested correctly. As shown in Figure 16.15, PFD_{LT} can eventually be calculated by using the approximated formula below.

$$PFD_{LT} = PFD_{avg} + PFD_{DU,PT} \qquad (16.29)$$

But: $PFD_{DU,\,PT}$ needs to be tested according to the requirements with regard to T_L. The formula is only valid for the case of:

$$PFD_{avg} \gg PFD_{DU,PT} \qquad (16.30)$$

This inequality can be rewritten as:

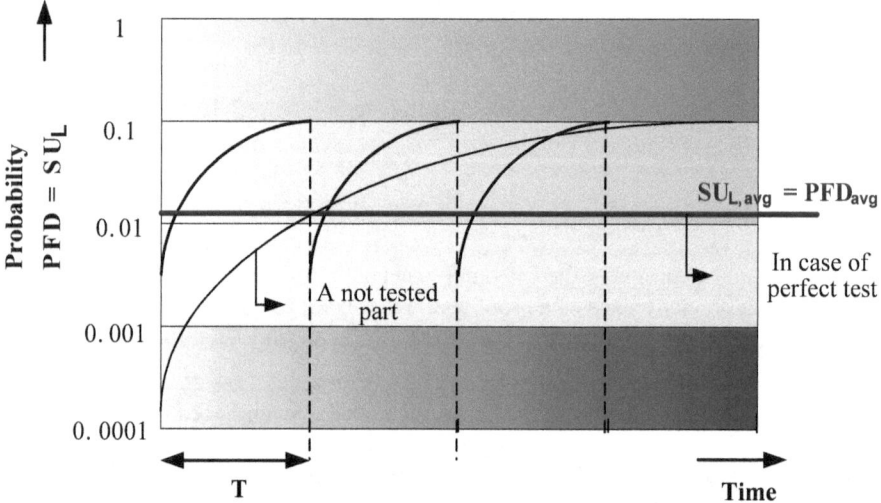

Figure 16.11: Correlation between the average $PFD_{L,avg} = SU_L$ for perfect proof tests and $PFD = SU$

16.8 Risk Reduction

$$\frac{1}{2} \cdot \lambda_D \cdot CF \cdot T \gg \frac{1}{2} \cdot \lambda_{DU,PT} \cdot T_L$$

$$\frac{1}{2} \cdot \lambda_D \cdot CF \cdot T \gg \frac{1}{2} \cdot \lambda_D \cdot (1-CF) \cdot T_L$$

$$CF \cdot T \gg (1-CF) \cdot T_L \qquad (16.31)$$

For example if $T = 1$ year and $T_L = 20$ years, then $CF \gg 0.95$. This result is not very promising because it means that the required CF-value comes very close to the ideal CF-value, which in practice is unrealistic.

Some users therefore apply an alternative approximation by introducing a complete inspection/repair/exchange-interval such as, for example, 5 years, assuming that the system can subsequently be regarded "as new". In other words, this can be regarded as a proof test with a high CF.

If one then applies equation 16.31 for $T = 1$ year and $T_L = 5$ years, the result is $CF \gg 0.83$. This is generally a very high coverage and can be proven for a well defined SIF. As mentioned above, a further improvement of $PFD_{DU,PT}$ by increasing the frequency of tests is hardly feasible since it again requires a higher CF.

17 Hardware of Safety-Related Systems

In safety-related systems, or in systems with functional safety requirements, hardware can generally be separated into the following groups:

- Electric (E)
- Electronic systems (E)
- Programmable electronic systems (PES)

All three groups are implemented nowadays in safety-related systems. In part, for certain fields of application or implementations the one or the other group is either required or excluded.

The same norms and standards apply to all E/E/PES, that are implemented in safety-critical applications.

17.1 Normative Architectural Specifications

Next to national laws and regulations, numerous norms and standards, that partially are legal in character, were developed for electronic installations world wide. The tasks and functions of norms are comparable to those of corporative norms. They are primarily designed to ensure a high degree of quality in safety with regard to the functionality of machines for the user. Furthermore, they should provide the distributors of electronic systems with planning and implementation safety regarding the procedure during development and manufacture of the installations.

17.1.1 Quality in Safety for Users of Safety-Critical Systems

According to the norms and standards, operators of safety-critical systems are required to develop their installations and machines according to the highest technological level. Therein, all aspects regarding new techniques, which are of importance in the manufacture and design of machines, must be described in detail in the individual norms and rulebooks. Through strict adherence to standards it can be ensured that the actual state of the art is achieved. Thereby, not only can the manufacturers of safety-critical installations fulfill their obligation to exercise diligence, but also, in the event of damage, they can refer to compliance with norms and standards. Figure 17.1 summarizes the relevant tasks and functions of norms and guidelines.

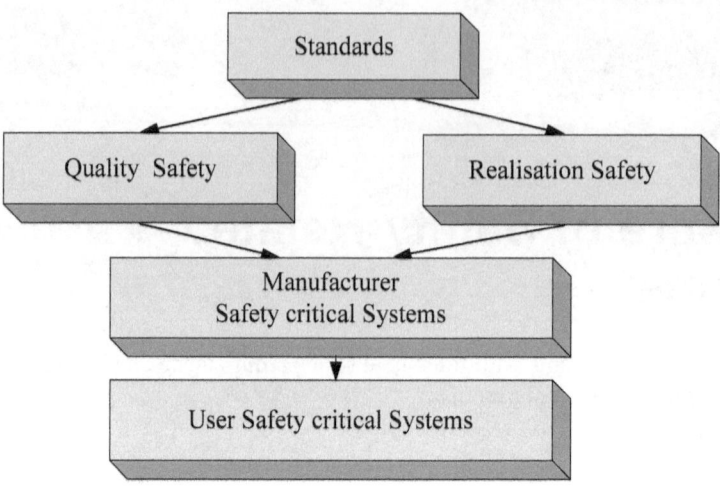

Figure 17.1: Tasks of norms and guidelines

17.1.2 Implementing Safety for Manufacturers of Safety-Critical Systems

Norms create the framework for a nationally and internationally binding procedure in specification, design and implementation of safety-critical installations. In them, rules and demands on the installations and machines are defined. The supplier has to prove, according to the respective applicable norms, that the end products meet the stipulated requirements. This proof is highly relevant not only from a legal but also from an economic point of view. Due to the immaterial and thereby also more complicated ex-ante evaluation possibility of the service, customers will often tend to place an order based on reputation. With this in mind, a damage event, that is demonstrably traced to non-compliance with relevant norms, can have considerable financial consequences for the manufacturer.

Thus, operators of safety-critical installations are highly interested, above and beyond norms and regulations, to take internal steps for ensuring functional safety.

Norms are developed in cooperation with a variety of stakeholders from economy, authorities and administration. The various representatives have a differentiated point of view which is also expressed in their norm work. It is therefore important to reflect all conceivable aspects and points of view of the members of the commission in the corresponding norm content. Before a norm can find national and international recognition though, the following principles must be fulfilled:

- The normwork should orient by the interests and the well-being of the public. The general acceptance of norms is the measure for achieving this goal.

- It must be possible to put norm contents into practice, that is, it must be possible to transfer and implement a norm in the daily routine of a business. This requirement is taken into account by the collaboration of various experts from economy, business, and diverse associations.

17.2 Hardware Safety Life Cycle

The hardware safety life cycle according to IEC/EN 61508, which must be applied to E/E/PE systems, is shown in Figure 17.2. It consists of six different implementation phases, in which different actions must be performed. Specification of the E/E/PES safety requirements occurs in phase 1. Herein, the necessary safety functions and the required safety integrity are stipulated in order to achieve the required functional safety. Phase 2 consists of planning the validation of the safety-related E/E/PE systems with regard to safety. The design of the safety-critical E/E/PE systems follows in phase 3, in due consideration of the requirements on the safety functions and the safety integrity. The integration of the developed safety-related E/E/PE-subsystems into the entire safety system occurs in phase 4. At this, test must be performed to prove that the safety-related E/E/PE system functions completely in compliance with the E/E/PES design. With the completion (implementation?) of phase 5 it is ascertained that procedures are developed so that functional safety is maintained during operation and during necesssary maintenance. Validation of the safety-related E/E/PE-system with respect to safety occurs is phase 6, in terms of safety of the required safety functions and safety integrity.

The individual phases are – with the exception of the specification of safety requirements in phase 1 – performed in the different developing departments/areas.

17.2.1 Safety Requirements Specification

In order to develop a safety-related E/E/PE-system, precise specifications of safety requirements must be available. The specification for the safety functions of the E/E/PES must contain accurate statements concerning how the E/E/PES should achieve and maintain the required safety. Therein, each safety function must contain the following descriptions:

- extensive, detailed requirements that are sufficient for the design and the development of the safety-related E/E/PE-system,

- the intended manner of how the safety-related E/E/PE-systems can achieve or maintain a safe state of the EUC (equipment under control, according to IEC/EN 61508-4: device, machine, apparatus or installation, used for the manufacture, matter conversion, transport, for medical or other applications),

- stipulation (predetermination?) whether, and for how long, continuous control (management) or regulation is required to achieve or maintain a safe state of the EUC, and

- stipulation whether the safety function is applicable to safety-related E/E/PE-systems which are operated in low demand mode or in high, respectively, continuous demand mode.

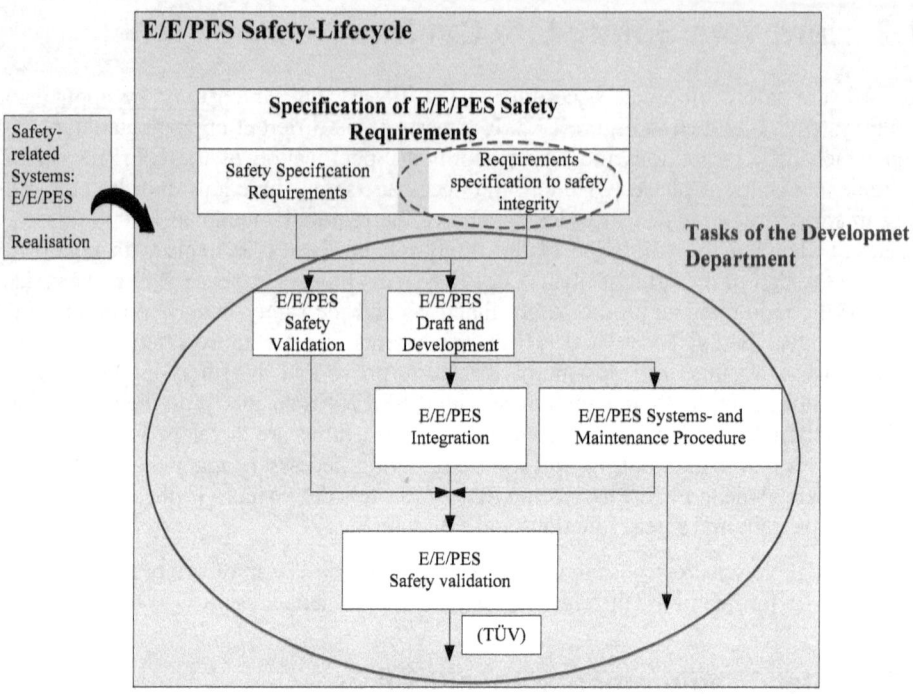

Figure 17.2: Hardware safety lifecycle

All relevant operating modes must be considered, such as:

- preparations for application, including settings and alignment,
- power-up phase, learning phase, automatic operation, manual operation, semi-automatic operation, stable operating phase,
- stable state without operation, resetting, powering-down, maintenance,
- reasonably predictable abnormal conditions

For certain operating modes (for example setup, alignment, or maintenance) additonal safety functions may be required to performe these actions safely. In addition, a specification for safety integrity must be produced, which contains statements concerning the safety integrity level (SIL) for each safety-relevant function. The intended electromagnetic compatibility must also be documented in the specification (the electromagnetic interference immunity threshold values are defined in IEC 61000-1-1, influences of an electromagnetic environment are described in IEC 61000-2-5).

17.2.2 Safety Validation Planning

All test procedure requirements with respect to test execution, test environment and criteria for existence/non-existence must be taken into account during planning. This is the only way in which a correct validation of all safety-relevant functions can be ensured.

17.2.3 Design and Development of the E/E/PES

The entire design of the E/E/PES must be in accordance with the documented safety specifications (Safety Requirement Specifications, SRS). The design and implementation of safety-related systems must fulfill the specified demands on the safety functions and the safety integrity. System design encompasses the entire hardware and software architecture, sensors, actuators, programmable electronics, embedded software, application software, etc.

The following general requirements must be fulfilled during the design of safety-related systems:

- requirements on hardware safety integrity, consisting of:
  - limitations of hardware safety integrity due to the architecture and
  - the requirements on the probability of dangerous, accidental hardware failures,
- requirements on systematic safety integrity, consisting of:
  - the requirements for avoiding failures and the requirements for controlling systematic failures, or
  - the proof that the production facilities are operationally well-tried;
- the requirements on system behavior upon detection of an failure.

The architectural requirements of hardware and software must be guided by the required SIL. The possible SIL is limited by hardware fault tolerance and the percentage of harmless failures.

17.3 Hardware Fault Tolerance

In the context of hardware safety integrity, the highest safety integrity level, which can be achieved for a safety function, is limited[150] by the hardware fault tolerance and the percentage of safe failures of subsystems which perform the safety function. Table 17.1 specifies the highest safety integrity level that can be exercised by a safety function which utilizes a subsystem. In doing so, hardware fault tolerance and the percentage of safe failures of this subsystem are examined. The requirements listed in Table 17.1 must be applied to every subsystem that executes a safety function, and therefore to each part of the safety-related E/E/PE-system.

---

[150] [IECb01]   IEC 61508-2

Table 17.1: Safety integrity of hardware: Limitations due to architecture and type of component used

Safe failure fraction (SFF)	Type A Hardware fault tolerance			Type B Hardware fault tolerance		
	0 Failures	1 Failure	2 Failures	0 Failures	1 Failure	2 Failures
< 60%	SIL 1	SIL 2	SIL 3	Not allowed	SIL 1	SIL 2
60% - <90%	SIL 2	SIL 3	SIL 4	SIL 1	SIL 2	SIL 3
90% - <99%	SIL 3	SIL 4	SIL 4	SIL 2	SIL 3	SIL 4
> 99%	SIL 3	SIL 4	SIL 4	SIL 3	SIL 4	SIL 4

Architectural constraints were incorporated into the IEC/EN 61508-2, in order to achieve a robust architecture allowing for the level of complexity of the subsystems. The hardware safety integrity level for the safety-related E/E/PE-system, derived from the implementation of these requirements, is the highest one to be used. This also applies to those instances where theoretically a higher safety integrity level can be determined, if a purely mathematical definition for the safety-related E/E/PE-system had been applied:

A subsystem can be classified as type A if the following conditions apply for the subunits that are required for achieving the safety function:

- The failure behavior of all covered components is known **and**
- the failure behavior is completely predictable **and**
- reliable failure data are available through field tests.

A subsystem must be classified as type B, if the following conditions apply for the subunits that are required for achieving the safety function:

- The failure behavior of at least one covered component is not sufficiently known **or**
- the failure behavior of the subsystem is not completely predictable **or**
- there are no reliable failure data from field tests.

If at least one component of the subsystem fulfills the conditions for a type B subsystem, then this subsystem must be considered as type B.

A 1oo1-system has a hardware fault tolerance of 0, i.e. if a system failure occurs, it can no longer perform its safety function on demand.

1oo2- and 2oo3-systems have a hardware fault tolerance of 1. If a failure occurs in systems with a 1oo2- or 2oo3-architecture, these systems can continue to fulfill their safety function. Only if another failure occurs in a second conduit, these systems can no longer control the failure, and they can no longer execute the requirements of the safety function correctly.

A 1oo3- and a 2oo4-system are 2-fault-tolerant. If two failures occur in these systems, they can still perform their safety function.

17.4 Constraints

17.4.1 Architectural Constraints

In safety-related E/E/PE-systems in which a safety function is executed by a single conduit (see, for example, Figure 17.3), the maximal hardware safety integrity level, which can be applied for the safety function under observation, must be determined by the subsystem which fulfills the lowest hardware safety integrity requirements (as determined according to Table 17.1).

Figure 17.3 demonstrates an architecture in which a specific safety function is performed by a single conduit through the subsystems (TS) 1 and 2. The subsystems meet the requirements of Table 17.1 as follows:

- TS 1 achieves a hardware fault tolerance of SIL1, with a specific percentage of safe failures;
- TS 2 achieves a hardware fault tolerance of SIL2, with a specific percentage of safe failures.

The maximum SIL of the complete system is determined by the subsystem with the lowest hardware integrity requirement.

Therefore, the TS 1 limits the hardware safety integrity level that can be utilized, in view of the hardware fault tolerance for the safety function observed, to SIL1.

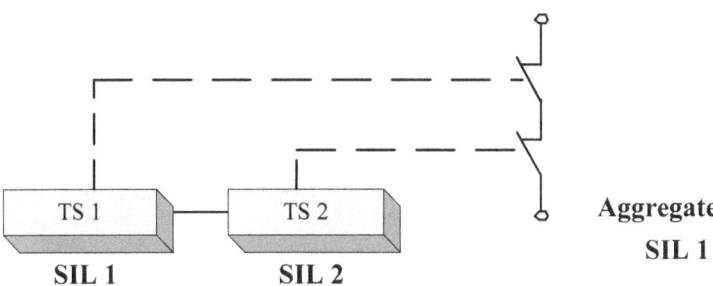

Figure 17.3: Architectural constraints for a 1oo1-system

Furthermore it must be considered that:

- If at least one of the components of the subsystem fulfills the conditions for a type B-subsystem, then the entire subsystem is considered type B.
- No other measures, which may control the consequences of failures, for example diagnostic devices, may be considered for determining hardware fault tolerance. For a 1oo1-system, the hardware fault tolerance equals zero failures.

17.4.2 General Concepts of Risk Reduction

Especially in complex systems, failures can only be fought effectively if design, development and maintenance are performed in a structured manner. The primary goal is to avoid failures in the first place. That type of strategy is implemented by a whole series of constructive, analytical and test methods in the course of the safety lifecycle. The phases of the safety lifecycle are described in norm IEC/EN 61508, through formulating essential requirements for each particular phase. Parts 2 and 3 of the IEC/EN 61508 contain details of the implementation phase of electric, electronic and programmable electronic (PES) systems. Therein, each phase is assigned concrete measures for failure avoidance, in order to implement the complex safety installation. These measures are contained in the appropriate supplements A and B of parts 2 and 3, where they are evaluated with respect to their efficiency, and they are eventually described in detail in part 7. The fundamental idea of a safety lifecycle is based on the fact that, especially with complex systems, safety can only be ensured in parallel with development over the entire system lifecycle. German testing agencies, such as the Technische Überwachungsverein (TÜV), have been following this path for a while in certifying computer-controlled systems. Certification by this type of testing agency starts with the so-called development-accompanying test in the customer requirement specifications phase, that is, in a very early phase. The test accompanies system development and operation, as well as system modification and maintenance.

The safety-related reliability of complex safety systems can only be ensured if constructive, analytical and test methods are combined. Methods vary in their complexity, depending on the minimization of risk to be warranted by the safety function.

Figure 17.4 shows the concept of risk minimization by technical appliances. Safety is considered to be achieved if, through any type of measure, whether technical or non-technical, the risk of a specific hazard to a machine or an installation has been reduced to an acceptable residual risk. When this is the case, then this machine or installation is considered safe. However, the question arises as to how the acceptable residual risk is defined.

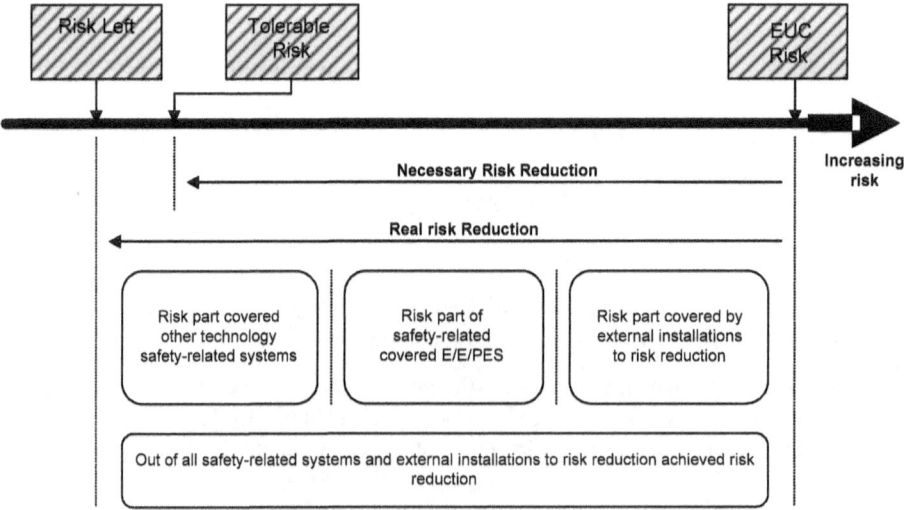

Figure 17.4: Risk reduction

17.4 Constraints

In the Federal Republic of Germany, the opinion has caught on that a generally tolerable residual risk can not be absolutely defined, merely that the necessary reduction of risks caused by dangers to technical installations can be achieved by drawing conclusion by analogy from the past. This implies that to reduce risk for an installation or a machine with cyclic access, during which, for example, irreversible injuries such as the loss of a hand and the like can occur, a very considerable effort must be expended towards failure avoidance and failure control.

A system can be considered unsafe if the system developers have neglected all safety considerations during development. The process of making a system safe can be considered risk management or risk reduction management. If, on the other hand, the risk to the system is so low that it is negligible, then no measures for risk reduction are required. If this is not the case though, the risk must be reduced to the point of becoming acceptable. This procedure is shown in Figure 17.4.

The measure of risk reduction necessary is defined by the distinctive safety requirements of various projects. For projects in which high-risk processes run, a high level of risk reduction is essential to achieve an acceptable risk. In order to achieve the high level of risk reduction, the risk management mechanisms applied must be highly reliable. For projects which demand a lower level of risk reduction, the safety system requirements are not so strict. The differing requirements of safety systems have led to the concept of integration levels for safety-critical systems. In this context, the term integrity is used in conjunction with safety integrity which can be defined as follows (from IEC/EN 61508-4):

"[Safety integrity is the] probability that a safety-related system fulfills the required safety functions under all specified conditions within a specified period of time according to requirements."

The basic idea is to assign a safety-critical system to a specifc level of safety integrity. This assignment can then be performed quantitatively or qualitatively.

The IEC/EN 61508 issues the following explanatory notes concerning the term safety integrity:

- The higher the safety integrity level of the safety-related systems, the lower the probability that they are incapable of performing their required safety functions.
- There are four levels of system safety intgegrity.
- For the determination of safety integrity, all failure causes (accidental hardware failures as well as systematic failures), which can lead to an unsafe state, must be considered, for example hardware failures, software failures, and failures caused by electric disturbances. Some of these failure types, in particular accidental hardware failures, can be quantified, for example by applying the failure rate of dangerous failures or the failure probability of a safety-related protective system on demand. Albeit, system integrity depends on many factors which can not be quantified accurately but must rather be examined qualitatively.
- Safety integrity encompasses hardware safety integrity and systematic safety integrity.

17.5 1oo1 System

The "single controller systems" with a CPU and a I/O block (Figure 17.5) represent a minimal system. In this system there is no fault tolerance. The circuits can be in safe state or dangerous state. Failures can be detected or undetected.[151]

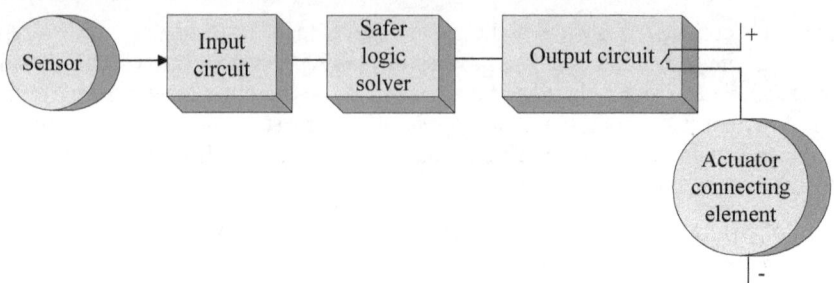

Figure 17.5: Block diagram of a single board controller

17.5.1 *PFD*-Fault Tree in 1oo1-Architecture

Figure 17.7 shows the fault tree of a 1oo1-system with dangerous failures. The system fails dangerously if dangerous detected (DD) or undetected (DU) failures are present. The reliability of a 1oo1-system is calculated with the following equation:

$$R(t) = e^{-\lambda \cdot t} \tag{17.1}$$

with

$$\lambda = \lambda_d + \lambda_s = const . \tag{17.2}$$

Herein, λ is the basic failure rate of the system, which is constant compared to time because we assume an exponential distribution of failures. For the calculation of the PFD_{avg}-value, only the failure rate for dangerous failures λ_D is required.

The failure probability of the system is calculated by

$$P(t) = 1 - R(t) . \tag{17.3}$$

If one applies equations 17.1 and 17.2 $\lambda = \lambda_D$ in equation 17.3, the following equation results for the 1oo1-system

$$P(t) = 1 - e^{-\lambda_D \cdot t} . \tag{17.4}$$

This equation is entered into the universally valid PFD_{avg}-equation

$$PFD_{avg}(T) = \frac{1}{T} \cdot \int_0^T P(t) \cdot dt \tag{17.5}$$

---

[151] [IECa] IEC 61508; *International Standard 61508 Functional Safety; Safety-Related Systems*

17.5 1oo1 System

The following solution[152] is obtained for a 1oo1-system

$$PFD_{avg}(T) = 1 + \frac{e^{-\lambda_D \cdot T} - 1}{\lambda_D \cdot T}.$$ (17.6)

This function can be developed to the power of T in place of $T = 0$ into a MacLaurin progression[153] For calculation of the *PFD*-value, the first three terms and the remainder R_3 are sufficient for the 1oo1-system. The remainder R_3 converges at point $T = 0$ towards a value of 0.

$$e^{-\lambda_D \cdot T} = 1 - \lambda_D \cdot T + \frac{\lambda_D^2 \cdot T^2}{2!} + R_3$$ (17.7)

If this equation is applied to a 1oo1-system, the result for the *PFD*-formula[154] is:

$$PFD_G(T) = \frac{\lambda_D \cdot T}{2}.$$ (17.8)

With

$$\frac{T}{2} = t_{CE} = \frac{\lambda_{DU}}{\lambda_D}\left(\frac{T_1}{2} + MTTR\right) + \frac{\lambda_{DD}}{\lambda_D} MTTR$$ (17.9)

and

$$\lambda_D = \lambda_{DD} + \lambda_{DU} \quad {}^{155}$$ (17.10)

the *PFD*$_{avg}$-equation for the 1oo1-system becomes:

$$PFD_{avg,1oo1} = \lambda_D \cdot t_{CE}.$$ (17.11)

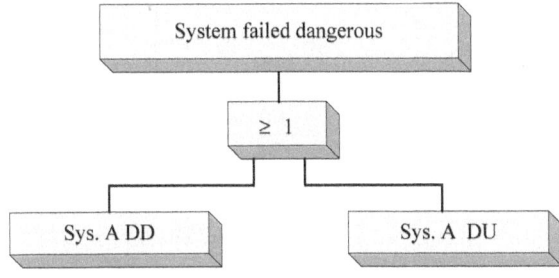

Figure 17.6: Fault tree of 1oo1-architecture for dangerous failures

---

[152] Derivation see [BÖRC04]
[153] Derivation see [BÖRC04]
[154] Derivation see [BÖRC04]
[155] The following variables represent:

λ_D	= dangerous failure rate	λ_{DD}	= dangerous detected failure rate
λ_{DU}	= dangerous undetected failure rate	λ_S	= safe failure rate ($\lambda_S = \lambda_{SD} + \lambda_{SU}$)
λ_{SD}	= safe detected failure rate	λ_{SU}	= safe undetected failure rate
t_{CE}	= channel equivalent mean down time		
T_1	= time interval	$MTTR$	= mean time to repair

17.5.2 Markov Model of 1oo1-Architecture

The 1oo1-architecture can be described by the Markov model, see Figure 17.7. In this context, the Markov model assumes that the status 0 represents the failure-free state. Starting from this state, the controller can assume three more states. Status 1 represents a safe state which the system achieves following safe detected (SD) or safe undetected (SU) failures. The transition rate is $\lambda_S = \lambda_{SD} + \lambda_{SU}$. The system enters status 2 if a dangerous detected failure (DD) occurs. Here, the transition rate is λ_{DD}. The system is in a dangerous state that is recognized through failure diagnosis. In status 3, the system has a dangerous, undetected failure (DU) which is not recognized by self-diagnosis. The transition rate is λ_{DU}. There are several possibilities for the system to achieve the failure-free status 0, from status 1, 2, or 3:

- The transition rate μ_R from status 1 into status 0 is $\mu_R = 1/\tau_{repair}$. This shows how much time is needed for repair and subsequent start-up of the system.

- The transition rate from status 2 is ($\mu_0 = 1/\tau_{Test}$). A 1oo1-system can not reach the safe status 1 at the end of the test interval time (τ_{Test}) because this system does not have any fault tolerance.

- When the system is in status 3, it can only return to status 0 at the end of its lifetime and subsequent exchange or repair. The transition rate in this case is $\mu_{LT} = 1/\tau_{LT}$.

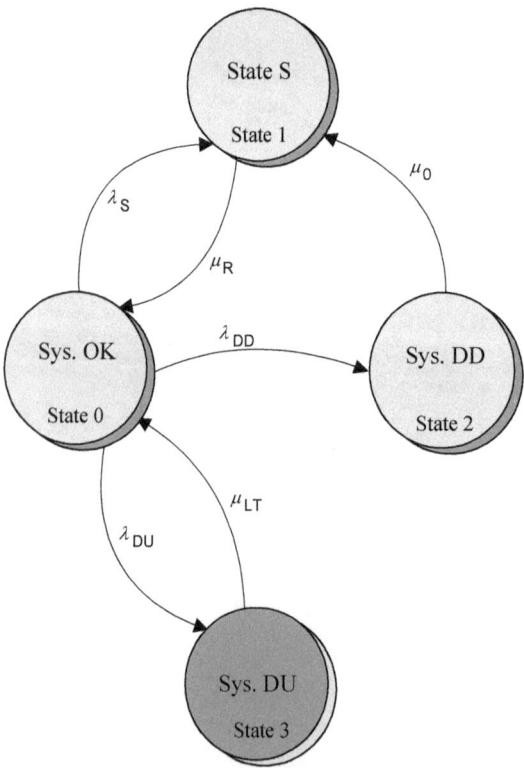

Figure 17.7: Markov model of 1oo1-architecture

17.5.3 Calculation of the MTTF-Value of a 1oo1-Architecture

The "transition matrix" **P** for a 1oo1-architecture is:

$$\mathbf{P} = \begin{bmatrix} 1-(\lambda_S + \lambda_D)\cdot dt & \lambda_S \cdot dt & \lambda_{DD} \cdot dt & \lambda_{DU} \cdot dt \\ \mu_R \cdot dt & 1-\mu_R \cdot dt & 0 & 0 \\ \mu_0 \cdot dt & 0 & 1-\mu_0 \cdot dt & 0 \\ \mu_{LT} \cdot dt & 0 & 0 & 1-\mu_{LT} \cdot dt \end{bmatrix}. \quad (17.12)$$

The **P**-matrix is formed as follows:

The elements of the **P**-matrix are probability densities from the different transition rates which were calculated in a Markov model. The individual elements P_{ij} of the **P**-matrix have the following significance: The first index shows the state from which the transition occurs. The second index shows the state to which the transition leads.

There are three different states in each system architecture:

- State ok if the system is fully functional
- Safe state S in which the system only has safe failures
- State DU in which the system only has dangerous unknown failures.

The last two states are also classified as absorbing states. An absorbing state is characterized by the fact that there is no failure-rate transition from this state into a different transition, but only transitions of the following type

$$\mu_R = \frac{1}{\tau_{repair}} \quad (17.13)$$

$$\mu_{LT} = \frac{1}{\tau_{LT}} \quad (17.14)$$

$$\mu_0 = \frac{1}{\tau_{Test}}. \quad (17.15)$$

The starting point of the following considerations is that each state has a probability density of 1. For probability, this means that a safe event exists. Since transitions occur from a status x into other status y_1, y_2, y_3 etc., one has to subtract each transition rate that describes state x from the reliability density of 1. The probability densities that describe the probability of occurrence of a status are in the main diagonal of the **P**-matrix. The other elements of the **P**-matrix describe the individual transitions from one status to another.

Since the first index in a **P**-matrix identifies the status from which the transition occurs, the elements of a line describe the different transition possibilities that the status x has. The elements of a column of the matrix show the transition rates which lead from the other states to status x.

In order to form the *MTTF*-value of a system based on a Markov model, or its corresponding **P**-matrix, respectively, several intermediate steps must be performed. First, the

Q-matrix is extracted from the **P**-matrix. The elements of the **Q**-matrix are formed by the probability densities in which the corresponding states fulfill the following criteria:

- system operational
- no absorbing state.

The following states are thereby excluded, among others:

- Safe status S in which the system has safe failures.
- Status DU in which all conduits have a dangerous undetected failure.

Since in a 1oo1-system only the 0 status ensures system functionality, and since all other possible states are absorbing, the **Q**-matrix of the 1oo1-system consists of only one term (the **Q**-matrix could also be called reliability matrix). It is formed by the element P_{11} of the transition matrix **P** and is called

$$\mathbf{Q} = 1 - (\lambda_S + \lambda_D) \cdot dt . \tag{17.16}$$

Once the **Q**-matrix is established it must be subtracted from the unity matrix **I**. The purpose of the subtraction becomes clear upon close scrutiny of the 1-dimensional **M**-matrix of the 1oo1-system. In order to determine the failure probability of a status, the safe event of a status must be excluded.

Therefore, the term for a 1oo1-system is:

$$\mathbf{I} - \mathbf{Q} = 1 - [1 - (\lambda_S + \lambda_D) \cdot dt] = (\lambda_S + \lambda_D) \cdot dt = \mathbf{M} \cdot dt . \tag{17.17}$$

Finally, the so-called **N**-matrix of a system must be defined in order to determine the *MTTF*-value of a system:

$$\mathbf{N} = \mathbf{M}^{-1} . \tag{17.18}$$

The **N**-matrix is the inverse matrix of the **M**-matrix. In a 1oo1-system, the **N**-matrix consists of one element, since the **M**-matrix itself only consists of one element. The **N**-matrix is calculated as:

$$\mathbf{N} = \frac{1}{\lambda_S + \lambda_D} . \tag{17.19}$$

The *MTTF*-value describes the mean time that passes between the occurrence of two failures. Herein, one proceeds from status 0, that is, the status in which the system operates failure-free. In this case, the reciprocal of the **M**-value is formed upon inversion, and therefore the failure rate becomes a time-dependent measure. This time-dependent measure describes the time that passes between two failures. To calculate the *MTTF*-value for the system one has to add the elements of the first line of the **N**-matrix. Therefore, the *MTTF*-value is

$$MTTF_{1oo1} = \frac{1}{\lambda_S + \lambda_D} . \tag{17.20}$$

17.6 Additional Architectures

The following segment describes a summary of results for frequently used architectures. A derivation of results is described, for example, in the book "Electronic Safety System"[156].

1oo2-System

The 1oo2-system is a safety system which consists of two independent conduits. For safety functions to be performed correctly, the conduits are connected in such a way that one of them is sufficient for triggering the safety function. One example of a 1oo2-system is the serial connection of two output circuits. In this arrangement, both systems must be flawed simultaneously, to achieve a dangerous state. The rise of the general failure rate is reflected in decreasing system availability. At the same time though, the occurrence of dangerous states is decreased which increases system reliability. The system can only enter a dangerous state if both systems (known or unknown) fail. Figure 17.8 shows the 1oo2-architecture of a SIS. Figure 17.9 shows a *PFD*-fault tree of a 1oo2-architecture. The system can fail dangerously in the following cases:

- A dangerous detectable failure is present in both circuits, which is caused by a common cause (common cause failure).

- A dangerous undetected failure is present in both circuits which is caused by a common cause dangerous undetected failure (common-cause-failure).

- A dangerous detected or undetected failure are present in each circuit which are not caused by a common failure.

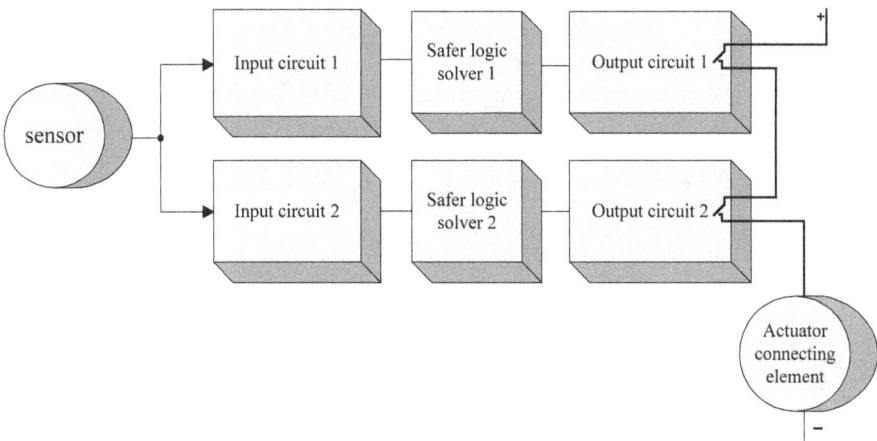

Figure 17.8: Block diagram 1oo2-board controller

---

[156] [BÖRC04] Börcsök, J., *Electronic Safety Systems*

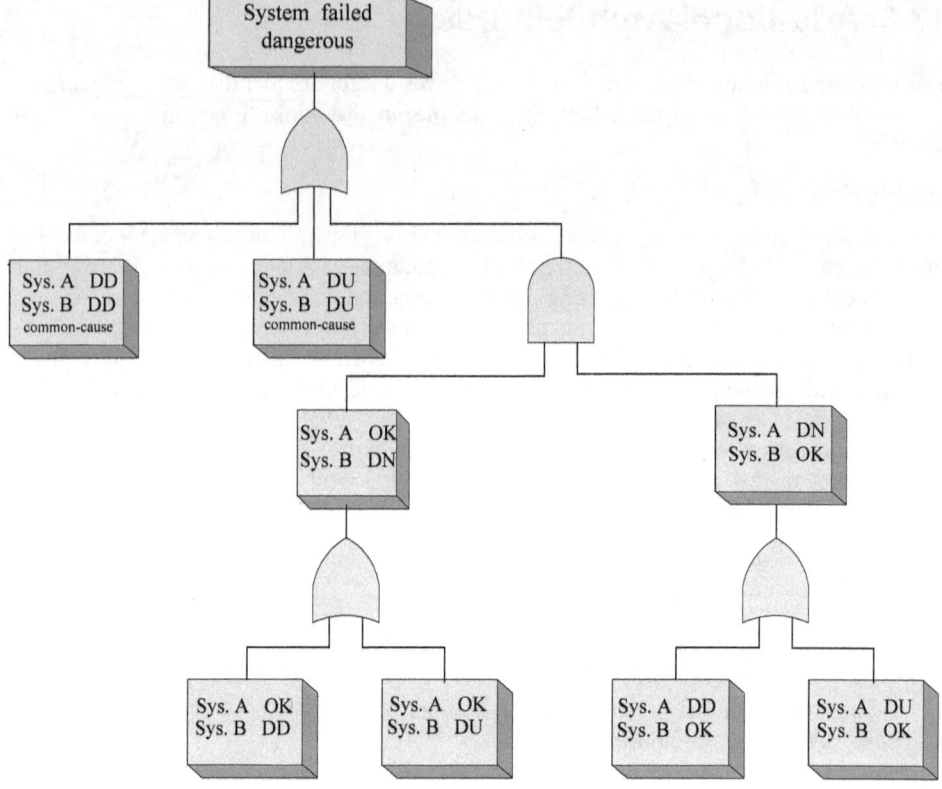

Figure 17.9: PFD-fault tree of 1oo2-architecture

If one considers the percentage of common-cause-failures, the mean failure probability PFD_{avg} for a 1oo2-system is:

$$PFD_{avg,1oo2} = 2 \cdot \left((1-\beta_D) \cdot \lambda_{DD} + (1-\beta) \cdot \lambda_{DU}\right)^2 \cdot t_{CE} \cdot t_{GE} + \beta_D \cdot \lambda_{DD} \cdot MTTR + \beta \cdot \lambda_{DU} \cdot \left(\frac{T_1}{2} + MTTR\right). \quad (17.21)$$

with

$$t_{CE} = \frac{\lambda_{DU}}{\lambda_D} \cdot \left(\frac{T_1}{2} + MTTR\right) + \frac{\lambda_{DD}}{\lambda_D} \cdot MTTR$$

$$t_{GE} = \frac{\lambda_{DU}}{\lambda_D} \cdot \left(\frac{T_1}{3} + MTTR\right) + \frac{\lambda_{DD}}{\lambda_D} \cdot MTTR$$

17.6 Additional Architectures

With:
λ_D rate of dangerous failures
λ_{DU} rate of dangerous undetected failures
λ_{DD} rate of dangerous detected failures
t_{CE} equivalent mean failure time of a circuit
t_{GE} equivalent mean failure time of a group
β weighting of undetected common cause failures
β_D weighting of detected common cause failures
T_1 proof-test interval and
MTTR   repair time

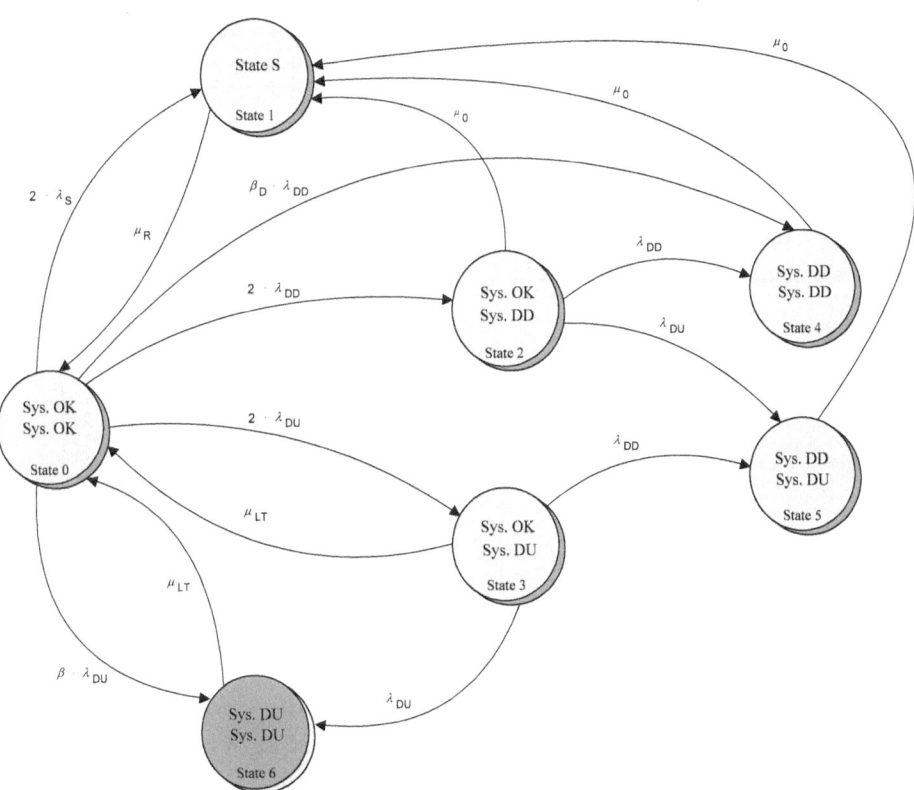

Figure 17.10: Markov model of 1oo2-architecture

The Markov model for a 1oo2-"single board system" is shown in Figure 17.10. Both controllers perform flawlessly in status 0. Status 1 represents the safe state which the system enters following a safe failure. The system is in an energy-free state. The transition rate from status 0 to status 1 is $2 \cdot \lambda_S$, because a safe failure can be present in both conduits.

The *MTTF*-value describes a mean time which passes between the occurrence of two failures. Herein, one sets out from status 0 in which the system was failure-free. In order to calculate the *MTTF*-value of the system, one must add the elements of the first line of the *N*-matrix[157]. The *MTTF*-term for a 1oo2-system is therefore as follows:

$$MTTF_{1oo2} = \frac{1}{A_1} + \frac{2 \cdot \lambda_{DD}}{A_1 A_2} + \frac{2 \cdot \lambda_{DU}}{A_1 A_3} \qquad (17.22)$$

with

$$\begin{aligned} A_1 &= 2 \cdot \lambda_S + \beta_D \cdot \lambda_{DD} + 2 \cdot \lambda_{DD} + 2 \cdot \lambda_{DU} + \beta \cdot \lambda_{DU} \\ A_2 &= \mu_0 + \lambda_{DD} + \lambda_{DU} \\ A_3 &= \mu_{LT} + \lambda_{DD} + \lambda_{DU}. \end{aligned} \qquad (17.23)$$

2oo2-System

Dual controller systems are developed to meet the requirements of high availability. The simplest architecture which fulfills these requirements is a 2oo2-system. Figure 17.11 shows a block diagram for a 2oo2-system. In this system, two circuits must perform failure-free.[158] Figure 17.12 shows the fault-tree of a 2oo2-architecture. It is apparent that the individual possible failure types are connected by an or-junction with each other. Due to the parallel connection, a dangerous failure in only one circuit is sufficient to make the whole system fail with a dangerous failure.

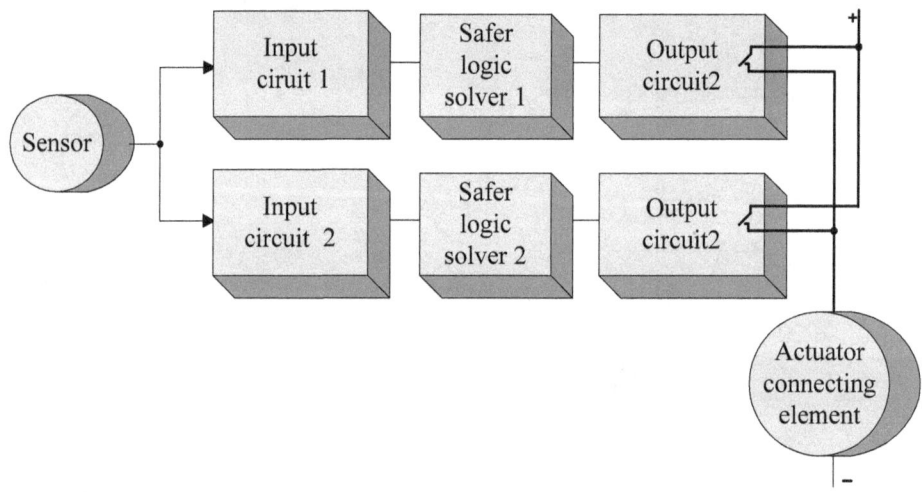

Figure 17.11: Block diagram of a 2oo2-board controller

---

[157] [BÖRC04] Börcsök, J. Electronic Safety Systems
[158] [IECa] IEC 61508; *International Standard 61508 Functional Safety: Safety-Related Systems*

17.6 Additional Architectures

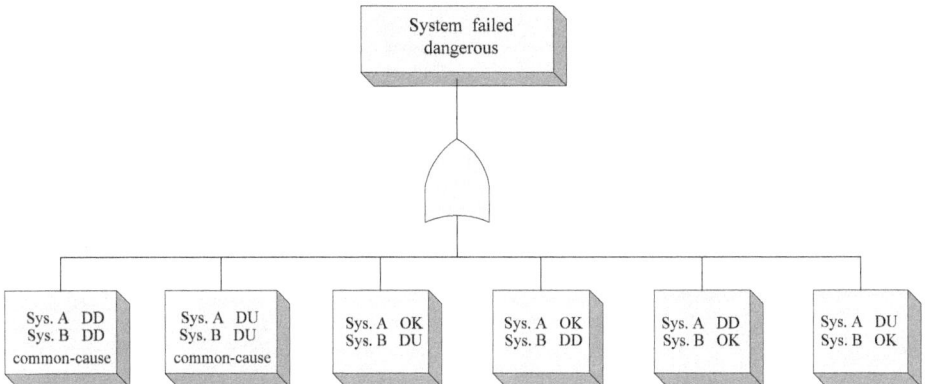

Figure 17.12: PFD-fault tree of 2oo2-architecture

The failure probability of a 2oo2-system is based on the failure probability of a 1oo1-system. A dangerous failure can lead to failure of either the first, the second, or both conduits.

The PFD_{avg}-equation of a 2oo2-system is:

$$PFD_{avg,2oo2} = 2 \cdot \lambda_D \cdot t_{CE} \quad \text{[159]} \tag{17.24}$$

with

$$t_{CE} = \frac{\lambda_{DD}}{\lambda_D} \cdot MTTR + \frac{\lambda_{DU}}{\lambda_D} \cdot \left(\frac{T_1}{2} + MTTR\right)^{160} \tag{17.25}$$

$$\lambda_D = \lambda_{DD} + \lambda_{DU}. \tag{17.26}$$

The Markov model of a 2oo2-"single board system" is shown in Figure 17.13. It is the same model as for the 1oo2-system. In conclusion, it can be documented that

Failures which elevate the 2oo2-system into status 2, 4, or 5 lead to the system returning to safe status 1, following a period of time which is smaller than $2 \cdot \tau_{Test}$. The transition rate from these states to status 1 is always $\mu_0 = 1/\tau_{Test}$.

The states 1, 4, 5, and 6 are absorbing states, that is, there are no further failure transitions to other states.

In states 0 and 3 the system is operational. This must be considered for the *MTTF*-calculation of the 2oo2-system.

To calculate the *MTTF*-value of the system, the elements of the first row of the *N*-matrix[161] must be added. The *MTTF*-value for a 2oo2-system therefore has the following form:

---

[159] Derivation see [BÖRC04]
[160] The following variables represent:
  t_{CE} = channel equivalent mean down time λ_D = dangerous failure rate
  λ_{DD} = dangerous detected failure rate λ_{DU} = dangerous undetected failure rate
  T_1 = time interval *MTTR* = mean time to repair

$$MTTF_{2oo2} = \frac{1}{A_1} + \frac{2 \cdot \lambda_{DU}}{A_1 \cdot A_3} \tag{17.27}$$

with

$$\begin{aligned} A_1 &= 2 \cdot \lambda_S + \beta_D \cdot \lambda_{DD} + 2 \cdot \lambda_{DD} + 2 \cdot \lambda_{DU} + \beta \cdot \lambda_{DU} \\ A_3 &= \mu_{LT} + \lambda_{DD} + \lambda_{DU}. \end{aligned} \tag{17.28}$$

The parallel structure of the 2oo2-system provides a higher availability than the 1oo1-system.

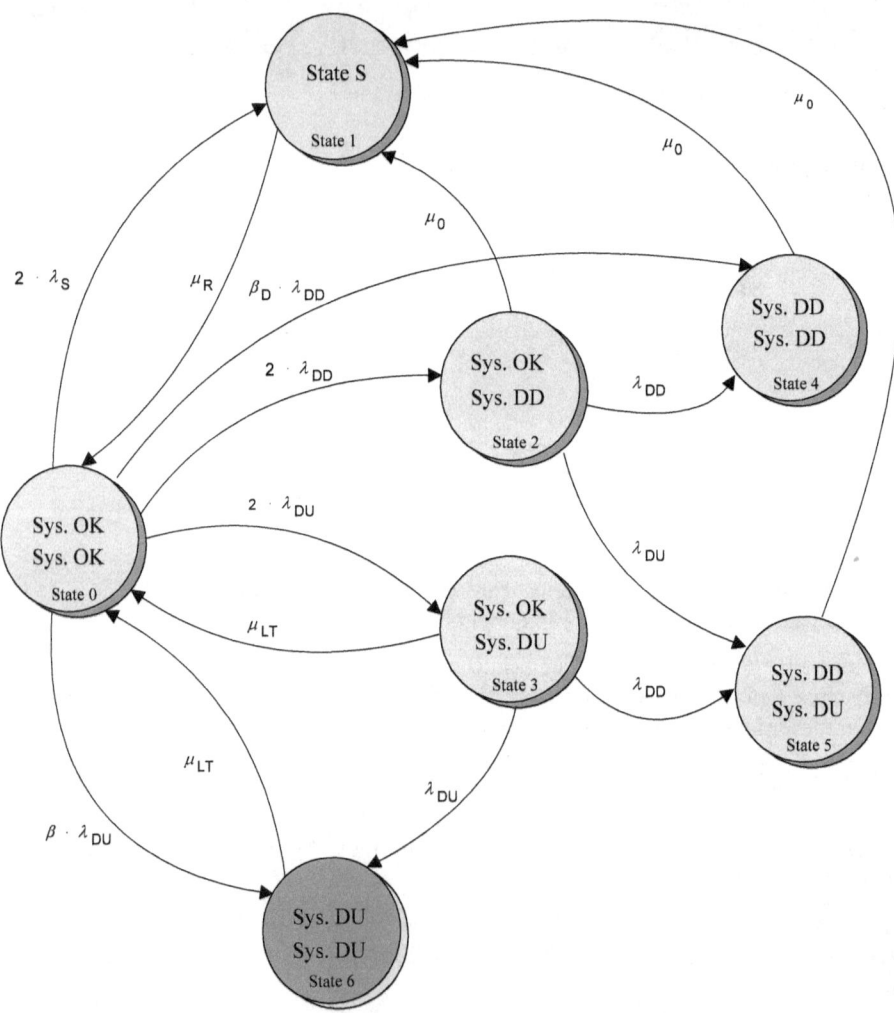

Figure 17.13: Markov-model of 2oo2-architecture

---

[161] [BÖRC04] Börcsök, J., *Electronic Safety Systems*

2oo3-System

The 2oo3-system is a safety system consisting of three independent circuits (Figure 17.14). The three circuits are connected such that two functional circuits are needed to trigger the safety function. The system enters a safe state if only one dangerous failure occurs in any circuit. It is said that "a 2oo3-system is 1-failure-safe".

The 1oo2-configuration reduces the probability of dangerous output states, whereas the 2oo2-architecture reduces the probability of harmless failures, respectively, increases the availability of the system. A 2oo3-architecture provides high safety as well as high availability.

A graphic representation of a 2oo3-system shows an output circuit in which each circuit has two switches. One switch each in one circuit is serially connected with a second switch on a second circuit. The following switch combinations are possible

A1-B1   A2-C2   B2-C1

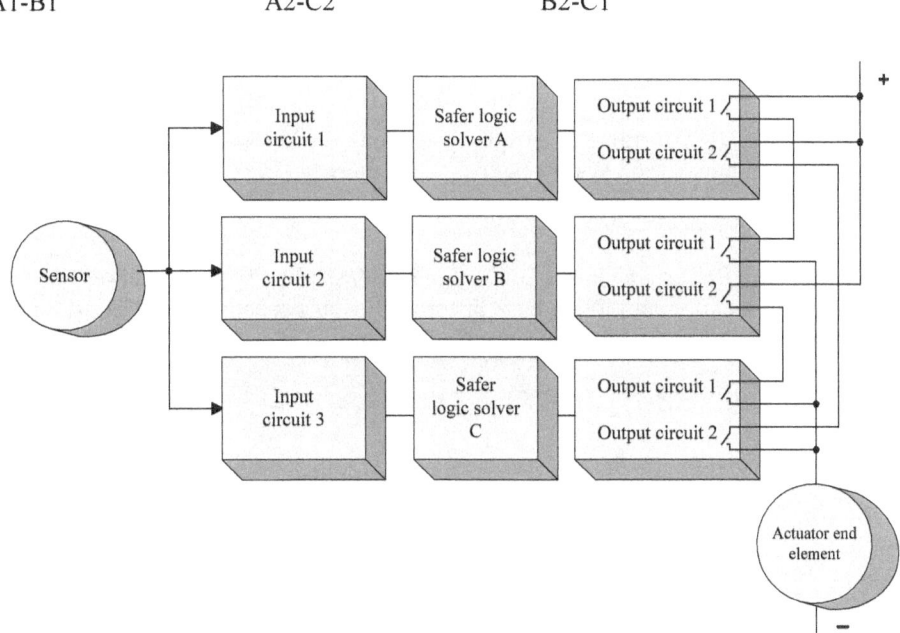

Figure 17.14: Block diagram of 2oo3-architecture

Herein, the notation A1-B1 means: Switch 1 of conduit A is connected in series with switch 1 of channel B. In addition, the three serial connections are connected in a so-called voting circuit which determines the actual starting position. In doing so, a majority decision is made.[162] A 2oo3-architecture is therefore comparable to three 1oo2-systems connected in parallel.

In theory, the 2oo3-system changes over to a safe state by majority decision when a dangerous failure occurs. In practise though, failure detection takes a certain amount of time. If a second failure occurs during this time, i.e., two failures are present, the system can

---

[162] [IECa]   IEC 61508; *International Standard 61508 Functional Safety; Safety-Related Systems*

not make a majority decision and remains in an undefined state. There are two possible solutions:

The system, respectively, the installation must be powered-down if no majority decision is possible. This results in an economic loss for the operator.

When a dangerous failure occurs, the 2oo3-system is degraded to a 1oo2-system, and the transition into a safe state can be defined.

Figure 17.15 shows the failure tree for a 2oo3-architecture. The system can fail dangerously in the following cases:

- In three conduits, a dangerous, detectable failure exists, which is caused by a common cause (common cause failure).
- In three conduits, a dangerous, undetected failure exists, which is caused by a common cause (common-cause-failure).
- In two out of three conduits, a dangerous detected or an undetected failure exists but are not caused by a common cause. The conduit pairs A-B, A-C, or B-C can fail dangerously.

For calculating the PFD_{avg}-value of a 2oo3-system one can refer to the derivation of the 1oo2 system. While in a 1oo2 system both conduits must fail in order to effect a dangerous system failure, in a 2oo3 system two of three circuits must fail to cause a dangerous system failure.

For calculating the PFD_{avg}-value of a 2oo3-system with simple failures, equation 17.21 of the 1oo2-system must be multiplied by a factor of 3. Consequently, the PFD_{avg}-formula with regard to common cause failures and simple failures:[163] is

$$PFD_{avg,2oo3} = 6 \cdot \left((1-\beta_D)\cdot \lambda_{DD} + (1-\beta)\cdot \lambda_{DU}\right)^2 \cdot t_{CE} \cdot t_{GE}$$
$$+ \beta_D \cdot \lambda_{DD} \cdot MTTR + \beta \cdot \lambda_{DU} \cdot \left(\frac{T_1}{2} + MTTR\right) \quad (17.29)$$

With

$$t_{CE} = \frac{\lambda_{DD}}{\lambda_D} \cdot MTTR + \frac{\lambda_{DU}}{\lambda_D} \cdot \left(\frac{T_1}{2} + MTTR\right)$$

$$t_{GE} = \frac{\lambda_{DD}}{\lambda_D} \cdot MTTR + \frac{\lambda_{DU}}{\lambda_D} \cdot \left(\frac{T_1}{3} + MTTR\right)$$
[164]
$$(17.30)$$

$$\lambda'_D = (1-\beta_D)\cdot \lambda_{DD} + (1-\beta)\cdot \lambda_{DU} \quad (17.31)$$

$$\lambda_D = \lambda_{DD} + \lambda_{DU}. \quad (17.32)$$

---

[163] Derivation see [BÖRC04]
[164] The following variables represent:
t_{CE}	= channel equivalent mean down time	t_{GE}	= group equivalent mean down time
λ_D	= dangerous failure rate	λ_{DD}	= dangerous detected failure rate
λ_{DU}	= dangerous undetected failure rate		
T_1	= time interval	MTTR	= mean time to repair

17.6 Additional Architectures

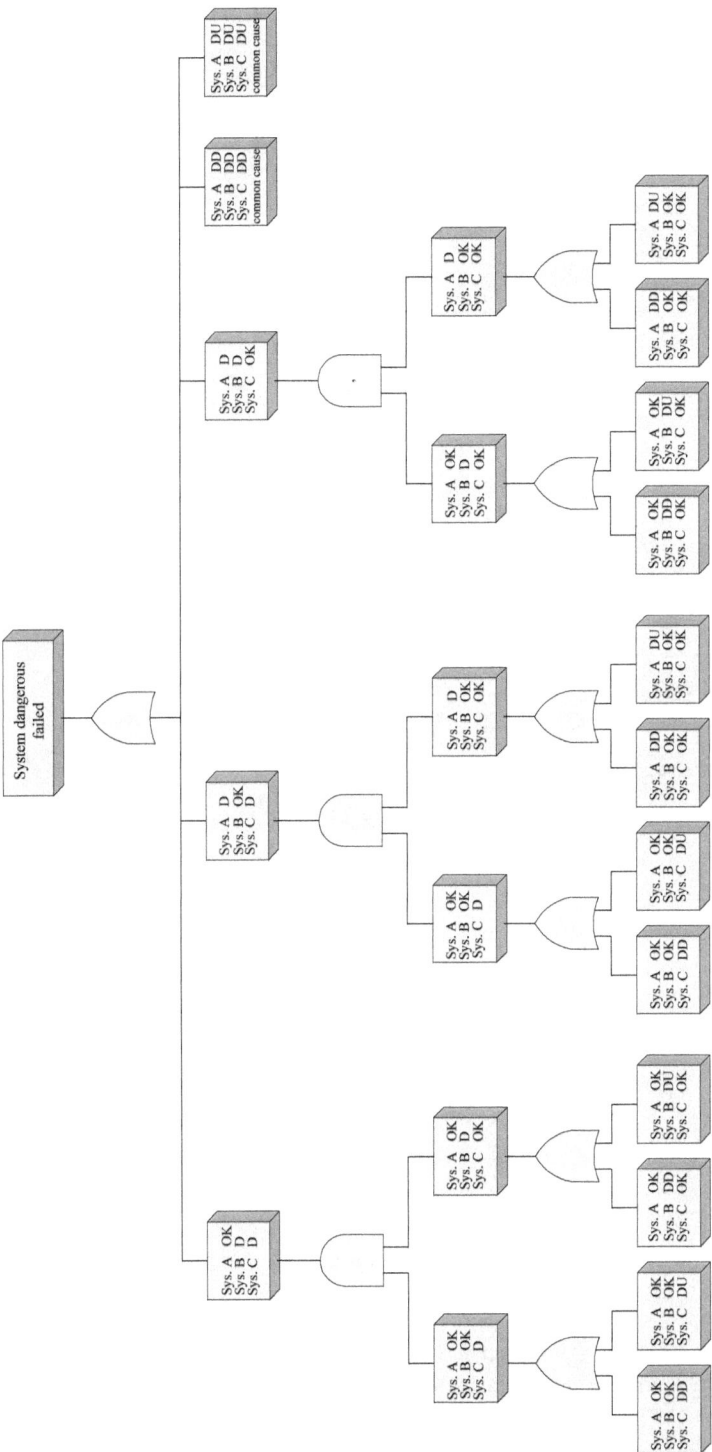

Figure 17.15: PFD-fault tree of 2oo3-architecture

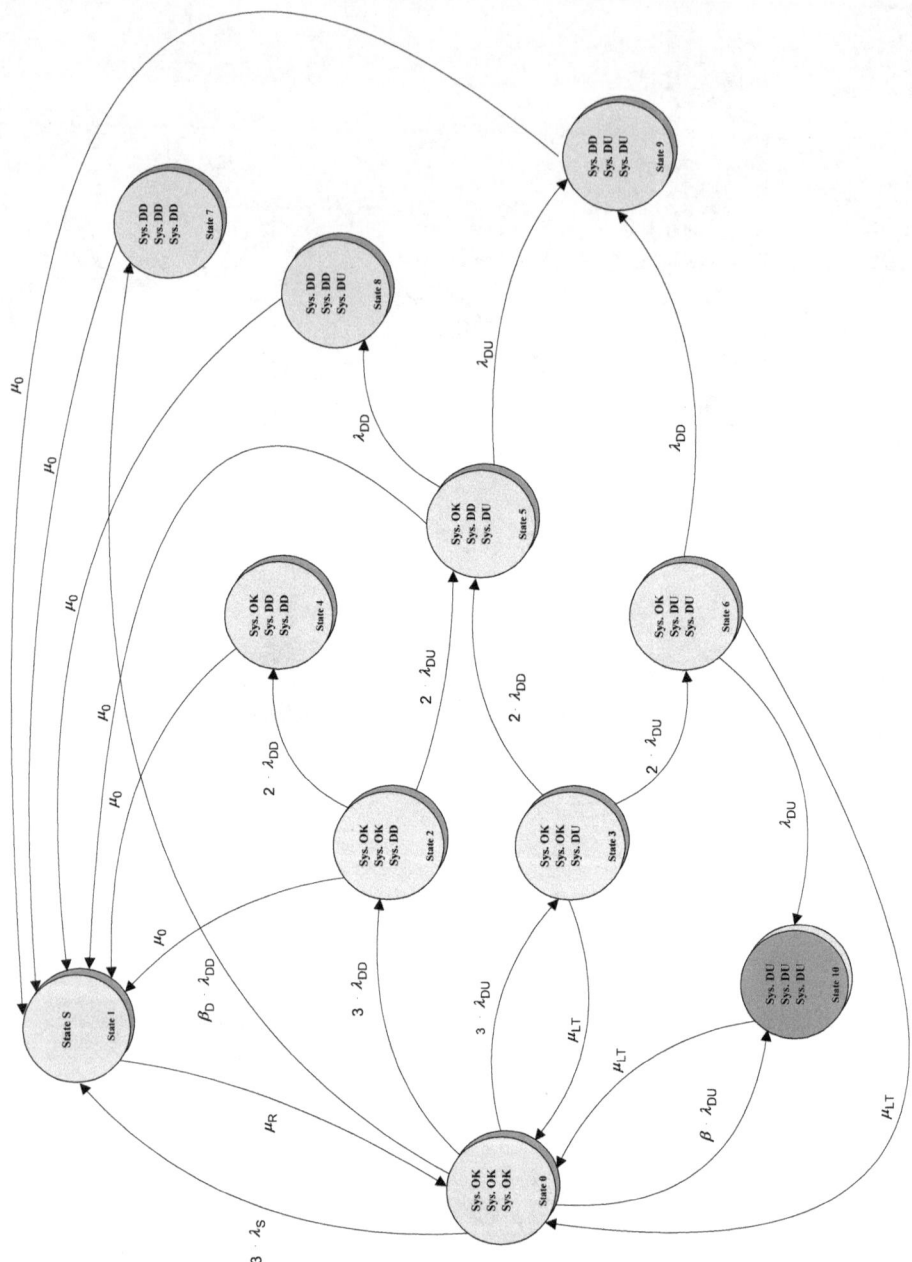

Figure 17.16: Markov-model of 2oo3-architecture

The Markov model for a 2oo3-"single board system" is shown in Figure 17.16. In status 0 all three conduits are in fault-free operating mode. Status 1 represents the safe state following transition after a safe failure. The system is in an energy-free state. The transition rate from status 0 into status 1 is $3 \cdot \lambda_S$, because a safe failure can exist in three conduits.

To calculate the MTTF-value for the system, the elements of the first line of the N-matrix[165] must be added. The MTTF-term for a 2oo3-system has the following form:

$$MTTF = \frac{1}{A_1} + \frac{3 \cdot \lambda_{DD}}{A_1 \cdot A_2} + \frac{3 \cdot \lambda_{DU}}{A_1 \cdot A_3} + \frac{6 \cdot \lambda_{DD} \cdot \lambda_{DU} \cdot (A_2 + A_3)}{A_1 \cdot A_2 \cdot A_3 \cdot A_5} + \frac{6 \cdot \lambda_{DU}^2}{A_1 \cdot A_3 \cdot A_6}$$

(17.33)

with

$$A_1 = 3 \cdot \lambda_S + 3 \cdot \lambda_{DD} + 3 \cdot \lambda_{DU} + \beta_D \cdot \lambda_{DD} + \beta \cdot \lambda_{DU}$$
$$A_2 = \mu_0 + 2 \cdot \lambda_{DD} + 2 \cdot \lambda_{DU}$$
$$A_3 = 2 \cdot \lambda_{DD} + 2 \cdot \lambda_{DU}$$
$$A_5 = \mu_0 + \lambda_{DD} + \lambda_{DU}$$
$$A_6 = \lambda_{DD} + \lambda_{DU}.$$

(17.34)

1oo2D-System

The 1oo2D-architecture is composed similarly to the 1oo2-architecture. In addition, it has a hardware-implemented diagnostic installation (Figure 17.17). This diagnostic installation allows the software to localize dangerous failures.[166]

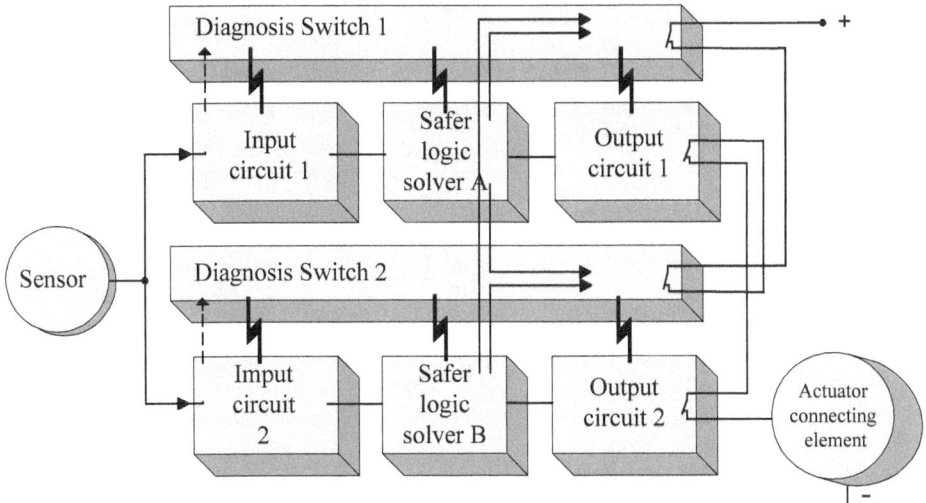

Figure 17.17: Block diagram of 1OO2D-architecture

The system is again composed of two independent conduits which are connected in such a way that one of the conduits is sufficient to bring the system into a safe state when a dangerous failure occurs in the second conduit. Contrary to a 1oo2-connection, the two

---
[165] [BÖRC04] Börcsök, J. Electronic Safety Systems
[166] [IECa] IEC 61508; *International Standard 61508 Functional Safety; Safety-Related Systems*

switches of the source circuit are not connected serially, but instead each output circuit switch is serially connected separately with a diagnostic switch.

In 1oo2D-architecture, a dangerous failure can only occur if both circuits produce a dangerous failure. Figure 17.18 shows the fault tree of a 1oo2D-architecture. The system can fail dangerously in the following cases:

Both conduits have dangerous detected failure which is caused by a common failure (common cause failure).

Both conduits have a dangerous, undetected failure which is caused by a common cause (common cause failure).

One conduit each has a dangerous detected or undetected failure which are not caused by a common cause. The one failure can be detected either by a software test or by hardware diagnosis.

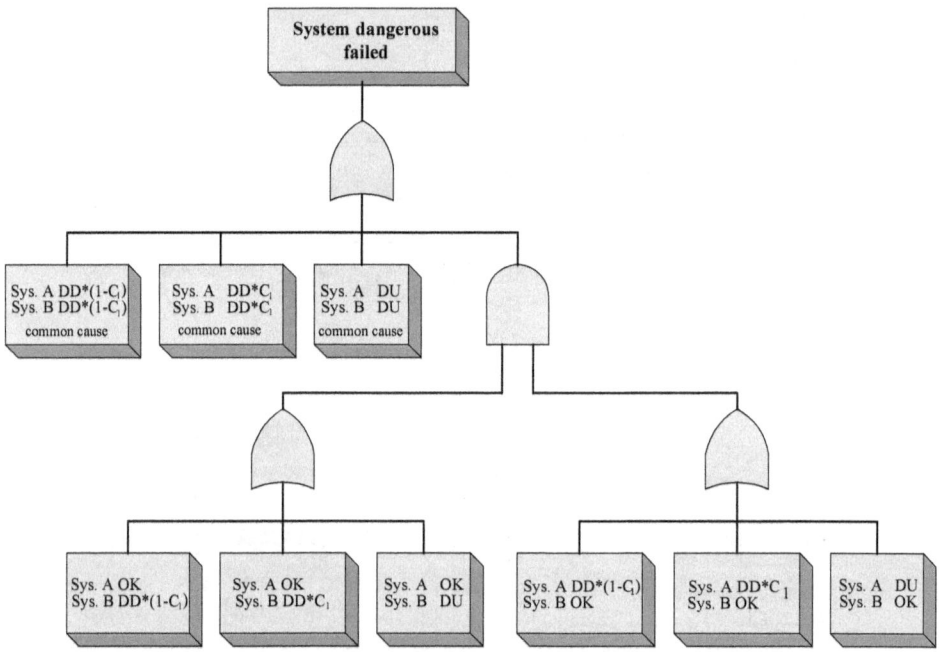

Figure 17.18: PFD-fault tree of 1oo2D-architecture

For the derivation of a PFD_{avg}-equation of a 1oo2D system one can refer to the derivation of the 1oo2-system. The difference is that, next to simple failures which can only be detected by software, there are also simple failures which are detected additionally by hardware diagnosis connection.

17.6 Additional Architectures

The following PFD_{avg}-equation:[167] is obtained for a 1oo2D system:

$$PFD_{avg,1oo2D} = 2 \cdot (1-\beta) \cdot \lambda_{DU} \cdot [(1-\beta) \cdot \lambda_{DU} + (1-\beta_D) \cdot \lambda'_{DD} + \lambda'_{SD}] \cdot \tilde{t}'_{CE} \cdot \tilde{t}'_{GE} +$$
$$\left[\beta \cdot \lambda_{DU} \cdot \left(\frac{T_1}{2} + MTTR \right) + \beta_D \cdot \lambda_{DD} \cdot MTTR \right] \quad (17.35)$$

with

$$\tilde{t}'_{CE} = \frac{\lambda_{DU} \cdot \left(\frac{T_1}{2} + MTTR \right) + (\lambda'_{DD} + \lambda'_{SD}) \cdot MTTR}{\lambda_{DU}' + \lambda'_{DD} + \lambda'_{SD}} \quad (17.36)$$

$$\tilde{t}'_{GE}' = \frac{\lambda_{DU} \cdot \left(\frac{T_1}{3} + MTTR \right) + (\lambda'_{DD} + \lambda'_{SD}) \cdot MTTR}{\lambda_{DU} + \lambda'_{DD} + \lambda'_{SD}} \quad (17.37)$$

Dangerous detected failures can be divided into two failure types by additional diagnosis connection. These dangerous detected failures are

- dangerous detected failures which are detected by diagnosis connection and
- dangerous detected failures which are not diagnosed by diagnosis connection.

The transition rate for the first type of failure equals

$$\lambda'_{SD} = DC \cdot \lambda_{DD} \quad (17.38)$$

and for the second type of failure equals

$$\lambda'_{DD} = (1 - DC) \cdot \lambda_{DD} \quad (17.39)$$

with

$$\lambda_{DD} = \lambda'_{DD} + \lambda'_{SD} . \quad (17.40)$$

The Markov-model of a 1oo2D-"single board system" is shown in Figure 17.19. In the state 0 both systems operate failure-free. State 1 represents the safe state in which the system transitions after a safe failure. The system is in an energy-free state. The transition rate from state 0 to state 1 is $2 \cdot \lambda_S$, because a safe failure can be present in both conduits.

- Failures which bring the 1oo2D-system into states 2 or 3 cause the system to be be in safe state 1 following a period of time which is smaller than or equal to τ_{Test}. The transmission rate from these states to state 1 is always $\mu_0 = 1/\tau_{Test}$.
- States 1, 5, 6, 7, 8, 9, and 10 are absorbing states.

In states 0, 2, 3, and 4 the system is operational. These states must be considered in the $MTTF$-calculation of the 1oo2D-system.

---

[167] [IECa] IEC 61508; *International Standard 61508 Functional Safety; Safety-Related Systems*

274   17 Hardware of Safety-Related Systems

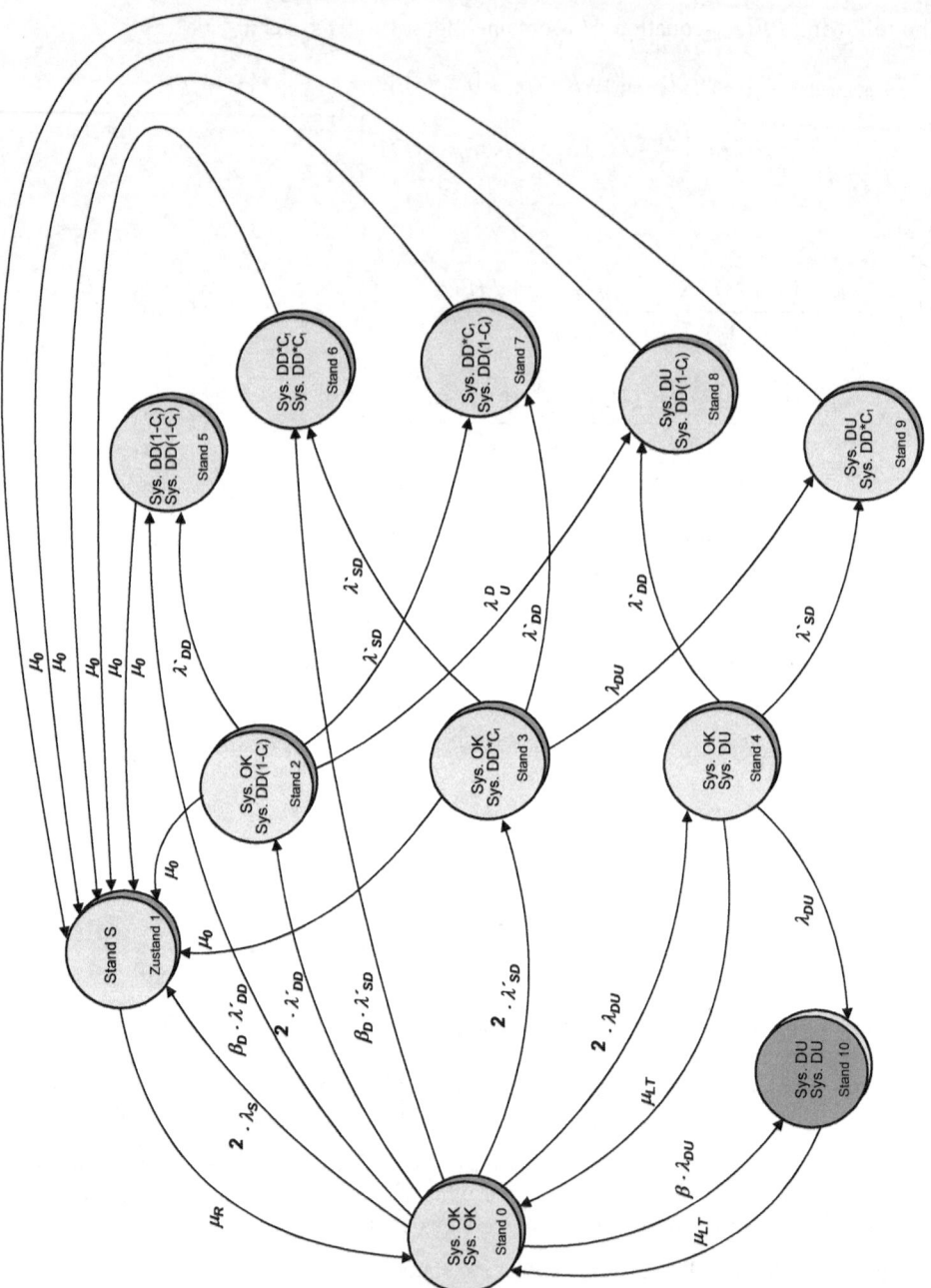

Figure 17.19: 1oo2D-Markov-model

17.6 Additional Architectures

To calculate the *MTTF*-value of the system, the elements of the first line of the *N*-matrix[168] must be added. The *MTTF*-term for a 1oo2D-system therefore has the following form:

$$MTTF = \frac{1}{A_1} + \frac{2 \cdot \lambda'_{DD}}{A_1 \cdot A_2} + \frac{2 \cdot \lambda'_{SD}}{A_1 \cdot A_3} + \frac{2 \cdot \lambda_{DU}}{A_1 \cdot A_4} \qquad (17.41)$$

with

$$A_1 = 2 \cdot \lambda_S + 2 \cdot \lambda'_{DD} + 2 \cdot \lambda'_{SD} + 2 \cdot \lambda_{DU} + \beta_D \cdot \lambda'_{DD} + \beta_D \cdot \lambda'_{SD} + \beta \cdot \lambda_{DU}$$
$$A_2 = \mu_0 + \cdot \lambda'_{DD} + \lambda'_{SD} + \lambda_{DU}$$
$$A_3 = \mu_0 + \lambda'_{SD} + \lambda'_{DD} + \lambda_{DU}$$
$$A_4 = \mu_0 + \lambda'_{DD} + \lambda'_{SD} + \lambda_{DU}.$$

$$(17.42)$$

---

[168] [BÖRC04] Börcsök, J., *Electronic Safety Systems*

18 Software requirements for a system with functional safety

18.1 Software in systems with functional safety

In order to operate computer-supported systems, it is necessary to have the proper hardware with the corresponding software. Developing this software is becoming increasingly difficult and complex. The following chapter describes what software really is, and why developing modern software is so complicated.

Generally, software and its programs are defined as a series of commands which has been designated for a specific computer system and which regulate communication between processor and periphery. The software therefore defines the processor's behavior in certain situations.

Definition: software

Programs, procedures, associated documentation, and data relevant to the operation of a computer system (IEEE 610.12).

From this definition it is apparent that there is far more to software than the mere program. The fact that only approximately 10 – 20% of the total effort expended in developing software goes into program writing further outlines this point. Compared to the entire life span of software, the resulting cost of all the effort invested into it drops to less than 10% of the overall cost of development (Figure 18.1). For this reason, expenditure estimates are often grossly skewed as only the actual program costs are being considered.

In contrast to materialized products, software does not have limitations such as properties of substance or physical laws. The only possible limitation presents itself in terms of accountability. This particular property of software has an effect on its behavior and reliability. Frequently, hardware developers are restricted to a much greater degree than software developers in regards to capabilities. For this reason, software programs can potentially contain more mistakes. Identifying mistakes in software presents an additional problem, whereas they are often relatively easy to identify in hardware. Extensive testing is often necessary to find software mistakes. However, software seems to be flexible and easy to alter; a few lines of additional or altered code often promise new and improved software functions. Nevertheless, in doing so one often neglects the fact that, in complex software projects, even small changes can have very significant effects. It is far more difficult to specify these alterations with respect to correct function and resultant mistakes than with hardware alterations.

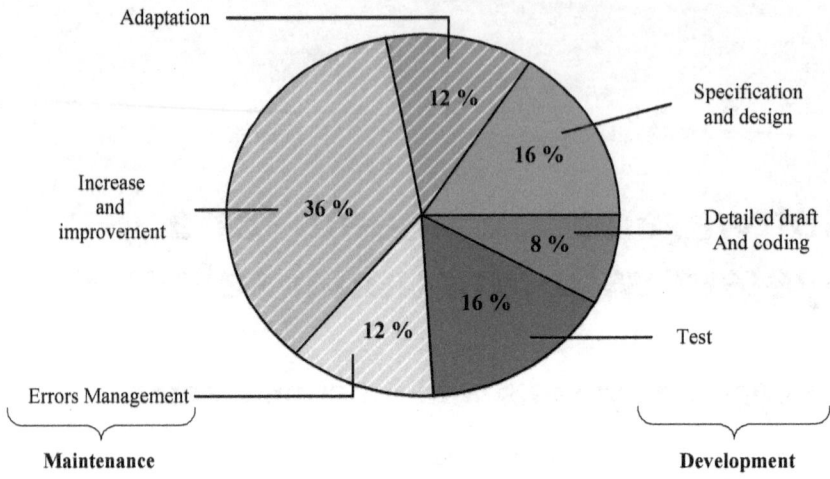

Figure 18.1: Relative percentage of overall effort during the entire lifetime of a product[169]

Table 18.1: Differences in developing smaller versus more expansive software

Smaller software	Extensive software
Programs from 1 to approx. 300 lines in length	Longer programs
For private use	For use by third party
Broad goal setting, product is its own specification	Precise goal setting, specification of requirements is requisite
One step only from problem to solution: the solution is programmed directly	Several steps required from problem to solution: requirement specification, solution concept, parts design, parts programming, parts composition, implementation
Validation and necessary corrections happen at the end product stage	Every developmental step must be followed by a testing step in order to minimize the risk of mistakes
One developer: no coordination, no communication necessary	Several developers working together: coordination and communication essential
As a rule, problem complexity is small, software structuring and keeping track (retaining the overview) is not difficult	Problem complexity is greater, explicit means for structuring and modulization are required
Software consists of few components	Software consists of many components which demand special measures of component management
In general, no documentation is drawn up (compiled)	Documentation is vital for economical operation and servicing of software
No planning and project organization required	Planning and project organization is crucial for goal-oriented economical development

Developing smaller software is markedly different from developing extensive software. Differences are shown in table 18.1. One of the major problems in software development is the fact that these differences are neglected. The development of expansive projects is simply carried out by many individual programmers whose contributions include the creation of smaller programs (Figure 18.2).

---

[169] [BOEH73]   Boehm, B., *Software and its Impact*

18.1 Software in systems with functional safety

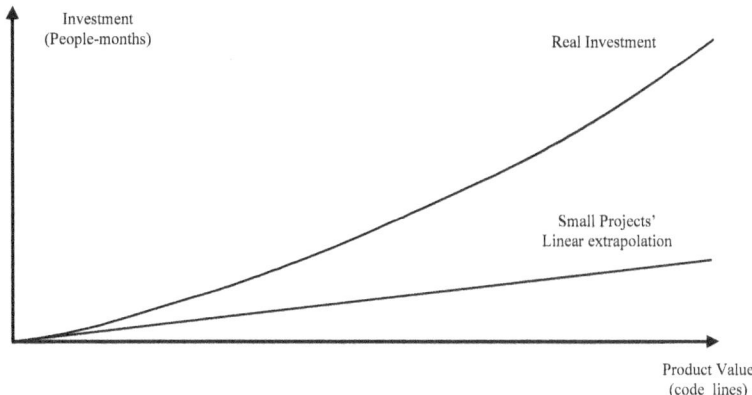

Figure 18.2: Miscalculations of effort in linear extrapolation

The effort expended in creating software increases extraproportionally with increasing product size. This connection is described in the following equation[170]

$$A = b \cdot P^m, \tag{18.1}$$

where A is effort, P is size of the software, b is a constant factor and m an exponent between 1.05 and 1.2. The causes for this extraproportional growth are found, among other things, in the growing extraproportional effort involved in communication between the developers.

The continuously changing expectations of the software, the so-called software evolution, are another problem in software development. New requirements arise during the developmental process which can often only be incorporated into the existing developmental process with utmost difficulties. In addition, software requires continuous care and expansion in order to remain useful for a period of time. In order to be able to make servicing as affordable as possible, a suitable design must already have been conceptualized during the developmental phase. Principally, four factors make software development a difficult task:

- The magnitude of problems to be solved. The bigger and more complex the software, the more elaborate and complicated the development.
- The fact that software is an immaterial product. This immateriality makes working with software more difficult than working with material technical products of comparable complexity. In addition, it is more difficult to recognize any risks.
- Developmental goals are continuously changing due to software evolution.
- Errors can be the result of estimations or perceptual mistakes.

The beginnings of software development were relatively problem-free. Programs were relatively small and independent of each other. In the mid-1960s necessary software became increasingly complex and expansive. In addition, developmental and service-

---

[170] [BOEH81]   Boehm, B., *Software Engineering Economics*

related costs of software skyrocketed, especially when compared with hardware (Figure 18.3). As a result, it became critical to implement a systematic procedure in software development, and the expression "software engineering" was created.

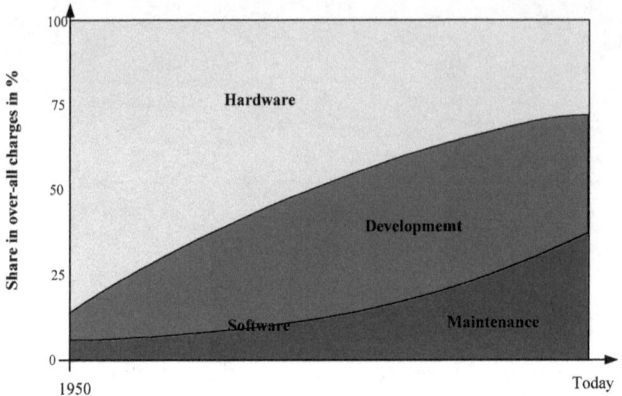

Figure 18.3: Development of hardware and software cost relationships[171]

Definition: software engineering

The application of a systematic, disciplined, and quantifiable approach to the development, use, and service of software, i.e., the application of principles of engineering to software (IEEE 610.12).

Software engineering pursues three fundamental goals:

- Productivity
- Quality
- Organisation

Productivity:

Productivity should be increased during the manufacturing process of software. In view of increasingly expensive software development, even minor productivity increases are sensible.

Quality:

Software quality should improve. This will result in advantages in competition and customer satisfaction. Service and further development of software should become more cost-efficient by an increase in quality.

Organization:

Improvements in punctuality and cost effectiveness can be achieved by the organization of software developmental projects.

---

[171] [BOEH81]   Boehm, B., *Software Engineering Economics*

18.1.1 Software requirements

It is the principal goal of software development to give the customer a guarantee that the software will have the desired functionality. En route to this goal there are many different problems to watch out for. For example, software development needs to ensure that servicing the system is as cost-effective as possible. On top of that, it should be possible to easily fix errors once they have been identified. These software requirements are independent of functionality and are therefore called non-functional requirements. The following paragraph outlines these non-functional requirements and explains how they are involved in the process of development.

18.1.2 Non-functional requirements

Non-functional requirements are requirements relating to the conditions under which the system should achieve its functionality. They contribute directly to a system's applicability. In order to minimize costs, with regards to the entire life cycle of the software, non-functional requirements must be recognized and incorporated into the developmental process early on in the beginning stages of software development. This requires accurate concepts of additional properties which the software should exhibit, next to a specification of functionality which is as accurate as possible. Therefore, systematic software development deems it necessary to set clearly defined goals.

18.1.2.1 Goal setting

Results of software development are random without the establishment of well-defined goals. Through goal setting it is possible, that participating persons work towards these goals together. The possible goals of software development are often in competition with each other. The decision, which goal to pursue, influences both the developmental process as well as the result. This is illustrated by an experiment performed in 1974 by Weinberg and Schulman (Table 18.2).

Table 18.2: Achievement of programming depending on preset goals (1 = best effort and 5 = worst effort)

Goal: Optimise	Production effort	Quality of results			
		Number of instructions	Storage needs	Program clarity	Output clarity
Production effort	1	4	4	5	3
Number of instructions	2-3	1	2	3	5
Storage needs	5	2	1	4	4
Program clarity	4	3	3	2	2
Output clarity	2-3	5	5	1	1

Five groups were assigned the task of developing the same software, but with different non-functional requirements or goals. Two characteristics soon became apparent:
- The quality of programs produced correlates strongly with preset goals.
- In order to achieve optimal success for these goals, other goals were neglected.

Some goals are in direct competition with each other. For this reason, targets must be tailored to both the customer's specific needs and pre-existing limiting conditions.

18.1.2.2 Goal control

Merely setting the goal is not sufficient. Strict adherence to the goals must be verified time and again during the entire course of the developmental process. Otherwise, it is possible for a small but steadily increasing misdevelopment to take place. By means of watchfully controlling a goal, any misdevelopment can be recognized, and counter-measures can be put in place before the costs of damage control become too high.

Control measures which consist of goal measurement are needed in order to achieve a set goal. For this purpose, measures have to be defined. For each goal,
- References, and
- Thresholds

can be defined. A reference value is the value at which 100% of the goal was achieved. Once the threshold has been reached, the result is close enough to the goal and therefore becomes acceptable. Some goals of software development can be measured directly, such as developmental costs, or runtime of the program. These measurements are called direct measurements. Other goals, such as how user-friendly the software is, require a lot of effort. It may also be that they can not be measured at all, as is the case with, for example, an absolute number of fields. Frequently, only empirical observations are available to determine these measurements. The following means of goal measurement are important:
- Effort
- Run times
- Work progress
- Developmental costs
- Error costs

18.1.3 Categories of non-functional requirements

According to the V-model[172], which has been formulated for the purpose of software development in public institutions, all non-functional requirements can be subdivided into six categories.

---

[172] [KBST05] KBSt, *V-Modell XT Release 1.1 Gesamtumfang*

1. **Pre-existing limiting conditions**

   These include, for example, legal limiting conditions such as the requirement to uphold legal requirements, or technical limiting conditions, such as hardware availability.

2. **System quality requirements**

   These include properties and qualities of the new system. The V-model utilizes a simplified scheme, according to ISO norm 9126:

   - Functionality
   - User friendliness
   - Reliability
   - Efficiency
   - Serviceability

   Additional quality properties which can be of importance in software projects are, for example, ease of alteration, reuseability of system components, or the ability to transfer to different hardware (details in software quality: section 18.2.10).

3. **Requirements of system safety**

   This includes requirements of data safety and data protection.

4. **Requirements during the developmental phase of the system**

   This includes developmental methods and quality standards which are dictated by the client, as well as regulations concerning the acceptance and introduction of the system.

5. **Logistic requirements**

   This includes demands on instructional documents and usage documentation.

6. **Demands on servicing and operation**

   This includes demands as to the qualifications of sales and service personnel.

Careful planning is required to meet the demands listed, from the beginning stages of software development on. To achieve this goal, all functional, as well as all non-functional, requirements must be developed in close collaboration with the client. These specifications are recorded in so-called customer requirement specifications. The effort required in finalising these requirements must not be underestimated. For example, 33% of the budget assigned to developing software for a harbor logistics and information system were spent on the analysis and modelling of required business processes.[173]

The advantages of such an elaborate procedure lie in a simplified implementation of software and a lower occurrence of errors. Design changes can be implemented with more ease, and the probability of failure of important functional requirements is minimized.

---

[173] [WEIL04]   Weilkiens, T., Schröder, C., *Praxisbericht: Erste Projekt-Erfahrungen mit der UML 2.0*

18.2 Software development

Modern software development of systems that are applied in environments where safety is of critical importance, much like in standard software development, consists of a series of individual stages which must be combined. Segmenting software development into separate steps, which are as independent from each other as possible, enables the production of complex and very elaborate projects. There are a number of models in which these steps and their interactions are both outlined and presented in detail. This segment shall provide a review of the individual phases and introduce some of the most well-known procedures in software development.

Individual stages of software development differ only marginally with regard to the variety of developmental models. Thus, there is a strong consensus concerning the character of certain stages. Differences are mainly encountered in connecting the separate steps, as well as in the details. The following steps should always be carried out when developing complex software.

1. **Requirements, stating specifications**

    This includes functional as well as non-functional requirements. It should eventually yield detailed and documented customer requirement specifications, which enumerate as accurately as possible that which the new software is expected to provide.

2. **Architectural design**

    During this stage, the fundamental architecture of the system is specified. The system is fragmented into separate components which are designed to be as independent of each other as possible. In addition, interaction between components is designated. This step also involves a crude specification of intersecting points within the system.

3. **Fine tuning design**

    Here, individual components are broken down into even smaller components until a structure develops which can be directly implemented. If you apply an object-oriented approach to programming, the end result of this step is a complete class structure. In the course of this step, all applied methods and attributes are defined. Furthermore, a complete definition of all interfaces is detailed.

4. **Implementation of individual components**

    Components are now implemented, according to predefined specifications. During this stage, programmers must adhere closely to the definitions drawn up as crude and fine design in prior stages in order for subsequent troubleshooting to be performed efficiently.

5. **Individual component testing, correction of errors**

    Here, all components are tested in order to ascertain that they perform according to customer requirement specifications. For this purpose, a precisely itemized examination procedure is necessary, one which is defined in the customer requirement specifications.

6. Integrating components into the total system

The total system is created by integrating all functional components into the system, one after the other. Following integration, each component is subject again to testing, in order to identify and rectify errors resulting from mutual interference. The total system is ready to be delivered to the client once it has passed a final test.

Each of these six steps has to be elaborately documented during the developmental process. The completion of one of these six steps is called a milestone. Advances in developmental process can effectively be controlled by milestones.

Definition: milestone

A milestone is a point in the process at which a planned result is achieved.

Milestones can therefore also be characterized by completion of segments. For planning a milestone, first one needs to define the goal or, respectively, the final outcome to be realized. Furthermore must be defined which resources (time, money, personal) are required to reach the target. A milestone is successfully reached when the planned event has been realized. Hereby it makes sense to compare the planned with the effective use of resources. Milestones only make sense if their result can be measured (Table 18.3).

Table 18.3: Suitability of intermediate results for milestones

Result during a software process	Suitability for milestones
Coding finished to 90%	useless because it can not be measured, only estimated
Component xyz has passed a certification test	qualified because it can be measured (by ways of the test acceptance protocol)
Requirement specification is available	qualified if content and form are tested

Prototypes are a central means for early recognition and solving of problems in software development.

Definition: prototype

A functional piece of software used for advance implementation of critical parts of a system in development. The development of prototypes is called prototyping.

There are three ways of prototyping:

- explorative
- experimental
- evolutionary

Explorative prototyping serves to clarify and define requirements by means of functional programs. On the one hand, it demonstrates the usefulness in principle and the feasibility of system ideas. On the other hand, it serves to create a checkable and criticizable model of a proposed solution. This kind of prototyping serves to examine the adequateness of requirements and their solutions.

Experimental prototyping is used for feasibility studies and testing of critical system components. Such prototypes, also termed laboratory prototypes, do not need to be documented. Their programming need not be pure but they cannot be utilized further.

In evolutionary prototyping, a type of software is developed which forms the nucleus of the total system. This software undergoes stepwise further development into the planned system by substitutions and expansions. Therefore, an evolutionary prototype must be carefully planned and developed.

18.2.1 Models of software development

It is important to utilize a model for the process of software development. Thus, developers can routinely refer to their model and are able to quickly recognize advances as well as errors in development.

18.2.1.1 Waterfall model

A relatively simple model is the so-called waterfall model (Figure 18.4). The developmental cycle is separated into sequential phases. A new phase can only be started when the previous one is completely finished.

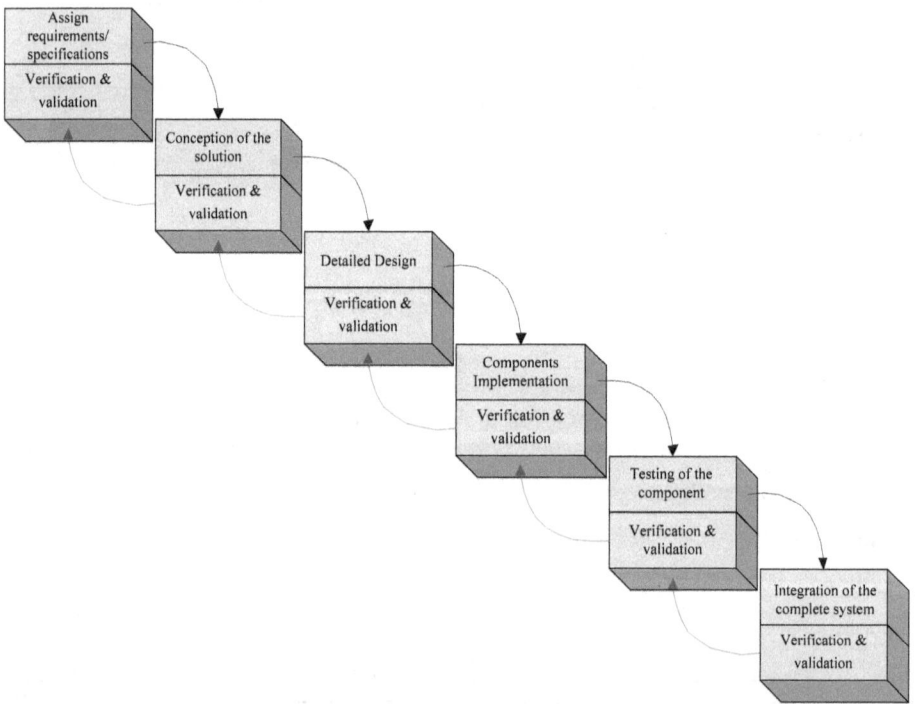

Figure 18.4: Waterfall model

An elaborate testing system should ascertain that the next phase can be entered into. In the meantime the findings of the phase are extensively documented. Even programming code is considered documentation. The new phase starts with the processing of results from the previous phase. If problems are encountered, retrograde steps into the previous phase are possible. The advantages of this model lie primarily in its simplicity. Other advantages are the straighforwardness and availability of interim results. Since errors, once they are detected, did not always occur in the immediately antecedent phase, frequent setbacks can occur. Thus, the number of iterations through the model will rise very high, and the entire developmental process will become very unclear. Another disadvantage is the very late testing which is responsible for highly increased costs for error detection in preceding phases. It is practically impossible to change requirements during the developmental process.

18.2.1.2 Spiral model

The spiral model is a progression of the waterfall model by Boehm[174]. In this model, the individual phases of the waterfall model are passed through in a cyclical manner (Figure 18.5). Each cycle consists of the following four elements:

1. Specifying targets, awareness of alternatives, and constraints
2. Evaluating alternatives and their risks, if necessary with the aid of prototypes
3. Development and testing the actual step
4. Planning the next cycle

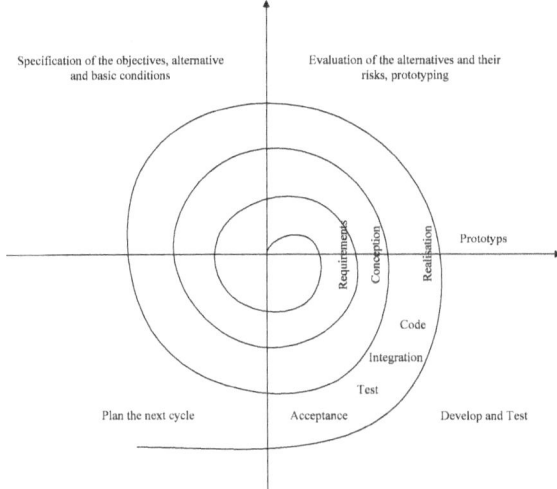

Figure 18.5: Spiral model[175]

Of special importance is the second step because the development of errors is avoidable through proper risk analysis. The above cycle will pass, at least, through the following steps:

---

[174] [BOEH88]   Boehm, B., *A Spiral Model of Software Development and Enhancement*
[175] [BOEH88]   Boehm, B., *A Spiral Model of Software Development and Enhancement*

1. Requirements
2. Conceptualisation of the solution
3. Realization (fine tuning the design, implementation, testing, integration)

A disadvantage of the spiral model is the unpredictable number of iterations necessary to achieve the desired solution.

18.2.1.3 V-Model

The V-model[176] was published in 1992 in terms of the "Software development standard of the German Army. It represents the developmental standard for IT systems of the German Confederation. Furthermore it represents the standard of software development and is used in quite a number of industrial firms.

System development is based on hierarchical subdividing of the system into ever smaller units. A clearly defined procedure exists for this partitioning. First, the requirements of the next higher stage are accepted; next the partitioning strategy is developed and executed. During the course of this procedure, new requirements arise which are passed on to the next stage. On every stage of this hierarchy, verification and validation are executed, in order to identify problems or wrong decisions early on.

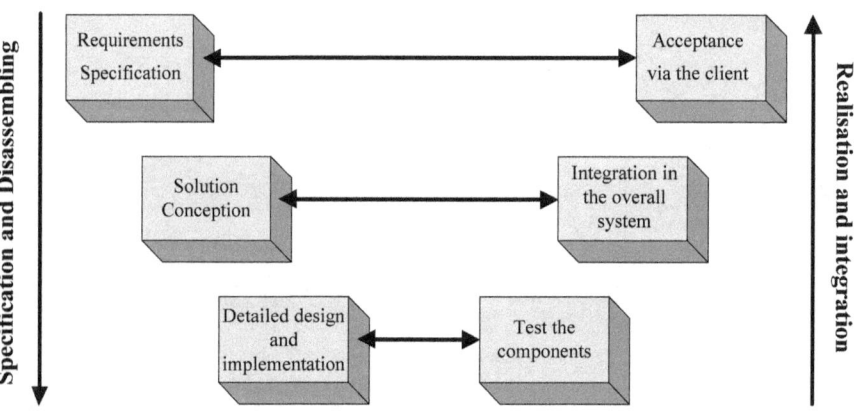

Figure 18.6: System development in the V-model

---

[176] [BALZ92]   Balzert, H., *Die Entwicklung von Software-Systemen*

18.2.1.4 Project planning

Project planning has a great influence on the successful completion of a project. All requirements and preconditions of the developmental process must be considered in the project plan. This includes functional and non-functional requirements, product quality, target deadline, and costs. In order to fulfill these requirements, a project plan must be created which specifies which participating partners in the project are assigned to which tasks, what kind of results are expectd, and what budget will be available.

Aspects of the project plan are not predefined, and are strongly dependent on the project. Detailed planning is essential. However, developers can be encumbered by strong restrictions written into the project plan, making design changes very costly.

18.2.2 Specification of requirements

The first step in software development consists of a specification of the functional requirements which is as comprehensive as possible. A large portion of subsequently possible problems can thereby be avoided. In this chapter, problems which occur during the formulation of requirements, as well as possibilities to display requirements for the system, are described. It is of crucial importance to formulate accurately all functional requirements of the system. If client and contractor are not entirely clear from the beginning of software development, what the purpose of the system is expected to be, high costs of re-design and troubleshooting are already preprogrammed into the process. Specifying non-functional requirements will lead to a substantial decrease in cost. For example, if it is foreseeable that some part of the functionality will also be required in a similar manner in other systems, then this functionality ought to be implemented in a reusable manner.

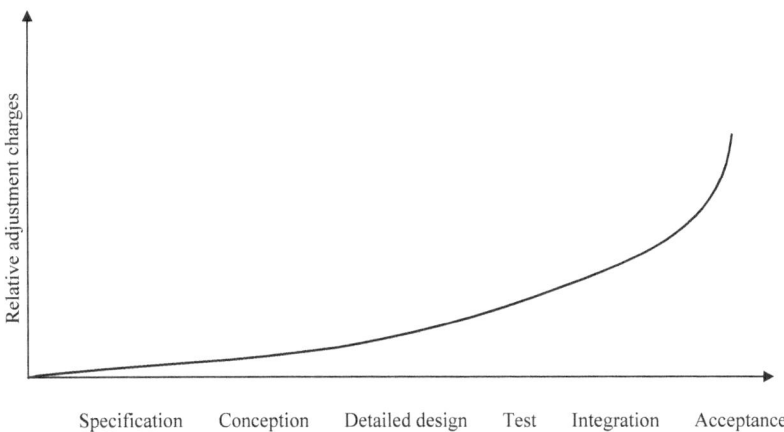

Figure 18.7: Cost of troubleshooting, dependent on the phase during which errors are detected[177]

---

[177] [BOEH81]   Boehm, B., *Software Engineering Economics*

Cost of requirement specifications, which can be considerable, is balanced out by benefit. Errors made during specification are extremely expensive because they are often not discovered until testing or installation of the system. The exponential rise of debugging during the course of development is shown in Figure 18.7.

18.2.2.1 Characteristics of a specification

A good requirement specification should safeguard (ensure) two attributes:

1. Few requirement errors
2. Existing errors should be recognized early

Therefore, each requirement specification should possess the following qualities:

1. Correctness

   Every requirement can be fulfilled through software.

2. Conclusiveness (Indisputability)

   Every requirement is explicit (definite), if applicable, term definitions are necessary

3. Completeness

   All requirements requested by the client are described. This also includes non-functional requirements. The behavior of unforeseeable inputs must be taken into account.

4. Consistency

   There must be no contradictory requirements.

5. Weighting of requirements

   There must be a clear weighting between target and performance requirements.

6. Testability

   It must be feasible to test each requirement through a clearly defined examination.

These requirements must be achieved during a customer-oriented specification process. This process proceeds repetitively and requires intensive communication between client and contractor.

When formulating these requirements, a variety of problems can arise. The biggest problem is language inaccuracies. Requirements described in natural language often result in ambiguities and contradictions. In addition, different representatives of the client have different expectations of the system to be specified. Often, requirements are described only as vaguely formulated system properties, because the clients are not able to formulate their expectations.

To solve these problems, a definition of terms should be drawn up if necessary. That way, conceptual ambiguities can be avoided. Next, all possible application scenarios are acted out, that is, all interactions between environment and system, and the reaction of the system to it, are modeled. The required information is ascertained through interview sessions or workshops.

18.2.2.2 Description of requirements

The *descriptive representation* regards the system to be specified as a black box. Requirements are described through connections between input and output. In descriptive representation only the exterior behavior of the system is described.

Advantages:

- The description concentrates on essentials: the connection between input data and result.

- The description gives the programmer complete freedom in formulating a solution.

Disadvantages:

- Descriptive presentation of big systems by means of text is very voluminous and confusing. Often connections between requirements are not clearly identified. Therefore the presentation becomes error-prone and difficult to test.

- The descriptive presentation of requirements by means of formal specifications is very difficult to construct and very hard for the client to understand.

For the above reasons, descriptive presentations are only useful in smaller systems.

The *constructive presentation* views the system to be specified as a quantity of interactive components. Requirements are described as the interaction of components and their resultant consequences.

Advantages:

- Requirements are easily reproducible because a descriptive model exists.

- The system can be fragmented into easy to manage partial tasks.

Disadvantages:

- Potentially, the requirement is specified from the viewpoint of implementation. The system is therefore merely fragmented into easy to implement components, thereby neglecting the specific requirement.

- It is equally disadvantageous if implementation occurs without previous adjustment of components.

Constructive demonstrations are especially irreplaceable in developing expansive systems.

18.2.2.3 Formality of requirements

In software development, requirements which have been formulated in natural, that is, informal language, must be translated into the completely formal code of a programming language. It is decisive for the developmental process which degree of formality the requirements have. Three degrees of formality are distinguished:

- **informal**

    Requirements are described in natural language. The advantages for this lie in ease of producibility. Disadvantages are ambiguities and contradictions which arise out of using natural language.

- **formal**

    Description of requirements follows as a result of mathematical models. Ambiguities and contradictions do not exist in a correct formal description. On the other hand, the cost of constructing and testing a formal specification is extremely high.

- **partially formal**

    Description of requirements takes place in descriptive formal models, for example, status machines which can also contain informal elements. Partially formal description is usually applied because the cost of formal specification is too high.

18.2.2.4 Customer requirement specifications

The requirement specifications are developed to completion and stored as customer requirement specifications. The customer requirement specifications describe "realisation precepts developed by the client" (according to DIN 69905). The customer requirement specifications contain a detailed description of the expected performance output to be achieved. This also includes technical prerequisites and requirements under which the system will be developed or implemented, respectively. Predefinition of quality measures, including, among others, the test conditions, as well as prerequisites for system certification are also formulated in the customer requirement specifications. For expansive projects, even time and resource planning are included. The exact composition of customer requirement specifications is not stated.

18.2.3 Software architecture

It is essential to conceptualize the solution, respectively the architectural design, in order to make implementation as simple and efficient as possible, and in order to guarantee reusability and changeability of the system, or of parts of the system. Without a structural solution concept, it becomes very difficult to partition developmental team effort. On top of that, understanding the solution without having an underlying concept becomes extremely difficult. Therefore, an easy to implement and reproducible solution must be carefully planned and designed from the beginning on. This chapter describes which elements of software development are contained in architectural design, and what effects they have. Conceptualization of a solution consists of the projection of architectural design, respectively, the crude design.

Definition: Architecture

The organizational structure of a system or of a component (IEEE 610.12).

Definition: Design

1. *The process of the definition of architecture, components, intersections and other characteristics of a system or a component.*
2. *The outcome of the process according to 1 (IEEE 610.12).*

18.2.3.1 Breakdown into components

The division of the complete system into individual components is a main characteristic of architectural design. Individual components make the developmental process easier to understand and manageable. Customarily, components are further broken down into smaller components until one particular requirement of the system can be solved. Without this kind of partitioning, the complexity of a big system is not manageable.

Components should have the following properties:

- Each component is a unit within itself.
- Utilizing the components does not require knowledge of their inner structure.
- Interactive communication of components follows clearly defined intersections.
- Changes made to one component, while leaving the intersection untouched (unchanged), have no effect on the whole system.
- A component can be tested independent of the whole system, respectively other components, for accuracy.

In order to achieve a good subdivision, it is essential to distinguish between the externally visible result of a component, and the internal calculation of this result. In architecture, it is important *what* the component achieves, not *how* it achieves it. A component's result is delivered to the outside via the intersection. How the result comes to be, is described in the design. There are four different elements:

- Realization of functions
- Representation of objects
- Construction of data structures
- Implementation of third party efforts

The partitioning of components should be reflected in the data to be manipulated. Ideally, a component only manipulates its own data, that is, no other component has access to these data. Data exchange occurs at the intersections. Furthermore, a component should implement a self-contained function, respectively, it should administer only one data object and its associated functions.

18.2.3.2 Intersections

Intersections describe the output range of components. The implementation of this output should remaining unknown to the outside. An intersection makes functions and data types available. It should be avoided to provide internal data directly because this would compromise the alterability of components. The supply of output through the intersection must be obligatory for the component. This is accomplished by these two postulations:

- **Prerequisites**

    These define the conditions which must be fulfilled before the component can start to function. This is important in order to avoid eventual malfunctioning of components.

- **Covenant**

    Here, the end effects of a component are clearly defined.

A well-defined interaction of components is facilitated by adhering to these postulates. In doing so, components can be interchanged or altered, so long as intersections remain unaffected.

18.2.3.3 Communication within the system

Crude communication processes between components must be determined during the architectural design. Any occurring messages and events must be delegated to the appropriate system part. During this process, fundamental decisions have to be made. If there is a central entity which coordinates the delegation of messages, then an independent component must be created for it. On the other hand it is also feasible that every component takes care of its own messages and sends them independently on to the recipient. This involves an increase of effort within the components.

18.2.3.4 Ability to test components

A qualitative valuable breakdown of the system into components also includes the ability to troubleshoot them. It must be possible to test each component individually. To do so, according to test procedures specified in the customer requirement specifications, each component is tested with regard to the output required of it at the system intersection. Furthermore it must be ascertained that requirements are met according to specification. The subdivision of the system into independent components must occur in such a way, that the above mentioned requirements can be fulfilled.

18.2.3.5 Additional quality characteristics

If parts of the system can be fashioned from already existing parts, or if parts of the system are to be re-utilized in subsequent projects, these parts must be identified and described during architectural design. Which parts can be produced must also be examined in this first step. For this purpose, the following considerations are essential:

- Is it possible to create a solution in completion, for example using standard software which can be configured accordingly?
- Is it possible to acquire partial systems, for example a data bank or a web link?
- Is it possible to implement individual components through acquisition, for example the calculation of mathematical functions through program libraries?

- Is it possible to utilize well-known and standardized architecture and design procedures for the actual project?
- Can and should the existing solution concept be modified in order to facilitate aquisition?

For all these considerations, cost effectiveness must be kept in mind. The cost of acquiring and integrating existing software must be lower than the cost of new development.

Likewise, the modification of solutions for reusable components must be examined with regard to cost/use relationship. The client must first conduct an internal test as to which components are qualified for re-use. Next, the costs of modification and implementation of reusable components, of maintenance and servicing, and of integration into other systems, must be determined. This is in contract to new development of future software projects.

18.2.3.6 Resources

Crude assignment of resources to conceptualized components is already essential during this phase of software development. Otherwise it is not possible to test whether demands can be fulfilled, and whether the actual design is technically feasible. The following resources must be assigned:

- Components to processes
- Processes to processors
- Data to data storage devices
- Process communication to communication technology, respectively communication media

The optimal workload of resources is not normally of vital importance. A clearly structured solution, which can be efficiently tested and modified, is more important.

In addition, economic resources must be incorporated into the design. Costs for developing the components and intersections must be known and effectively allocated. Furthermore there should be a timetable for development of components, as well as assignment of project participants to individual tasks.

18.2.3.7 Quality of the solution

The architectural design should be documented extensively. The result of this documentation must be correct, that is, all requirements must be fulfilled, as described in the customer requirement specifications. Furthermore, resources such as time and money must be allocated to the individual parts of the system, according to the requirement specifications. The conceptual solution must be easy to understand, error-tolerant, and alterable. Error tolerance and alterability should ensure low costs for a potential re-design.

18.2.4 Possible architectural styles

The type of components used, their apportionment, and their interaction determine the architecture. If these three characteristics are in tune with in each, a uniform architectural style can be applied.

18.2.4.1 Functional orientation

A functionally oriented design, also called structured design[178] focusses its attention on the system algorithms. The system is considered a black box whose task it is to solve a problem. This black box is subdivided into component parts which themselves solve partial tasks independently from each other. This procedure is utilized until the components merely calculate functions which can be transferred into the selected programming language with ease. The result is a hierarchy of functions, where complex functions are on top and elementary functions on the lower stages (Figure 18.8).

For each desired task, participating components must be identified. In addition it must be obvious how these components interact with each other in order to solve the problem. Connections between the partial solutions must therefore be specified. Only on the bottom stage should details of individual solutions be examined. On the higher stages, only the partitioning into partial solutions and the connection between these partial solutions is essential.

Desired data structures are first duly appropriated in the lower stages. Previously, data abstractions are used. Data transfer usually takes place via parameter transfer, or via shared storage space.

Components which utilize shared data can be summarized only with difficulty, or not at all, because each component should only implement elementary functions. This leads to an abstruse design, because logically connected parts of the system are evaluated separately from one another. This also makes the changeability of the solution more difficult, because one alteration can cause effects in many different places.

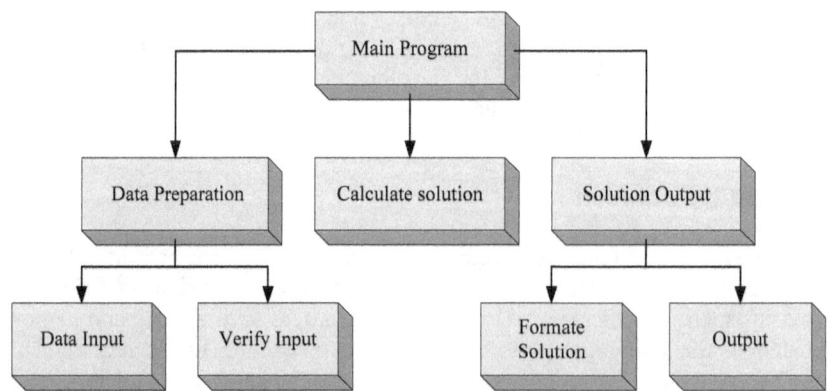

Figure 18.8: Example of a functionally oriented design

---

[178] [PAGE88] Page-Jones, M., *The Practical Guide to Structured Systems Designs*

18.2.4.2 Object orientation

In contrast to function-oriented design, object-oriented design concentrates on data and their associated operations, instead of on algorithms. Data which belong together, and the operation applied to these data, are called objects. Certain types of objects are combined into classes. These form the components of architectural design. Of particular importance is now the identification of matching classes, and of the corresponding communication between the objects of these classes.

18.2.5 Reusable architectural structures

18.2.5.1 Design patterns

A design pattern[179] describes a specific solution for frequently occurring problems. They facilitate therefore re-use of structured and field-tested solutions, and are particularly suitable for object-oriented software development. Design patterns are subdivided into three classes:

- **Structural design**

  Here, classes and objects are combined into larger structures. Class designs combine intersections and implementations, whereas object designs combine different objects.

  Example: adapter

  An adapter bridges one intersection to another. This enables communication between two intrinsically incompatible intersections. By using an adapter, classes can be reutilized without altering their intersections.

- **Production design**

  Production designs encapsulate the production process of objects. Thereby the system becomes more independent in interacting with applied objects.

  Example: prototype

  Prototypal entities are used to produce new objects. The prototype is copied and adjusted to match the necessary attribute. The advantage lies in the rapid production of complex objects.

- **Pattern of behavior**

  Patterns of behavior described the interaction between objects and classes.

  Example: iterator

  An iterator enables a sequential access to the elements of a complex object. At the same time the structure of the object is unknown to the iterator.

---

[179] [GAMM95]   Gamma, E., Helm, R., Johnson, R., Vlissides, J., *Design Patterns: Elements of Reusable Object-Oriented Software*

18.2.5.2 Frames

Frames or frameworks provide base frames for particular problem classes. Therefore they facilitate re-use of entire architectures. An object-oriented frame predefines the essential classes and objects of software. The utilization of certain frames is normally restricted to a particular application. One application example, which can be developed by use of a framework, is a web-based warehouse. Many operations and intersections are standardized and occur in all applications of this type, so that a framework can be used as a scaffold.

18.2.5.3 Architectural design

Architectural designs, like developmental designs, provide solutions for frequently recurring problems. However, architectural designs apply to the entire system, not just individual parts. One example of an architectural design is the model-view-controller (MVC) design. This particular system is structured into three units. The model puts into practice the whole of the application logic and has no knowledge of the view layer, respectively, the navigation layer. The view layer registers with the model and is kept up to date with messages about every change. Its task is to represent implementation logic data to the outside. The navigation layer processes user input and regulates the actual procedure of the application. This design separates application logic from navigation logic. This allows a simplified breaking up into components, because dependencies are minimal, as described above.

18.2.6 Programming convention

Unequivocal and project-wide programming conventions are indispensable for software development in a team. "A well written program is sensibly structured, has a clean layout, uses sensible names, is elaborately annotated and utilizes constructs of language in such a way that maximal safety and legibility of the program are obtained"[180].

18.2.6.1 Documentation and appearance of source text

An elaborate and uniform documentation, respectively, annotation, of source text is indispensable for comprehension, and for subsequent program testing. Developers must agree upon a uniform language commentary, respectively, German or English.

Introductory comments at the beginning of the source text should delineate the purpose of the data set. Furthermore, statements about author, creation date, licences used, etc. should be made at the beginning. Use and purpose of all attributes and methods of a class must be documented. If a tool for the automatic generation of documentation is used, tool-specific conventions must be followed. Important parts of the source text, for example the use of an external (exclamation?), or of complicated methods, must be extensively supported with commentaries, so that other developers can immediately understand what exactly it is that happens at this point.

---

[180] [SOMM87]   Sommerville, I., *Software Engineering*

The layout of source text files must be uniform throughout the entire project, for example:

1. Data file commentary heading with statements about name the file, author, date etc.
2. Include instructions of the external system (usually in < >)
3. Include instructions of the system proper (usually in "...")
4. Definition of data types, global variants, ...

In order to make source texts more legible one should take care not to make them too long. If source texts are very long, an attempt should be made to differentiate them into several logically connected files. The same applies to very long methods.

Indentation depth, as well as the position of brackets, in method definition and control instructions must also be clearly regulated. For indentation, one must avoid the use of tabulators, if several different program editors are employed during the course of development. Instead, a fixed number of empty spaces should be used.

18.2.6.2 Naming convention

A uniform style contributes considerably to better understanding of source texts, when allocating attribute and method names. One possible example of naming convention is shown in Table 18.4. It is important that one convention is agreed upon, and that this convention is consequently used; the convention's appearance is but of secondary importance.

Hungarian notation

In Hungarian notation, an attribute name consists of base type, prefix, and identifier. Different variables are represented by abbreviations, the so-called base types, for example *b* for variables of type BOOL. Unique abbreviations have to be found for self-defined attributes. The prefix describes the purpose of a variable, for example *a* for array or *p* for pointer. The identifier stands for the expressive name of the variable.

Table 18.4: Example of an application of naming convention

Type	Convention
Constants	Constants are written in capital letters. Individual word parts are separated by underscore. The name of the constant contains an abbreviation of the designated use (intended purpose), for example: MAX_NUM
Classes	Class names are written as a combination of capital and small letters without underscore. The new part of a class name begins with a capital letter, for example: BankCustomer
Methods	Method names always start with a capital letter. Individual word parts are also separated by capital letters, for example: NewBankCustomer(). Methods which identify the attribute of an object start with Get, for example: GetName(). Methods which alter the attributes of an object start with Set, for example: SetName().
Attribute	Attribute names are written in small letters and contain a meaningful name, for example: secretnumber. Excempt are simple number variables such as i, j.

- *Advantages*

    Error avoidance is increased because, for example with assignments, it is immediately possible to see whether they are executable. Comprehension of code is dramatically improved because type and purpose of each variable are directly obvious.

- *Disadvantages*

    Hungarian notation can lead to uninformative attribute names because identifiers are simply left out. Furthermore, the existence of many different versions of Hungarian notation enormously hampers intelligibility.

18.2.7 Software development with UML

The object-oriented premise for developing complex software consists of two steps, analysis and design. The following part orients itself on the Object Engineering Process (OEP [181]).

18.2.7.1 Object-oriented analysis

Object-oriented analysis uses elements of object orientation in order to simplify requirements and problems of the developmental process. For this, the above described techniques of abstraction, enclosure, and heredity are applied. Object-oriented analysis is a relatively elaborate method to describe the system-in-development as completely as possible. The goal is to make up for the effort expended on analysis during design and implementation.

System description

System description is the first step into analysis. System description contains initially only vague ideas of the system to be developed. These should be briefly documented in written form.

In the next step, all possible interest holders should be identified. Interest holders are all persons who have a particular interest in output or requirements of the system to be developed. To these belong, among others:

- System users
- Clients
- Developers
- Service personnel
- Customers

The variety of desires and demands of interest holders must be evaluated according to their relevance. Not all interest holders will be 100% satisfied, so that a choice has to be made in advance between the musts, the shoulds, and the coulds of requirements.

---

[181] [OEST05]   Oesterreich, B., *Analyse und Design mit UML 2*

Subsequently, different use cases can be developed. In order to do so, individual use cases are modeled by means of appropriate diagrams. Accordingly, descriptions are crude and do not need to make any statements as to the interior of the system.

Next, use cases are appropriated by considering environmental conditions and technical facts. For this purpose, individual use cases are put into temporally correlated system application cases. This creates a temporal sequence and a definable system state for each time point. Individual steps must be formulated as abstractly as possible.

Identifying subject classes

Herein, the most important subject matters are identified and placed in a structural context. This results in a crude class diagram. Connections and dependencies between the classes are defined through the use of association and inheritance resources.

All essential terms should be defined as soon as subject classes are known. This vastly facilitates communication and comprehensibility between different departments or developers.

Modeling the system processe

The individual steps of system application cases are further split up until they can be shown in an activity diagram. At first, the standard process is described, that is, the planned sequence without exceptions and variants. Following that, all conditions of exceptions and possible branching points are described. Now, black boxtest cases (see paragraph 18.2.14.2) can be deduced from the completed activity diagram, which perform a comparison between desired results and actual results. There is a possibility of implementing these test cases already at the present point in time.

Now, all as yet developed application cases are presented in an application case model. To do so, all application cases with a logical connection are combined into one package.

Finding redundant application cases

Redundancies often occur during the systematic procedure of application case development. Certain steps are executed multiple times. These steps should be isolated and modeled as a stand-alone application case.

Describing system intersections and developing prototypes

Each application case has an intersection for incoming and outgoing data, objects, and messages assigned to it. That is, all intersections of the future system, which will lead to the outside, are described. It is therefore essential to find all intersections included in the system.

Once system intersections have been identified, it is possible to develop explorative prototypes in order to test the suitability of application cases.

18.2.8 Object-oriented design

The outcome of object-oriented analysis is a detailed model of the application field. It serves as the foundation for object-oriented design. In practice, the transition between analysis phase and design phase is often fluctual. Object-oriented design takes structures which were developed during analysis, and inspects them from a software developer's point of view. The interrelationships of classes are examined in more detail and made more abstract. The design serves as the base of implementation. The goal of object-oriented design is the construction of an architecture, the modeling of special procedural structures from the general procedural structures, and the conversion of a logical structure into an implementable structure.

18.2.8.1 Architecture

Prior to drawing up the design, the fundamental system architecture (see section 18.2.3) must be defined. This results in the necessary classes and intersections which the system must contain. In order to describe architecture, three models are customarily required.

Layer model

The elements of the layer model must be recognized. This includes individual layers, such as client/server layer, components, and intersections. These are presented in a component diagram. Furthermore, any dependencies between the individual layers, as well as the direction of access, are formulated. In order to safeguard a uniform application of architecture, the software to be developed can be embedded in a frame. Communication between the layers, respectively, the components, can then proceed along predefined classes of the framework. In the process, communicative relationships should be modeled such that communication takes place, for the most part, inside a layer.

Distribution model

Herein, existing architecture is customized to hardware requirements. For this purpose, existing nodes are defined. Next, individual components are assigned to these nodes. Likewise, the technical prerequisites of communication are described, and a distribution diagram created.

Subsystem model

Next, based on the previously created class model, and the architecture applied, a subsystem model is developed. A subsystem consists of a multitude of components or classes. Classes, respectively, components, which have a logical connection, are defined and combined into a subsystem.

18.2.8.2 Assigning procedural structures

The hereto developed procedural structures, respectively, the individual activities, have one component assigned to them each. For this purpose, the activity diagrams of the application cases are examined, and each activity is assigned a component. The assignment must take place in such a manner that dependencies between the components are minimized. This reduces the necessary communication and strengthens the inner coherence of the components.

18.2.8.3 Developing design classes

During the analytical phase a class model was developed, which describes the application cases and thereby the problem. Now, on the basis of this model, classes are defined which facilitate the solution of these problems. Potentially, analysis classes must be combined or separated for this purpose. Furthermore, tasks and properties of all classes of a component must be described.

18.2.8.4 Describing component intersections

Starting from the system description, intersections for all identifiable application cases, and thus for their associated components, are described. During this step a decision has to be made as to which intersections provide the component, and which intersections are needed by the component. For this, the activities of the activity diagram are examined, and the component intersections are described according to the dependencies which emerge from the diagram.

18.2.8.5 Specializing status models

The potential statuses of different objects are specified and described in status diagrams. For this, all operations which could potentially bring about a change in object status, must be identified and named. In addition, status-dependant operations must be described. For all statuses, any possible consecutive statuses, as well as necessary warranties and prerequisites, must be defined.

18.2.8.6 Object flow of activity models

Objects contained in activity models are described. For each step of an activity diagram, the incoming and outgoing objects, their status and data are modeled. The result is an activity diagram which has been broadened by integration of participating objects and status changes, and is called object flow diagram.

18.2.8.7 Modeling interaction models

For each use case, a sequence or communication diagram is modeled. To do so, the standard procedure, without exceptions and variants, is described. All participating objects, and their communication among each other, are examined. The participating messages are recorded along association lines. Messages are numbered according to their chronological order of occurrence. Weaknesses and design errors of the standard procedure can now be identified. Next, the most notable exceptions and variants are documented. Eventually, the missing properties of affected classes, such as missing classes, associations, or operations, are identified.

18.2.8.8 Developing tests

For each use case, respectively, its associated process, component tests are developed which test all possible processes. Ordinarily, this allows the identification of still missing properties, which are required in the development of automated tests. That means that, potentially, operations have to be redefined or restructured. In order to facilitate automated component tests, the design has to be properly construed right from the start. Because only component interfaces have been defined so far, and the actual content of the components is still missing, these tests can not run successfully at the beginning. In spite of this fact, the tests should be implemented because they provide a way of determining whether all operations have been defined. The advantage of component tests is that they provide new insights because, through component testing, a different angle of the system is shown.

In the following step, class tests are developed. To this end, it must be defined which classes are responsible for performing which operations of component intersections. The prerequisites and results of these operations are specified and correspondent class tests are defined.

18.2.8.9 Specifying attributes

Once the individual class tests are defined, classes must be examined for their requisite attributes. For each attribute it must be determined whether it belongs in a class of its own. Attributes are distinguished according to their type, that is, it must be specified whether an attribute is a primitive data type, an enumeration, or an ordinary attribute, etc.

Once all participating classes, their relationships, as well as the attributes and operations contained in them, are defined, the system, respectively, the components, can be implemented and tested. The weaknesses of design, respectively, errors, which occur in the process are critically examined and incorporated into the design process. This procedure is repeatedly iterated until the system meets the requirements.

18.2.9 The use of CASE tools

CASE (Computer Aided Software Engineering) tools aid in the development of software. These tools should support the entire developmental process and provide aids for every phase. Not all CASE tools are in a position to warrant this. The demands made on CASE tools are listed in Table 18.5.

Table 18.5: Demands made on CASE tools

Property	Value
Complete support of at least one modeling method	Complete concentration of developers on analysis and design feasible
Generation of code	Avoidance of errors in transferring the design into the source code
Preparation of documentation	Central administration of documentation, uniform appearance of documentation and thus improved comprehension
Support of round trip engineering (see section 18.2.9.1)	Avoidance of errors in redesign
Syntax test	Spotting of formal errors in individual components
Semantic tests	Prevention of inconsistencies within or in between components
Qualified and fast support	Fast and competent help when problems occur
Modern operation	Shorter training periods

18.2.9.1 Round trip engineering with CASE tools

The code generation from computerized UML diagrams is called forward engineering. Through the use of forward engineering, an error-free and uniform scaffold is formed which has to be filled with content. If code design changes occur during the implementation phase, for example new operations or associations, these requirements should be taken up directly into the design documents. This process is called reverse engineering. The cyclic progression of forward and reverse engineering is called round trip engineering.

Round trip engineering is, in principle, a specialization of the spiral model. The four phases are:

1. Analysis of demands

   Demands are analysed and described by means of UML diagrams.

2. Design

   The design is created. In doing so, UML diagrams are refined and specialized.

3. Implementation

   From the UML diagrams, code is generated and filled with content.

4. Test

   The described code is tested with respect to requirement specifications.

Alterations to the code, respectively the design, usually result following the fourth phase. These can be performed and tested directly on the code. After the test, design changes are directly incorporated through reverse engineering into the analysis, respectively, the design documents, and can be examined as to their effects on the whole system. The developmental process can begin again.

An important requirement of modern CASE tools is the support of round trip engineering. Thereby, balancing of design and implementation can be automated. The result is better software quality, fewer errors, and an optimized implementation of requirement specifications.

18.2.9.2 MDA

In model-driven software development MDA (Model-Driven Architecture), the division of software into platform-independent and platform-specific models is prioritized. Platform-independent models describe the factual aspects of software. This includes business processes and applications, which are independent of the technology applied. Platform-specific models contain elements of implementation which are dependent on technology.

Actual performance requirements remained virtually unchanged over time. Thus, sorting algorithms or transaction concepts are still implemented nowadays. Applied and supported technologies, on the other hand, are in constant change. There is, for example, a continuous development of programming languages, or new concepts, such as web connection through a server-client architecture.

MDA software development provides for a better reusability and increased lifespan of software production due to its separation between application and technology.

18.2.9.3 Comparison of UML CASE tools

Table 18.6: Features of UML CASE tools[182]

Name	Rational Rose	Software through Pictures (StP)	Together
Manufacturer	IBM	Aonix	Borland
Supported UML diagram types	all	all	all
Supported operating systems	Windows, Linux, Solaris, HP-UX, IRIX	Windows, Solaris, Linux, HP-UX	Windows, Linus, Solaris, HP-UX, Mac Os X
Round trip engineering	yes	yes, via StP/OO-Re	yes
Supported languages for round trip engineering	C/C++, Java, C#, .NET	C++, IDL, Java, TOOL libraries	C++, Java Visual Basic, C#, IDL
MDA (Model Driven Architecture)-support	yes	yes, via StP/ACD	yes
Integration of IDEs	Visualstudio, Eclipse, JBuilder	Visualstudio	own IDE, JBuilder
XMI (XML metadata interchange)-auppoer	yes	yes	yes

---

[182] [MÜLL03]   Müller, F., *Entwicklungshelfer, iX*
   [JECK05]   Jeckle, M., *Unified Modeling Language (UML) Tools*
   [IBM05]   IBM, *Rational Software*
   [AONI05]   Aonix, *Software through Pictures*
   [TOGE06]   Together 2006 für Eclipse

18.2.10 Software quality

In the following section, software quality is evaluated by inspecting the definitions and properties of software quality. Furthermore, procedures are considered with which software quality can be determined. There are different formulations for defining software quality:

- Quality is the adherence to requirements *(Philip B. Crosby)*

- Quality is the sum total of attributes of a unit with regard to its qualification to fulfill defined and provided requirements. *(ISO 8402)*

- Quality is the sum total of characteristics and attributes and of a product, or of an activity which refers to its qualification to fulfill predefined requirements. *(DIN 55350, part 11)*

- Quality should aim at fulfilling the present and future wishes of the customer. *(W. Edwards Deming)*

The most important components of software quality features are:

- Reliability
- Comprehensibility
- Portability
- Maintainability
- Efficiency
- Reusability
- Correctness
- Safety
- Robustness
- Modularity
- Testability
- Flexibility
- Usefulness

Some properties of quality can be directly contradictory, and not every property of quality is relevant for every software.

In order to ensure the desired properties of software quality, a systematic procedure is needed. This is summarized as quality management. Quality management attempts to achieve quality through appropriate course of action, aids, and methods.

Definition: quality management

All actions of collective executive function which determine politics of quality, goals,, and responsibilities, as well as implement them by means of quality planning, quality control, and quality improvement within the framework of the quality management system (ISO 8402).

- **Politics of quality**

    The politics of quality describes the quality targets of the client, for example, production of high quality products or good customer service.

- **Quality management system (QMS)**

    QMS contains the organization of composition and process, the means and responsibilities of handling quality management.

- **Quality planning**

    At this, process requirements and product volume are determined. Process and product characteristics must be selected, classified, and weighted.

- **Quality plan**

    The quality plan contains all quality measures of a project.

- **Quality control**

    In order to accomplish the goal, the actualization of developmental steps is controlled, monitored, and corrected.

- **Quality testing**

    Testing determines which requirements are met by the software system, or by a part of the system.

18.2.10.1 Quality plan

For each specific software project, a quality plan is prepared, which contains measures which are relevant to quality. Normally the quality plan refers to other project documents and project plans, as shown in Figure 18.9.

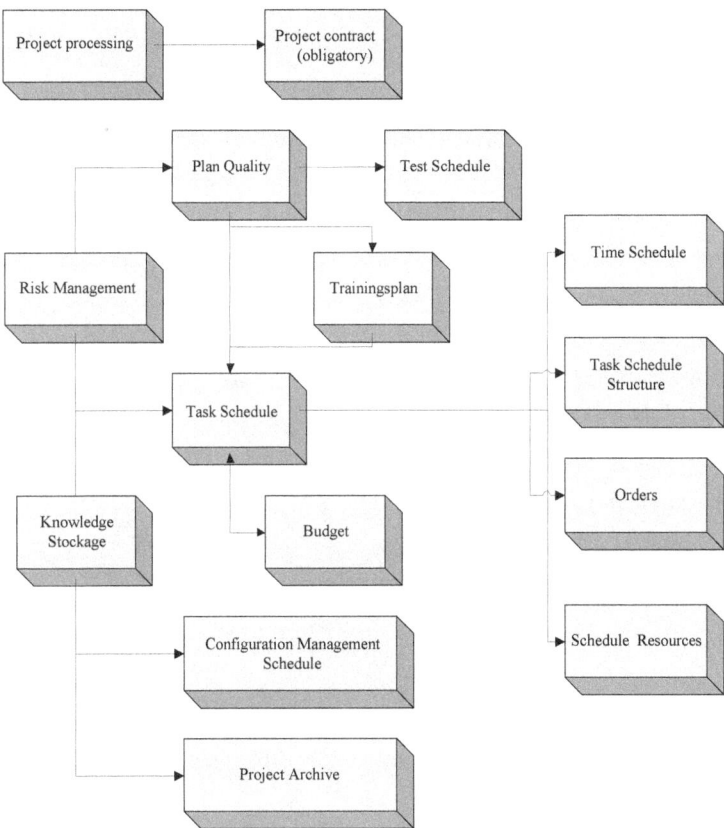

Figure 18.9: Correlation of various project documents

The following points are examined in the quality plan, starting from predetermined risks, project type, and complexity of the project:

- Certification and suitability of tools
- Reviews
- Testing process
- Error management
- Certification of completed tasks
- Qualification of resources, aids, etc.
- Documentation and its control

18.2.11 Software reliability

One of the most important quality targets is malfunction-free and error-free availability of software. In order to be able to describe malfunctions and errors in software systems, their reliability must be examined.

Definition: software reliability

Reliability is defined as the probability of a failure free system for a specified purpose, during a specified time span, and in a specified setting.

According to this definition, the setting, as well as the purpose of software, is significant. That is, software must behave appropriately and according to specifications in an examined setting.

Definiton: software defect, software error, software failure

1. *A software defect occurs as the result of errors made by the software developer during the prgramming phase.*
2. *A software error is a manifestation of a software defect.*
3. *Software failure occurs when an error prevents the software from performing its function according to specification.*

Software defects occur due to a great number of factors. The most important factors are:

- Incomplete or inaccurate requirements list
- Incomplete or incorrect developmental document
- Program error
- Incomplete tests
- Inaccurate solutions

As a result of these factors, most software systems are not failure-free on average, each 1,000 lines of source text contain six errors, and the average effectiveness of a solution is 99.5%.

18.2.11.1 Measurements of reliability

The most important measurements of reliability are:

Relative functional period (availability):

$$\frac{t_F}{(t_F + t_A)} \tag{18.2}$$

Relative downtime:

$$\frac{t_A}{(t_F + t_A)} \tag{18.3}$$

with: t_F = functional period
t_A = downtime

While the time the system is running, representative data can be collected and calculated during a predetermined functional period. Most commonly though it is the case that the client sets terms with regard to reliability data which have to be met by the contractor. For this purpose, appropriate procedures are employed to collect averages from the individual components. These measures are called meantime between failure *MTBF*, and the downtime is called meantime to repair *MTTR*. Decrease time describes the medium time period necessary for repairing or replacing a failed component. The *MTBF* is calculated as the reciprocal of the failure rate λ of the component to be examined.

Average of failure-free functional period

$$MTBF = \frac{1}{\lambda}. \tag{18.4}$$

Failure rates of individual components are added in order to obtain the failure rate of the total system. However, it is assumed that λ always remains constant. Thus, the availability *A* of the system can be calculated as follows:

$$A = \frac{MTBF}{MTBF + MTTR} \tag{18.5}$$

18.2.11.2 Differences between hardware and software reliability

Software reliability is different from hardware reliability in the following attributes:

- Software does not wear out
- Software is updated
- Problems with software can be new and unexpected
- Software tests are incomplete as a result of the complexity of software
- Software requires methods of tolerating errors that are different from hardware

Developing a software failure rate is fundamentally different from that of hardware failure rate. There are three types of failures in electronic systems:

- **Early failures**

  Failure rate decreases over time, because the system is revised repeatedly in order to avoid and correct errors, and to make system functional.

- **Accidental failures**

  Accidental failures occur while the system is fully functional and operating normally. Failure rate here is constant. This failure rate serves as the basis for calculating the reliability formulas.

- **Wearout failures**

  Wearout failures, or late failures, occur towards the end of the life span of an electronic system. In this phase, the failure rate increases constantly, until the system ultimately becomes unusable.

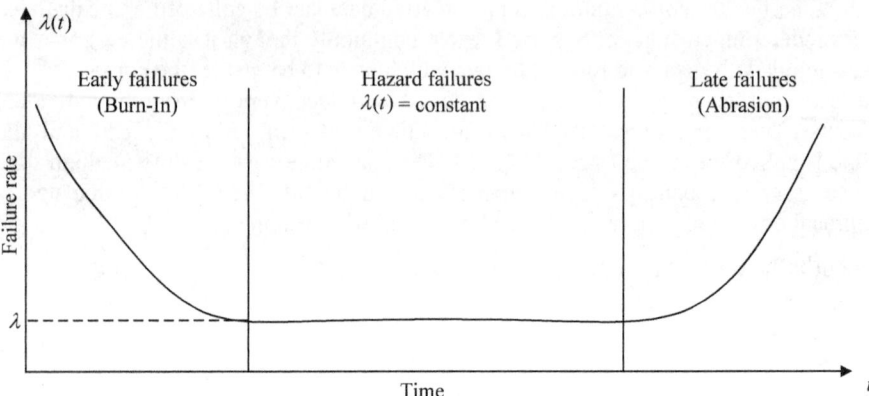

Figure 18.10: Bath tub curve of hardware reliability

A curve of these failures over time displays the bath tub curve shown in Figure 18.10.

Software reliability can also be divided into three phases (Figure 18.11):

- **Testing phase**

  Failures during this phase are caused by errors in implementation, respectively, design. Intensive testing and improvement reduces the failure rate until the software is operational.

- **Application phase**

  Analogous to hardware reliability, the system is fully functional and operates normally during this phase. Upgrades are functional upgrades which increase the functionality of software. Without upgrades, failure rate is constant.

- **Aging phase**

  Failure rate is constant. Albeit, software functionality is no longer adequate to meet increased demands. A new development is necessary.

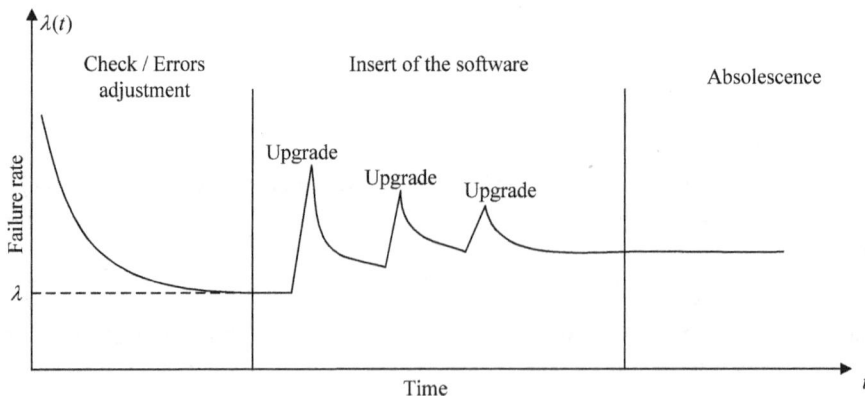

Figure 18.11: Software reliability over time

18.2.11.3 Increase in reliability by verification and validation

Verification and validation (V&V) is an important phase of software development. They are mechanisms (processes) which are applied during the entire developmental process. Verification and validation should examine results and advances in process according to specifications. The following statements can be used for differentiation:[183]

- **Validation:** Do we build the correct product?
- **Verification:** Do we build the product correctly?

One can deduce from the following statements that during verification the system is tested for functional and non-functional requirements of specification, whereas validation involves determining whether clients' expectations are met.

In V&V process the following methods of system testing and analysis are distinguished:

- Software inspections

    These analyse the entire appearance of the system. For example, request document, blueprint diagrams, and source code are examined.

- Software testing

    Herein, implementation processes and software behavior during operation are examined.

Figure 18.12 shows the phases during which software inspections and program testing can be applied. Software inspection has the advantage that it can be applied to each phase of the developmental process. Specially designed prototypes are necessary to test software.

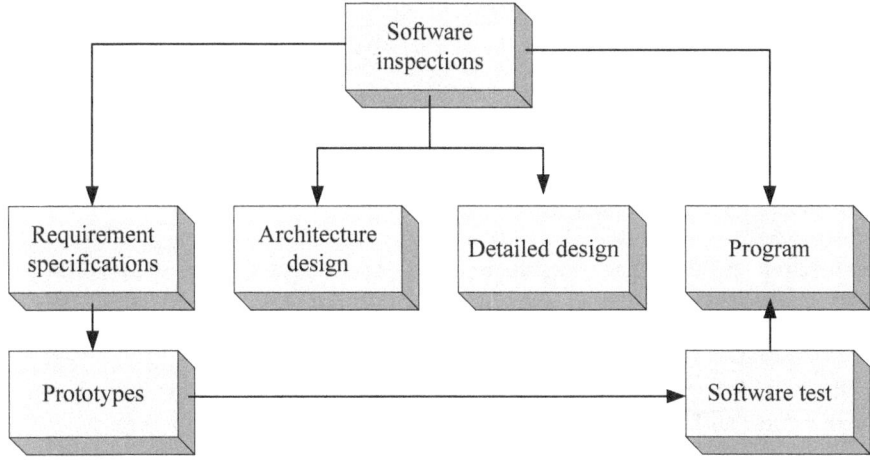

Figure 18.12: Structure of verification and validation

---

[183] [BOEH79]   Boehm, B., *Software Engineering*

When testing software, a real program is run with data. There are two types of software tests:

- Tests to find failures

    With these tests, contradictions between programs and requirement specifications should be identified. The tests are constructed such that they not only simulate the program, but also actively look for errors, for example by entries in the marginal zone of what's allowable.

- Statistical tests

    These tests give the client additional information about the software, for example, performance and reliability of the product. These data can only be evaluated statistically and demand a simulation of the program under varying conditions. By means of the number of system failures, a reliability prediction of the system can be made (see section 18.2.11.1). For performance, the speed of program execution and the speed of data processing are measured, among other things.

Verification and validation are aimed at proving that the software fulfills the client's expectations, and its purpose. This does not simultaneously imply that all errors are detected.

18.2.11.4 Validation of reliability

For the specification of reliability data, for example the failure rate λ, specific measurements are needed. The individual phases of reliability evaluation are shown in Figure 18.13.

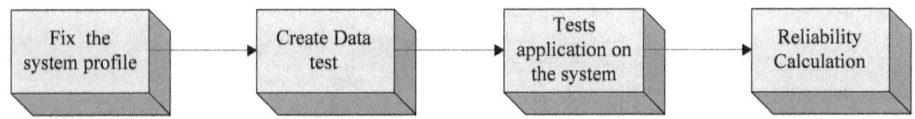

Figure 18.13: Procedure of reliability evaluation

- The operating profile characterizes the frequency and the classes of system inputs.
- Test data are created for examining the system. These can originate from a test generator or older data from the data bank.
- Test data are fed into the system, and the number of failures are observed. Time points of errors and failures are recorded.
- Reliability is calculated using statistical methods and recorded data.

In practice, it is not easily possible to develop a specific operating profile. Some systems have several users, who potentially utilize differing system functions and therefore require different test data. Some test data and inputs can not be generated automatically and must therefore be produced manually, at high cost.

The longer and more frequently a system is tested, the more accurate is the reliability observation. Elaborate testing procedures aim at lowering the failure rate by minimising errors, or through other methods. If the failure rate is low enough though, no accurate statements can be made concerning reliability, because the basis for statistical evaluation is too small.

18.2.11.5 Proof of reliability

Already during the developmental phase, informal certification procedures can be applied. With these methods, errors can be detected early on, which then clearly results in an increase in quality. These informal procedures include inspection, review, and walkthrough. Another fundamental method of increasing reliability is to conduct tests. A variety of testing procedures are discussed in section 18.2.14.

- **Inspections**

    During an inspection, the actual status of the software project is examined by means of check lists. Herein, the correctness of the program is paid less attention to, instead, the formal and factual correctness of documents is examined. Documents from different phases of the project are compared with each other, in order to identify errors, respectively, undesirable developments.

- **Review**

    In a review, quality and status of the project under development is examined. In addition, suggestions are made for improvement. Usually, a review occurs at the end of a developmental phase. Hitherto accomplished targets are examined and compared with specifications.

- **Walkthrough**

    In walkthrough, the program is examined at the functional or source code level. Here, elements are inspected for errors and inconsistencies.

18.2.12 Measuring software quality

To measure software quality, interesting characteristics are observed and quantitatively recorded. These observations are sorted into a scale by means of a formula. The transformation of measurement, that is, the transformation between an empirical field and the number field of the scale, is called homomorphism.

Definition: Homomorphism

Let A and B be two non-empty sets on which a two-part connection is defined. It follows that the transformation $f : A \to B$ is a homomorphism if

$$f(x * y) = f(x) * f(y) \qquad (18.6)$$

is valid for all $x, y \in A$ l.

A homomorphism also forms parts of a structure of "equal-meaning" parts of another structure. Through this, measurable properties of the system become comparable. For the purpose of evaluation, different operations can be performed on the measurements. Depending on the relations present in these measurements, only certain operations are permitted on scale values. Five scale types are distinguished, according to operations allowed (see Table 18.7).

Examples of diverse scale types:

- **Nominal scale:**   students' matriculation number
- **Ordinal scale:**   school grades
- **Interval scale:**   temperature
- **Relation scale:**   lengths, mass, volume, number of program lines
- **Absolute scale:**   probabilities and frequencies

Table 18.7: Scale types

Scale type	Operation	Description
Nominal scale	= ≠	Pure categorising of values. Scale values are neither comparable, nor can they be connected.
Ordinal scalea	= ≠ < >	Scale values are ordered and comparable with each other. Determinating the median value is possible.
Interval scale	= ≠ <> distance	Values are ordered. The distance between two scale values can be determined. The zero point of the scale is arbitrarily chosen. Average and variance are definable.
Rational scale	= ≠ <> distance, (+,-), multitude, %	Values are ordered and, as a rule, additive. Multitudes and percent values are definable. The scale has an absolute zero point which represents the total lack measured properties.
Absolute scale	= ≠ <> distance, (+,-), multitude, %	Scale values are absolute values, that is, they can not be converted into other scales. Apart from that, the same properties are valid as for the rational scale.

In software development, for counting and measuring, the type of scale must be defined for all observed measures. Unfortunately, a large number of presently known measurements lack well-proven statements concerning the type of scale.

Software measures are often subdivided into process and product measures, where a process measure is a quantifiable attribute of the software process or the developmental environment, for example, the experience which a developer gains over the years. A product measure is a quantifiable measure, which is applied to the product. It does not make any statements about how the product was created. Product measures are, for example, product size, complexity, or applicaton field.

The measure of program complexity is essential in the evaluation of software reliability. This makes it easier to estimate the effort expended in program development, as well as troubleshooting. Beyond that, lower complexity increases reliability.

Below, we differentiate between static measures and developmental measures. Static measures, which measure product quality at a particular time point, are divided into:

- Measures which encompass the size of the product, for example, the number of instructions, or the number of operators and operands.
- Measures which measure the program control structure of a product, for example, nesting depth
- Measures which concern data structure

Developmental measures measure product quality over a certain period of time.

Many measures can be utilized for evaluating static program complexity, where each of these measures regards different elements, as referring to complexity. For example, the number of applied operators and operands can be utilized as a measure of complexity, or the number of linear independent paths through the program. Distinct measures can not be compared with each other.

18.2.12.1 Lines of code (LoC)

The simplest measure of evaluating program complexity is counting the number of program lines in the source code. The measured value is a result of eliminating empty lines and comment lines.

By means of the number of program lines, respectively, lines of code, a variety of properties can be estimated. Thus, for example, time, effort, or storage space can be estimated. Indirect measures, such as error density, can also be deduced.

Lines of Code can deliver an approximation but is too crude for more accurate analysis. Thereby, different programming languages or programming styles have a direct effect on the number of code lines without changing functionality.

18.2.12.2 McCabe measure

This measure is not directly applied to programming code but refers to program flow diagrams. This measure elucidates the linear independent program paths and weighs them as a measure of complexity. Thereby predictions can be made about the expected test effort.

McCabe represents a program in form of a directed graph, where nodes represent instructions, and borders represent control flow. McCabe says that program complexity is dependent on the number of main paths of this program flow graph. A linear independent program path, respectively main path, is the number of border lines V, which are necessary, without intersection, as a minimum for the creation of border lines by combination. The formula is:

$$V(g) = e - n + 2p \qquad (18.7)$$

with:
- e = number of borders in the program flow graph
- n = number of nodes in the program flow graph
- p = number of connection components

McCabe's measure serves to find a minimal number of test cases by using the number of linear independent program paths. According to McCabe, programs with a value of $V(g) > 10$ are prone to error.

18.2.12.3 Halstead measures

The length of software systems is called on as a measure for determining production effort. On the basis of easy to determine values, the multitude of different results can be calculated.

These four values result from the number of operators, that is, the symbols and key words which characterize an action, and the number of operands, that is, the symbols which represent the data.

n_1	number of different operators
n_2	number of different operands
N_1	total number of all operators
N_2	total number of all operands
$n = n_1 + n_2$	number of different symbols
$N = N_1 + N_2$	total number of all symbols

n is named vocabulary of the program, and N is program length. Program length is an intuitive measure for the size of a program.

The following equations can be articulated from the above variables:

Program volume:

$$V = N \log_2 n \tag{18.8}$$

Level of difficulty:

$$D = \frac{n_1 \cdot N_2}{2 \cdot n_2} \tag{18.9}$$

Programming effort:

$$E = D \cdot V \tag{18.10}$$

The above measures can be reexamined during the entire programming process, in order to estimate the growth of complexity. Unfortunately, Halstead measures only consider the source code and neglect other properties, which influence complexity, such as stacking depth. On top of that, the computed findings reflect, at best, a trend, and deliver no accurate results.

18.2.12.4 Usefulness of formulas

Software formulas can be used to measure different aspects of software complexity. This is helpful in estimating the testing effort and error proneness. In addition, information regarding the expected service and maintenance efforts can be gained. Certain parts of software quality can therewith be measured and compared.

Yet, all software formulas only consider parts of software development and are overall merely crude estimates. They can not replace elaborate testing and review procedures.

18.2.13 Failures in software systems

Failures are defined as the non-implementation of requirement specifications. For quality assurance of a product, the following must be known:

- Which errors can occur?
- Which consecutive reactions can these errors trigger?
- Whch error behavior can be observed over a longer period of time?

The distinction of error types occurs through the use of standardized norms[184] (see Table 18.8).

Table 18.8: Standardized norms for error description

Failure	**fault:** Unallowable property which can lead to failure
	defect: Unallowable deviation, respectively, non-implementation of predefined requirements
	error: Deviation between calculated and correct value
	mistake: Human action with undesired consequences
(Hardware) failure	Violation of at least one failure criteriion
Failure criterion	Marginal condition for the deviation of allowable measures of property following start of use
Specification errors	Deviation between specified and intended function
Programming error	Error in applying the specifications to the program
Program error	Deviation between realised and intended program function
Fault	Behavior of an implementation unit (IU) which does not match the specified function
Consecutive failure	Failure caused by the failure of another IU
Systematic multiple failure	Multiple IUs fail due to the same cause
Partial failure	Failure of a part of the function of an IU
Critical failure	Failure of at least one absolutely essential function of the IU
System failure	Sytem behavior does not conform to intended function
Disturbance	Impairment of a function

---

[184] [NTG82] NTG-Empfehlung 3004: *Zuverlässigkeitsbegriffe im Hinblick auf komplexe Software und Hardware.*

The following are critical errors in software development:

- Specification errors

    While transferring client demands into a formal or informal specification, deviations, incompleteness, or contradictions can arise.

- Implementation errors

    Implementation can deviate from specification due to slips of the pen or misinterpretations. Programming errors are especially typical here.

- Documentation errors

    Faulty documentation of an implementation can lead to improper use or maintenance errors. In error analysis, all deviations between system and documentation must be included.

The point in time of error recognition is of fundamental importance for the entire developmental process. Errors which are recognized too late, for example upon activation, require very high costs and an elaborate re-design, whereas errors which are identified during the specification phase can be corrected with adequate effort. Figure 18.14 shows the fundamental occurrence of errors in different life phases of a combined hardware-software system.

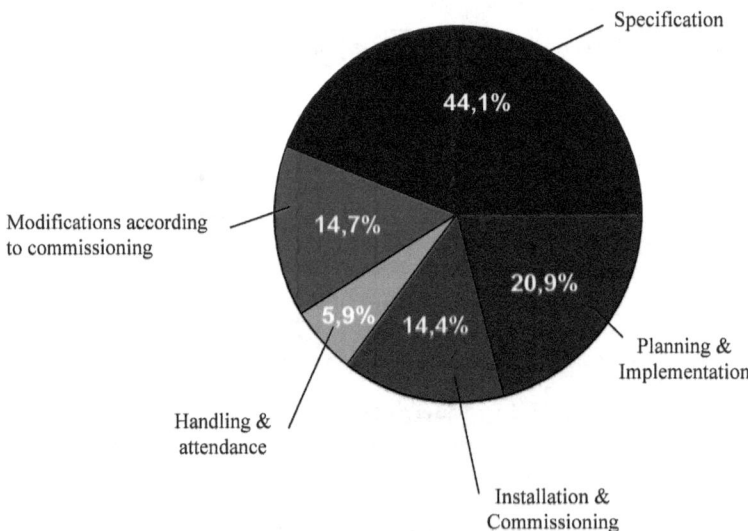

Figure 18.14: Error percentages in different life phases of a system

18.2.13.1 Error tolerance and error avoidance

There are two different methods for reducing the effects of errors on the system.

- **Error avoidance**

   At this, reliability is improved by perfecting the process of software development. The occurrence of errors should be avoided from the start. Through careful planning, tests, and improvements prior to activation, the number of errors should be minimised.

- **Error tolerance**

   At this, the attempt is to design a system which safeguards proper operation if a certain number of components should fail. This can be achieved by redundant development.

Cost effectiveness must be considered during the application of this method, or a combination thereof. Indeed, extremely elaborate testing is desirable but usually not possible due to cost factors.

18.2.14 Testing procedure

Generally, testing is differentiated into static or dynamic testing. Static tests were described in section 18.2.11.5 above. It includes examination of the source code. With that, it evaluates whether programming guidelines have been adhered to, whether documentation was done in a precise manner, and whether programming was performed in a tidy manner.

In dynamic tests, on the other hand, program flow is examined. Suitable data are entered into the program, followed by a systematic examination of the program for errors. The following section discusses dynamic testing intensively.

18.2.14.1 Testing procedure

Within the framework of the developmental process, several testing tasks present themselves. The minimally requisite testing tasks are defined in norm IEEE 1012-1986.

- Module testing
- Integration testing
- System testing
- Acceptance trial testing

Table 18.9: Standardised activities and results of a testing task

Activities	Results
Test planning: Generation of a document which contains the targets, the extent, the methods, the resources, a time table, as well as the responsibilities of the actual testing task	Testing plan
Test design: Test methods are formulated in detail. It is established which methods will be applied, and in which manner, in order to reach the predetermined target. The criteria which decide success or failure of a test are determined.	Test design specifications
Test case determination: It is specified, which test object will fail in which test case and with which input, and which task it is expected to deliver	Test cases
Test procedure planning: Steps for the process of test cases are determined, that is, requirements for the execution of tests and their sequence	Test procedure specifications
Test conduct: Specified tests are conducted. The outputs of individual tests, as well as problems and incidents during the testing procedure, are recorded.	Testing protocol Test incident record
Test evaluation: Test results are analysed. The test is evaluated and a decision is made whether another test should be executed.	Test report

The norm suggests standardised activities and results for solving the test tasks.

The standardization of testing protocol enables more accurate time and cost planning, and diminishes the risk of uncontrolled testing. Two distinct methods are possible for test case determination:

- Black box test

  The test object is considered a black box, that is, the tester does not have any knowledge about the inner workings of the test object.

- White box test

  The structure, respectively, the inner workings are known to the tester. This information is used for designing the test and is incorporated into the test.

18.2.14.2 Black box test methods

In this segment, four different black box test methods are explained.

- Coverage of functions

  Functions of test objects are identified by means of concrete application cases. For each function, an input and output specification is specified. Test cases are generated from these specifications. The tests are conducted in order to determine whether functions are present and practicable. Test cases do not normally take into consideration (make allowances for) peripheral or exceptional situations.

- Equivalency class method

    A test case is regarded as ideal, if it detects a class of errors, for example, the processing of floating point variables in a particular class. This way, the number of test cases can be reduced. An equivalency class is a multitude of qualities of size. It is assumed for all values of an equivalency class, that the test will deliver the same results for one quality of this class, for example error type or error number, as for all other qualities of this class. When designing test cases, the value ranges of input and output sizes are allotted (graduated). One particular quality of this class is selected and used in testing the equivalency class. The allotment of classes must first be examined with regard to completeness.

- Marginal value analysis

    Of special importance in testing is the treatment of marginal situations. Many errors can occur here. In the analysis of marginal value, test cases are generated which examine just that one particular marginal situation. In this, marginal situations are used as test cases for input and output sizes, or their close approximation.

- Cause/effect graph

    Herein, program specification is translated into graphic formalism. The nodes on the graph represent causes and effects. A border connects a cause with an effect. Causes can be, for example, input conditions and their effect, for example, output conditions. The variety of causes and their resulting effects are examined during testing. These graphs are very difficult to design and relatively error-prone for complex specifications.

To be able to perform a black box test, all functional requirements on system and component must be known. Even then, it is still difficult to develop a complete test plan, because often times a very large number of functions has to be tested. If only informal requirement specifications are available, it takes very high effort to generate the corresponding black box tests.

18.2.14.3 White box test methods

In white box tests, tests are extrapolated from the code. The goal is, to cover certain criteria which refer to the code, see Table 18.10.

To test software using the white box method alone is not recommended because requirements are not tested. A program can therefore be absolutely error-free, but still does not perform what the client expects of it.

Table 18.10: Coverage measures in white box tests

C_0	Run through of all instructions
C_1	Run through of all bifurcations
C_2	Run through of all conditions in bifurcations and loops
C_3	Multiple run through of all conditions in bifurcations and loops
C_4	Run through of all paths

Instruction coverage test/ C_0-test

In the C_0 test, all instructions of the source text are run through at least once. But in doing so, control structures, and data dependencies between program parts, are not taken into account. With merely 18%, this test has the lowest rate of identifying existing errors[185]. It should therefore only be used in connection with other tests.

Branch coverage test/ C_1-test

In the C_1 test, all branches or leaps are run through. Loops, combinations, and complex conditions of branches are not tested sufficiently because a single runthrough is satisfactory. Error identification rate is at 34%[186]. This test is known as the minimal test criterion.

Condition coverage/ C_2-, C_3-test

In condition overlap, all terms within conditions are overlaid at least once with TRUE and once with FALSE. These procedures do not guarantee complete branch coverage. In multiple runs of condition coverage tests C_3, all possible combinations of conditions are tested. This includes branch coverage. However, an enormous effort is necessary because n terms result in 2^n test cases. For this reason, the test is hardly practicable.

Path coverage/ C_4 test

At this, all paths of a component are run through at least once. The number of paths grows exponentially if repeat instructions are present and in the absence of a fixed repetition number. For that reason, this procedure is practically insignificant.

18.2.14.4 Intuitive test case determination

Intuitive test case determination is based on a person's perceptive abilities and on heuristic procedures. Thus one often has a sure feeling where errors are hidden. Furthermore, there are error categories which very occur very frequently, and in many applications.

---

[185] [BALZ97] Balzert, Helmut, *Lehrbuch der Software-Technik*
[186] [BALZ97] Balzert, Helmut, *Lehrbuch der Software-Technik*

18.2.15 Testing in practice

In practice, one can not avoid using a combination of white box and black box test methods. Black box tests probe the requirements according to specification. It is necessary to know the inner structure of the system in order to localize errors that have occurred. For this purpose, white box tests are required. A step-wise procedure is recommended for testing a module:

- Probing specifications with the cause/effect graph method
- Equivalency method
- Boundary value analysis
- Quality improvement of test cases by intuitive test case determination
- Completion of test cases with white box methods

Proof of complete accuracy of a program is not possible using the methods described above.

18.2.16 Integration

The above described test methods are primarily applicable to program components. Following the successful completion of their respective tests, these components are combined into a software system. This so-called integration process must be probed systematically. During integration tests, interaction of components with each other is tested. As to which tests must be performed, can be derived from system specification. Of particular difficulty is the exposure (detection) of errors in the system. The interaction of individual components in the system is often very complex and not straightforward not easy to understand as a whole). At first, only a minimal system configuration should be tested. Following successful completion of these tests, further components can be added to this minimal configuration and subsequently tested. There are a number of different procedures in integrating new modules: top-down integration, and bottom-up integration, as well as outside-in integration, a mixed variant.

18.2.16.1 Top-down integration

In the top-down method, primary components are tested first. A main program is created and tested. Dummy programs, that is, temporary substitutions for a component which is needed for the actual test, are placed in the location of missing program parts. An advantage is the early generation of demonstrable models. Intersection errors are identified early on, making it easier to localize errors. A disadvantage is the additional creation of dummy programs. Furthermore, critical base modules are designed and tested late in the process. The collaboration of system software, hardware, and of software-in-testing is examined only late.

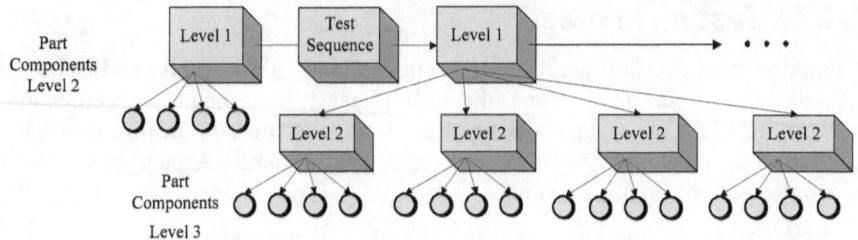

Figure 18.15: Top-down integration

18.2.16.2 Bottom-up integration

With bottom-up integration, secondary components are integrated first. First, base modules are developed and tested. Next, further modules are layered on from bottom to top and tested. A disadvantage is the late development of a demonstrable product. Intersection errors are only detected late yet are easy to localize. The bottom-up method requires additional test actuators, that is, special test intersections, in order to test system components. It is advantageous that several modules can be worked on simultaneously. Furthermore, the collaboration of system software, hardware, and of software-in-testing is examined early on.

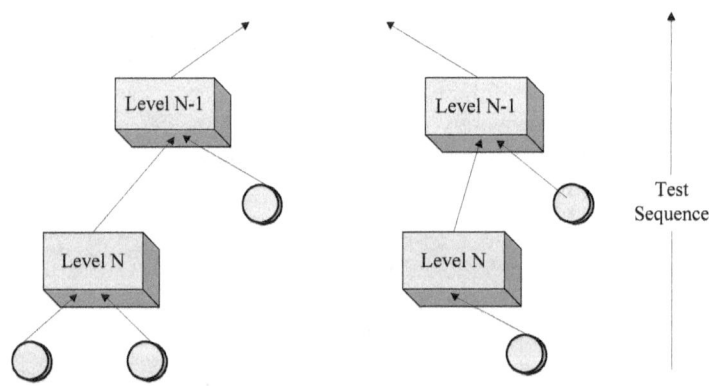

Figure 18.16: Bottom-up integration

18.2.16.3 Outside-in integration

In practice, most often a mixture of top-down and bottom-up integration is used. This is called the outside-in procedure. Integration is started simultaneously from top and bottom, and works towards the middle. The target of this is to combine the advantages of top-down and bottom-up methods, and to minimise the disadvantages of these procedures.

18.2.17 System and certification test

The system test is the finishing test of a software system without the client. It is conducted under realistic conditions by the software developer and quality control personnel. The foundation for the test is provided by the requirement specifications which have been recorded as customer requirement specifications.

- Completeness

    Functional and non-functional requirements, according to specification, are tested

- Performance

    It is determined whether the system will remain functional within specified response times, while under load.

- Suitability

    User-friendliness of the system is tested. An evaluation is performed, from the point-of-view of the end user, whether the software is easy to learn and to understand.

- Documentation

    System documentation is examined. This includes installation, service, and user documentation. Incomprehensibilities and contradictions must be found and corrected.

Aside from servicing and overhaul work, software development is brought to a close by a client certification test. Herein, the software is tested and evaluated under actual conditions in the client's hands. Conditions of the certification test are similar to those of the system test.

19 Application examples

19.1 Practical Implementation of the IEC 61508 Safety Standard

The IEC 61508[187] is an international norm which was issued by the international electrotechnical commission (IEC) in 1999. This international standard is the first of its kind, in that the Functional Safety of electrical/electronical/programmable safety-oriented systems plays a central role.

It is vital for the owner of an installation to set up a sufficient safety management system. If, in the event of an accident, the safety-related system does not conform to the preset standard, the company could be incriminated and held liable for neglecting proper safety guidelines. Accordingly, adequate provisions for deactivation and an appropriate technique, which conforms to the international standard, are necessary and probably essential, for downstream maintenance and for external companies.

The term Safety Instrumented System (SIS) was introduced into the safety standards. Included in it is all basic hardware, from sensors to logic solvers, to the terminating actuators. This procedure of risk reduction is essential for the smooth and safe operation of the installation.

The IEC 61508 safety norm commences with a determination of the risks inherent in the process, and their analysis, followed by risk assessment. All required safety functions must be specified and documented in a so-called "Safety Requirement Specification", which includes all sensor and valve configurations. The supplier of the safety system starts with the design of the SIS, based on predetermined specifications.

The norm also covers all safety validations subsequent to installation, as well as requested or periodic manual proof tests, and unavoidable modifications during operation. In the process, it is essential that all decisions, actions, and results are noted, in order to leave a verifiable trace behind. This TI (Technical Information) represents the practical, consecutive steps and measures which the user must take, and which must be adhered to similarly by the suppliers, respectively, external companies, in order to conform to the standard. The selection and scope of the input sensors, as well as the type of logic actuator and safety valves, is also mentioned.

---

[187] [IECa]   IEC 61508, *International Standard 61508 Functional Safety: Safety-Related System*

19.1.1 IEC 61508 Norm

The IEC 61508 represents a systematic method of compiling all related process risks and their definitions, and of taking measures. Extraordinary emphasis is placed on the design and the validity declaration of the safety-related system. The norm presents a life cycle concept, including a method, of producing the required safety integrity level (SIL) for the particular kind of installation. The focus is generally on built-in safety and a strategy of risk management; in addition, the demands placed on the SIS are considered the last layer for containing a dangerous situation.

This new approach requires expert knowledge, which must be procured in the course of apprenticeship, or during a training program. Through industry contacts it has become apparent that interest in consultation, respectively, support in implementing standards IEC 61508 and IEC 61511, has increased.

The IEC 61508 norm encompasses seven parts:

- General requirements
- Requirements for the E/E/PES safety-related system
- Software requirement
- Definitions and abbreviations
- Methods for establishing guidelines for the SIL
- Guidelines for the application of parts 2 and 3
- Overview of techniques and methods

Parts 1, 2, 3, and 4 are normative, parts 5, 6, and 7 are for information only.

The "safety life cycle", which is fundamental to the IEC 61508, is shown in Figure 19.1. It covers not only the development of a system, but all main phases of its existence. The approach to the total safety life cycle requires a repeated evaluation of the safety of measuring instruments through all participating companies, for example, contractors, distributors, integrators, and users.

The emphasis of the norm is placed on four main aspects:

- Total life cycle
- Management of functional safety
- Evaluation of quantitative safety
- Pipe to pipe approach

Each of these aspects is discussed briefly in the following section.

19.1 Practical Implementation of the IEC 61508 Safety Standard

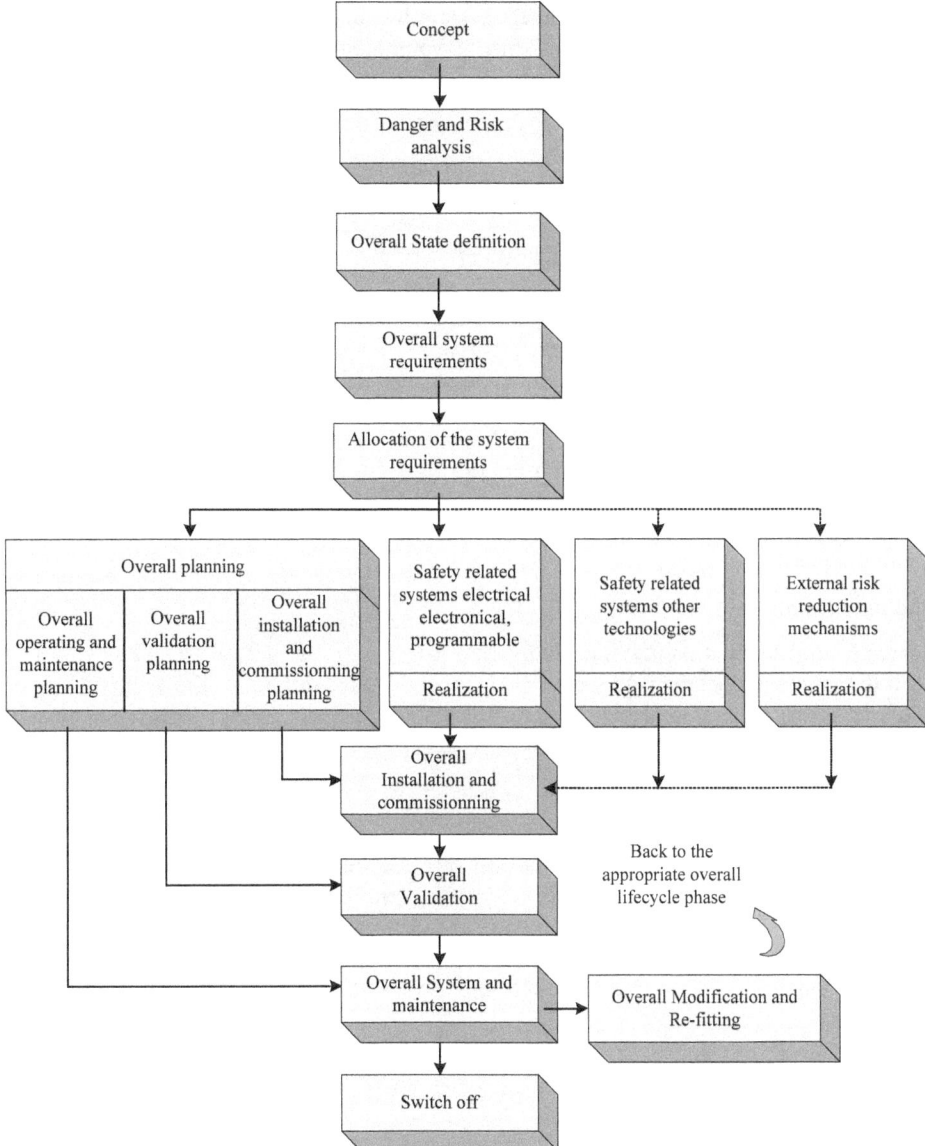

Figure 19.1: The safety life cycle[188]

It must be pointed out that, in safety management, it is emphasized that every step in the life cycle is documented in full. Included in this is the competence of the organization and of the participating persons, who are incorporated into this life cycle.

---

[188] [IECa99]   IEC 61508-1, Figure 2

In some companies, a certain management system for functional safety is already in place. These generally conform to IEC requirements, yet the aforementioned life cycle is not always given the same level of significant attention.

The following chapter describes all steps of the total life cycle. Figure 19.2 shows the practical consecutive steps, including those which determine safety functions. In addition, the design and implementation of safety loops is shown.

19.1.1.1 Functional Safety Management

This step can prove to be the most difficult and most underestimated part of the norm. All companies which are involved with every possible step of the life cycle, must be accomodated in the Functional Safety Management system (FSM). All manipulation and technical activities, which are essential in order to achieve the necessary functional safety, should be specified. Furthermore, all procedures applied during the life cycle are shown, the competence of the persons, departments, and organizations in charge, and the manner in which testing and the declaration of validity are handled. All of these events require exact implementation, so that they can be verified, and all decisions are understandable and open to scrutiny.

Documentation

Normative documentation measures in the new safety norm are very strict and extensive. It is a requirement to document all steps which are undertaken during the life cycle (in written or digital form). The entire design, and technical decisions and their justifications, must be documented diligently. All activities and results must be carefully organized to be well traceable.

Competence

All organizations which are incorporated in the life cycle must be able to prove that they are competent to perform the tasks for which they are responsible. All managers, project leaders, and engineers, who are involved in the design, technique and integration phase, must prove to possess a sufficient level of experience and competence. Operating norms for education and training must be established, and regular training sessions must also be scheduled.

19.1 Practical Implementation of the IEC 61508 Safety Standard

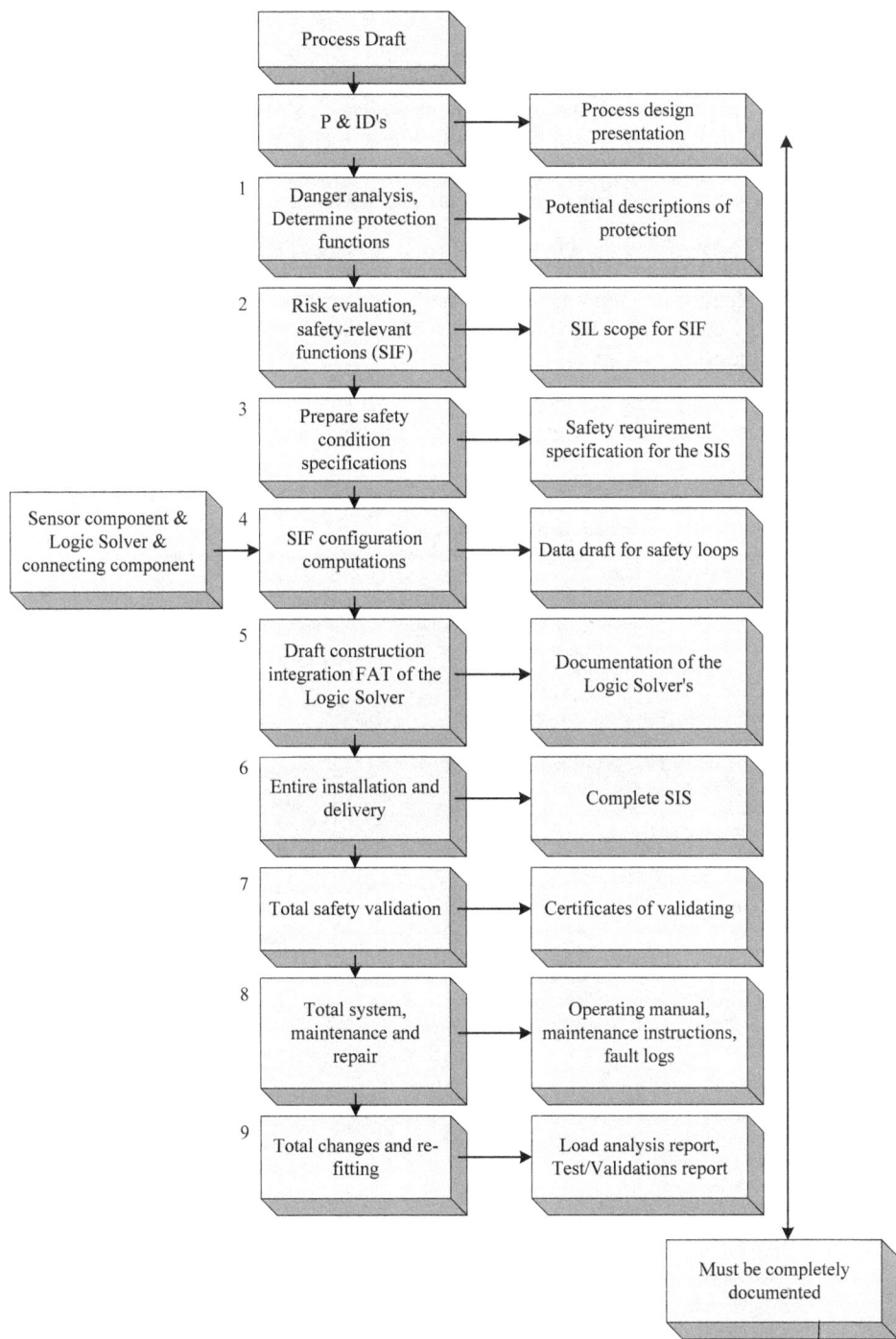

Figure 19.2: Practical flow chart

19.1.1.2 Pipe to Pipe Approach

The norm prescribes that the complete safety loop must be regarded. Normally, the loop contains sensors, logic elements, terminal elements (valves, contact makers), as seen in Figure 19.3. At the present time, some users still believe that they have fulfilled the safety requirements of their system by buying a safety-oriented controller, and they disregard other safety-technical systems (SIS Safety Instrumented System).

The norm requires a pipe to pipe approach, in which all parts of the safety loop are regarded parts of the safety life cycle. This is a very comprehensive approach, since sensors, inputs, logic elements, outputs, and terminal elements all represent vital connections, and are part of the same safety chain. It follows that firmly assigned sections of the standard specification enumerate the requirements for field mechanisms and field intersections.

In practice, valves contribute the most to instability (50 %), followed by sensors (35 %). Logic elements in an average configuration contribute approximately 15 %, often less.

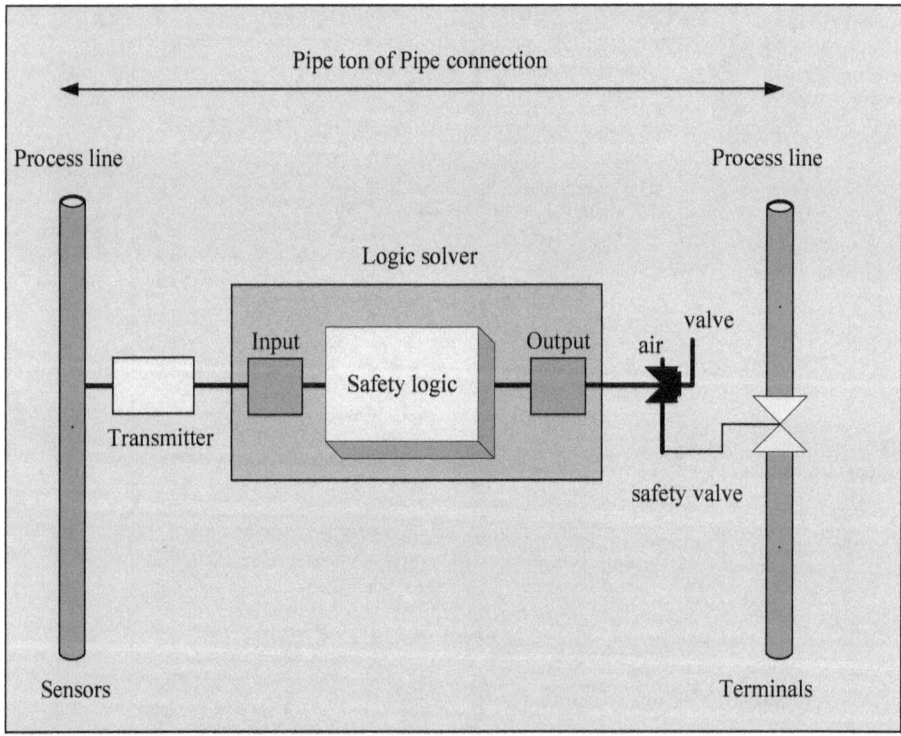

Figure 19.3: A safety-oriented function

19.1.1.3 Quantitative Safety Evaluation

In the course of hazard and risk analysis, all possible dangers are identified and classified. The results of these studies are defined in safety integrity functions (SIF). This risk reduction is called safety integrity level (SIL).

Before a safety system can be activated, it must be shown, through calculations, that the system delivered conforms to the SIL.

19.2 Determining the SIL of a Processor Based System

Below, a PFD-calculation for a processor system with periphery and subsequent SIL determination is conducted. The system is subdivided into the following subsystems:

- Oscillator
- Processor
- Flash
- SRAM
- Watchdog

The individual subsystems have failure rates λ and DC-factors as shown below in Table 19.1.

Table 19.1: Failure rate and DC-factor for the subsystems of a processor system

	Oscillator	Processor	Flash	SRAM	Watchdog
Failure rate λ	$30 \cdot 10^{-9}$	$450 \cdot 10^{-9}$	$100 \cdot 10^{-9}$	$5 \cdot 10^{-9}$	$150 \cdot 10^{-9}$
DC-Factor	0.99	0.99	0.999	0.999	0.99

Figure 19.4 shows the individual subsystems, which lie in sequence and are interconnected as 1oo1-architectures.

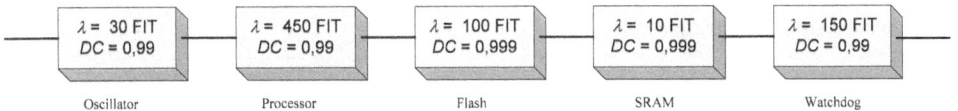

Figure 19.4: Processor system with periphery

The following limiting conditions pertain to the calculation of all subsystems:

- The proof-test interval T_1 is 10 years
- Repair time $MTTR$ is 8 hours
- The factor for safety-critical failures S is 50 %
- It is a type B subsystem

Table 19.2 shows the differentiated failure rates, PFD-, SFF-, and MTTF-values of the individual subsystems.

Table 19.2: Differentiated failure rates, PFD-, SFF-, and MTTF-values for the subsystems of a processor system

	Oscillator	Processor	Flash	SRAM	Watchdog
λ_S in 1/h	$1.5 \cdot 10^{-8}$	$2.25 \cdot 10^{-7}$	$5.0 \cdot 10^{-8}$	$5.0 \cdot 10^{-9}$	$7.50 \cdot 10^{-8}$
λ_D in 1/h	$1.5 \cdot 10^{-8}$	$2.25 \cdot 10^{-7}$	$5.0 \cdot 10^{-8}$	$5.0 \cdot 10^{-9}$	$7.50 \cdot 10^{-8}$
λ_{DD} in 1/h	$1.485 \cdot 10^{-8}$	$2.2275 \cdot 10^{-7}$	$4.995 \cdot 10^{-8}$	$4.995 \cdot 10^{-9}$	$7.425 \cdot 10^{-8}$
λ_{DU} in 1/h	$1.5 \cdot 10^{-10}$	$2.25 \cdot 10^{-9}$	$5.0 \cdot 10^{-11}$	$5.0 \cdot 10^{-12}$	$7.5 \cdot 10^{-10}$
PFD_G	$6.69 \cdot 10^{-6}$	$1.0035 \cdot 10^{-4}$	$2.59 \cdot 10^{-6}$	$2.59 \cdot 10^{-7}$	$3.345 \cdot 10^{-5}$
SFF	99.50 %	99.50 %	99.95 %	99.95 %	99.50 %
MTTF in years	3805.18	253.68	1141.55	11415.53	761.04

As can be seen from the last-but-one line in Table 19.2 above, every individual subsystem has an SFF-value greater than 99 %.

19.2.1 SIL Requirements

A complete system must fulfill a variety of requirements before it can be applied in a SIL 3 application, in the event of a low request mode. The requirements concerning PFD-value and architectural constraint are specified in the IEC/EN 61508. The latter restraints are decisive. This means, that even if the calculated PFD-value shows a higher SIL-value, it is the SIL-value after architectural constraints which is decisive. (This means, that the SIL-value after architectural constraints overrides the calculated PFD-value even if the latter shows a higher SIL-value.). These architectural constraints take into account the fault tolerance of hardware, and the percentage of safe and dangerous-to-detect errors with regard to the total error percentage, as expressed in the SFF-parameter. These connections are summarized in number form in Tables 19.3 and 19.4.

Table 19.3: SIL classification for PFD- and PFH-value during operation with low and high demand rate

Safety Integrity Level (SIL)	Operation with low demand (average probability of failure on demand)	Operation with high demand (probability of one dangerous failure per hour)
4	$\geq 10^{-5}$ to $<10^{-4}$	$\geq 10^{-9}$ to $<10^{-8}$
3	$\geq 10^{-4}$ to $<10^{-3}$	$\geq 10^{-8}$ to $<10^{-7}$
2	$\geq 10^{-3}$ to $<10^{-2}$	$\geq 10^{-7}$ to $<10^{-6}$
1	$\geq 10^{-2}$ to $<10^{-1}$	$\geq 10^{-6}$ to $<10^{-5}$

Table 19.4: Architectural constraints, SFF and hardware fault tolerance

Safe Failure Fraction (SFF)	Type A Hardware fault tolerance			Type B Hardware fault tolerance		
	0 error	1 error	2 errors	0 error	1 error	2 errors
< 60 %	SIL 1	SIL 2	SIL 3	Not allowed	SIL 1	SIL 2
60 % - < 90 %	SIL 2	SIL 3	SIL 4	SIL 1	SIL 2	SIL 3
90 % - < 99 %	SIL 3	SIL 4	SIL 4	SIL 2	SIL 3	SIL 4
> 99 %	SIL 3	SIL 4	SIL 4	SIL 3	SIL 4	SIL 4

In order to classify a system as fit for use in SIL 3, it must have a PFD value between 10^{-4} and 10^{-3}, during the operation type "low demand". Ultimately though it is critical, that for example in a system which consists of subsystems classified as type B with a 0-error hardware fault tolerance, each subsystem must have an SFF of greater than 99 %.

19.2.2 Determining the SIL of a Processor Unit with Processor Periphery

For the processor system, the following total PFD value can be calculated from Table 19.2:

$$PFD_G = PFD_{G,Oszi} + PFD_{G,Proc.} + PFD_{G,Flash} + PFD_{G,SRAM} + PFD_{G,WD}$$
$$= 6.69 \cdot 10^{-6} + 1.00 \cdot 10^{-4} + 2.59 \cdot 10^{-6} + 2.59 \cdot 10^{-7} + 3.35 \cdot 10^{-5}$$
$$= 1.43 \cdot 10^{-4}$$

This value is sufficient (see Table 19.3) for a PFD-equivalent SIL 3 classification of the system.

Relative to the SFF parameter, the individual subsystems satisfy the requirements for SIL 3 (see Table 19.2). If you calculate the SFF for the entire processor system, you obtain

$$SFF = \frac{29.85 \; FIT + 447.75 \; FIT + 99.95 \; FIT + 9.995 \; FIT + 149.25 \; FIT}{30 \; FIT + 450 \; FIT + 100 \; FIT + 10 \; FIT + 150 \; FIT}$$
$$= 99.57\%$$

Finally, when calculating the MTTF value in order to be able to compare with the selected proof-test-interval T_1 of 10 years, the following result is obtained:

$$MTTF = \frac{1}{\sum \lambda_i} = \frac{1}{30 \; FIT + 450 \; FIT + 100 \; FIT + 10 \; FIT + 150 \; FIT}$$
$$= 154.26 \; years$$

It is obvious that the MTTF value is much greater than the selected proof-test-interval with $T_1 = 10$ years. The aforementioned approximations with regard to the calculation of the PFD value are therefore valid and can be used for calculation.

The result for the processor system reads as follows: The system can be used for applications up to SIL 3.

19.2.3 DC-Measures for a Processor Unit with Processor Periphery

19.2.3.1 Processor Units

In order to achieve a high degree of coverage for the diagnosis of a processor unit, according to IEC/EN 61508-2, faults which are described in the DC fault model for data and addresses, or faults which can occur in storage cells as a result of dynamically contradictory influences, must be detected, using suitable measures, with a certain probability. The IEC/EN 61508-2 specifies the following high-quality measures for this purpose:

- Comparator
- Majority decider
- Mutual comparison through software
- Coded processing

The first three points can only be implemented by multichannel architecture. As for the last point, coded processing, which is possible inside of a one-channel system, the IEC/EN 61508-7, A.3.4, contains the following statement:

"Processing units [this includes a processor unit] can be designed with special error-recognizing or error-correcting circuit modes. Up until now, these procedures have only been used in relatively simple circuits, and their use is not widespread, but future developments should not be excluded."

Since a processor unit for the SIL 3 requirement is a highly technical, complex system with complex circuits, during its development, HW, and SW, a redundant architecture should be provided for doing so will ensure that a safety function with a SIL 3 requirement does indeed fulfill this requirement.

One additional aspect in favor of redundant processor architecture is the fact that, for the units oscillator, flash, SRAM, and watchdog, a large number of test and diagnosis data accrues. These data must be compared and analysed within a specified period of time, in order to fulfill the appropriate safety requirements. A redundant processor unit is best suited for this purpose.

The so-called "walking bit test" is conducted in order to identify errors in the physical memory or in the command decoder of the processor as early as possible. This self-test is implemented exclusively by additional software functions, according to a data model, for example the walking bit model.

19.2.3.2 Read-Only Memory

According to IEC/EN 61508-2, in order to achieve a high degree of diagnostic coverage for an unalterable memory unit (for example flash), all errors which have an influence on memory data must be revealed with certain probability, using appropriate measures. The IEC/EN 61508-2 lists high-quality measures for doing so:

- Signature with double word width (16 bit → 32 bit)
- Block repetition

The IEC/EN 61508-7, A.3.4 contains the following statement concerning these two points:

- *About signature with double word width: This procedure calculates a signature by using a CRC-algorithm (cyclic redundancy check), although the result comprises at least two words. The broadened signature is stored, calculated anew, and compared. In the event of a difference between the stored and the newly calculated value, a failure notice is generated.*
- *About block repetition: The address space is deposited in two memories. The first memory is operated as usual. The second memory contains the same information and is addressed in parallel with the first. The outputs are compared, and if a difference is detected, a failure notice is generated. In order to identify certain types of bit aberrations, the data must be deposited in an inverse fashion in one of the memories, to be inverted upon reading.*

At least regarding the last aspect this implies that the flash module must also be present in redundancy.

19.2 Determining the SIL of a Processor Based System

19.2.3.3 Alterable Memory

According to IEC/EN 61508-2, in order to achieve a high degree of diagnostic coverage for "volatile memory" (for example SRAM), all errors which are described through the DC error model must be uncovered with certain probability, using proper methods, both for data as well as for addresses. The DC error model comprises the following failure types:

- Stuck-at error
- Stuck open
- Open exits or exits with high impedance, as well as
- Short circuits between signal cables

Further detection concerns errors which are caused by

- dynamic crosstalk between storage cells
- no, erroneous, or multiple addressing, or by an
- exchange of information, caused by transient errors (soft errors) in DRAMs of 1 MBit size and greater.

These errors can be detected, according to IEC/EN 61508-2, using the following high-quality measures:

- RAM test "Galpat" or "transparent Galpat"
- RAM test "Abraham"
- Monitoring the RAM with a modified Hamming-code, or identification of data errors by the use of error detection and correction codes (EDC), or
- Double RAM with hardware or software comparison and read/write test.

The IEC/EN 61508-7, A.5.3 to A.5.7 contains the following statement concerning these points:

The goal of the "Galpat test" and of the "transparent Galpat test" is the identification of static bit errors and of the bulk of dynamic couplings.

- In the "Galpat" RAM test, the selected memory area is first uniformly preallocated (that is, all zeros or all ones). Then, the first memory cell to be tested is inverted and all remaining cells are examined, in order to ensure that the contents are correct. Following read access to one of the remaining cells, the inverted cell is examined as well. This procedure is repeated for every cell in the selected memory area. A second run is performed with the opposite preallocation. Each difference produces an error message.
- The "transparent Galpat" test is a variation of the above procedure. Instead of preallocating all cells in the selected memory area, the existing content is left unchanged, and instead signatures are used to compare the contents of the cell selection. The first cell to be tested in the selected area is chosen, and the signature S1 of all remaining cells in this particular area are calculated and stored. The cell to be tested is then inverted, and the signature S2 of all remaining cells is again calculated. (Following every read access on one of the remaining cells, the inverted cell is tested as well). S2 is compared with S1, and each difference produces an error message. In order to restore the original content,

the cell which the test is based on is then again inverted, and the signature S2 of all remaining cells is again calculated and compared with S1. Each difference produces an error message. All memory cells in the selected area are tested in the same manner.

In the RAM test "Abraham", all stuck-at errors and couplings between the memory cells should be identified.

- The proportion of identified errors surpasses the number obtained in the "Galpat" RAM test. The number of operations necessary for a complete memory test is 30 n, where n represents the number of memory cells. The test can be performed transparently during the operational cycle, in that the memory can be subdivided and each part tested in different time segments.

In the course of RAM monitoring with a modified Hamming code, or recognition of data errors with error-detection-correction-codes (EDC-codes), all odd-numbered bit errors, all two-bit errors, some three-bit errors, and some multiple-bit errors should be identified.

- Every memory word is expanded by several redundant bits, in order to achieve a modified Hamming code with a Hamming distance of at least 4. By testing redundant bits during each read access, it can be determined whether corruption has taken place. If a difference is found, an error message is produced. The procedure can also be used to detect addressing errors, in that redundant bits are calculated through the nexus of a data word with its address.

When performing a diagnostic detection by "double RAM with hardware or software comparison and read/write test", all bit errors should be identified.

- The address space is deposited in two memories. The first memory is operated as usual. The second memory contains the same information and is addressed in parallel with the first. Outputs are compared, and if a difference is identified, an error message is produced. In order to recognize certain kinds of bit aberration, the data must be deposited inversely in one of the two memories and then re-inverted for reading.

19.3 Determining the SIL of a Safety Function

Below, a SIL assessment for a safety function is to be performed. This safety function should meet requirements for SIL 3 at low demand level. In the first step, a SIL assessment is performed for a simply constructed safety function. This is followed by two modifications of the safety function, which demonstrate different possibilities for a safety function to fulfill the requirements imposed on it. With reference to the PDF value, and the SFF at a given hardware error tolerance, SIL 3 requirements are identical to the SIL 3 requirements for the processor system in section 19.2 (refer to Tables 19.3 and 19.4). The safety function consists of the following type B subsystems:

- Signal preparation
- CPU
- Safety switch (abbreviated: SIS)

The safety function is constructed in single-channel fashion, as seen in Figure 19.5. Because the system does not possess a diagnosis, the DC factor for all subsystems equals zero.

19.3 Determining the SIL of a Safety Function

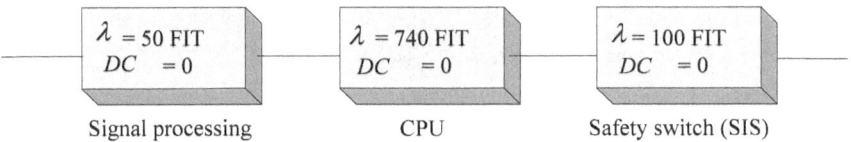

Figure 19.5: Single-channel safety function

Furthermore, it should apply to all subsystems that, for the safety-relevant factor S:

$$S = \frac{\lambda_D}{\lambda_S + \lambda_D} = 0.5$$

This means that 50 % of all occurring errors are dangerous. In addition, the following parameters should apply to the subsystems:

- $T_1 = 10$ years
- MTTR = 8 hours
- Type B subsystems

Table 19.5 shows the failure rates, PFD-, SFF-, and MTTF-values for the individual subsystems.

Table 19.5: Differentiated failure rates, PFD-, SFF- and MTTF-values for the subsystems of a safety function

	Signal preparation	CPU	SIS
λ in 1/h	$5.0 \cdot 10^{-8}$	$7.40 \cdot 10^{-7}$	$1.0 \cdot 10^{-7}$
λ_S in 1/h	$2.5 \cdot 10^{-8}$	$3.70 \cdot 10^{-7}$	$5.0 \cdot 10^{-8}$
λ_D in 1/h	$2.5 \cdot 10^{-8}$	$3.70 \cdot 10^{-7}$	$5.0 \cdot 10^{-8}$
λ_{DD} in 1/h	0	0	0
λ_{DU} in 1/h	$2.5 \cdot 10^{-8}$	$3.70 \cdot 10^{-7}$	$5.0 \cdot 10^{-8}$
PFD_G	$1.0952 \cdot 10^{-3}$	$1.6209 \cdot 10^{-2}$	$2.1904 \cdot 10^{-3}$
SFF	50.00 %	50.00 %	50.00 %
MTTF in years	2283.11	154.26	1141.55

As can be seen from the last-but-one line in Table 19.5, each individual subsystem has a SFF-value of exactly 50 %.

19.3.1 Determining the SIL of a Safety Function

Since the individual subsystems of the safety function exhibit a 1oo1-architecture, the PFD equation

$$\begin{aligned} PFD_{G,1oo1} &= (\lambda_{DU} + \lambda_{DD}) \cdot t_{CE} \\ &= \lambda_D \cdot t_{CE} \quad (19.1) \\ &= \lambda_{DU} \cdot \left(\frac{T_1}{2} + MTTR\right) + \lambda_{DD} \cdot MTTR \end{aligned}$$

can be applied to each subsystem (for values see Table 19.5).

The total PFD-value for the safety function can be deduced from Table 19.5 by addition of the individual PFD-values.

$$PFD_G = PFD_{Signal\ proc.,1oo1} + PFD_{CPU,1oo1} + PFD_{SIS,1oo1}$$
$$= 1.0952 \cdot 10^{-3} + 1.6209 \cdot 10^{-2} + 2.1904 \cdot 10^{-3}$$
$$= 1.9495 \cdot 10^{-2}$$

According to the SIL requirements table (see Table 19.3), this value implies that this safety function can be applied in a SIL 1 application.

Regarding the SFF parameters, the individual subsystems have a value of 50 %. In order to reach SIL 1 for a type B component with hardware error tolerance of 0 errors, the SFF must be between 60 % and 90 %, according to the SIL requirements table (see Table 19.4). In the above described safety function the SFF is only 50 %, since no diagnosis is given. Therefore, this value is not sufficient for SIL 1 requirement.

If you calculate the SFF for the complete safety function, the result is

$$SFF = \frac{2.50E(-08) + 3.70E(-07) + 5.00E(-08)}{5.00E(-08) + 7.40E(-07) + 1.00E(-07)} = 50\%$$

Finally, in order to be able to compare with the selected proof-test-interval T_1 of 10 years, if you calculate the MTTF value, you receive the following result:

$$MTTF = \frac{1}{\sum \lambda_i} = \frac{1}{5.00E(-08) + 7.40E(-07) + 1.00E(-07)} = 128.26\ years$$

The MTTF-value is conspicuously greater than the selected proof-test-interval with $T_1 = 10$ years. Therefore, the above mentioned approximations regarding the calculation of the PFD-value are valid and can be applied to the calculation.

The result for the safety function is: According to IEC/EN 61508, the safety function does not represent a safety function and can not be integrated into a safety system.

19.3.2 Modification of the Architecture of a Safety Function

The safety function consists, again, of the following type B subsystems:

- Signal preparation
- CPU
- Safety switch (abbreviated: SIS)

But only the signal preparation is single-channel, that is, it continues to have a hardware error tolerance of 0. The two other subsystems, the CPU, and the SIS, have a 1oo2-architecture, that is, they are 1-error-tolerant. The design of these safety functions is shown in Figure 19.6. Because the system does not possess a diagnosis, the DC factor continues to equal zero for all subsystems.

19.3 Determining the SIL of a Safety Function

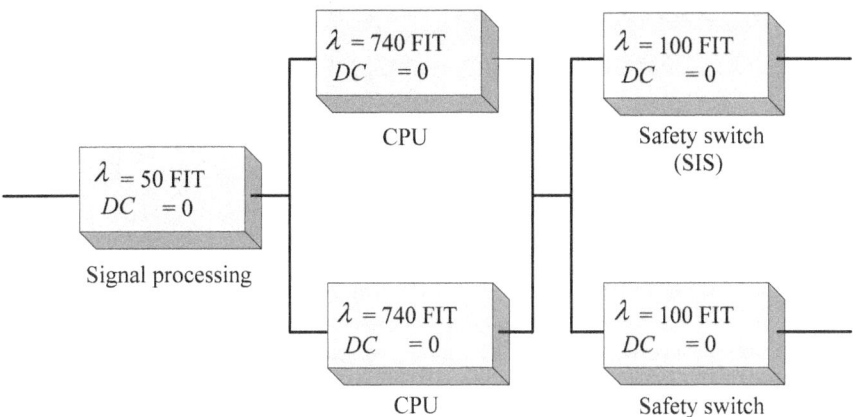

Figure 19.6: Modified safety function, 1oo1- and 1oo2-architecture

For all subsystems, the safety-relevant factor S should again be:

$$S = \frac{\lambda_D}{\lambda_S + \lambda_D} = 0.5$$

This means, that 50 % of all occurring errors are dangerous. In addition, the following parameters should again apply to the subsystems:

- $T_1 = 10$ years
- MTTR = 8 hours
- Type B subsystems

Since the subsystems CPU and SIS have a 1oo2-architecture, PFD calculations must, in addition, take into account "common cause" errors. While in a 1oo1-system only so-called simple errors can occur, 1oo2-, 1oo3-, 1oo4-architectures etc. can have errors which can be traced to a common cause. Such errors are weighted by means of the β-factor. Thereby, one distinguishes between the β-factor for dangerous, but non-detectable errors, and the β_D-factor for dangerous but detectable errors. The values for both β-factors are calculated by means of tables in the IEC/EN 61508-6. For this, among other things, experience values of the applied components, experience of the development personnel, or design diversity are taken into account. In the current case, the following values, which are actually determined in practice, were selected for both subsystems:

- $\beta_D = 0.01$
- $\beta = 0.02$

Table 19.6 shows the failure rates, PFD-, SFF- and MTFF-values for the individual subsystems.

Table 19.6: Differentiated failure rates, PFD-, SFF- and MTTF-values for the subsystems of a safety function

	Signal preparation	CPU	SIS
λ in 1/h	$5.0 \cdot 10^{-8}$	$7.40 \cdot 10^{-7}$	$1.0 \cdot 10^{-7}$
λ_S in 1/h	$2.5 \cdot 10^{-8}$	$3.70 \cdot 10^{-7}$	$5.0 \cdot 10^{-8}$
λ_D in 1/h	$2.5 \cdot 10^{-8}$	$3.70 \cdot 10^{-7}$	$5.0 \cdot 10^{-8}$
λ_{DD} in 1/h	0	0	0
λ_{DU} in 1/h	$2.5 \cdot 10^{-8}$	$3.70 \cdot 10^{-7}$	$5.0 \cdot 10^{-8}$
PFD_G	$1.0952 \cdot 10^{-3}$	$6.6064 \cdot 10^{-4}$	$4.9952 \cdot 10^{-5}$
SFF	50.00 %	50.00 %	50.00 %
MTTF in years	2283.11	$1.30 \cdot 10^{+7}$	$7.13 \cdot 10^{+8}$

As can be seen from the last-but-one line in Table 19.6, each individual subsystem has an SFF-value of exactly 50 %.

19.3.3 Determination of the SIL of a Modified Safety Function

For subsystem signal preparation, the 1oo1-PFD equation continues to be valid (for value see Table 19.6):

$$PFD_{G,1oo1} = (\lambda_{DU} + \lambda_{DD}) \cdot t_{CE}$$

$$= \lambda_D \cdot t_{CE} \quad (19.2)$$

$$= \lambda_{DU} \cdot \left(\frac{T_1}{2} + MTTR\right) + \lambda_{DD} \cdot MTTR$$

For the subsystems CPU and SIS, the 1oo2-PFD equation (for values see Table 19.6) must be applied:

$$PFD_G = 2\{(1-\beta_D)\lambda_{DD} + (1-\beta)\lambda_{DU}\}^2 t_{CE} t_{GE} + \beta_D \lambda_{DD} MTTR + \beta \lambda_{DU} \cdot \left(\frac{T_1}{2} + MTTR\right) \quad (19.3)$$

with

$$t_{CE} = \frac{\lambda_{DU}}{\lambda_D} \cdot \left(\frac{T_1}{2} + MTTR\right) + \frac{\lambda_{DD}}{\lambda_D} \cdot MTTR$$

$$t_{GE} = \frac{\lambda_{DU}}{\lambda_D} \cdot \left(\frac{T_1}{3} + MTTR\right) + \frac{\lambda_{DD}}{\lambda_D} \cdot MTTR$$

The total PFD-value for a modified safety function can be calculated from Table 19.6 by addition of the individual PFD-values:

$$PFD_G = PFD_{Signal\ proc.,\ 1oo1} + PFD_{CPU,\ 1oo2} + PFD_{SIS,\ 1oo2}$$

$$= 1.0952 \cdot 10^{-3} + 6.6064 \cdot 10^{-4} + 4.9952 \cdot 10^{-5}$$

$$= 1.8058 \cdot 10^{-3}$$

According to the SIL requirements table (see Table 19.3), this value indicates that this safety function can be applied in a SIL 2 application.

If one takes into account that the percentage of a logic part system is 15 % of the total safety loop (see section 5.1), the following becomes apparent: for a SIL 1 application, the PFD

19.3 Determining the SIL of a Safety Function

value for the logic part system can be between 1.5E-03 and 1.5E-02. For the selected example this means that it can be categorized as a SIL 1, according to its PFD value.

The individual subsystems continue to have a value of 50 % with regard to their SFF parameters. In order to achieve SIL 1 for a type B component with a hardware error tolerance of 0 errors, the SFF must be between 60 % and 90 %, according to the SFF requirements table (see Table 19.4). The above described safety function, for the subsystem signal preparation, only achieves a SFF of 50 % due to the absence of a diagnosis. This value does not meet SIL 1 requirements.

The same can be said for the other two subsystems, CPU and SIS: For a type B component with an error tolerance of 1 error, the SFF must be less than 60 % according to the SIL requirements table (see Table 19.4) in order to achieve SIL 1. Both subsystems achieve this value with SFF = 50 %.

When calculating the SFF for the total safety function, the value of

$$SFF = \frac{2.50E(-08) + 3.70E(-07) + 5.00E(-08)}{5.00E(-08) + 7.40E(-07) + 1.00E(-07)} = 50\%$$

is again achieved. When the MTTF value is calculated, in order to have a comparison with the selected proof-test-interval T_1 of 10 years, one gets the following result (using the part-count method):

$$MTTF = \frac{1}{\sum \lambda_i} = \frac{1}{5.00E(-08) + 7.40E(-07) + 1.00E(-07)} = 128.26 \text{ years}$$

As can be seen, the MTTF value is clearly greater than the selected proof-test interval with $T_1 = 10$ years. The above mentioned approximations regarding the calculation of the PFD value are therefore valid and can be used for calculation.

The result for the safety function is: According to IEC/EN 61508, the safety function does not represent a safety function due to its SFF value being too low, and can not be integrated into a safety system.

19.3.4 Modification of the Safety Function

The safety function again consists of the following type B subsystems:

- Signal preparation
- CPU
- Safety switch (abbreviated: SIS)

The signal preparation is again single-channel in that it continues to have a hardware error tolerance of 0. The two other subsystems, the CPU and the SIS, have a 1oo2-architecture, that is, they are 1-error-tolerant. The structure of these safety functions is shown in Figure 19.7. The difference to the first modification is, that all subsystems have a diagnosis. The degree of diagnostic coverage DC = 99 % for the signal preparation, whereas for the other two subsystems, CPU and SIS, the DC = 90 %. A definite number of dangerous errors is detectable by these diagnostic measures.

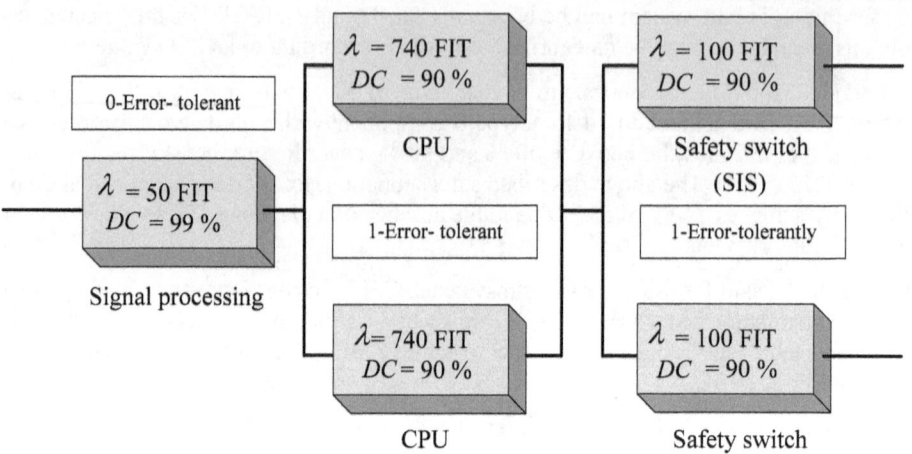

Figure 19.7: Modification of the safety function by diagnostic measures

For all subsystems, the safety-relevant factor S should continue to be:

$$S = \frac{\lambda_D}{\lambda_S + \lambda_D} = 0.5$$

This means that 50 % of all occurring errors are dangerous. In addition, the following parameters are again valid for the subsystems:

- $T_1 = 10$ years
- MTTR = 8 hours
- Type B subsystems

Same as for the first modification, because of the 1oo2-architecture of the two subsystems CPU and SIS, the common-cause error must be taken into account. In the example shown, the same values are used for the two β-factors, β and β_D, as in the previous example:

- $\beta_D = 0.01$
- $\beta = 0.02$

Table 19.7 shows failure rates, PFD-, SFF-, and MTTF-values for the individual subsystems.

Table 19.7: Differentiated failure rates, PFD-, SFF-, and MTTF-values for the subsystems of a safety function

	Signal preparation	CPU	SIS
λ in 1/h	$5.0 \cdot 10^{-8}$	$7.40 \cdot 10^{-7}$	$1.0 \cdot 10^{-7}$
λ_S in 1/h	$2.5 \cdot 10^{-8}$	$3.70 \cdot 10^{-7}$	$5.0 \cdot 10^{-8}$
λ_D in 1/h	$2.5 \cdot 10^{-8}$	$3.70 \cdot 10^{-7}$	$5.0 \cdot 10^{-8}$
λ_{DD} in 1/h	$2.475 \cdot 10^{-8}$	$3.33 \cdot 10^{-7}$	$4.50 \cdot 10^{-8}$
λ_{DU} in 1/h	$2.5 \cdot 10^{-10}$	$3.70 \cdot 10^{-8}$	$5.0 \cdot 10^{-9}$
PFD_G	$1.1150 \cdot 10^{-5}$	$3.5885 \cdot 10^{-5}$	$4.4472 \cdot 10^{-6}$
SFF	99.50 %	95.00 %	95.00 %
MTTF in years	2283.11	$1.30 \cdot 10^{+7}$	$7.13 \cdot 10^{+8}$

19.3 Determining the SIL of a Safety Function

As can be seen from the last-but-one line in Table 19.7, the subsystem signal preparation has an SFF value of 99.50 %. The two subsystems CPU and SIS each have a SFF value of 95.00 %.

19.3.5 Determining the SIL of a Safety Function with Diagnosis

For the subsystem safety preparation, the 1oo1-PFD equation continues to be valid (for value see Table 19.6):

$$PFD_{G,1oo1} = (\lambda_{DU} + \lambda_{DD}) \cdot t_{CE}$$
$$= \lambda_D \cdot t_{CE} \quad (19.4)$$
$$= \lambda_{DU} \cdot \left(\frac{T_1}{2} + MTTR\right) + \lambda_{DD} \cdot MTTR$$

For the subsystems CPU and SIS, the 1oo2-PFD-equation (for values see Table 19.6) needs to be applied:

$$PFD_G = 2\{(1-\beta_D)\lambda_{DD} + (1-\beta)\lambda_{DU}\}^2 t_{CE} t_{GE} + \beta_D \lambda_{DD} MTTR + \beta \lambda_{DU} \cdot \left(\frac{T_1}{2} + MTTR\right)$$
$$(19.5)$$

with

$$t_{CE} = \frac{\lambda_{DU}}{\lambda_D} \cdot \left(\frac{T_1}{2} + MTTR\right) + \frac{\lambda_{DD}}{\lambda_D} \cdot MTTR$$

$$t_{GE} = \frac{\lambda_{DU}}{\lambda_D} \cdot \left(\frac{T_1}{3} + MTTR\right) + \frac{\lambda_{DD}}{\lambda_D} \cdot MTTR$$

The total PFD-value for the modified safety function can be calculated by adding the individual PFD-values from Table 19.7:

$$PFD_G = PFD_{Signal\ proc.,\ 1oo1} + PFD_{CPU,\ 1oo2} + PFD_{SIS,\ 1oo2}$$
$$= 1.1150 \cdot 10^{-5} + 3.5885 \cdot 10^{-5} + 4.4472 \cdot 10^{-6}$$
$$= 5.1483 \cdot 10^{-5}$$

According to the SIL requirements table (see Table 19.3), this value indicates that, according to its PFD-value, this safety function can be utilized in a SIL 4 application. On the other hand, if one considers that the logic part system represents 15 % of the total safety loop (see section 5.1), this signifies the following: the PFD-value for the logic part system of a SIL 3 application can be between 1.5E-5 and 1.5E-4. For the selected example, this implies a SIL 3 classification, based on the PFD value.

With regard to the SFF parameters, the individual subsystems have the following values (see Table 19.7):

- Signal preparation: SFF = 99.50 % (hardware error tolerance = 0)
- CPU: SFF = 95.00 % (hardware error tolerance = 1)
- SIS: SFF = 95.00 % (hardware error tolerance = 1)

In order to achieve SIL 3 for a type B component with hardware error tolerance of 0 errors, the SFF must be greater than 99 %, according to the SIL requirements table (see Table 19.4). With the safety functions described above, the subsystem signal preparation will achieve a SFF of 99.50 %. This value is sufficient for a SIL 3 requirement.

For the other two subsystems, CPU and SIS, the following result is obtained: For a type B component with hardware error tolerance of 1 error, the SFF must be between 90 % and 99 %, according to the SIL requirements table (see Table 19.4), in order to achieve SIL 3. Each of the two subsystems achieves a value of 95 %. This value is sufficient for SIL 3 requirement.

Calculation of the SFF for the total safety function results in

$$SFF = \frac{2.50E(-08) + 2.4750E(-08) + 3.70E(-07) + 3.33E(-07) + 5.00E(-08) + 4.50E(-08)}{5.00E(-08) + 7.40E(-07) + 1.00E(-07)}$$

$$= 95.25\%$$

Finally, calculating the MTTF-value, in order to have a comparison with the selected proof-test interval T_1 of 10 years, yields the following result (when using the part-count method):

$$MTTF = \frac{1}{\sum \lambda_i} = \frac{1}{5.00E(-08) + 7.40E(-07) + 1.00E(-07)} = 128.26 \ years$$

It is obvious that the MTTF-value is much greater than the selected proof-test interval with $T_1 = 10$ years. The previously mentioned approximations with regards to the calculation of the PFD-value are therefore valid and can be applied for the calculation.

The result for the safety function is as follows: The safety function represents a safety function according to IEC/EN 61508 and can be integrated in a safety system for a SIL 3 application.

19.4 Determining the SIL of a Safety Loop

The two preceding sections described the development and evaluation of a safety processor system and its subsequent integration into a safety function. In the following segment, the safety of a safety loop consisting of sensor, logic part system, and actuator will be assessed. The goal is to design a loop which can be used in a SIL 3 application with low requirement. The SIL 3 requirements with regard to the PFD-value and the SFF with given hardware error tolerance is identical to the SIL 3 requirements for the processor system in section 19.2, Tables 19.3 and 19.4. The safety loop consists of the following type A and type B subsystems:

- Sensors (type B subsystem)
- Logic part system (type B subsystem)
- Actuators (type A subsystem)

19.4 Determining the SIL of a Safety Loop

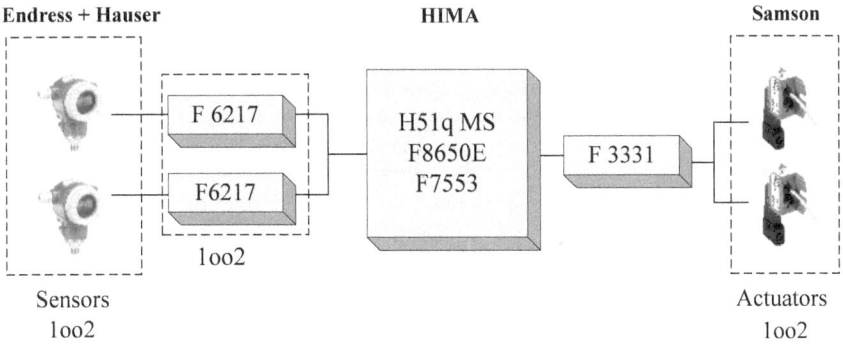

Figure 19.8: Safety loop

As shown in Figure 19.8, the safety loop consists of the following architectures:

- Sensors: 1oo2-architecture, DC=82.0%
- Analog input module (AI): 1oo2-architecture, DC=99.6%
- CPU logic module: 1oo1-architecture, DC=99.6%
- Digital output module (DO): 1oo1-architecture, DC=99.4%
- Actuators: 1oo2-architecture, DC=22.4%

For all subsystems, the safety-relevant factor S should be:

$$S = \frac{\lambda_D}{\lambda_S + \lambda_D} = 0.5$$

This means that 50 % of all occurring errors are dangerous. In addition, the following parameters apply to the subsystems:

- $T_1 = 3$ years
- MTTR = 8 hours

Because of the 1oo2-architecture of the subsystems sensor, analog input module, and actuator, the common-cause errors must be taken into account. In the example on hand, the following values are assumed for the two β-factors, β and β_D:

Table 19.8: β-factors for the subsystems of a safety loop

	β_D	β
Sensor	0.05	0.05
AI	0.01	0.02
Actuator	0.05	0.05

For the individual subsystems, the resulting failure rates, PFD-, SFF-, and MTTF-values are shown in Table 19.9.

Table 19.9: Differentiated failure rates, PFD-, SFF-, and MTTF-values for the subsystems of a safety function

	Sensor	AI	CPU logic module	DO	Actuator
λ in 1/h	$1.4174 \cdot 10^{-6}$	$5.934 \cdot 10^{-7}$	$1.9564 \cdot 10^{-6}$	$2.8389 \cdot 10^{-7}$	$1.4050 \cdot 10^{-7}$
λ_S in 1/h	$7.0872 \cdot 10^{-7}$	$2.967 \cdot 10^{-7}$	$9.7820 \cdot 10^{-7}$	$1.4190 \cdot 10^{-7}$	$7.025 \cdot 10^{-8}$
λ_D in 1/h	$7.0872 \cdot 10^{-7}$	$2.967 \cdot 10^{-7}$	$9.7820 \cdot 10^{-7}$	$1.4190 \cdot 10^{-7}$	$7.025 \cdot 10^{-8}$
λ_{DD} in 1/h	$5.810 \cdot 10^{-7}$	$2.955 \cdot 10^{-7}$	$9.7430 \cdot 10^{-7}$	$1.41104 \cdot 10^{-7}$	$1.5470 \cdot 10^{-8}$
λ_{DU} in 1/h	$1.2760 \cdot 10^{-7}$	$1.199 \cdot 10^{-9}$	$3.913 \cdot 10^{-9}$	$8.517 \cdot 10^{-10}$	$5.4510 \cdot 10^{-8}$
PFD_G	$8.7500 \cdot 10^{-5}$	$3.393 \cdot 10^{-7}$	$5.9240 \cdot 10^{-5}$	$1.233 \cdot 10^{-6}$	$3.6460 \cdot 10^{-5}$
SFF	91.00 %	99.79 %	99.80%	99.70%	61.20 %
MTTF in years	40.27 (1)	96.12 (2)	53.35 (2)	402.11 (2)	406.25 (1)
Note: (1):	MTTF is calculated according to part count				
(2):	MTTF is also calculated with non-secure components according to part count.				

As can be seen from the last-but-one line in Table 19.9, the sensor subsystem has a SFF-value greater than 90 %. The AI, CPU, and DO logic subsystems each have a SFF-value of greater than 99 %. The type A actuator subsystem has a SFF-value of greater than 61%.

19.4.1 Determining the SIL of the Safety Loop

For the CPU-logic module and the DO subsystems, the following 1oo1-PFD equation is valid (for value see Table 19.9):

$$PFD_{G,1oo1} = (\lambda_{DU} + \lambda_{DD}) \cdot t_{CE}$$
$$= \lambda_D \cdot t_{CE} \qquad (19.6)$$
$$= \lambda_{DU} \cdot \left(\frac{T_1}{2} + MTTR\right) + \lambda_{DD} \cdot MTTR$$

For the sensor, AO, and actuator subsystems, the 1oo2-PFD-equation must be applied (for values see Table 19.9):

$$PFD_G = 2\{(1-\beta_D)\lambda_{DD} + (1-\beta)\lambda_{DU}\}^2 t_{CE} t_{GE} + \beta_D \lambda_{DD} MTTR + \beta \lambda_{DU} \cdot \left(\frac{T_1}{2} + MTTR\right) \qquad (19.7)$$

with

$$t_{CE} = \frac{\lambda_{DU}}{\lambda_D} \cdot \left(\frac{T_1}{2} + MTTR\right) + \frac{\lambda_{DD}}{\lambda_D} \cdot MTTR$$

$$t_{GE} = \frac{\lambda_{DU}}{\lambda_D} \cdot \left(\frac{T_1}{3} + MTTR\right) + \frac{\lambda_{DD}}{\lambda_D} \cdot MTTR$$

The total PFD value for the modified safety function can be calculated by addition of the individual PFD-values from Table 19.9:

$$PFD_G = PFD_{Sensor,1oo2} + PFD_{F6217,1oo2} + PFD_{H51q\_MS,1oo1} + PFD_{F3331,1oo1} + PFD_{Actor,1oo2}$$
$$= 8.750 \cdot 10^{-5} + 3.393 \cdot 10^{-7} + 5.9240 \cdot 10^{-5} + 1.233 \cdot 10^{-6} + 3.646 \cdot 10^{-5}$$
$$= 1.8427 \cdot 10^{-4}$$

According to the SIL requirement table (see Table 19.3), this value indicates that, according to its PFD-value, this safety function can be applied in a SIL 3 application. In addition, it is

19.4 Determining the SIL of a Safety Loop

apparent that the weighting of the sensor, logic, and actuator subsystems, in relation to the total PFD-value, correspond to the percentages described in section 19.1.1.2 (sensor: 35 %, logic: 15 %, actuator: 50 %). The distribution in this example, in terms of the maximum possible PFD value of 1E-03 for an SIL 3 loop, is as follows:

- Sensor: 8.77 %
- Logic: 6.08 %
- Actuator: 3.64 %

The total PFD value for the loop is therefore 14.36 % of the maximum permissible PFD-value for a SIL 3 application.

By increasing the proof-test interval T_1, the total PFD-value can be increased until it reaches the maximum available PFD-value for a SIL 3 application. If one takes into account that the sensor percentage must be no more than 35 %, the logic percentage no more than 15 %, and the actuator percentage no more than 50 %, the following PFD-values (calculated using SILence) result for a proof-test interval of 10 years and 10 months:

- Sensor: $3.4410 \cdot 10^{-4}$ equals 34.41 %
- Logic: $2.3441 \cdot 10^{-4}$ equals 23.44 %
- Actuator: $1.3630 \cdot 10^{-4}$ equals 13.63 %
- Total: $7.1481 \cdot 10^{-4}$ equals 71.48 %

(The percentages refer to the SIL 3 value of 1E-03).

Without taking into account the individual subsystem percentages, and solely based on the requirement that the total PFD-value must be lower than 1E-03, the following maximum possible proof-test interval of 18 years and 8 months, and the following PFD-values with their corresponding percentages are obtained (calculated using SILence):

- Sensor: $4.8545 \cdot 10^{-4}$ equals 65.92 %
- Logic: $3.1310 \cdot 10^{-4}$ equals 39.96 %
- Actuator: $1.8760 \cdot 10^{-4}$ equals 24.68 %
- Total: $9.8520 \cdot 10^{-4}$ equals 98.52 %

Regarding the SFF parameters, the individual subsystems have the following values (see Table 19.9):

- Sensor: SFF = 91.00 % (hardware error tolerance = 1)
- AO: SFF = 99.79 % (hardware error tolerance = 1)
- CPU logic: SFF = 99.80 % (hardware error tolerance = 0)
- DO: SFF = 99.70 % (hardware error tolerance = 0)
- Actuator: SFF = 61.20 % (hardware error tolerance = 1)

To achieve SIL 3 for a type B component with hardware error tolerance of 0 errors, according to the SIL requirements table (see Table 19.4), the SFF must be greater than 99 %. In the safety loop described above, a SFF of greater than 99 % is achieved for the H51q MS and F3331 subsystems, each of which have type B components and a 1oo1-architecture. This value is sufficient for SIL 3 requirement.

For the two sensor and AO subsystems, which also contain type B components, but which have a hardware error tolerance of 1 – due to their 1oo2-architecture – the SFF must be between 90 % and 99 % to fulfill the requirements for SIL 3, according to the SIL requirements table (see Table 19.4). Both subsystems reach a value of greater than 90 % (*Note:* For the F6217 subsystem, a SFF value of greater than 99 % can be calculated. This would, de facto, permit SIL 4 classification. Yet, according to IEC/EN 61508, not only the hardware must comply with SIL requirements, but also the software applied. In this case, the applied software is classified as SIL 3, and therefore the AO can "only" be utilized in SIL 3 applications.)

The actuator subsystem is exclusively constructed with A components. Additionally, it has a 1oo2-architecture with a hardware error tolerance of 1. Accordingly, this subsystem requires a SFF value of between 60 % and 90 % (see Table 19.4) in order to achieve SIL 3. With a SFF value of greater than 61 %, the actuator subsystem fulfills the SFF requirements for use in a SIL 3 application.

Calculation of the SFF for the total safety function results in

$$SFF = \frac{7{,}0872E-07 + 5{,}8086E-07 + 3{,}5546E-07 + 2{,}3698E-07}{1{,}4174E-06 + 5{,}9340E-07 + 1{,}9564E-06 + 2{,}8389E-07 + 1{,}4050E-07} +$$

$$\frac{1{,}0308E-06 + 9{,}2200E-07 + 1{,}6039E-07 + 1{,}2274E-07 + 8{,}600E-08}{1{,}4174E-06 + 5{,}9340E-07 + 1{,}9564E-06 + 2{,}8389E-07 + 1{,}4050E-07}$$

$$= 95{,}73\%$$

$$SFF = \frac{7.0872E(-07) + 5.810E(-07) + 2.9670E(-07) + 2.955E(-07)}{1.4174E(-06) + 5.9340E(-07) + 1.9564E(-06) + 2.8389E(-07) + 1.4050E(-07)}$$

$$+ \frac{9.7820E(-07) + 9.7430E(-07) + 1.4190E(-07) + 1.4110E(-07)}{1.4174E(-06) + 5.9340E(-07) + 1.9564E(-06) + 2.8389E(-07) + 1.4050E(-07)}$$

$$+ \frac{7.0250E(-08) + 1.5470E(-08)}{1.4174E(-06) + 5.9340E(-07) + 1.9564E(-06) + 2.8389E(-07) + 1.4050E(-07)}$$

$$= 95{,}73\%$$

Finally, calculation of the MTTF value, for the purpose of comparison with the selected proof-test interval T_1 of 10 years, yields the following result (calculated with part-count method, only the given failure rates λ are considered.

$$MTTF = \frac{1}{\sum \lambda_i} = \frac{1}{\sum \frac{1}{MTTF_i}}$$

$$= \frac{1}{2 \cdot 1.4174E(-06) + 2 \cdot 5.9340E(-07) + 1.9564E(-06) + 2.8389E(-07) + 2 \cdot 1.4050E(-07)}\, h$$

$$= \frac{1}{6.54319E(-06)}\, h$$

$$= 17.44\, h$$

It is apparent that the MTTF value is clearly greater than the selected proof-test interval with $T_1 = 3$ years. The above mentioned approximations concerning the calculation of the PFD value are therefore valid and can be used for the calculation

The result for the safety loop is as follows: The safety loop fulfills the requirements of a safety function, according to IEC/EN 61508, and can be integrated into a safety system for a SIL 3 application.

19.5 Examples of Reliability Analysis

In the following section, a variety of situations are described for which a reliability analysis is subsequently performed.

19.5.1 Example 1 (Chemical Installation)

In a chemical installation, different chemicals flow down three pipes into a kettle (see Figure 19.9). Electronic valves are installed on the three pipes. All three valves are controlled by a central microcontroller MSP430 by Texas Instruments. It compares its values with those of another MSP430 microcontroller. If the values are identical, the appropriate valve is opened; if they are not identical, the valves stay closed. Likewise, if one of the microcontrollers fails, no valve is opened. The same thing happens if one of the three valves fails.

Control of the valves is necessary because it must be avoided that all three valves open at the same time. It would be risky if, through all three pipes, chemicals were to flow into the kettle at the same time. Down each pipe, a chemical must flow only during a specific time frame. These time frames are specified and controlled by MSP430. In addition, a person is responsible for each kettle. The plant contains 20 kettles.

If at least two valves are opened at the same time, a dangerous situation occurs. The following segment describes just such a case. All other danger situations are not considered in the risk graph (see Figure 19.9).

19.5.1.1 Risk Graph

The risk graph arising out of the arrangement is shown in Figure 19.10. The fact, that risks and dangers are high in chemical processes, requires the contemplation of a high extent of damage. Complications with one kettle, and the resulting consequences, would not only endanger the person responsible for that particular kettle, but could also endanger the lives of several people in the vicinity (D3).

Furthermore, it must be taken into account that it is the person's responsibility to observe the kettle permanently. For this reason, they spend a very long time in the vicinity of the kettle and thus in the danger zone (E2).

The probability of such an accident happening, with dire consequences for people and environment, is very low in this case, because the installation is controlled by two microcontrollers, and the probability that both microcontrollers fail is nearly zero (P1).

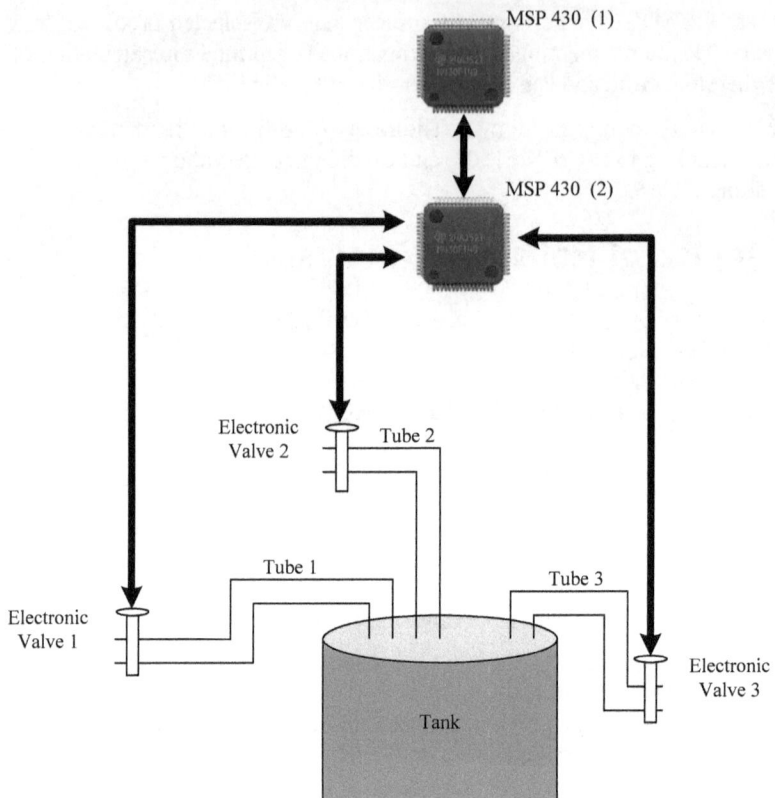

Figure 19.9: Layout of the installation

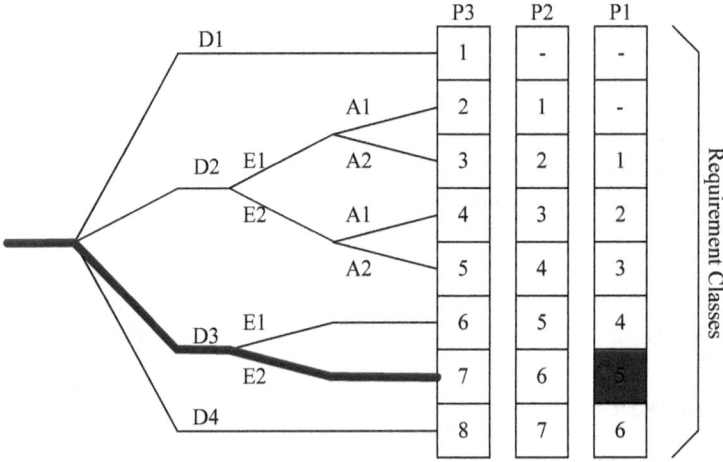

Figure 19.10: Risk graph for example 1

19.5.1.2 Event Tree

The event tree for this example is shown in Figure 19.11.

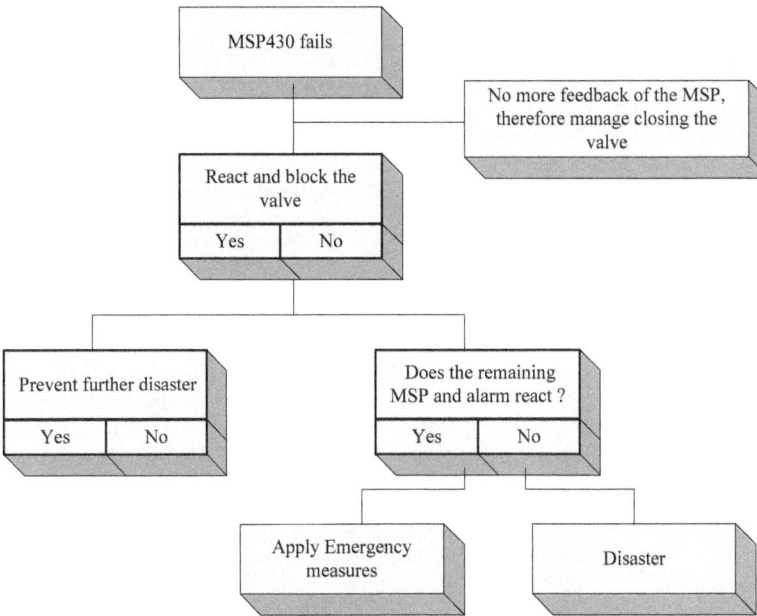

Figure 19.11: Event tree after the failure of one of the microcontrollers

If one of the two microcontrollers fails, and the remaining control can sound an alarm, then it is possible to prevent the catastrophic situation. If neither of the two microcontrollers reacts, that is, the valves cannot be closed, the resulting situation can lead to a catastrophe.

19.5.1.3 Error Tree Analysis

Figure 19.12 shows the error tree analysis for the first example. The undesired event is the simultaneous opening of more than one valve. One level down, this event is subdivided into three combinations. Another level down, the individual events for an open valve are shown.

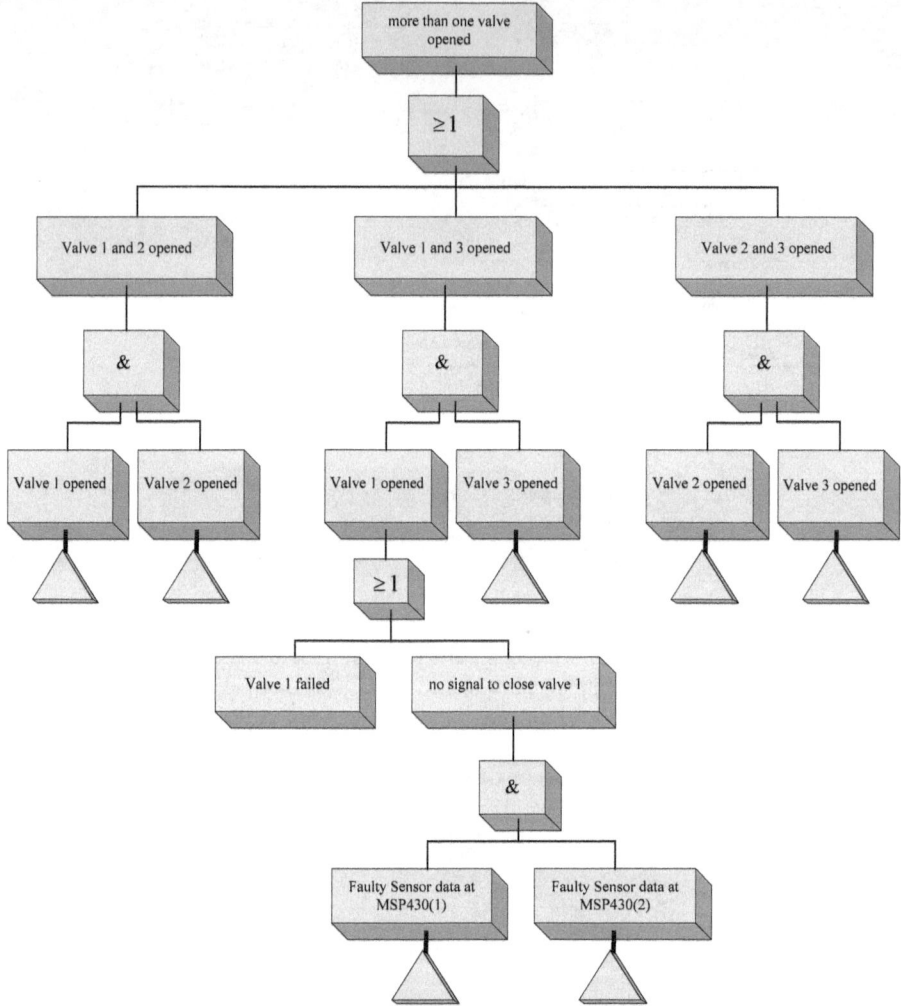

Figure 19.12: Error tree analysis in the event that more than one valve is open

Here, only one exemplary event was developed, the others are analog to it. One valve remains open if it is defective or if no signal is sent to close the valve. A defective valve is a primary failure. This primary failure is analysed within its own error tree. The signal to close the valve remains absent if both microcontrollers contain inaccurate sensor data. These events are not resolved further here, as would, naturally, have to be done in a complete analysis.

19.5.1.4 Reliability Block Diagram

For example 1, reliability block diagram should be drawn. Since all units must function in order for the system to function, a serial structure is chosen in which the parameters have the following significance:

$x_i = 1 \Rightarrow$ „Function"

$x_i = 0 \Rightarrow$ „Failure"

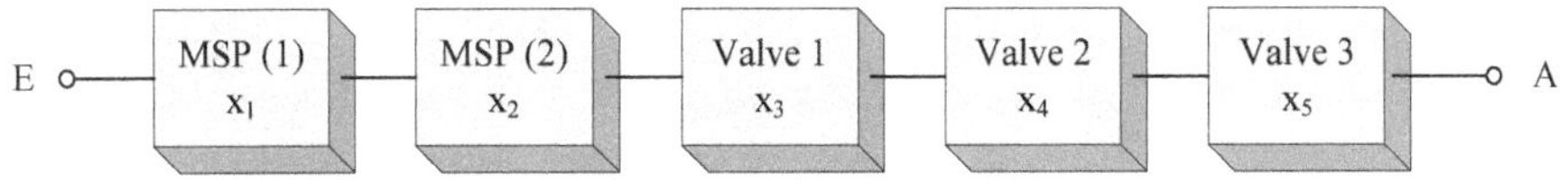

```
E o—[ MSP (1) x₁ ]—[ MSP (2) x₂ ]—[ Valve 1 x₃ ]—[ Valve 2 x₄ ]—[ Valve 3 x₅ ]—o A
```

Figure 19.13: Reliability diagram for example 1 (serial structure)

With the failure probabilities of the units (q_i), the failure probability of the system P is

$$P(q_1,...,q_5) = 1 - ((1-q_1)(1-q_2)(1-q_3)(1-q_4)(1-q_5))$$

The general failure probability with exponential distribution is given as

$$P(t) = 1 - e^{-\lambda t}$$

$$\Rightarrow q = e^{-\lambda t} \qquad \text{for } \lambda t \ll 1 \Rightarrow q \approx \lambda t$$

The following failure rates are specified for the units:

MSP = 150 FIT = 150 10^{-9} 1/h

Valve = 50 FIT = 50 10^{-9} 1/h

Thus, the individual failure probabilities over the course of 1 year (8760h) come to

$$MSP = 8760h \cdot 150 \cdot 10^{-9} \frac{1}{h} = 0{,}001314$$

$$Ventil = 8760h \cdot 50 \cdot 10^{-9} \frac{1}{h} = 0{,}000438$$

And thereby the failure probability of the total system can be calculated

$$P(q_1,...,q_5) = 1 - ((1-0{,}001314)(1-0{,}001314)(1-0{,}000438)(1-0{,}000438)(1-0{,}000438))$$
$$= 0{,}00396 = 0{,}396\%$$

19.5.2 Example 2 (Driver-Side Airbag)

In the second case, we examine the control of the driver-side airbag in a car. In a car, one control logic is in charge of all airbags. Once the control electronics (MSP430) has recognized an accident which will be dangerous for the passengers, based on data from the so-called airbag sensors, a solid propellant is ignited which causes a nylon cushion to be inflated extremely rapidly (approximately 10 – 40 thousandths of a second).

In this case, we will design the risk graph for two different situations. We will start out with different speeds in the two situations and examine the differences.

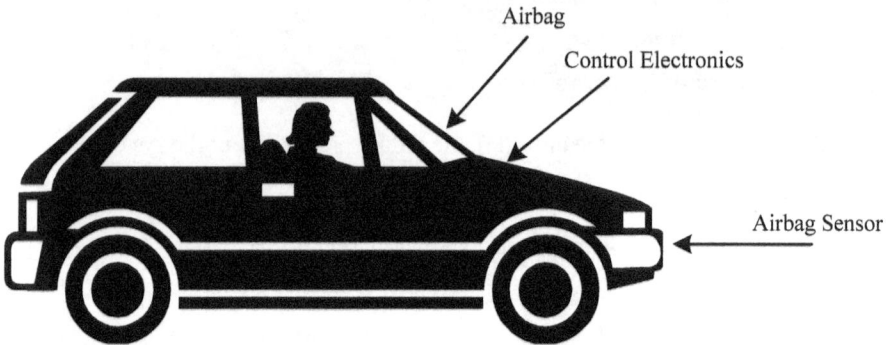

Figure 19.14: Car with airbag, control electronics, and airbag sensors

19.5.2.1 Risk graph

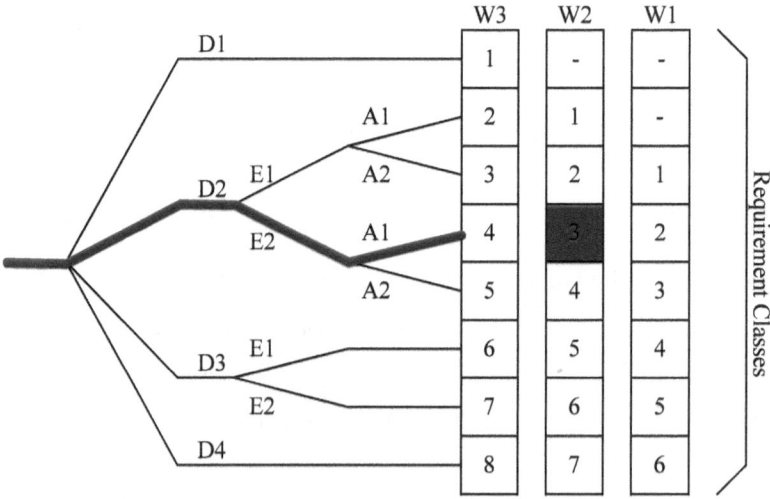

Figure 19.15: Risk graph for example 2 (speed of up to 70 km/h)

Since one can assume that each airbag in the car has its own controller, the extent of damage must be restricted to one person. For this reason, the worst consequence in this particular situation can be the death of a person (D2).

19.5 Examples of Reliability Analysis

The "worst case" scenario must be assumed, because the person's length of stay in the danger zone depends on the length of the drive, and in this case no specific travel time is given; therefore the stay in the danger zone is considered frequent to permanent (E2).

Although, in this case, danger avoidance depends on individual abilities (reflexes, ability to concentrate, ...) and on the dangerous situation itself, it can be assumed that at speeds of up to 70 km/h, danger avoidance is possible under certain circumstances (A1).

At low speeds, the probability that an event occurs which has dire consequences for a person (W2), is highly unlikely. Naturally, it must be added that this would depend on the dangerous situation.

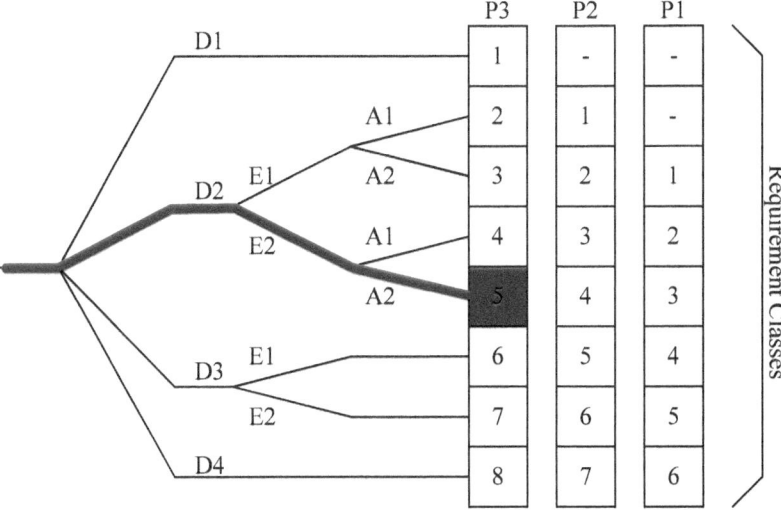

Figure 19.16: Risk graph for example 2 (speed higher than 70 km/h)

The difference regarding the first part of the situation is, that at higher speeds it is hardly possible to avoid imminent danger (A2). Higher speed also increases the probability of occurrence of a dangerous situation, and thereby also the risk (P3)

19.5.2.2 Event Tree

If, following an accident, the airbag sensor stop functioning, the person at the wheel can be seriously injured and, as a consequence of this injury, the accident can prove to be fatal. In order for the driver to come away with only minor injuries, the airbag sensors and control electronics must function flawlessly. In all other cases, the driver will be seriously hurt.

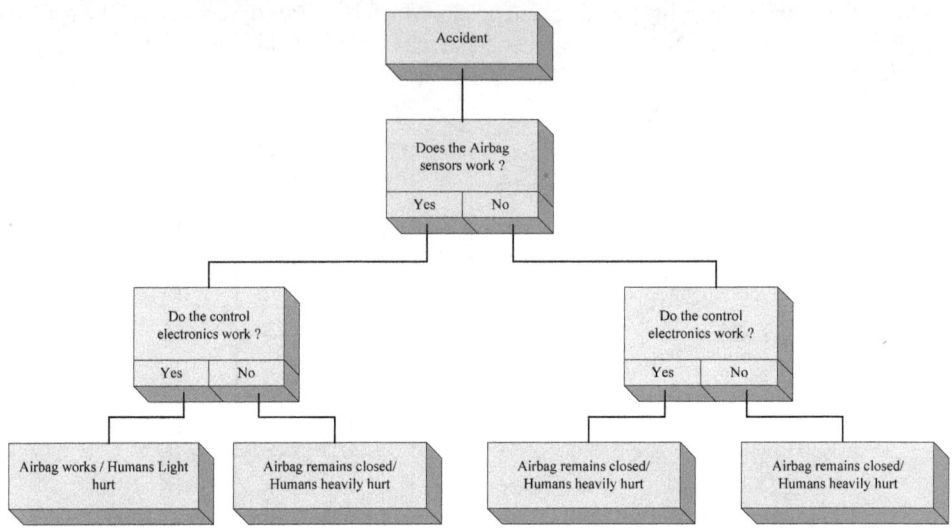

Figure 19.17: Event tree after an accident

19.5.2.3 Error Tree Analysis

Here, the undesired event is a non-functional or only partially functional airbag. This event can be triggered by one of the underlying primary failures, which are separately analyzed each in their own error tree. Since even the occurrence of only one primary failure will precipitate the undesired event, the probability of failures must be reduced to a minimum.

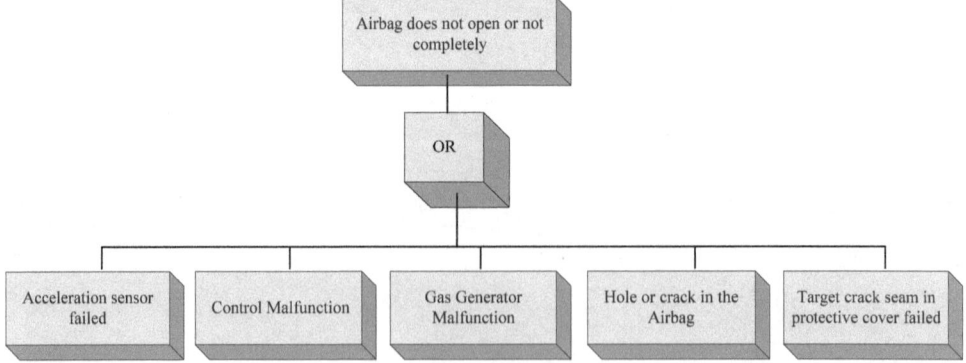

Figure 19.18: Error tree analysis in case of the airbag not opening

19.5.2.4 Reliability Block Diagram

Next, the reliability block diagram is created. Here one must also assume a serial structure, because all units (sensor, control electronics, and airbag) must function in order for the whole system to function.

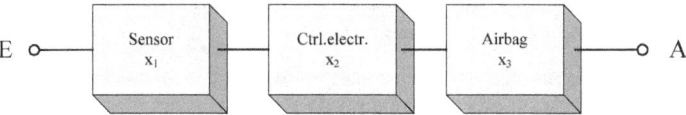

Figure 19.19: Reliability diagram for example 2 (serial structure)

With the failure probabilities of the units (q_i), the failure probability of the system P is

$$P(q_1,...,q_3) = 1 - ((1-q_1)(1-q_2)(1-q_3))$$

With the failure probabilities of the units (q_i), the failure probability of the system P is

$$P(q_1,...,q_3) = 1 - ((1-q_1)(1-q_2)(1-q_3))$$

The following failure rates are specified for the units:

- Sensor = 35 FIT = $35 \cdot 10^{-9}$ 1/h
- Control electronics = 200 FIT = $200 \cdot 10^{-9}$ 1/h
- Airbag = 5 FIT = $5 \cdot 10^{-9}$ 1/h

For the failure probability of individual units in a 1 year timeframe we get

$$P_{Sensor}(t) = 8760h \cdot 35 \cdot 10^{-9} \frac{1}{h} = 0.0003066$$

$$P_{Cont.elect.}(t) = 8760h \cdot 200 \cdot 10^{-9} \frac{1}{h} = 0.001752$$

$$P_{Airbag}(t) = 8760h \cdot 5 \cdot 10^{-9} \frac{1}{h} = 0.0000438$$

Therefore, the failure probability of the entire system comes to

$$P(q_1,...,q_3) = 1 - ((1 - 0.0003066)(1 - 0.001752)(1 - 0.0000438)) = 0.00210$$
$$= 0.21\%$$

19.5.3 Example 3 (Airplane)

Example 3 examines a 4-engine aircraft. As seen in Figure 19.20, each wing has 2 engines.

In order for the airplane to fly unrestrictedly, all 4 engines must be functional. If one of the four engines fails (for example A), or in the event of one engine failing on each wing (for example A and C), the flight can only be continued in a limited manner. Accident risk is increased but an emergency landing is not mandatory.

If two engines on one single wing fail, or if 3 engines fail, continuation of the flight is severely handicapped and an emergency landing should be prepared for immediately.

If all four engines fail, flying is no longer possible (gliding). An emergency landing is the only possibility of saving passenger's lives, and thereby avoiding a catastrophe.

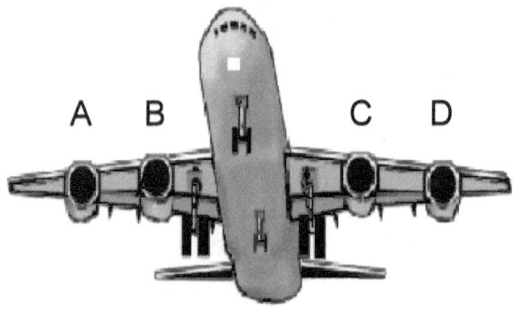

Figure 19.20: Airplane with 4 engines

19.5.3.1 Risk Graph

In the following segment, a risk graph is designed for the case of one engine failing, or of two engines failing, that is, one engine on each wing.

While this situation restricts the flight, it has no effect on the passengers. Today's airplanes are technically well equipped to continue the flight in such a situation, without disadvantages. The passengers are hardly aware of these technical mishaps. For this reason, the extent of damage can be classified as very low (D1).

19.5 Examples of Reliability Analysis

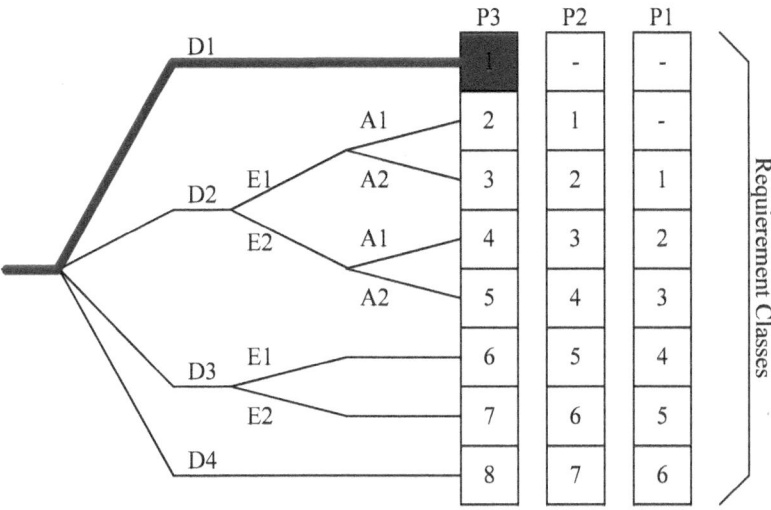

Figure 19.21: Risk graph for example 3 (failure of one engine, or of one engine on each wing)

Next, a risk graph is drawn for the event of failure of two engines on one wing, or of three engines.

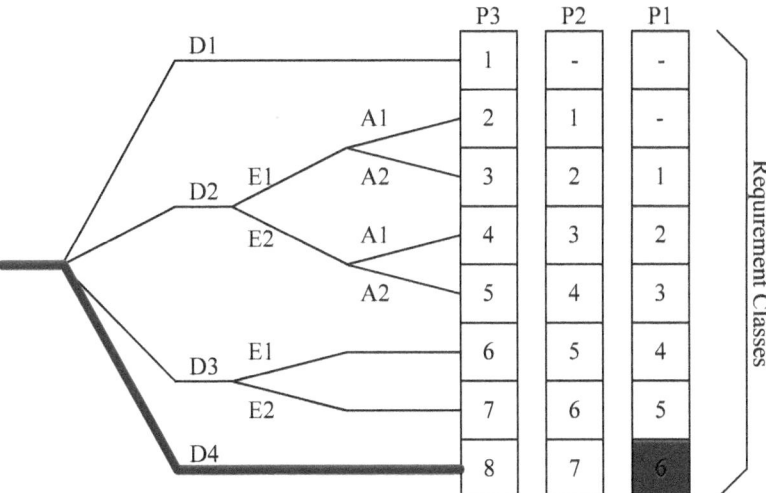

Figure 19.22: Risk graph for example 3 (failure of two engines on one wing, or failure of three engines)

In such a situation, the flight of the airplane is severely restricted so that it must perform an emergency landing. In the worst case scenario, with four failed engines, the pilot must even attempt an emergency landing by gliding. This, of course, has major effects on the passengers (D4) because, in the worst case, a crash is imminent and thus a catastrophe with very many people killed. The probability of such a situation occurring is very small in today's airplanes, because they are equipped with a variety of technical safeguards (P1).

19.5.3.2 Event Tree

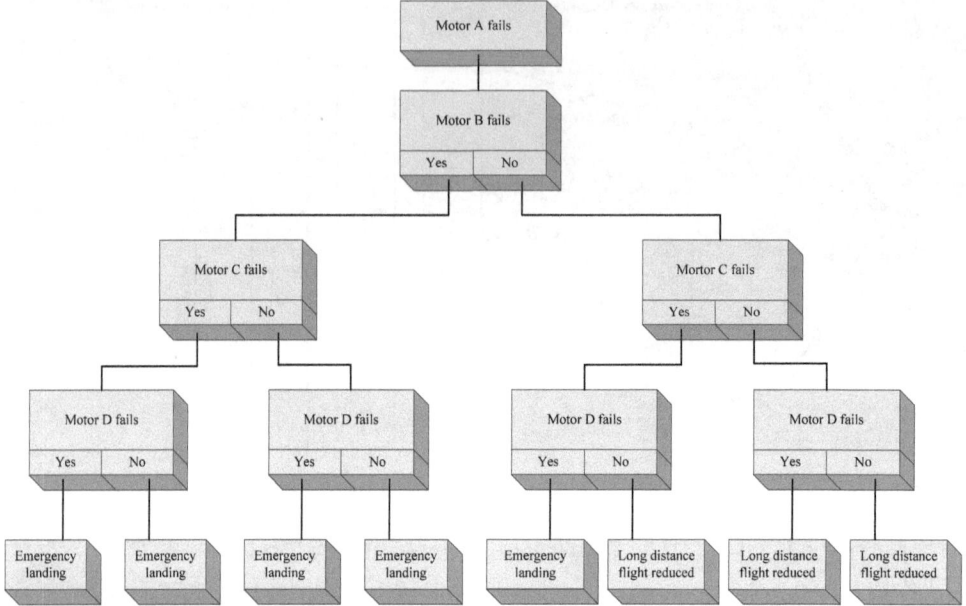

Figure 19.23: Event tree after engine failures

As soon as two engines fail on one side (for example A and B), or a total of more than two engines fail (for example A, C, and D), an emergency landing must be performed because the situation can lead to a catastrophe. If, on the other hand, only one engine fails (for example A), or if only one engine fails on each side (for example A and D), the flight can be continued within limits.

19.5.3.3 Fault Tree Analysis

In this example, the possibilities of fault tree simplification are readily apparent. The three fault trees (see Figures 19.24 to 19.26) are equivalent and can be transformed among themselves. The first one is based on the statement that the failure of two engines on one wing, or the failure of more than two engines, independent of which, requires an emergency landing. The 3-out-of-4 trellis represents the failure of more than three engines. If one resolves the 3-out-of-4 trellis, one first obtains the more expansive, second error tree, where all failure combinations can be recognized. If one crosses out all failure combinations which contain other failure combinations, then one obtains the third, most compact error tree, which demonstrates minimal cuts.

19.5 Examples of Reliability Analysis

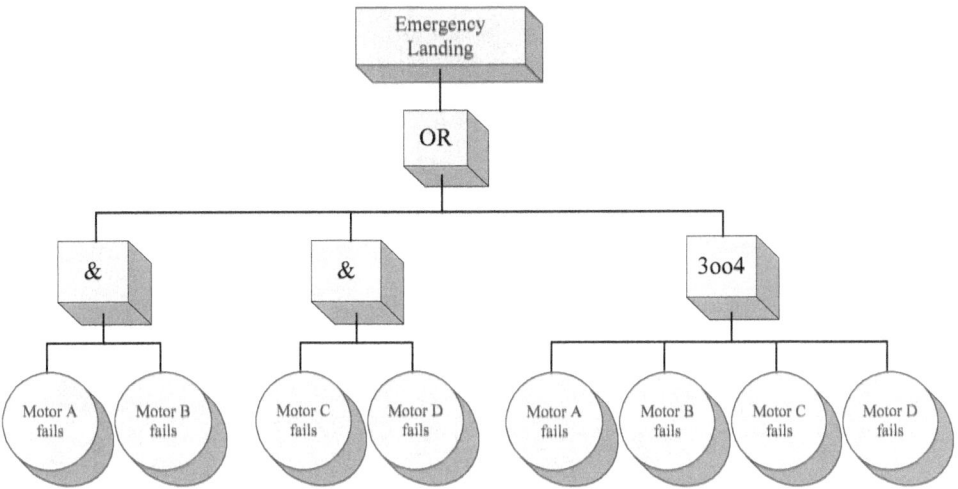

Figure 19.24: Fault tree 1 for the event of an emergency landing

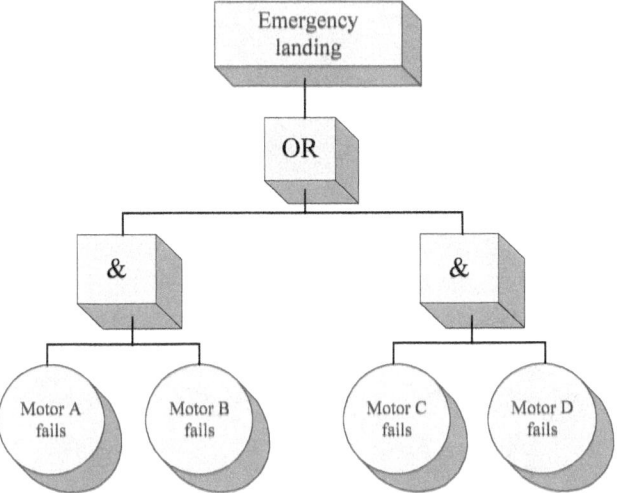

Figure 19.25: Fault tree 2 for the event of an emergency landing

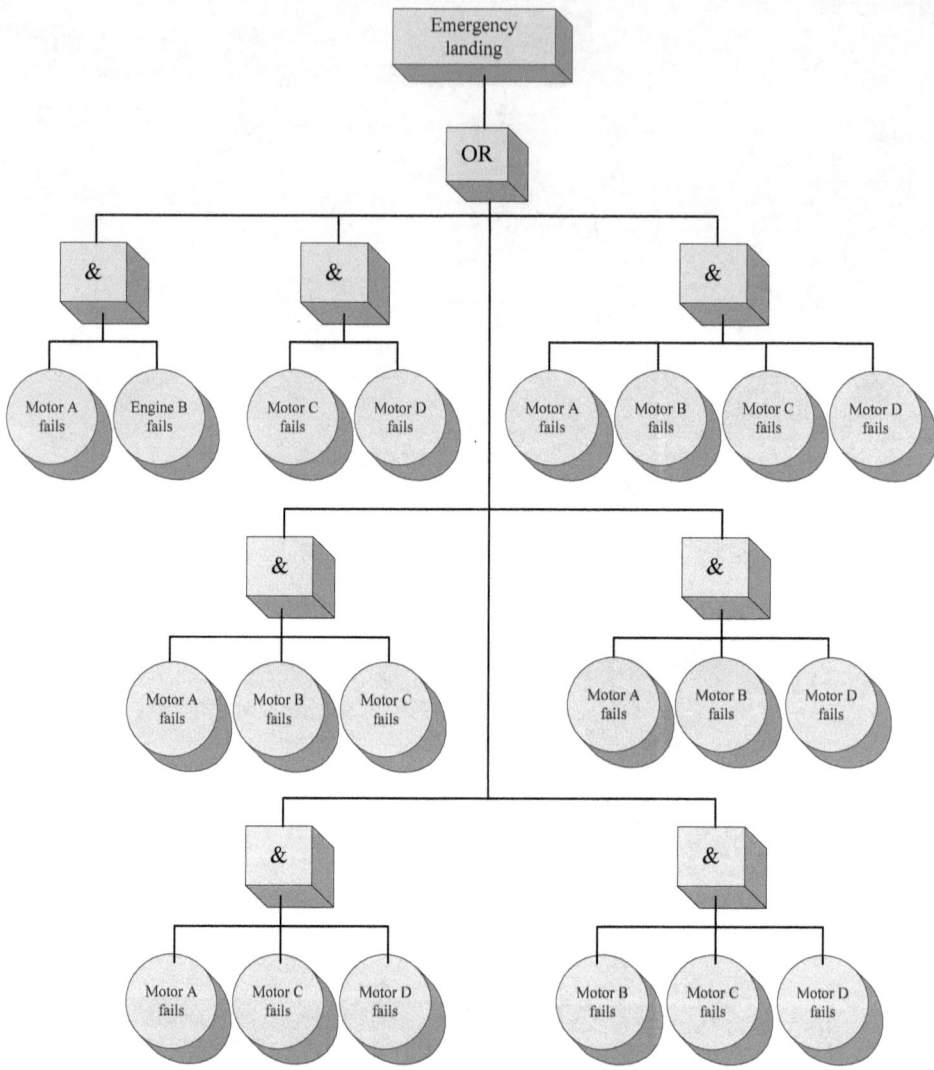

Figure 19.26: Fault tree 3 for the event of an emergency landing

19.5.4 Example 4 (Pipeline)

In this example concerns the transport of oil via a pipeline (see Figure 19.27). The distance from the oil drill rig to the oil refinery is 20 km. Control stations are established every 5 km, which utilize pressure sensors to check the pipeline for leaks. The control stations have built-in F35 controls by HIMA, which transmit the data received to the head stations. Communication between control stations and head stations takes place through glass fiber cables.

Individual segments can be opened and closed with valves, which allow a safe status in the event of an oil spill. In such a situation, the first valve to close is the one which is responsible for the segment where the spill occurred. Next, all other control stations are notified and take steps to close the valves on the remaining segments as well.

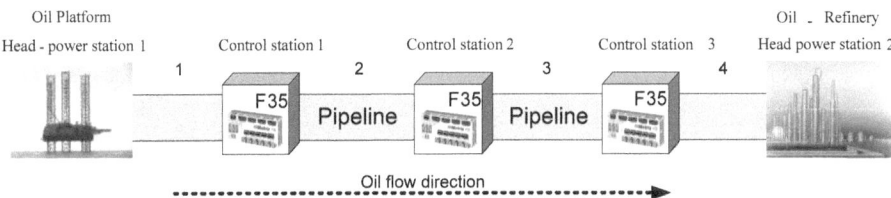

Figure 19.27: Pipeline between oil drill rig and oil refinery

19.5.4.1 Risk Graph

First, a risk graph is created for the event that, during an oil spill, the proper valve in the segment concerned does not close. Communication continues to be functional and thereby all other valves can be accessed.

Because communication is not affected by the failure of one valve, and because all other valves are still accessible, the oil spill will not be very big. The consequences for humans and for the environment are therefore limited (D1).

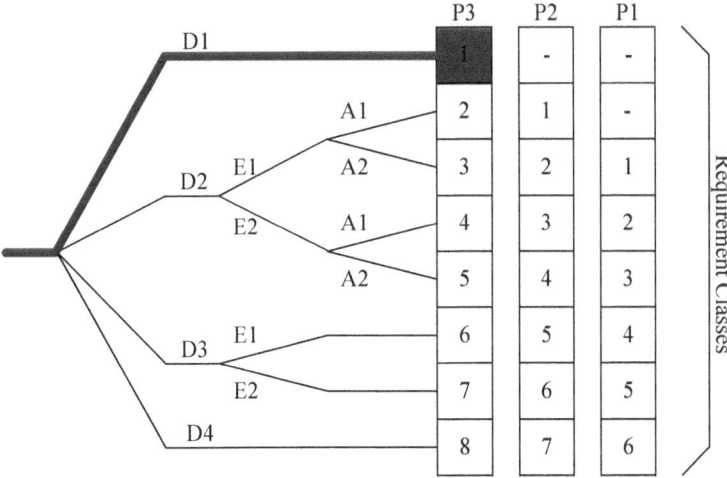

Figure 19.28: Risk graph for example 4

19.5.4.2 Event Tree

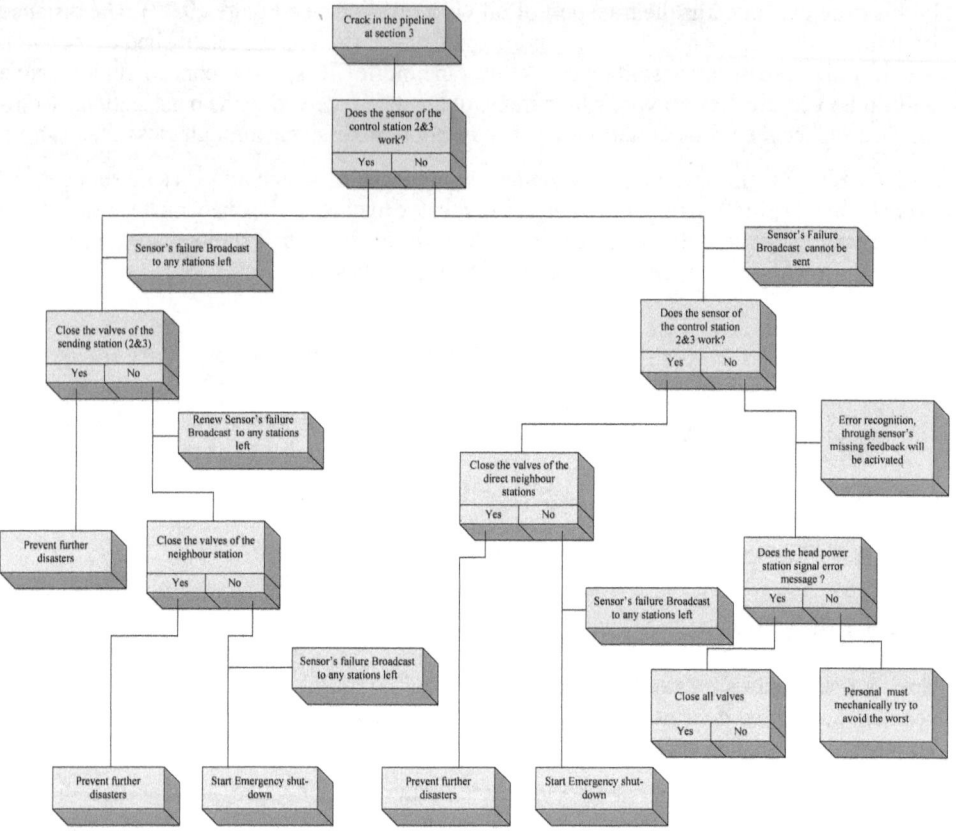

Figure 19.29: Event tree following a tear in the pipeline

19.5.4.3 Error Tree Analysis

If the valves of the transmitting station and the valves of the neighboring stations can not be closed following a tear in the pipeline, an emergency shut-off must be initiated, in order to avoid a catastrophe. As soon as the valves of the transmitting station, or the valves of the neighboring stations, are closed, a catastrophe can be avoided.

The undesired event in this case is the oil spill which will happen if there is a tear in the segment and if all valves of the segment concerned are open. In the following example, only one open valve is further analysed, all other valves would be developed analogously.

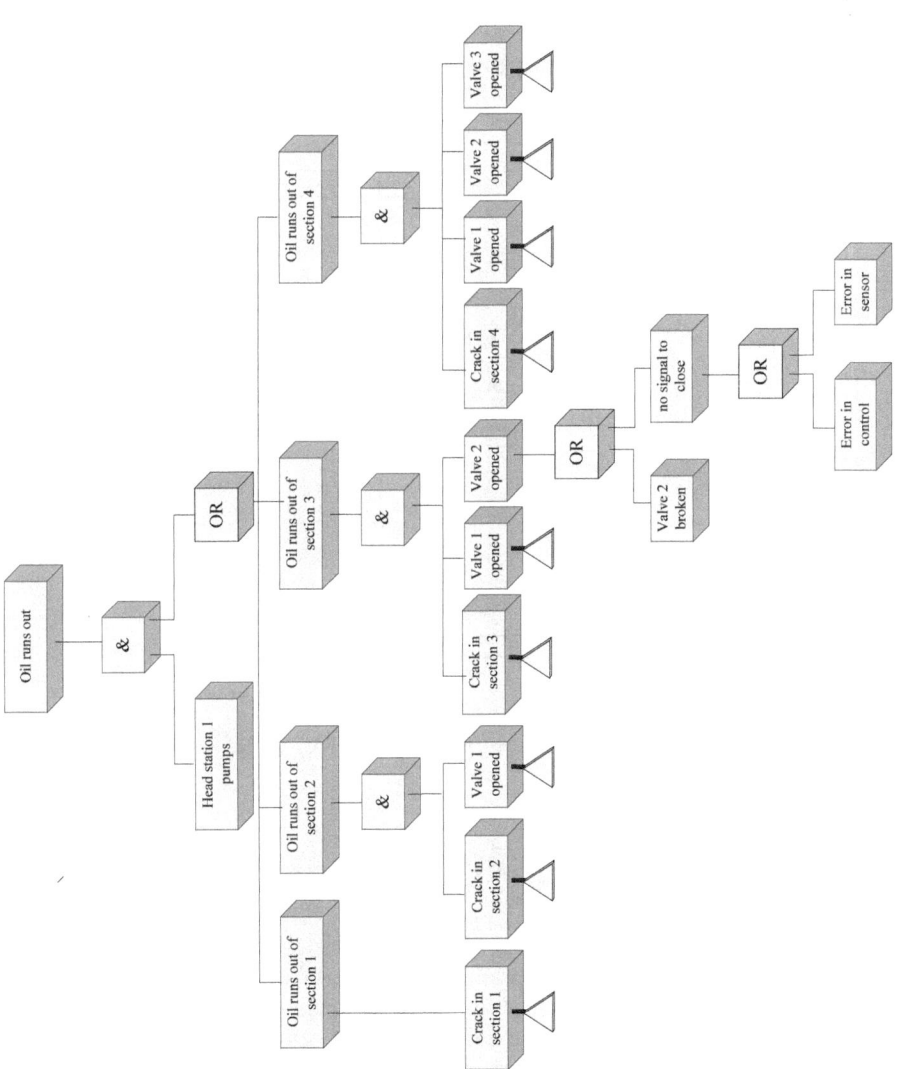

Figure 19.30: Error tree analysis in the event of an oil spill

19.5.5 Example 5 (Coliseum)

As a final example, we will consider a coliseum with a seating capacity of 10,000 people. The coliseum is equipped with safety devices to detect smoke and fire danger.

The coliseum is equipped with, among others, three smoke sensors (R_1-R_3) and three temperature sensors (T_1-T_3). A built-in controller directs the ventilation and sprinkler system. Surveillance of the control is performed by a HIMatrix F30 controller manufactured by HIMA. This device has purely a supervisory function and no control function. Both controls are "low activity". As soon as one of the two controls fails, all of the outputs are switched to "low".

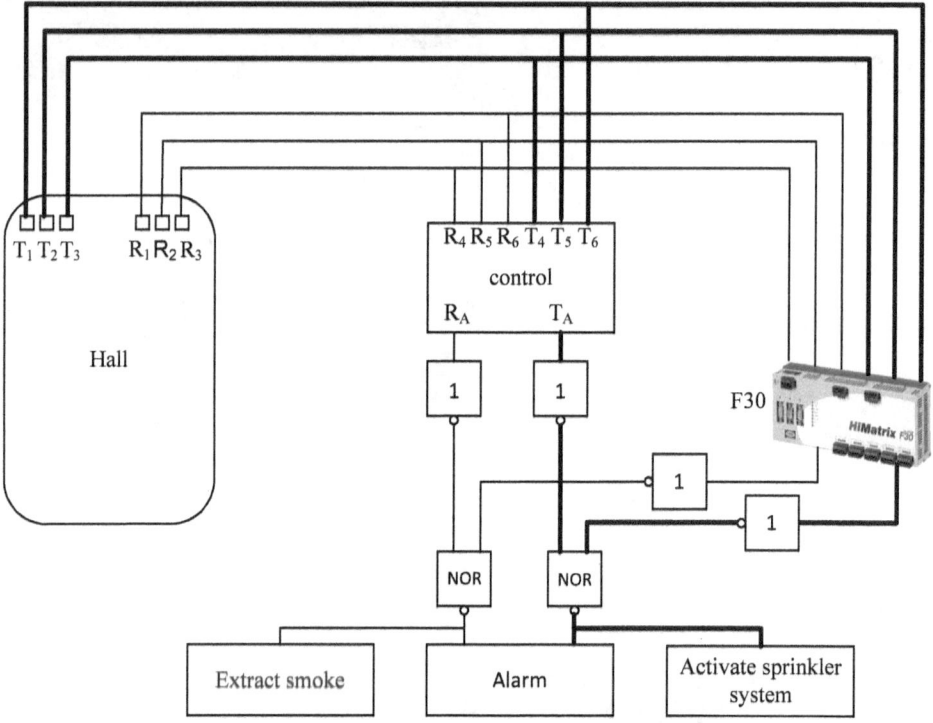

Figure 19.31: Safety devices in the coliseum

The sensors transmit data to the controller and to the F30. As soon as both controllers receive dangerous values, an alarm is triggered and, depending on the level of danger in the particular situation, either the smoke is sucked off, or the fire is extinguished.

Two of the three values transmitted by the sensors must match for the appropriate action (sucking off the smoke or activating the sprinkler system) to take place in a danger situation (2oo3-system). If one of the three sensors fails, the values transmitted by the two remaining sensors must match. As soon as two sensors fail, a very dangerous situation exists because action can no longer be taken. Two cases of sensor failure need to be distinguished:

- Sensor has failed and remains in danger status
- Sensor has failed and remains in non-danger status

19.5.5.1 Risk Graph

When two sensors fail and remain in non-danger status, it means that in the event of a fire, no reaction is possible because two sensors transmit non-danger values. This could potentially lead to a catastrophe (D4). The probability of such an event is low (P2).

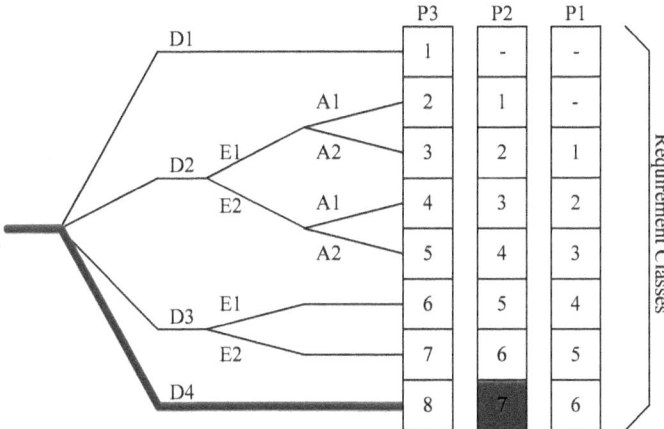

Figure 19.32: Risk graph for example 5

19.5.5.2 Event Tree

In this case, a catastrophe occurs if, in the event of a fire in the coliseum, the control devices do not register a deviation of the failed sensors. The catastrophe can only be averted if the control devices recognize the sensor failure, and if appropriate actions are taken, either automatically through the safety system or manually by a person, as soon as they have been alerted.

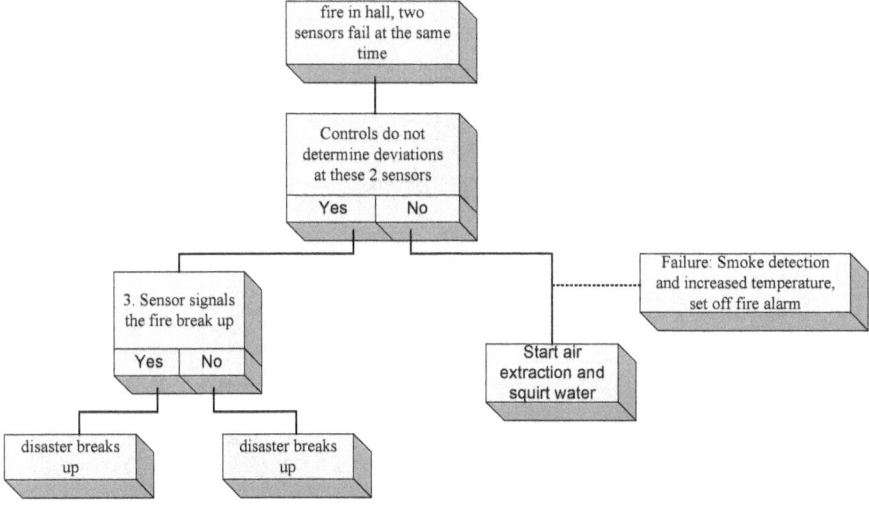

Figure 19.33: Event tree following the simultaneous failure of two sensors

19.5.5.3 Error Tree Analysis

In this example, the undesired event is a fire which does not get extinguished. This can happen if a fire breaks out and the sprinkler system fails to operate. This in itself can occur if both the control output, as well as the output of the HIMatrix, remain on HIGH (see flow schematics in Figure 19.31: The actions "suck off smoke", "alarm', and "activate sprinkler system" are LOW-active.). If a fire breaks out, it is consequently not extinguished if both controls fail, or if at least two sensors transmit incorrect data.

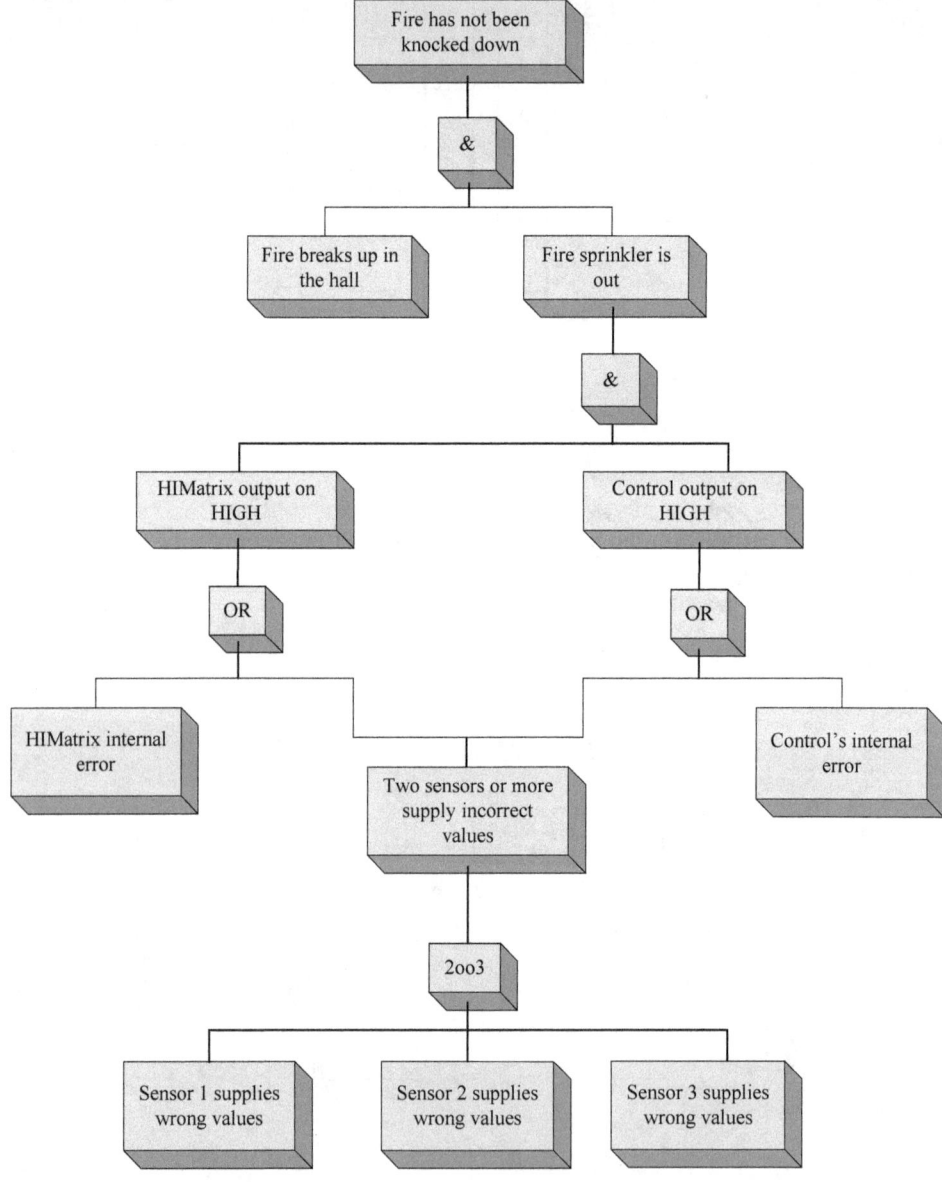

Figure 19.34: Error tree analysis in the event of a fire breaking out

20 IEC/EN 61508

Throughout the world, there is a multitude of various committees which have dedicated themselves to the specification of norms and standards, namely in regards to the procedures used in the development, verification, and implementation of systems which are crucial for safety. In the implementation of critically important safety systems, provisions are made in order to maintain a standard for the operation of such installations: to certify a system through a supervising authority, according to the standards and guidelines that are relevant to its implementation.

The application of safety-oriented computer systems in various industrial sectors has lead to a wealth of experience and knowledge in this area. This knowledge is mostly present in compact form, as standards and guidelines, which are designed to assist the designing engineer in achieving a uniformly high standard of quality and safety in prospective systems. Depending on their exact purpose, safety-oriented systems must fulfill certain standards in order to determine developmental methods and system requirements. Certification of the final total process, including the safety system which regulates this process, depends on compliance with these standards, and with the criteria that they define. Nowhere else in industry is strict adherence to standards as strictly regulated as in the implementation of safety-oriented measurement and control[189]- and ESD[190]-technology.

Industry-specific standards and guidelines, such as in the chemical, petrochemical, airline, atomic, and mining industry, each describe the principal dangers inherent in these industries. The remaining risks, which ultimately go hand-in-hand with the demands on reliability and availability, are regulated and defined by the advisement of the supervising authorities and by the industry itself. Resulting from this are the safety requirement standards regarding systems of a specific sector of industry. General standards and guidelines, on the other hand, refer to all industry sectors.[191, 192]

The IEC[193] develops and administers standards in collaboration with national committees. In doing so, work groups are put in place which deal with a special subject at a time; for by example, the Technical Committee 65, which is responsible for „measuring and regulating in industrial processes". This committee created the document „IEC 61508: Functional Safety: Safety related Systems" which is described in more detail below. The IEC 61508, also labelled basic safety standard, describes the fundamental, complete life cycle of safety-oriented systems. It is subdivided into seven parts, where parts 1, 2, 3, and 4 are

---

[189] MSR-Technik: Messungs-, Steuer- und Regelungstechnik
[190] ESD-Technik: Electronic-Shut-Down-Technik
[191] [HSE-87] HSE, *Programmable Electronic Systems in Safety-Related Applications*; An Introductory Guide. Health and Safety Executive. London: Her Majesty's Stationery Office
[192] [IECa02] IEC 61508, *Functional Safety; Safety-Related Systems*.
[193] International Electrotechnical Commission

also known as basic safety standards and are referred to by technical committees during the creation of standards according to IEC guide 104, and ISO/IEC guide 51.

Aside from the IEC 61508, the next chapter also describes the IEC 61511. The group title of this standard is: „Functional safety: Safety Instrumented Systems for the process industry sector" [194], in German: „Funktionale Sicherheit: Sicherheitstechnische Systeme für die Prozessindustrie"[195]. This international standard deals with the application of safety-instrumented systems in the process industry. It is the application of the IEC 61508 for process engineering and requires the execution of a hazard and risk analysis. According to this analysis, a specification of the safety-instrumented systems can be designed.

20.1 IEC/EN 61508-1

20.1.1 Outline and Field of Application

It is imperative for the owner of an installation to arrange for adequate safety management. If, in the event of an accident, the safety-oriented system does not conform to the standard, the firm could be held liable for neglecting proper safety guidelines. As such, a proper safety technology, which is in compliance with international standard, is required for all downstream servicing and for outside contractors. The IEC 61508 is an international standard which was issued by the International Electrotechnical Commission (IEC) 1999. It refers to all aspects that are connected to the use of E/E/PES (Electrical / Electronical / Programmable Electronic Systems) for safety-relevant functions and applications. They are fundamentally applicable to all safety-related E/E/PES, in particular during the absence of a special safety standard for an application area. The IEC 61508 standard comprises the following seven parts (61508-1 through 61508-7), see also Figure 20.1:

- Part 1: General requirements
- Part 2: Demands on safety-related electric/electronic/programmable electronic systems
- Part 3: Software requirements
- Part 4: Terms and abbreviations
- Part 5: Examples for determining the level of safety integrity
- Part 6: Application guideline for IEC 61508-2 and IEC 61508-3
- Part 7: Application instructions for procedures and methods

---

[194] [IEC-02] IEC 65A/324/FDIS 2002, *Functional safety: Safety Instrumented Systems for the process Industry sector*
[195] [DIN-03] DIN IEC 61511, Teil 1 bis 3, (VDE 0810 Teil 1), *Funktionale Sicherheit: Sicherheitstechnische Systeme für die Prozessindustrie*

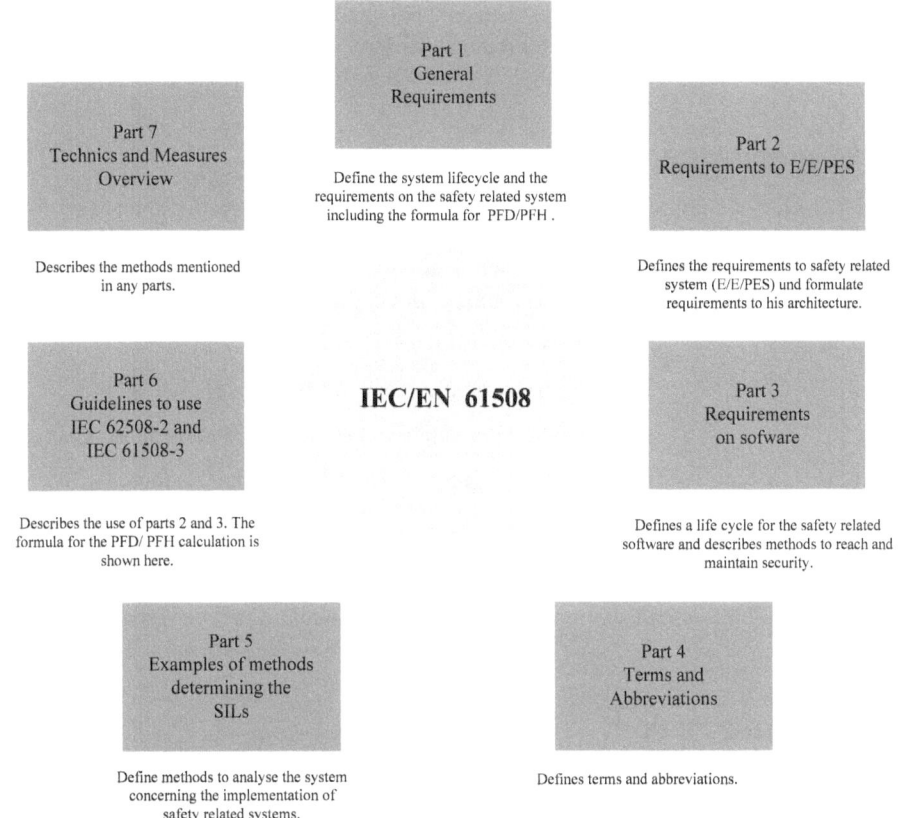

Figure 20.1: Outline of the parts of the IEC/EN 61508

This standard facilitates a systematic, risk-based approach to safety-relevant problems. Part 1 of this standard specifies the general requirements which are applicable to all other parts. Parts 2 and 3 define additional requirements of system hardware and software. Part 4 explains the definitions and abbreviations used in this standard. Part 5 provides guidelines for the application of part 1, and part 6 provides guidelines for the application of parts 2 and 3. Finally, part 7 contains an overview of procedures and methods.

This International Standard regards all relevant safety-related phases of the total concept of E/E/PES and of the software safety lifecycle, from the concept, through development, execution, implementation and maintenance until deactivation. It allows for the compilation of application-specific international standards, which are concerned with safety-related E/E/PES. Furthermore, it provides a method for developing the specifications of safety requirements, which is necessary in the achievement of stipulated functional safety for the safety-related E/E/PE system. It utilizes the safety integrity level for specifying the goal of the safety integrity of related functions. In order to determine the demands placed upon the safety integrity level, it applies a risk-based approach to the problem in which criteria for numeric failures are set.

The standard is universally valid and applicable to all safety-related systems.

Figure 20.2 shows the assignment of parts of the standard to the requirements with regards to the safety lifecycle. The requirements differentiate between technical and other requirements (for example, documentation and management of functional safety).

20.1.2 Compliance with this Standard

In order to achieve compliance with this standard, it must be proven that the requirements are fulfilled and of accord with the specified criteria. Excempt are systems of low complexity, for which reliable field experience is at hand. Their requirements were, and are, integrated into the safety standards of various application fields.

CD: IEC 62061	Machine safety
CDV: IEC 61511	Safety-instrumented systems for the process industry
EN 50126	Rail industry applications: Specification and proof of safety
EN 50128	Software for rail guidance and supervision systems
EN 50129	Safety-relevant electronic systems for signal technology
pr EN 50156	Electrical equipment of furnace devices

The IEC/EN 61508 is applicable both as a stand-alone standard, as well as a basic standard.

20.1.3 Documentation

The entire documentation must contain sufficient information to make it possible for all phases of the safety lifecycle- as well as all occurring design and verification tasks. safety management, and safety inspection- to be executed efficiently. Documentation should be accurate and concise, making it easy to understand. Furthermore, it must be accessible and serviceable at all times. The individual segments of this standard specify exactly which information has to be documented.

20.1.4 Safety Management

Persons or organizations responsible for individual phases, or for the entire lifecycle, must specify all technical and management activities. These ensure that the E/E/PES will achieve and maintain the required functional safety. Among these activities are, most notably, the organization of the internal information flow, the strategy for achieving functional safety, and the qualifications of the personnel involved. It is mandatory to create a plan that includes all of the aforementioned activities in order to achieve safety.

20.1 IEC/EN 61508-1

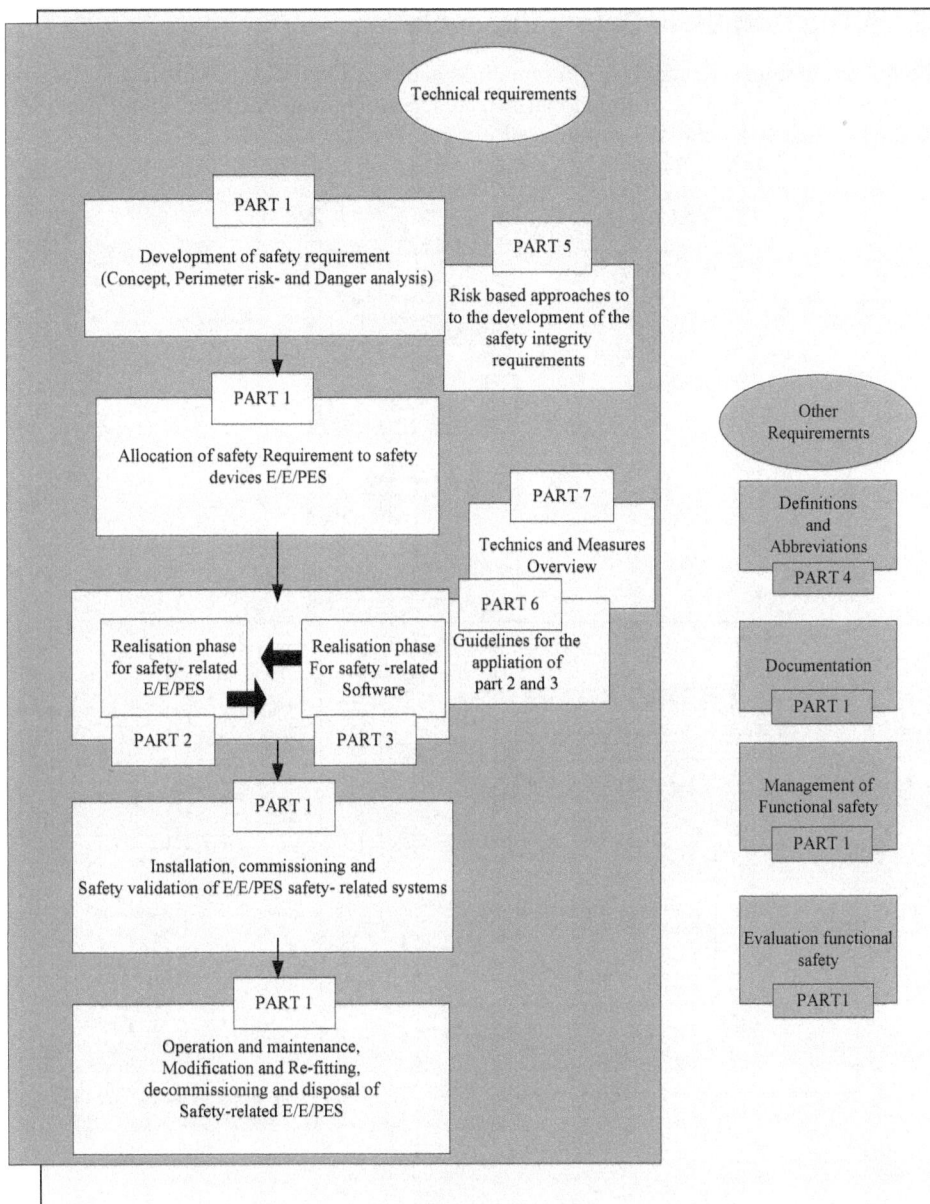

Figure 20.2: Complete structure of the IEC/EN 61508

20.1.5 The Complete Safety Lifecycle

The following safety lifecycle (see Figure 20.3) is defined in order to achieve a systematic approach to the problems of functional safety. Upon application of the safety lifecycle, the safety integrity level (SIL) is achieved.

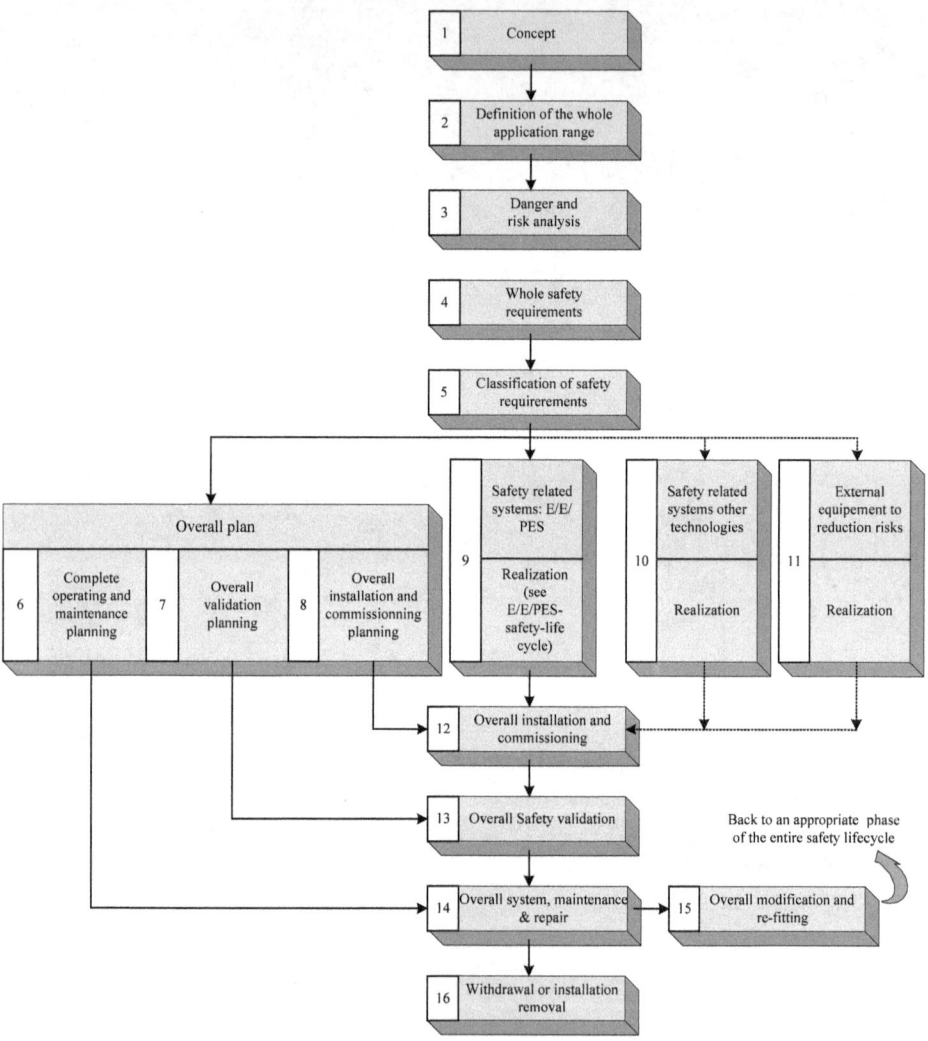

Figure 20.3: Entire safety lifecycle

The safety lifecycle contains 3 methods to minimize risk: phases 9, 10, and 11. In the following segment, the individual phases from Figure 20.3 are illustrated in more detail.

In the *concept* phase, a sufficient understanding of the EUC installation (plant, machine, etc.) and its environment must be achieved. This phase also includes observations regarding probable safety hazards and legal provisions. In the following phase, the *entire appli-*

cation field is defined, including the limits of the system and of the EUC, the hazards, and any external events. Furthermore, a *hazard and risk analysis* must be performed, in which the information from phase 2 is incorporated. Thereby, all hazards and potentially dangerous events of the EUC must be incorporated as reasonable and foreseeable circumstances. Likewise, the probabilities and potential consequences of the hazardous events must be determined. In phase 3, in order to achieve the necessary functional safety, the *entire safety requirements*, containing both the safety functions and their safety integrity, must be specified. In addition, risk reduction through external systems and systems from other technologies must be determined. In the *assignment of safety requirements* phase, certain safety-related systems are specified in order to achieve the necessary functional safety. A variety of safety integrity levels (SIL) are also specified, each of which must be assigned to an individual safety function (see tables at the end of this chapter). This is followed by the *planning of the complete installation and of the complete maintenance* of the E/E/PES. A plan is drawn up for this purpose which contains, as needed, the standard routines for upkeep of functional safety and of the safety-related E/E/PES. Likewise, measures for preservation of safety during servicing activities must be planned. In the *planning of complete safety validation*, a plan for the validation of the system must be created, with consideration of different operating conditions of the system, predictable disturbances, validation strategies, and criteria for existence, all in respect to nonexistence. One plan for each measure must be created for the controlled *complete installation and complete commissioning* of the system. In particular, the following aspects must be taken into consideration: time, responsibilities, procedure, and criteria, all of which declare the completion of this installation. In phase 9, the *materialization of the safety-related systems* is planned. Hereafter, the safety-related systems which meet the specified safety requirements must be compiled. This is where the development of hardware and software of a E/E/PES for safety applications is performed. For the development of hardware and software, the lifecycle and the requirements from parts 2 and 3 of the IEC 61508 apply. For the second phase of materialization, certain safety tasks must be specified for *safety-related systems from other technologies, and external devices for risk reduction*. Such systems are not discussed, however, in this standard. In phase 12, *total installation and total commissioning*, the E/E/PES is installed and commissioned. Strict adherence to the plans constructed in phase 8 is required. In *total safety validation*, it must be ascertained that the installed E/E/PES conforms to safety requirements. This must be achieved by following the plan drawn up in phase 7. All of the measuring instrumentation needed in validation must be calibrated according to a reproducible standard. In the *complete operation, complete maintenance and repair* phase of the safety lifecycle, it is explained that the E/E/PES must be run, serviced, and repaired in such a manner that the required functional safety is preserved. In *total modification and total upgrade*, it must be ascertained that functional safety of the E/E/PES is preserved both during, as well as after, the modification. As such, all activities concerning the modification must be carefully planned with any effects considered. Before *deactivation or deinstallation* can begin, all effects of the deactivation of the E/E/PES on the functional safety of other systems, which are in connection to the system to be dismantled, must be analyzed. Only after this analysis is complete can deactivation be authorized.

20.1.6 Verification

A verification plan must be designed, which contains the criteria and methods used to perform the verification. Such a plan must be created for each phase of the safety lifecycle, and, with it, the verification of each individual phase can be performed

In the following tables, the safety integrity level is defined as dependent upon the failure probability, both for an operating level at low demand (see Table 20.1) as well as for an operating level at high demand (see Table 20.2) or continuous operation.

Table 20.1: SIL low demand rate

Safety Integrity Level (SIL)	Low demand operating level (medium probability of the failure of the safety function during operation)
4	$\geq 10^{-5}$ to $< 10^{-4}$
3	$\geq 10^{-4}$ to $< 10^{-3}$
2	$\geq 10^{-3}$ to $< 10^{-2}$
1	$\geq 10^{-2}$ to $< 10^{-1}$

Table 20.2: SIL high demand rate

Safety Integrity Level (SIL)	High demand operating level (medium probability of the failure of the safety function during operation)
4	$\geq 10^{-9}$ to $< 10^{-8}$
3	$\geq 10^{-8}$ to $< 10^{-7}$
2	$\geq 10^{-7}$ to $< 10^{-6}$
1	$\geq 10^{-6}$ to $< 10^{-5}$

20.1.7 Evaluation of Functional Safety

The actual achieved functional safety of the E/E/PES must be evaluated. Therein, aspects concerning the degree of independence and competence of the implementing persons must be taken into account.

20.2 IEC/EN 61508-2

20.2.1 Field of Application

This part of standard IEC/EN 61508 applies to all safety-related systems, which fall under the definition in part 1. It makes additional demands on the hardware. Furthermore, requirements are specified for all activities, which are necessary during the developmental and manufacturing process; this does not include existing system software, which is discussed in part 3.

20.2.2 The E/E/PES Safety Lifecycle

So as to achieve compliance with the standard, the following safety lifecycle must be applied:

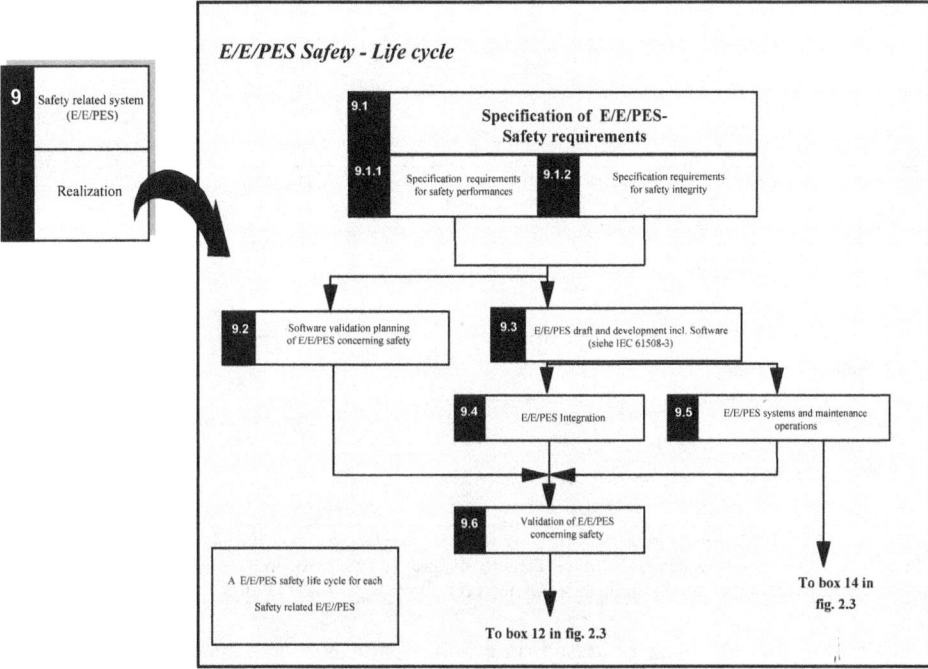

Figure 20.4: E/E/PES safety lifecycle

If a different hardware safety lifecycle is to be applied, it must be one which has been specified during the planning phase of functional safety, see segment 6 of the IEC 61508-1. The individual phases from Figure 20.4 are explained in more detail below.

A *specification for the safety function of the E/E/PES* must be created, which must contain specifics on how the E/E/PES will achieve and maintain the required safety. In addition, all relevant operating levels must be considered. Also, a specification for the safety integrity must be created, which contains information about the safety integrity level for each safety-relevant function. In particular, and among others, the intended electromagnetic compatibility must be included. For *planning the validation of the E/E/PES with regards to safety,* all demands on test procedures concerning execution, test environment, and criteria for the existence/non-existence must be paid attention to, so that a correct validation of all safety-relevant functions is possible. The entire *design and development of the E/E/PES* must be in accordance with the safety specifications developed in phase 9.1 from Figure 20.4. The architectural demands on hardware and software must abide by the required SIL. The possible SIL is limited by hardware failure tolerance, and by the SFF (safe failure fraction). The highest SIL which can be chosen for a safe function, is specified in Tables 20.3 and 20.4. Subsystems can be subdivided into type A and type B. With type A, in contrast to type B, the behavior in the event of an failure is completely known. For both types, the following requirements apply to safety integrity:

Table 20.3: Hardware safety integrity type A

Safe Failure Fraction	Hardware fault tolerance (see Note 2)		
	0	1	2
< 60 %	SIL1	SIL2	SIL3
60 % – < 90 %	SIL2	SIL3	SIL4
90 % – < 99 %	SIL3	SIL4	SIL4
> 99 %	SIL3	SIL4	SIL4

Note 1: See 7.4.3.1.1 to 7.4.3.1.4 in IEC/EN 61508 for detailed explanations of table
Note 2: A hardware fault tolerance of N means, that N+1 faliures can cause the loss of the safety function.
Note 3: See Appendix C for details regarding calculation of the percentage of safe failures

Table 20.4: Hardware safety integrity type B

Safe Failure Fraction	Hardware fault tolerance (see Note 2)		
	0	1	2
< 60 %	Not permitted	SIL1	SIL2
60 % – < 90 %	SIL1	SIL2	SIL3
90 % – < 99 %	SIL2	SIL3	SIL4
> 99 %	SIL3	SIL4	SIL4

Note 1: See 7.4.3.1.1 to 7.4.3.1.4 in IEC/EN 61508-2 for detailed explanations of the table
Note 2: A hardware fault tolerance of N means, that N+1 failures can cause the loss of the safety function
Note 3: See Appendix C for details on the calculation of the percentage of safe failures

Furthermore, the plan must be based on a subdivision into subsystems with their own specific design. Other components of the design and developmental process are estimates of the failure probability of the safety functions of components through accidental hardware failures. For this, the architecture of the safety-related E/E/PES, the estimated failure rate of subsystems, the probability of detecting a failure by routine diagnosis, and the length of time required for repair of the detected failure must be taken into account. In phase 4, the *E/E/PES integration,* the E/E/PES is integrated according to the specified design. The integrated system with all software and hardware modules is now tested specifically. The test must show that all modules interact correctly, and that they fulfill all intended functions. Next comes the *E/E/PES operation and maintenance procedure.* In this phase, procedures are specified which enable the safe operation and maintenance of the E/E/PES. The *validation of the E/E/PES relating to safety* must follow a preconceived plan, in which each specific safety function is validated by means of tests or analysis.

To avoid negative effects on the safety of the E/E/PES, each modification must be validated and documented diligently. The results of each phase of the lifecycle must be reviewed for accuracy. This is referred to as verification.

20.2.3 Techniques and Measures for Control of Failures during Operation

This addendum provides an overview of techniques and measures for the control of failures during operation. For each safety integrity level, there are techniques and measures for the control of accidental hardware failures, as well as for systematic, environment-related, and operation-related failures. In Table A.1 of the IEC/EN 61508-2, appendix A, the errors and failures are listed which must be recognized during operation. More detailed information is given in additional tables.

20.2.4 Methods for Avoiding Systematic Errors during Different Phases of the Lifecycle

In addendum B of the IEC/EN 61508-2, methods are indicated for avoiding systematic errors during different phases of the lifecycle, for each safety integration level. Accordingly, the individual methods are evaluated in the form of tables, with regards to their suitability for a particular SIL.

20.3 IEC/EN 61508-3

20.3.1 Field of Application

This part of standard IEC/EN 61508 refers to each software system which is either a part of a safety-related E/E/PES or an application used during its development.

20.3.2 Quality Management System of Software

The requirements meet the demands specified in section 6.2 of the IEC 61508-1. Additionally, the planning of functional safety must define strategies for the acquisition, development, integration, verification, validation, and modification of the software.

20.3.3 Software Safety Lifecycle

For the development of software, a safety lifecycle must be specified, in accordance with section 6 of the IEC 61508-1 (see Figure 20.5).

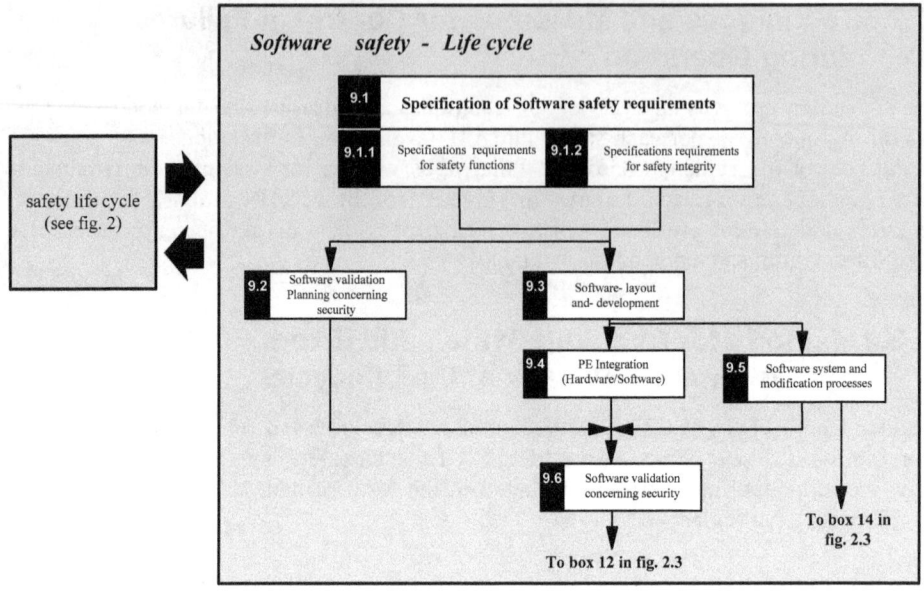

Figure 20.5: Software safety lifecycle

The *safety requirements of the software* must be specified in detail, if that has not already been done. In this, the specification must the tested by the software developer, in order to ascertain that all requirements are sufficiently described. The specification must contain, among others, instructions for self-monitoring of the software, and hardware monitoring. During the *planning of software validation with regards to safety*, all demands on the test procedure, the persons who implement it, the environmental conditions, as well as the criteria for existence and non-existence must be specified. The third phase of the software safety lifecycle is the *software design and software development.* Here, software must be developed which meets the specified safety requirements. The developed software must be analyzable and verifiable in order to guarantee that the required safety integrity level is achieved. Tools and compilers used for software development must be carefully selected. For its development, the software should be dissembled into modules or similar partitions, which reduce the complexity. Thereby, each module must already be specifically tested during the design. The software must be integrated into the target hardware of the programmable electronics. This part is referred to as *PE integration*. Tests must be performed to ascertain the compatibility of software and hardware, so that the required safety integrity level can be achieved. The *software operation and modification procedure* is explained in segment 9.5 of the IEC 615080-3. The integrated system must fulfill the specific demands on software safety. On account of this, next comes the *validation of software with regard to safety,* which must proceed according to the safety plan constructed. Validation will, for the most part, be made through tests.

In the event that the software must be modified, one has to ensure that the safety integrity level is not compromised. For this purpose, the effects of planned modifications on the functional safety of the E/E/PES must be analyzed. The results of each phase of the software lifecycle must be evaluated for accuracy. Therein, verification must be planned simultaneously with development. Each individual software component must be verified, that which includes, among others, code, data, modules, and architecture.

20.3.4 Evaluation of Functional Safety

The requirements regarding the evaluation of functional safety are specified in segment 8 of the IEC 61508-1. For the purpose of evaluation, the results of activities, as listed in Table A.10 in appendix A of the IEC/EN 61508-3, can be applied.

20.3.5 Appendix A – Guidelines for the Selection of Techniques and Methods

Appendix A gives an overview of possibly applicable procedures and measures, which must be included in the individual design phases. It evaluates them according to the requirements of the respective safety integrity level. The individual tables of appendix A of the IEC/EN 61508-3 refer to the different phases of the life cycle:

- A.1: Specification of safety requirements
- A.2-A.5: Design and development of software
- A.6: Integration of the programmable electronics
- A.7: Safety validation of software
- A.8: Modification
- A.9: Verification
- A.10: Evaluation of functional safety

20.4 IEC/EN 61508-4

In part 4 of standard IEC/EN 16508, all terms used in parts 1 – 7 are explained, as well as a variety of abbreviations defined. Some terms from this standard were singled out in order to explain them briefly. The definitions are helpful for understanding the standard.

20.4.1 Terms Regarding Safety

- *risk:* combination of the probability, with which damage occurs, and the extent of the damage
- *tolerable risk:* risk which is tolerable, based on the actual corporative value proposition in a given context
- *remaining risk:* risk which remains despite protective measures
- *safety:* freedom from indefensible risks
- *functional safety:* part of the complete system, referring to the EUC installation and the EUC guidance or control system, which depends on the correct functioning of the E/E/PE safety-related systems, safety-related systems of other technologies, and external devices for risk reduction

20.4.2 Terms relating to Devices and Equipment

- *Software:* Intellectual product consisting of programs, procedures, data, and all associated descriptions, which is part of a data processing system.

- *EUC (equipment under control):* devices, equipments, apparatuses, or installations which are used for the manufacturing, substance conversion, transportation, or for medical and other activities

- *PE (programmable electronics):* based on computer system technology, they can consist of hardware, software, and of input and/or output units

- *E/E/PE (electrical/electronic/programmable electronics):* based on electric (E), electronic (E), and/or programmable electronics (PE)

20.4.3 System Terms

- *System:* A multitude of elements, which are in an interactive relationship according to a certain design. Simultaneously, a system element can also be a separate system, a so-called subsystem. The latter can be either a controlling or a controlled system containing hardware, software, and human interaction.

- *Programmable electronic system (PES)* : A system for control, protection, or supervision, based on one or more programmable electronic devices. This contains all elements of the system, for example, energy supply, sensors and other input devices, data connections, and other communication pathways, as well as activators and other output devices (Figure 20.6)

- *Safety-related system:* System which provides both the necessary safety functions for the EUC, as well as the necessary safety integrity, if applicable, in combination with other systems.

- *Logical system* Part of a system which executes the functional logic, although it does not contain sensors or actuators.

- *E/E/PES (electrical/electronic/programmable electronic system)* System for controlling, protection, or supervision, based on one or more electric/electronic/programmable electronic devices. This contains all system elements, for example energy supply, sensors and other input devices, data connections, and other communication pathways, as well as activators and other output devices (see Figure 20.7).

20.4 IEC/EN 61508-4

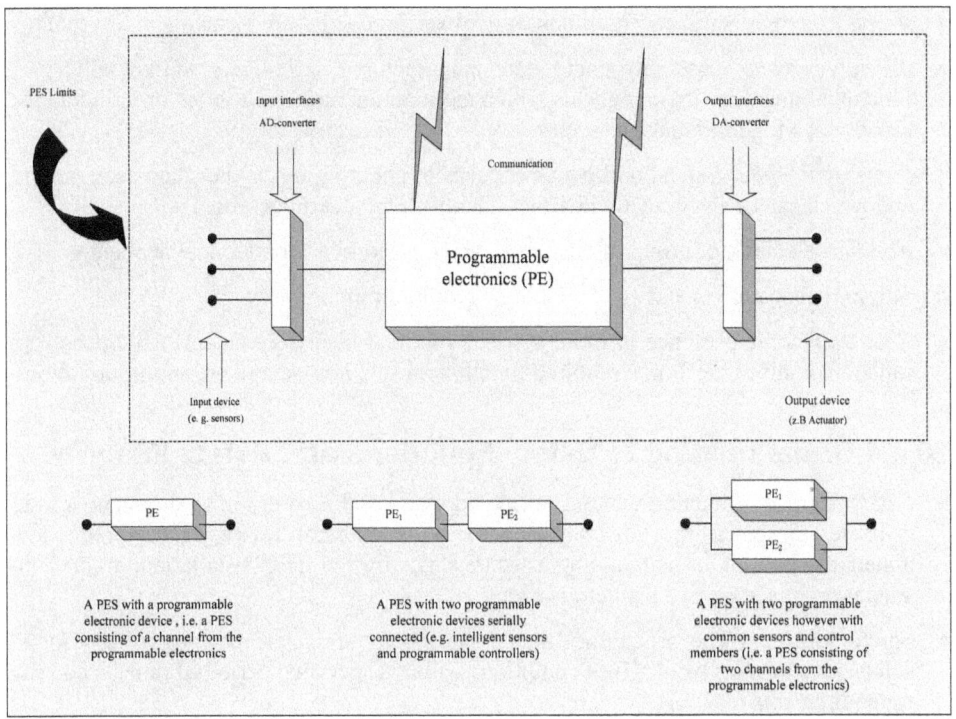

Figure 20.6: PES structure and terms

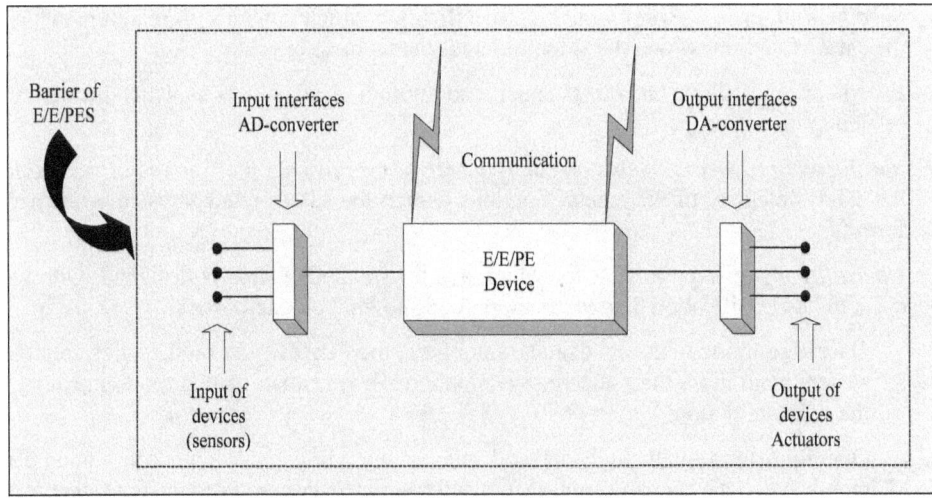

Figure 20.7: E/E/PES structure and terms

- *Architecture:* Specific configuration of hardware and software elements in a system
- *Assembly group:* assembly component manufactured serially, or individually, or a functional quantity of components which are manufactured as a series or individually, all of which together make an entity
- *Software module:* logical unit which consists of subprograms and/or data declarations, and which can be in contact with other, identical, logical units
- *Conduit:* element, or group of elements, which perform a function independently
- *Diversity* disparate means for executing a required function
- *Redundancy* the presence of more than the minimal necessary means, which enable a functional unit to perform a required function, or to allow data to present information

20.4.4 Terms relating to Safety Functions and Safety Integrity

- *Safety function:* Function which is executed by a E/E/PE safety-related system, a safety-related system from a different technology, or external devices for risk reduction, aimed at achieving or maintaining a secure status for the EUC installation, in due consideration of a specified dangerous event.
- *Safety integrity:* The probability that a safety-related system performs the required safety functions under all given conditions within a specified period of time, according to requirements
- *SIL safety integrity level:* One of four discrete steps for specification of the requirements for the safety integrity of the safety functions, which are assigned to the E/E/PE safety-related system, wherein safety integrity level 4 is the highest degree of safety integrity, and safety integrity level 1 is the the lowest.
- *Specification of safety requirements:* Specification which contains all requirements on the safety function, which the safety-related system must execute
- *Specification of safety functions:* Specification of all safety functions which the safety-related system must perform
- *Specification of requirements on safety integrity:* Specification of the requirements on the safety integrity of the safety functions, which the safety-related system must perform
- *Operating mode:* Application for which a safety-related system is designed, with respect to its classification degree, to adopt the following characteristics:
  - Operating mode with low demand rate: here, the demand rate on the safety-related system is no more than once per year, and no greater than double the frequency of the re-examination
  - Operating mode with high demand rate, or operating mode with continuous demand: here, the demand rate on the safety-related system is more than once per year, or is greater than double the frequency of the re-examination.

20.4.5 Terms relating to Errors, Failure, and Deviation

- *Error:* abnormal condition which can cause the impairment or the loss of ability of a functional unit to perform a requested function

- *Error avoidance:* Application of techniques and procedures aimed at avoiding the occurrence of errors during each phase of the safety lifecycle of the safety-related system

- *Failure:* Termination of the ability of a functional unit to perform a required function.

- *Systematic failure:* A failure, for which the cause can clearly only be eliminated by changing the design or the manufacturing process, the manner of operation, the operating instructions, or other influential factors.

- *Dangerous malfunction:* A malfunction with the potential of placing the safety-related system in a dangerous or inoperative state

- *Harmless malfunction:* malfunction which lacks the potential of putting the safety-related system in a dangerous or inoperative state

- *Human error:* A person's action, or failure to act, which leads to an undesired event

- *Deviation:* Mismatch between computational results, observed and measured values or qualities, and the corresponding true, specified, or theoretically accurate values or qualities

20.4.6 Terms relating to Lifecycle

- *Safety lifecycle:* Necessary actions within the framework of the implementation of safety-related systems, in the course of a timeframe which begins with the conceptual phase of a project, and ends when all E/E/PE safety-related systems, safety-related systems of other technologies, and external devices for risk reduction are no longer available to the application

- *software lifecycle:* Action during a period of time, which begins with software development and ends when the software is no longer used

- *Effect analysis:* Action to determine the influence which the alteration of a function, or of a system component part, can have on other functions or system component parts, or on other systems

20.4.7 Terms relating to Verification of Safety Measures

- *Verification:* Affirmation, based on an examination and on demonstration of proof, that the requirements were fulfilled
    - Examples: Examination of results, design examination, performed tests on developed products, execution of integrity tests

- *Validation:* Affirmation, based on an examination and on demonstration of proof, that the special requirements for a special intended usage were fulfilled; the standard contains three validation phases:
  - the safety total validation,
  - the E/E/PES validation and
  - the software validation
- *Evaluation of functional safety:* an examination based on proof, which evaluates the functional safety, which is achieved by one or more E/E/PE safety-related systems, safety-related systems of other technologies, or external devices for risk reduction
- *Degree of diagnostic coverage:* Partial reduction of the probability of hazardous hardware malfunctions, due to the administration of automatic diagnostic tests
- *Testing facility:* Facility which is capable of simulating the operational environment of the software or the hardware during the development, by subjecting the software to test cases, and by registering the replies.

20.5 IEC/EN 61508-5

20.5.1 Field of Application

Part 5 of the standard IEC/EN 61508 provides an overview of the underlying risk concepts, and the relationship between risk and safety integrity. Furthermore, methods are made available for determining the safety integrity level.

20.5.2 Appendix A – Underlying Concepts

In appendix A, the required risk reduction, the role of the E/E/PES, the safety integrity, as well as the safety integrity level are defined and described. Figure 20.8 shows the principle of risk reduction down to the remaining residual risk. The residual risk must always be smaller than the tolerable risk.

The tolerable risk depends on many factors:

- Severity of injury,
- Number of persons who are exposed to the danger,
- Frequency and
- Duration of exposure.

20.5.3 Appendix B – ALARP and the Concept of Tolerable Risk

In appendix B, the ALARP principle (As Low As Reasonably Practicable) is elucidated, see Figure 20.9.

See Table 20.5 for corresponding defined risk categories.

Table 20.6 shows an example for the classification of an accident.

Table 20.5: Interpretation of risk categories

Risk Category	Interpretation
Category I	Risk not tolerable
Category II	Undesired risk, which is only tolerable if risk reduction is not possible, or if the cost of reduction is unproportionally high when compared to the degree of improvement achieved
Category III	Tolerable risk, if the costs of risk reduction surpass the improvement achieved
Category V	Negligible risk

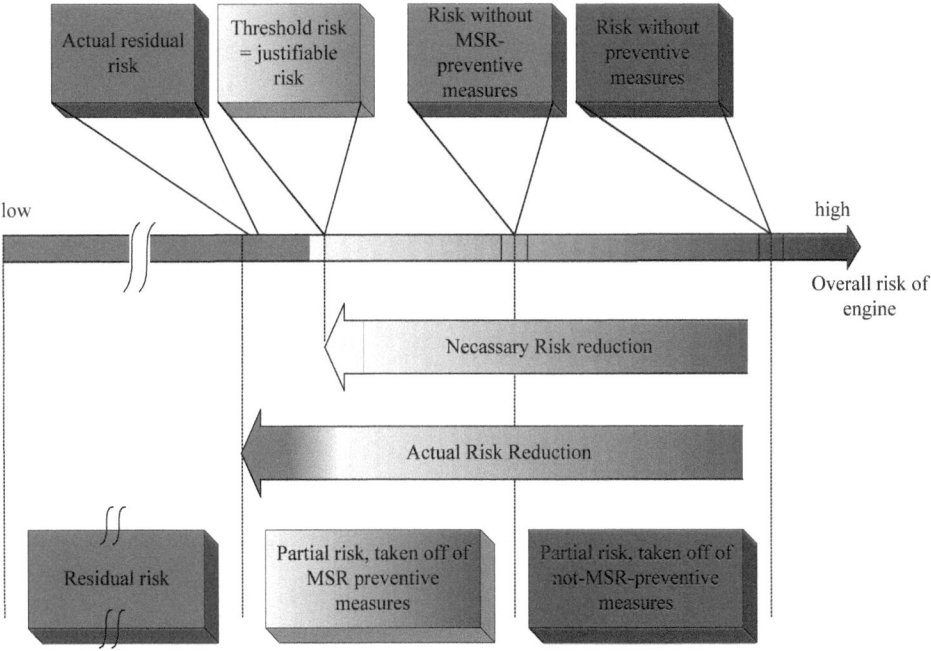

Figure 20.8: General concepts of risk reduction

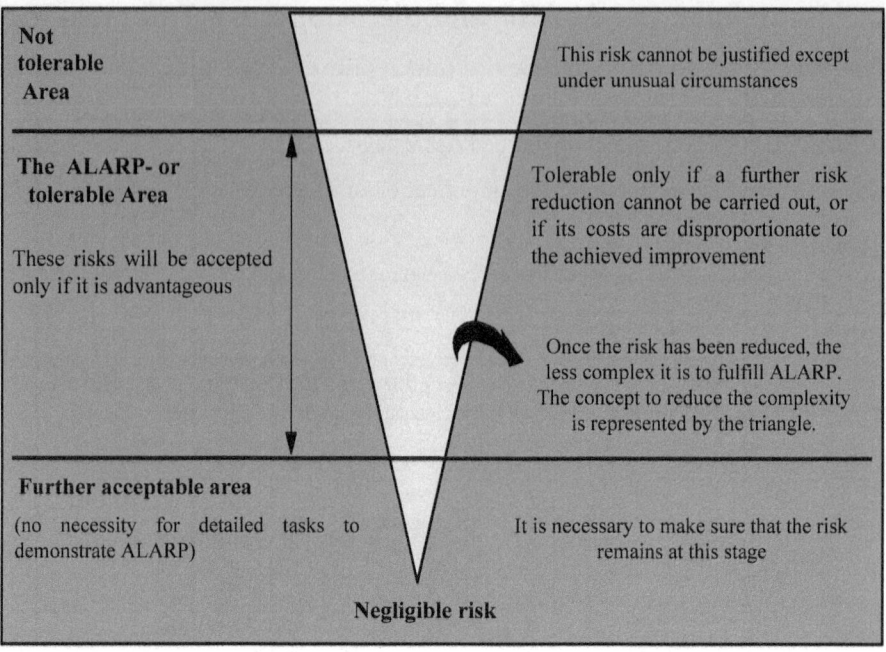

Figure 20.9: Tolerable risk and ALARP

Table 20.6: Examples for the risk classification of accidents

Frequency	Consequence			
	catastrophic	critical	limited	minimal
frequent	I	I	I	II
probable	I	I	II	III
occasional	I	II	III	III
rare	II	III	III	IV
improbable	III	III	IV	IV
not plausible	IV	IV	IV	IV

20.5.4 Appendix C – Quantitative Methods for Determining the Safety Integrity Level

This appendix describes, how the safety integrity levels can be determined, with the aid of a quantitative method (see Figure 20.10). For this, the tolerable risk is linked to the risk of the EUC in a systematic manner, in order to determine the required risk reduction:

$$PFD_{avg} \leq \frac{F_t}{F_{np}} = \Delta R$$

With the definitions being:

PFD_{avg}: medium failure probability

F_t: frequency of the tolerable risk

F_{np}: classification grade of the safety-related protective device

ΔR: required risk reduction

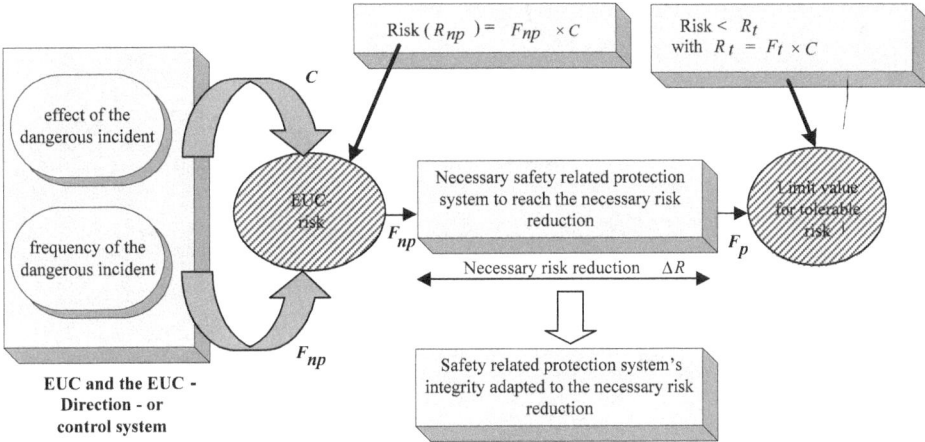

Figure 20.10: Assignment of safety demands

The safety integrity level for the hereby determined PFD_{avg} can be determined from Table 20.1.

20.5.5 Appendix D – Qualitative Methods for Determining the Safety Integrity Level (Risk Graph)

Appendix D describes how, aided by a qualitative method, the so-called risk graph (see Figure 20.11), the safety integrity level can be determined based on knowledge of the risk factors. Here, the risk graph was taken up from the DINV 19250, where it depicts the therein described procedure from AK[196]1 to AK8, on SIL 1 to SIL 4.

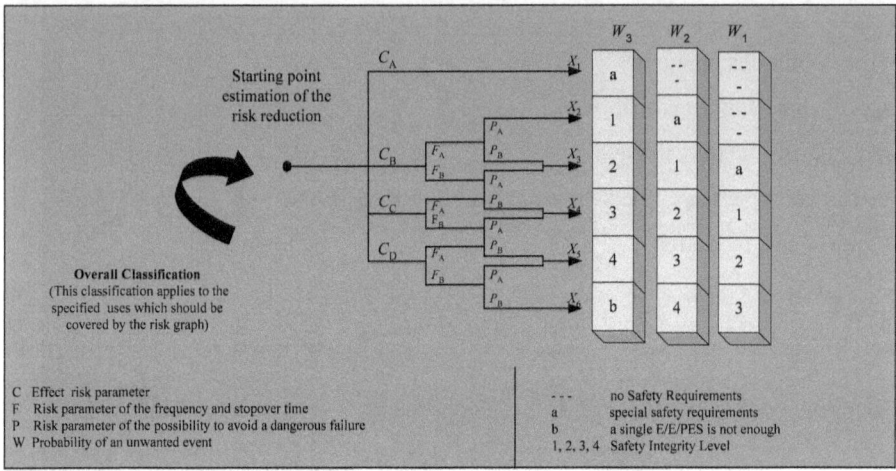

Figure 20.11: Risk graph: general design

---

[196] AK Anforderungsklasse

20.5.6 Appendix E – Specification of the Safety Integrity Level A Qualitative Procedure – Matrix of the Extent of a Dangerous Event

In the event that the method described in appendix C can not be applied, for lack of quantitative knowledge of the risk factors, the qualitative method of the matrix of extent (see Figure 20.12) of the dangerous events, as described in appendix E, can be applied.

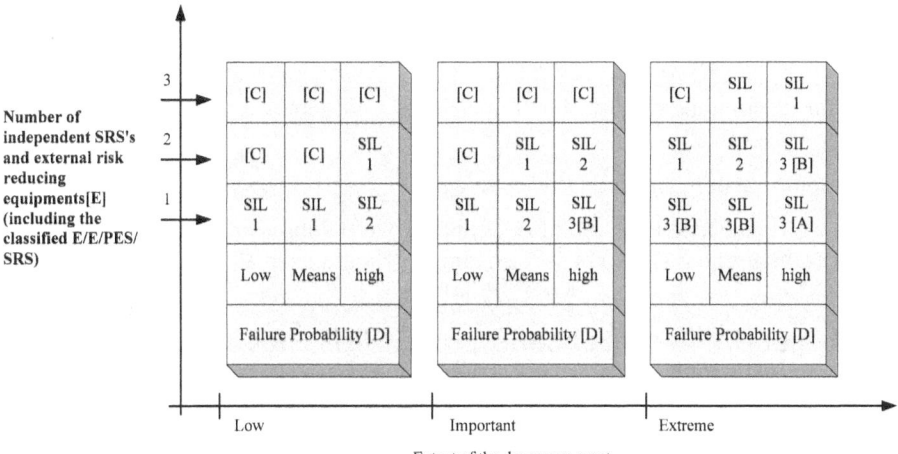

[A] A SIL 3 E/E/PES SRS does not supply sufficient risk reduction for this risk level. Further risk reducing measures are necessary

[B] A SIL E/E/PES SRS would not supply any sufficient risk reduction for this risk level. Further risk reducing measures could be necessary.

[C] An independent E/E/PES SRS is probably not necessary.

[D] Probability of the event [D] is the probability that the dangerous event occurs without any SRS or external risk reduction mechanism.

[E] SRS = safety related system. The event probability and the number of independent Protection equipment will be set up in dependently of the application.

Figure 20.12: Matrix of the extent of a dangerous event

20.6 IEC/EN 61508-6

20.6.1 Field of Application

Part 6 of the standard IEC/EN 61508 makes available information and guidelines for the application of parts 2 and 3 of this standard. Next to parts 2 and 3, it is very important for some developmental steps, because it contains detailed instructions for the quantitative calculation of the E/E/PES. They are:

- Block diagram and formula for PFD (Probability of Failure on Demand) calculation and PFH calculation
- Table for determining the ß-factor (ß: percentage of unknown malfunctions resulting from the same cause)
- Tables for estimating the diagnostic coverage
- Tables with calculated PFD and PFH (Probability of Failure per Hour) numbers for all system configurations shown in the standard, with variants in all relevant parameters

20.6.2 Appendix A – Application of IEC/EN 61508-2 and -3

Appendix A contains subchapters A.1, A.2, and A.3:

- A.1: Application of IEC/EN 61508-2 and -3
- A.2: Application of IEC/EN 61508-2
- A.3: Application of IEC/EN 61508-3

Segment A.1 again gives a short overview of the contents of parts 1-3 of the IEC/EN 61508. In segment A.2, the individual functional steps for the application of part 2 of the standard for hardware development are elucidated, in A.3 those for software development.

20.6.3 Appendix B – Exemplary Procedure for Determining Hardware Failures

To begin with, foundations and assumptions are listed, around which the calculations of the following segments are built. There are several methods for the analysis of safety integrity of safety-related E/E/PES. Some of the more widespread methods are reliability block diagrams and Markov models. Both methods produce similar results, when applied correctly, but reliability block diagrams in the case of complex PES yield fewer exact results than Markov models.

The medium probability of failure of a safety function of the E/E/PES, for the *low demand* operation mode, is determined by calculating the medium probabilities of the failure and adding:

$PFD_{sys} = PFD_S + PFD_L + PFD_{FE}$

With:

- PFD_{sys} the medium probability of the failure of a safety function during demand
- PFD_S the medium probability of failure during demand on a safety function
- PFD_L the medium probability of failure on demand for the logic subsystem
- PFD_{FE} the medium probability of failure on demand for the output subsystem

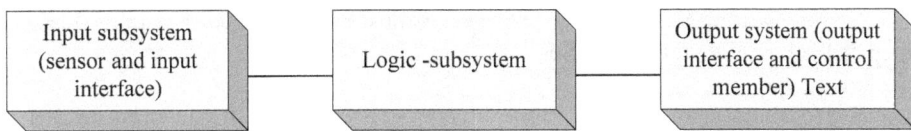

Figure 20.13: Subsystem structure

In order to calculate the medium failure probability for each subsystem, the following must be determined (see Figure 20.13):

- the underlying architecture (for example, 1oo2D, 2oo3)
- the diagnostic coverage of each conduit
- the failure rate per hour ℓ for each conduit
- the β-factors β and β_D for failures resulting from the same cause

Below, five architectures are described for a low demand operation mode:

- 1oo1
- 1oo2
- 1oo2D
- 2oo2 and
- 2oo3

The 1oo1-architecture (see Figure 20.14) consists of a single conduit, in which every dangerous error leads to a failure of the safety function.

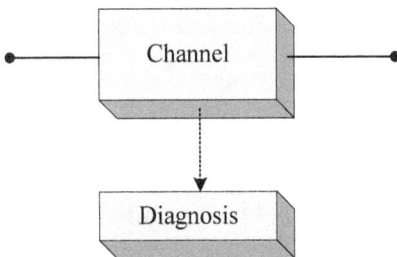

Figure 20.14: 1oo1 block diagram

For this architecture, the medium probability of failure during demand is:

$$PFD_{avg} = (\lambda_{DD} + \lambda_{DU}) \cdot t_{CE}.$$

This means

λ_{DD} the rate of dangerous failures and

λ_{DU} the rate of undetected dangerous failures

t_{CE} the equivalent medium failure duration of a conduit.

This *1oo2-architecture* (see Figure 20.15) contains two conduits in parallel, each of which can perform the safety function.

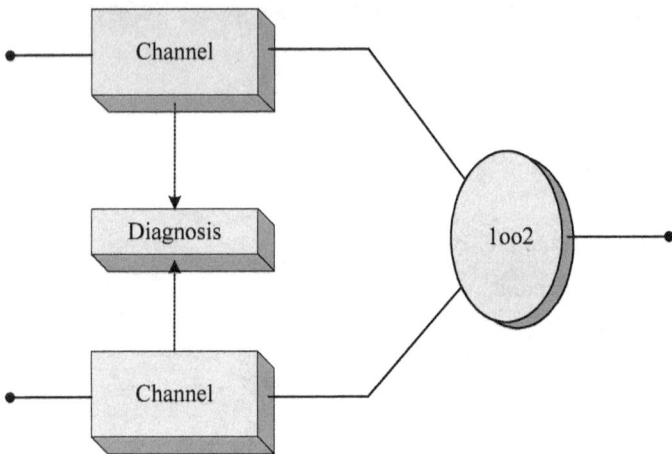

Figure 20.15: 1oo2 block diagram

The medium failure probability during demand is:

$$PFD_{avg} = 2\{(1-\beta_D)\lambda_{DD} + (1-\beta)\lambda_{DU}\}^2 t_{CE}t_{GE} + \beta_D\lambda_{DD}MTTR + \beta\lambda_{DU}\left(\frac{T_1}{2} + MTTR\right)$$

With *1oo2D-architecture* (see Figure 20.16), the two parallel-switched conduits are monitored, in addition, by a diagnosis unit. If it detects an error in one of the conduits, the total result is determined exclusively through the intact conduit. Only if both conduits are flawed, or if an error can not be associated with a particular conduit, does the output enter a secure state.

20.6 IEC/EN 61508-6

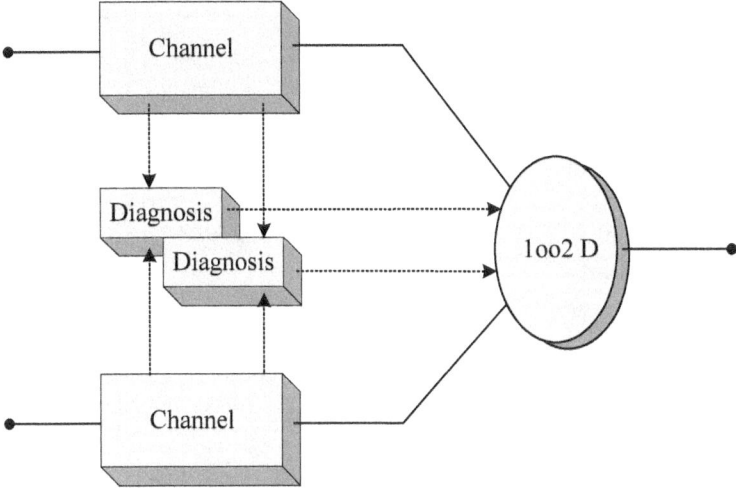

Figure 20.16: 1oo2D block diagram

The medium failure probability is:

$$PFD_G = 2(1-\beta)\lambda_{DU}\{(1-\beta)\lambda_{DU} + (1-\beta_D)\lambda_{DD} + \lambda_{SD}\}t'_{CE}t'_{GE} + \beta_D\lambda_{DD}MTTR + \beta\lambda_{DU}\left(\frac{T_1}{2} + MTTR\right)$$

This *2oo2-architecture* (see Figure 20.17) consists of two parallel-switched conduits, both of which have to request the safety function, before it can be executed.

The medium failure probability on demand is:

$$PFD_G = 2\lambda_D t_{CE}$$

The *2oo3-architecture* contains three parallel-switched conduits and a provision for arrangement. The output state is specified by at least two concurring conduits. The medium failure probability on demand for this arrangement is:

$$PFD_{avg} = 6\{(1-\beta_D)\lambda_{DD} + (1-\beta)\lambda_{DU}\}^2 t_{CE}t_{GE} + \beta_D\lambda_{DD}MTTR + \beta\lambda_{DU}\left(\frac{T_1}{2} + MTTR\right)$$

The procedure for calculating the *failure probability per hour* for a E/E/PES with *high demand rate* is identical to the method for systems with low demand rate:

$$PFD_{sys} = PFD_s + PFD_L + PFD_{FE}$$

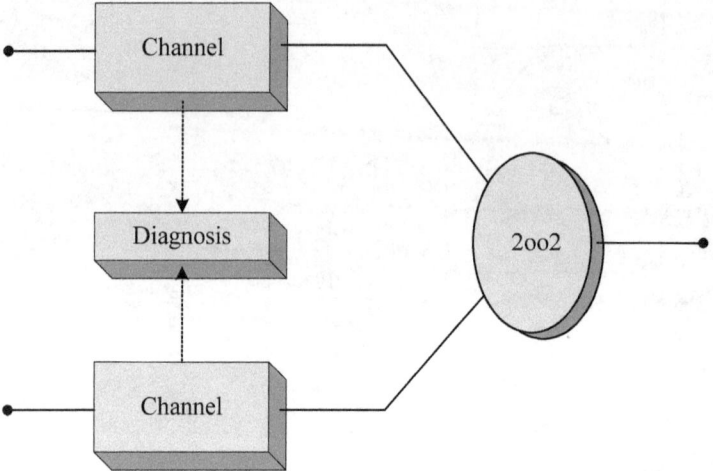

Figure 20.17: 2oo2 block diagram

For systems with high demand rate, the above mentioned architectures are again discussed. The block diagram figures are identical to those for low demand rate. The formulas for calculating the *PFH*-numbers are as follows:

- 1oo1

$$PFH_G = \lambda_{DU}$$

- 1oo2

$$PFH_G = 2 \cdot \{(1-\beta) \cdot \lambda_{DU} + (1-\beta_D) \cdot \lambda_{DD}\}^2 \cdot t_{CE} + \beta_D \cdot \lambda_{DD} + \beta \cdot \lambda_{DU}$$

- 1oo2D

$$PFH_G = 2 \cdot (1-\beta) \cdot \lambda_{DU} \cdot \{(1-\beta) \cdot \lambda_{DU} + (1-\beta_D) \cdot \lambda_{DD} + \lambda_{SD}\} \cdot t'_{CE} + \beta_D \cdot \lambda_{DD} + \beta \cdot \lambda_{DU}$$

- 2oo2

$$PFH_G = 2 \cdot \lambda_{DU}$$

- 2oo3

$$PFH_G = 6 \cdot \{(1-\beta) \cdot \lambda_{DU} + (1-\beta_D) \cdot \lambda_{DD}\}^2 \cdot t_{CE} + \beta_D \cdot \lambda_{DD} + \beta \cdot \lambda_{DU}$$

For a simple evaluation, appendix B of the IEC/EN 61508-6 contains tables with *PFD*- and *PFH*-numbers, which can be applied if the structure of the system exactly corresponds to the data in the standard. Thereby, a variety of architectures can be compared rapidly. For a more in-depth analysis, a detailed calculation is imperative.

20.6.4 Appendix D – Methods for Quantifying the Consequences of Hardware Failures due to the Same Cause in E/E/PES

This method is applied to calculate β, a factor which is frequently used in model building of failures caused by the same reason. With it, the failure rate, evidenced as a consequence of the same cause, of two or more systems working in parallel can be determined. This is done by means of the rate of accidental hardware malfunctions. Finally, the entire method for calculating the factor is explained.

20.7 IEC/EN 61508-7

20.7.1 Field of Application

Part 7 of this standard contains an overview of different safety-related procedures and measures, which are relevant to the application of parts 2 and 3. This part is applied during the planning phase and serves as a reference text for all questions, which occur while working with the standard.

20.7.2 Appendix A – Overview of Procedures and Measures for E/E/PES: Control of Accidental Hardware Failures

For the following components, a variety of measures is described to help avoid accidental hardware malfunctions:

- Electric
- Electronic
- Central Processing Units (CPUs)
- Variable/constant memory depth
- Variable memory
- E/A-units and intersections
- Data pathways
- Voltage supply
- Temporal and logical program runtime supervision
- Ventilation and heating
- Communication and mass memory
- Sensors
- Actuators

One typical application for each of the most important technologies is described below.

Example 1 Electric: Supervision of relay contacts. Relays with positively driven contacts are arranged such that their contacts are firmly connected. One can assume that contacts which are connected to each other always have the same status. Thereby, for example, a contact can be utilized to perform the actual switch function, while the second contact, which is connected to this contact, can be used for supervision.

Example 2 Electronic: Supervised redundance. The safety function is performed by at least two conduits. The results of these conduits are supervised, and a safe status is introduced as soon as just one conduit conveys an aberrant result.

Example 3 Central Processing Units (CPUs): Self-test through software. „Walking Bit" (one conduit). In order to identify errors in the physical memory or in the command decoder of the processor as early as possible, the so-called „walking bit test" is conducted. This self-test is exclusively implemented through additional software functions. In it, a data pattern is applied (for example walking bit pattern), to test the physical memory (data and address registry) and the command decoder.

Example 4 Constant memory areas: Signature. The content of a memory block is compressed to one word, respectively two words, by means of a CRC-algorithm, a polynom division is performed, and the left-over division remainder is stored as signature. Later on, the signature is re-calculated and compared with the stored signature. In case of a discrepancy, an error message is generated.

Example 5 Variable memory area: double RAM with hardware or software comparison and read-/write test. The address space is deposited in two storage areas. The first one is operated normally, the second one contains the same information and is accessed in parallel to the first. The outputs are compared, and an error message is generated in case of a discrepancy. In order to identify certain types of bit errors, the information in one of the two memories must be stored in an inverted manner and must be re-inverted upon reading.

Example 6 E/A-units and intersections: Test pattern. This test is a cyclic test of the E/A-units. It utilizes a defined test pattern to compare input values with expected values. The test pattern information, test pattern reception, and test pattern evaluation must be independent of each other. The EUC must not be unduly influenced by the test pattern.

Example 7 Voltage supply: Surge suppressor with safety switch: Potential power surges are recognized in time, all outlets are brought to a safe state by the switch-off routine.

Example 8 Temporal and logical program runtime supervision: Watchdog with separate time basis and time slot. External timers with separate time basis are periodically triggered, to test the behavior of the computer and the plausibility of the program sequence. Herein, it is important that the trigger moments are correctly placed in the program. If the program sequence over- or undershoots a specified upper or lower tolerance limit, an emergency measure is performed.

Example 9 Ventilation and heating: Temperature sensor. A temperature sensor supervises the critical parts of the E/E/PES. Before the temperature leaves the specified zone, a safety-related action is triggered.

Example 10 Communication and mass storage: Separation of energy conduits and information conduits. Electrical supply lines are separated from information conduits, to prevent crosstalk between the two lines. The electric field, which could induce power spikes in the data conduits, decreases as distance increases.

Example 11 Actuators: Supervision. The operation of the actuator is supervised. The redundancy introduced by this supervision can be applied towards triggering a safety-related action.

20.7.3 Appendix B – Overview of Techniques and Measures for Prevention of Systematic Failures

An important measure for prevention of systematic failures is project management. For quality assurance, an organisational model is designed which is documented in the quality assurance handbook. Furthermore, rules and measures are specified which are required for the creation and validation of safety-related systems. To prevent errors and impairment of system safety, each step is documented during development. One further important measure is the separation of safety-related systems from non-safety-related systems. Thereby is prevented that non-safety-related systems do not influence safety-related systems in an undesired manner. One other method of reducing the probability of failures is the utilization of different types of building blocks for the diverse conduits of a system.

It must be shown during assessment, that the development was performed diligently. For this purpose, and in addition to EDV support, specific guidelines must be introduced, for example:

- Computer supported administration of documentation
- Definition of standardized documentation modules
- Guidelines for content and form of the documentation
- Functional Safety Managament

Computer-supported specification tools and checklists, among others, serve as tools for specification of the safety requirements on the E/E/PES. Computer-supported specification tools generate a specification in form of a data bank, which can be examined automatically for completeness and consistency. Generally, this technique can be applied not only for specification, but also for other phases of the lifecycle, for example design. Checklists contain all important aspects of the system. The usefulness of checklists depends mainly though on the interpretation of the engineer who prepares the checklist, who therefore should always carefully document and substantiate his decisions. The goal is, to fulfill the checklist as thoroughly as possible.

The most important method for design and development of the E/E/PES is the simulation. Herein, individual parts of the circuit are simulated to safely evaluate function. For this purpose, software with models of the building blocks is put in place, to describe their behavior.

For operation and maintenance procedures, the goal is to develop procedures which help prevent failures during operation and maintenance of the safety-related system. Of particular importance is, that operation is performed by experienced personnel only who are specifically trained with regard to the requirements of the system. This also includes background knowledge of the production process and the EUC.

One further goal is the prevention of malfunctions during the integration phase, and the identification of failures which occur during this and previous phases. Of particular usefulness for this purpose is, for example, the black box test. The functions of a system are tested in a specific environment with specified test data, which were deduced from the specification. This illustrates the behavior of the system and enables a comparison with the specification. In doing so, no knowledge of the inner structure of the system must be used in testing. The goal is to find out whether the system performs all functions required by the specification correctly.

The E/E/PES must be validated in regard to safety. One possibility of validation is the function test under environmental conditions. For validation, a test is performed under environmental conditions (for example, according to IEC 60068 climate and environment, or IEC 61000 EMV), which evaluates the function and reliability of the safety function. Another possibility of validation is the failure modes and effects analysis FMEA. Hereby, each individual building block of a system is analyzed, to find the quantity of its failure types, their causes, and their effects, and to find procedures for recognizing them.

20.7.4 Appendix C – Overview of Techniques and Measures for Achieving Safety Integrity of Software

Software is developed based on mathematical methods. The mathematical model of the system can be subjected to analyses, to uncover a variety of errors. This analysis can also be performed by a computer. There are several formal methods, of which OBJ is described here in more detail. OBJ is an algebraic language in which the user specifies the requirement in the form of algebraic equations. This type of implementation makes it possible to validate the system directly on the specification.

Prior to an expansion or modification of the system, an analysis of the consequences of expansion or modification on the software must be performed. For this purpose, influence analysis is useful. Furthermore, it must be determined which software system and module are concerned. After finishing the analysis, it must be decided whether re-verification is necessary.

The following possibilities come to mind:

- only the changed module is re-verified
- all modules concerned are re-verified
- the complete system is re-verified.

The decision depends on a number of criteria, to which belong, among others, the significance of the module, the quantity of modules concerned, and the type of change. The system is now examined for functional safety. As graphic models for combinations of events or conditons, which are necessary in the error-free operation of a system, reliability block diagrams are applied. In a block diagram, each block represents an event or a condition in an interconnected fashion. Next, a path is searched for from one side of the diagram to the other. If the path comes upon a block, then the event described in the block must have taken place; that is, that particular condition must be true in order for any continuation of the path. If at least one path goes through the diagram, the system can be said to function correctly.

21 IEC 61511

The IEC (International Electronic Commission) is a worldwide standards organization which encompasses all national electrotechnical committees. Its goal is the advancement of international collaboration in all matters of standardization in the field of electrical engineering and electronics.

The international standard IEC 61511[197] is of significance for the process industry. It was designed by subcommittee 65A of the IEC TC 65 and consists of the following parts:

- Part 1: General, terms, system, hardware, and software requirements
- Part 2: Instructions on the application of IEC 61511-1
- Part 3: Instruction on determining the required safety integrity level

In the process industry, safety-related systems have been in use for a long time. They include all components and subsystems which are required for the implementation of safety-related functions. The process control techniques applied in safety-related functions must thereby fulfill certain minimal requirements and performance levels. The safety-related systems described below apply to electric (E), electronic (E) and programmable electronic (PE) technologies. Safety can often be best ensured by means of safe processes, whereby the safety-related systems must be tested, if possible, in combination with other protective systems. This test determines the comprehensive safety requirements by hazard and risk evaluation, and assigns them to individual systems. In addition, certain actions, such as safety management, are performed. These requirements and methods of operation are prerequisites in this international standard. It discusses every critical period within the safety lifecycle. These periods are concept, design, execution, operation, maintenance, and decommissioning. Other, already existing standards for the process industry can harmonize with this standard.

21.1 Scope of Application

This international standard determines requirements on the entire field of work of a safety-technical system (SIS), so that a process can be set into a safe state, or can be maintained in a safe state. It was designed as the application of international standard IEC 61508 and concerns devices which are integrated into a system used by the process industry, and application software which was designed for safety-technical systems with limited language complexity or with fixed programs, see Figures 21.1 and 21.2.

---

[197] [IECb00]   IEC 61511, Functional Safety: Safety Instrumented Systems for the Process Industry Sector

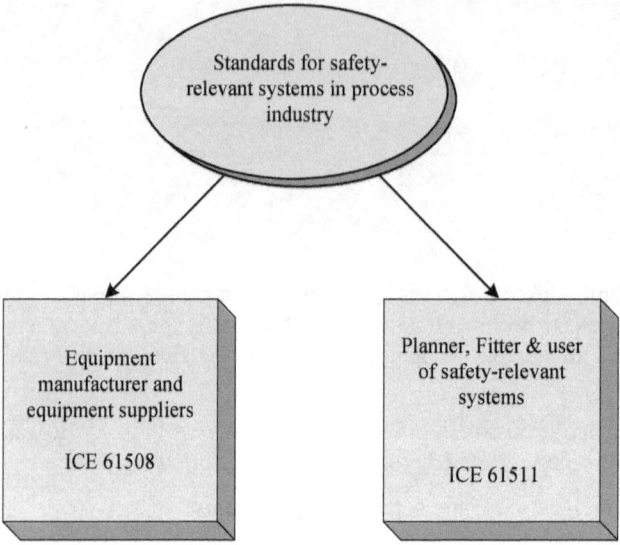

Figure 21.1: Relationship between IEC 61508 and IEC 61511

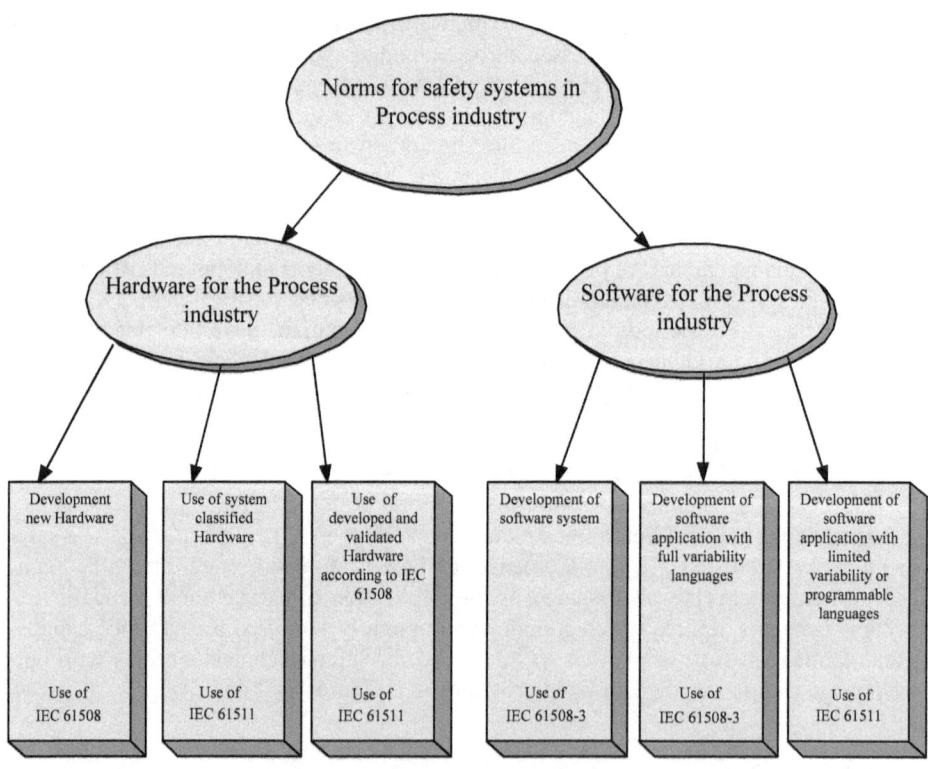

Figure 21.2: Relationship between IEC 61508 and IEC 61511

Its field of application lies in the process industry, chemical industry, paper manufacturing, gas delivery, in refineries, etc. This standard is fulfilled if functional safety for the protection of coworkers and of the environment is achieved. It also serves to protect economic goods.

In this segment, the correlation between safety-technical and non-safety-technical functions is explained, see Figure 21.3. The connection between system, hardware, and software is shown in Figure 21.4.

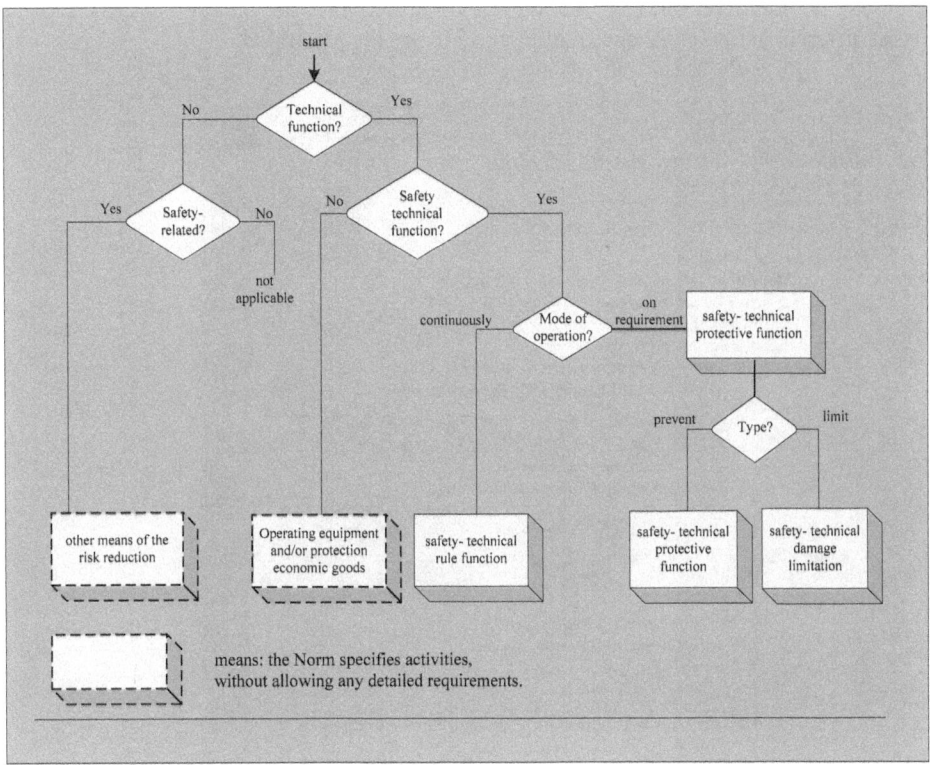

Figure 21.3: Correlation between safety-technical and other functions

21.2 Subdivision of Standard IEC 61511

The international standard IEC 61511 is subdivided into three parts. The following segment gives a brief description of the individual parts.

The first part, the IEC 61511-1, is described in the following chapters. It explains the terms and abbreviations used in the standard, and subsequently describes system requirements and procedures of ensuring the safety of hardware and software.

The second part, the IEC 61511-2, contains an instruction with regard to how the safety-related system and its safety-technical functions, as defined in the part 1, are to be specified, planned, built, operated, and maintained. The structure of the chapter and the subchapter mirrors that of the IEC 61511-1. In a more expansive appendix, a variety of examples are shown in which failure probabilities are calculated, and architectures of the safety-technical system and application software are constructed, according to the standard.

The IEC 61511-3 is the third part of the standard and describes an instruction for determining the required SIL (safety integrity level). The basic concept of risk, and its relationship to safety integrity, is discussed. In the appendix, the ALARP model is introduced, which can be used to determine tolerable risk. Furthermore, the underlying principles of several different methods for determining the SIL are demonstrated.

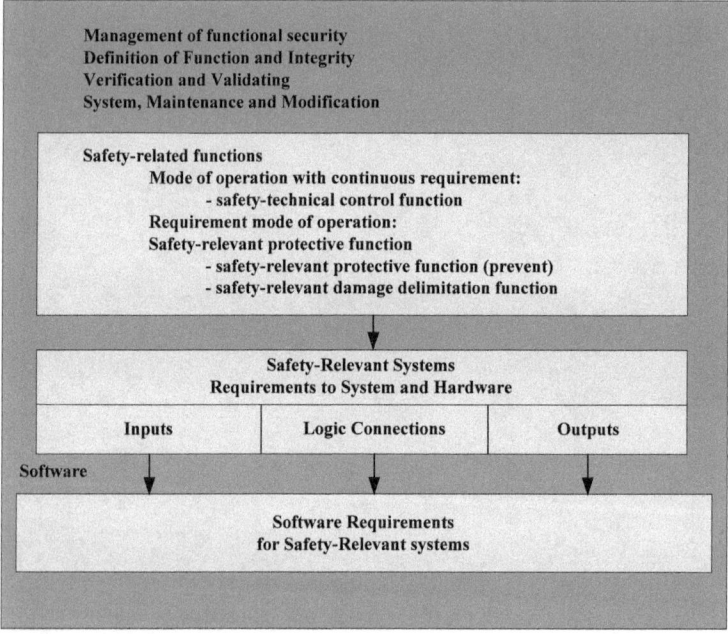

Figure 21.4: Correlation between system, hardware, and software in the IEC 61511-1

21.2 Subdivision of Standard IEC 61511

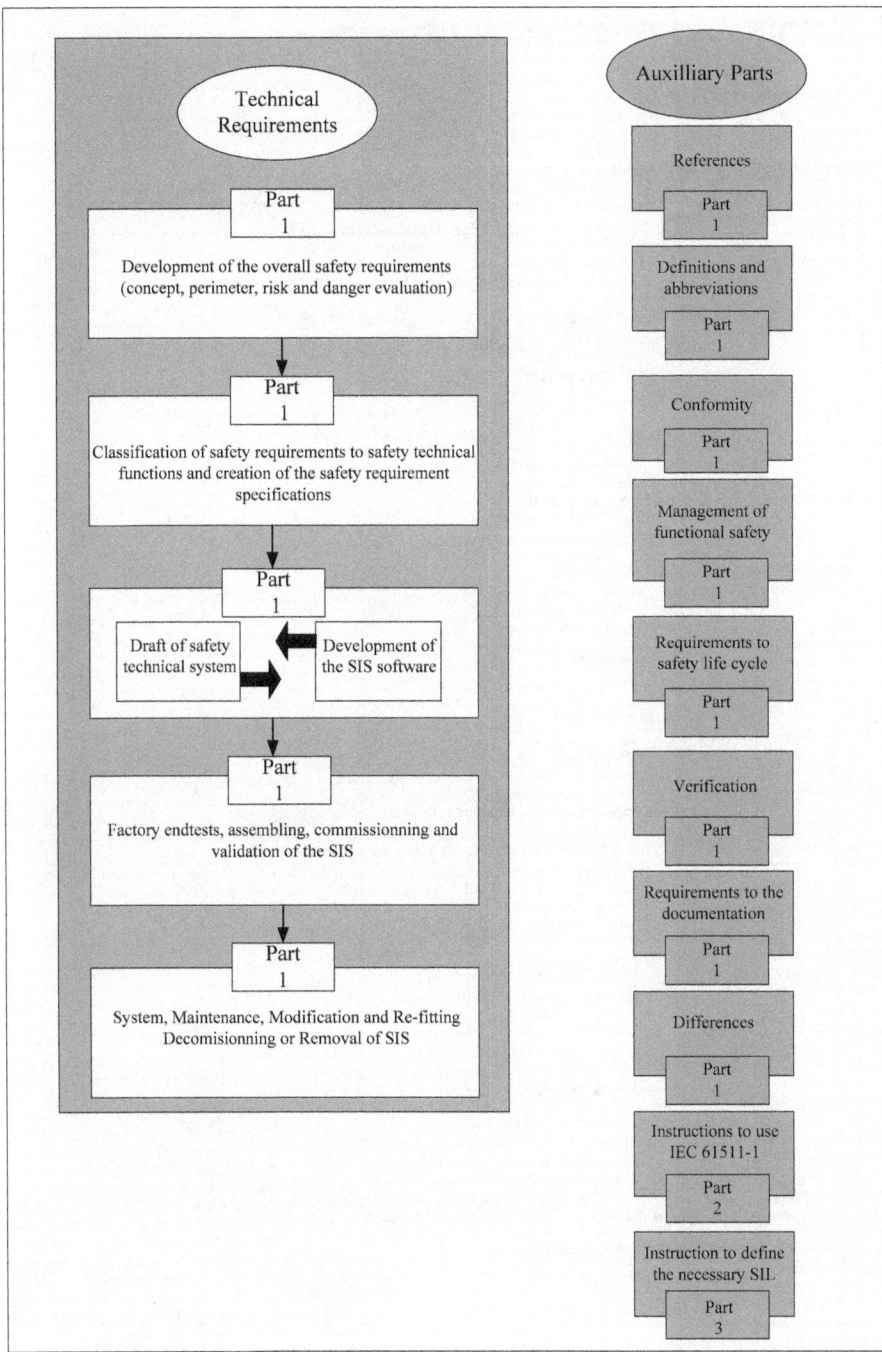

Figure 21.5: Overall framework of this standard

21.3 Terms and Abbreviations

21.3.1 Abbreviations

Table 21.1: Abbreviations used in the IEC 61511

Abbreviation	Meaning
1vN	One out of N
AC/DC	Alternating current/direct current
ALARP	As low as reasonably practicable
ANSI	American National Standards Institute
BPCS	Basic process control system
DC	Diagnostic coverage
E/E/PE	Electrical/electronic/programmable electronic
E/E/PES	Electrical/electronic/programmable electronic system
EMC	Electro-magnetic compatibility
FAT	Factory acceptance testing
FPL	Fixed program language
FTA	Fault tree analysis
FVL	Full variability language
H & RA	Hazard & risk analysis
H/W	Hardware
HFT	Hardware fault tolerance
HMI	Human machine interface
HRA	Human reliability analysis
IEC	International Electrotechnical Commission
IEV	International Electrotechnical Vocabulary
IPL	Independent protection layer
ISA	Instrument Society of America
ISO	International Organization for Standardization
LVL	Limited variability language
NPPE	Non-programmable programmable electronics
PES	Programmable electronic system
PFD	Probability of failure on demand
PFD_{avg}	Average probability of failure on demand
PLC	Programmable logic controller
S/W	Software
SAT	Site acceptance test
SFF	Safe failure fraction
SIF	Safety instrumented function
SIL	Safety integrity level
SIS	Safety instrumented system

21.3.2 Terms

- *Architecture/Configuration*

  the arrangement of hardware and/or software components in a system such as, for example
  - the combination of safety-technical subsystems;
  - the inner structure of a safety-technical subsystem;
  - arrangement of software programs

- *Protection from property damage*

  the function associated with system design whose purpose it is to prevent property damage

- *BPCS*

  System which responds to input signals from the process and its associated technical devices, from other programmable systems, and/or from a user, and which generates output signals which regulate the process and its associated technical devices in the desired manner. The BPCS does not, however, assume any safey-technical functions with an SIL≥ 1.

- *Conduit*

  Element or group of elements, which execute a function independently

- *Coding/Programming*

  Design, construction, and test of a series of program commands for solving a task, or for data processing

- *Failure as a result of common causes*

  Failure which is the result of one or more events, which cause failures of two or more separate conduits in a multichannel system, and which lead to system failure

- *Failure as a result of identical failure type*

  Identical failure of two or more conduits, which generate the same incorrect event

- *Component*

  Part of the system or subsystem or of a device which performs a specific function

- *Configuration management*

  Procedure for identification of components of a (hardware and software) system under development, for the purpose of controlling changes in these components, and for maintaining the continuity and retraceability during the lifecycle.

- *Control system*

  System which reacts to input signals from the process and/or from the user, and which generates output signals which cause the process (facility for process engineering) to work in the desired manner

- *dangerous failure*

  Failure which has the potential of putting the safety-technical system into a dangerous or inoperative state

- *dependent failure*

  Failure whose probability can not be expressed as the simple product of the probabilities of the individual events which caused it

- *recognized, conspicuous*

  In connection with hardware failures and software errors recognized through diagnostic tests or the usual operation

- *Installation/device*

  Functional unit, consisting of hardware or software, or a combination of both, which serves a special purpose (for example, field devices, or devices connected to the input and output cards of the SIS. These devices also include field wiring, sensors, actuators, logic systems, and user interfaces, which are wired to the input/output level of the SIS)

- *Degree of diagnostic detection*

  Percentage of the failure rate of errors detected by diagnostic tests compared to the overall failure rate of the component or the subsystem. The degree of diagnostic detection does not contain errors found during repeat tests

- *Diversity*

  Presence of disparate methods for executing a required function

- *electric/electronic/programmable electronic (E/E/PE)*

  Based on electric/(E) and/or electronic (E) and/or programmable electronic (PE) technology

- *Deviation*

  Non-correspondence between computational results, observed or measured values or states, and the corresponding true, specific or theoretically correct values or states

- *External device for risk reduction*

  Method of reducing or alleviating risks which is separate and different from the SIS

- *Failure*

  Cessation of the ability of a functional unit to perform a required function

- *Error*

  Abnormal state which can cause the reduction or loss of the ability of a functional unit to perform a required function

- *Error avoidance*

  Application of techniques and procedures aimed at avoiding the occurrence of errors during each phase of the safety lifecycle of the SIS

21.3 Terms and Abbreviations

- *Error tolerance*

  Ability of a functional unit to continue performing a required function in the presence of errors or deviations

- *Actuator*

  Part of the safety-technical system, which performs the procedures in the process necessary to reach a safe state

- *Functional safety*

  Part of the overall safety which refers to the process and the BPCS, and which depends on the designated function of the SIS and other safety levels

- *Evaluation of functional safety*

  Analysis that is based on evidence which evaluates the functional safety achieved by one or more protection levels

- *Audit of functional safety*

  Systematic and independent examination which determines whether the techniques for specifying the functional safety requirements agree with the planned procedures, are conducted efficiently, and are suitable for achieving the

- *Functional unit*

  Hardware or software unit, or both, which are suitable for implementation of the specified task

- *Hardware safety integrity*

  Part of the safety integrity of the safety-technical function which refers to accidental hardware failures with dangerous consequences

- *Damage*

  Physical injury of, or damage to, a person's health, either directly or indirectly as the result of damage to property or environment

- *Endangerment*

  Potential source of harm

- *Human failure, error*

  Action or act of omission by people, which lead to an undesired result

- *Influence analysis*

  Action to determine the influence which altering one function or one component of the system can have on other functions or components in the system or on other systems

- *Independent department*

  Department which is separated and not in connection with the departments who are responsible for actions which take place during a certain phase of the safety lifecycle, which is the object of evaluation of functional safety or validation

- *Independent organization*

    Organization which, because of its management and its other means, is separated and not connected with the organizations which are responsible for the actions which take place during a special phase of the safety lifecycle, which is the object of evaluation of functional safety or validation

- *Independent person*

    Person who is separated and not tied into the actions which take place during a special phase of the safety lifecycle, which is the object of evaluation of functional safety or validation, and who takes no direct responsibility for these actions

- *Entry function*

    Function that supervises the process and the corresponding equipment, and makes this information available to a logic function

- *Instrument*

    Instrument that is used for executing an action (typically in systems for process control techniques)

- *Logic function*

    Function that generates the transformation of input information (made available by one or more entry functions) to output information (applied by one or more output functions). Logic functions therefore effect the transformation of one or more input functions to one or more output functions

- *Logic system*

    Part of the BPCS or SIS that executes one or more logic functions

- *Maintenance and engineering interface*

    The maintenance and engineering interface describes that part of hardware and software which is designed for the maintenance or modification of an SIS. It can also include instruction or diagnostic methods in the form of software, programming equipment, diagnostic tools, indexes, bridging operations, testing and calibration equipment.

- *Damage control method*

    Method for reducing the effect(s) of hazardous events

- *Operating mode*

    Mode in which a safety-technical function is run

- *Safety-technical function in demand mode*

    In cases where the specified action (for example, the closing of a valve) is initialized in response to process states or other requirements. In the event of a hazardous failure of the safety-technical function, a potential endangerment only occurs if the process or the BPCS fails.

- *Safety-technical function in interrupted mode*

  In cases where, when a safety-technical function fails, a potential endangerment occurs without any further failure, if no measure is taken to prevent it.

- *Module*

  Self-contained unit made of hardware components, for implementation of a specific hardware function (for example, digital input module, analog output module), or reusable application code (within a program, or a number of programs), which performs a specific function, for example that part of a computer program which performs a specific function

- *Requisite risk reduction*

  Risk reduction to ensure that the acceptable risk (risk threshold) is not surpassed

- *Non-programmable system*

  A system that is not based on computer technology, meaning, a system that is not based on programmable electronics (PE) or software.

- *User interface*

  User interfaces are devices (for example, monitors, illuminated displays, push buttons, horns, alarms) which serve in the exchange of information between the user and the SIS. The user interface is also called man-machine-communication

- *Alternative technology safety system*

  Safety system that is not based on electric, electronic, or programmable electronic technology

- *Basic function*

  Function that controls the process and the associated devices according to the control signals specified by the logic function

- *Phase*

  Period of time in the safety lifecycle during which actions take place that are described in this standard

- *Safeguard*

  Measure designed to reduce the frequency of occurrence of a hazardous event

- *Process engineering risk*

  Risk that results from process states which are caused by extraordinary events (including malfunctions of the BPCS)

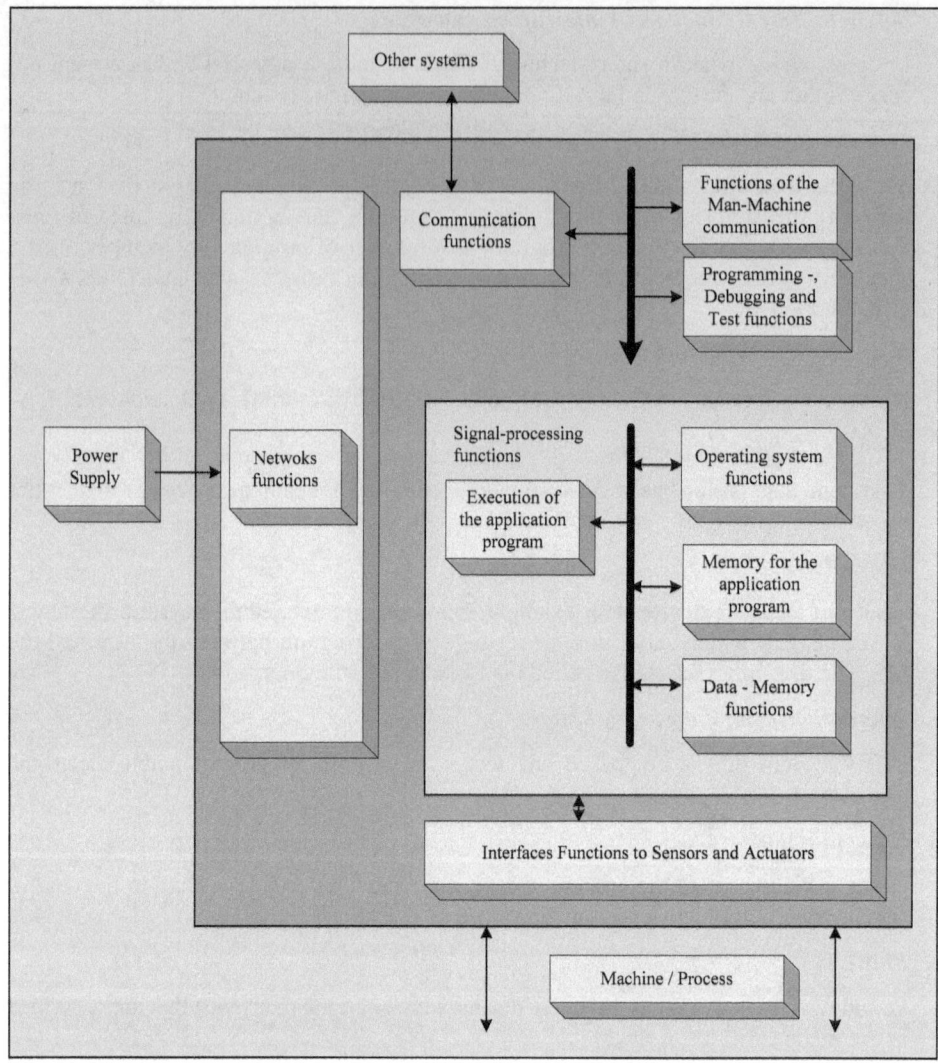

Figure 21.6: Structure and functions of an SPS (from IEC 611319)

- *Programmable electronic*

    Electronic component or device within a PES that is based on computer technology. The term encompasses hardware, software, input and output devices

- *Programmable electronic system (PES)*

    System for control, protection, or supervision, based on one or more programmable electronic devices, including all elements of the system, such as energy supply, sensors, and other input devices, data connections and other communicative pathways, as well as activators and other output devices

- *Repeat test*

  Test to detect hidden errors in a safety-technical system, so that, if necessary, the system can be returned to a state where it can fulfill its scheduled function

- *Protection layer*

  Every independent measure that reduces risk by controlling or supervising protective or damage control measures

- *Operationally approved*

  A component is operationally approved if an examination with proper documentation has shown that appropriate evidence from previous operations attests that the component is suitable for use in a safety-technical system.

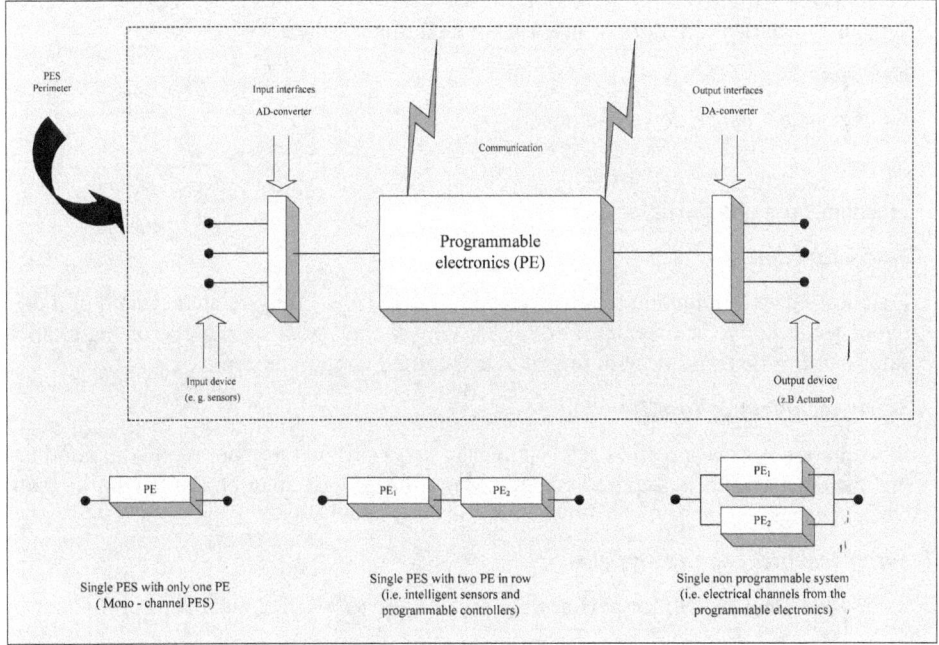

Figure 21.7: Programmable electronic system (PES), structure and terms

- *Quality*

  Multitude of qualities of a unit with regard to its suitability to fulfill specified and predetermined requirements

- *Accidental hardware failure*

  Failure that occurs at an accidental time and results from one or more possible mechanisms in the hardware, which leads to impairment of component properties

- **Redundancy**

    The use of several elements or systems to perform the same function. Redundancy can be carried out with identical elements (homogeneous redundancy) or with different elements (diversity)

- **Risk**

    Combination of the probability of occurrence of damage and of the severity of damage

- **Harmless failure**

    Failure which lacks the potential to put the safety-technical system into a hazardous or inoperable state

- **Percentage of harmless failures**

    Percentage of the overall failure rate for accidental failures of a device that either results in an accidental failure or in a known hazardous failure

- **Safe state**

    Process state where safety is achieved

- **Safety**

    Freedom from indefensible risks

- **Safety function**

    Function for risk reduction that is executed by an SIS, a safety-related system of a different technology, or of external devices, with the set goal to achieve or maintain a safe state for the process, with regard to a specified dangerous event

- **Safety-technical control function**

    SIS function with a specified safety integrity level (SIL) in an operation with continuous demand, which is required for avoiding a hazardous state and/or for limiting its consequences

- **Safety-technical control system**

    Safety-technical system for performing one or more safety-technical control functions

- **Safety-technical function**

    Function with specified safety integrity level (SIL) that is required for achieving functional safety. A safety-technical function can be either a safety-technical protective function or a safety-technical control function

- **Safety-technical system**

    Safety-technical system for execution of one or more safety-technical function. A SIS consists of sensor(s), logic system, and actuator(s)

- **Safety integrity**

    Medium probability that a safety-technical system executes the required safety-technical functions, according to all specified requirements, under all specified conditions within a specified time frame

- *Safety integrity level (SIL)*

  One of four discrete steps in the specification of requirements for the safety integrity of safety functions, which are assigned to the safety-technical system, where safety integrity level 4 is the highest degree of safety integrity, and safety integrity level 1 the lowest degree

- *Specification of safety integrity requirements*

  Specification in which the requirements made on the integrity of the safety-technical functions of the safety-technical system(s) are specified

- *Safety lifecycle*

  Necessary actions within the framework of implementing safety-technical functions during a period of time which begins with the conceptual phase of a project and ends when all safety-technical functions are no longer available for use

- *Safety manual*

  Manual that describes the safe use of a device, subsystem or systems

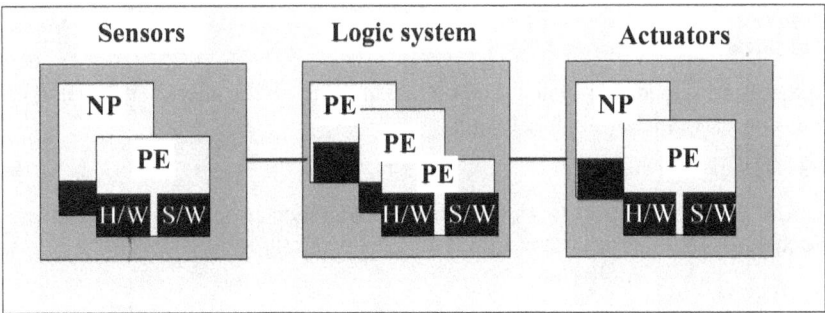

Figure 21.8: SIS architecture

- *Safety-related software*

  Software that is required for performing safety-technical functions

- *Specification of safety requirements*

  Specification that contains all safety-technical function requirements, which are performed by the safety-technical system

- *Safety software*

  Software in a safety-technical system with application, system, or assistant software

- *Software*

  Intellectual product consisting of programs, processes, data, rules, and the corresponding documentation which refers to the operation of a EDV system

- *Fixed program language (FPL)*

  With this language, the user can only set few parameters (for example, the measuring range of a pressure transmitter, alarm threshold values, network addresses)

- *Limited variability language (LVL)*

  This kind of language is understandable for users in the process industry and allows the combination of predefined, application-specific library functions, thereby fulfilling the safety requirement specifications. The functions of a LVL correspond to a high degree to functions required for the application.

- *Full variability language*

  This language was designed for application by computer programmers. With it, a wide array of functions and applications can be covered.

- *Application software*

  Software for a specific user application. In general, it contains the chain of procedures, conditions, threshold values and expressions that influence the corresponding inputs, outputs, calculations, decisions, in order to satisfy the requirements of the safety-technical functions

- *Embedded software (firmware)*

  Software that is delivered as a system component and can not be altered by the end user. Firmware is also called system software

- *Supplemental software*

  Software tool for the creation, maintenance and documentation of application programs. Such software tools are not required for the operation of the SIS

- *Software lifecycle*

  Actions during a certain time period that begins with software development and ends when the software can no longer be used permanently

- *Software safety integrity*

  The probability that a software fulfills its safety-technical functions under all conditions, in a programmable electronic system, over a certain period of time

- *System/subsystem*

  Number of elements that interact with each other. One element of the system can at the same time be another system, a so-called subsystem, which is a controlling or controlled system and can contain hardware, software and human intervention

- *Systematic failure*

  Systematic malfunction/failure that can be conclusively traced back to a cause, which can only be eradicated by a modification of the design or of the manufacturing process, the manner of operation, the operation instructions, or other factors

- *Systematic safety integrity*

  Part of the safety integrity of a safety-technical function that refers to systematic malfunctions/failures with dangerous failure types

- *Failure threshold value*

  Targeted probability of dangerous failures that is specified, according to the requirements for safety integrity, as the medium failure probability of the function during demand, or as the probability of a dangerous failure of the SIF per hour (in continuous demand operation mode)

- *Component, software component*

  A structured, non-specific part of the application software that is easily adjustable while maintaining the original structure, to perform certain functions, for example, an interactive screen model. This component regulates the sequence of the application views but is not fixated on the information presented. A programmer uses this generic component in order to create new screen masks for users, in the course of function-specific revisions.

- *Acceptable risk*

  Risk that is acceptable, within a given concept and based on the valid value propositions of the company

- *Not recognized, hidden*

  In connection with hardware and software errors that are not recognized by diagnostic tests or by normal operation

- *Validation*

  Proof that the examined safety-technical functions and the safety-technical system, following montage, fulfill the specifications of the safety requirements in every respect

- *Verification*

  Theoretical or practical proof for every phase of the appropriate safety lifecycle, that, for certain inputs, the results fulfill the goals and requirements of the phase concerned in every respect

- *Watchdog*

  Combination of diagnosis and output device (typically a switch) that supervises the error-free operation of the programmable electronic (PE) and initiates an action in the event of erroneous operation

- *MvN system*

  Does not require all conduits which it consists of, in order to perform a safety-technical function flawlessly

21.4 Management of Functional Safety

21.4.1 Goal

In this segment, actions for achieving functional safety are listed.

21.4.2 Requirements

Safety management must be in place to allow safety-technical systems to put a process engineering device in a safe state and to maintain it there. Hereto, an evaluation method must be determined and published in the organization.

Every person, whose task touches upon the scope of the safety lifecycle, must be named and educated as to responsibilities. It is a prerequisite that the persons are capable of fulfilling these tasks.

Existing risks must first be named and estimated. According to segment 8 of the IEC 615110-1, steps must be taken to reduce these risks.

Safety planning sessions must be conducted as frequently as possible with the persons, departments, organizations, or other units concerned. Herein, all measures to be performed must be established.

Procedures must be established so that recommendations from the different sources of danger analysis and risk evaluation, evaluation actions and audits, verification and validation actions can soon be worked up and implemented satisfactorily.

Each manufacturer must introduce an appropriate quality management system for the delivery of their products or services, by which the customer can ascertain that they meet his requirements.

In order to test safety-technical systems with regard to safety requirements, evaluation procedures must be developed. At the same time, procedures are required that are designed to recognize and avoid systematic failures, and that clarify whether the previous assumptions correspond to the probability of dangerous failures.

21.4.3 Evaluation, Auditing, and Revisions

In order to evaluate functional safety, a procedure is necessary which also requires the appointment of an advisory team that possesses technical know-how and operational experience with regard to the installation. This team must have at least one competent person who is not involved in project design.

The time points of the safety life cycle which are to be evaluated must be specified during safety planning.

At least one evaluation of functional safety must be performed, which ensures that every hazard is controllable. At least one of them must be performed prior to operation, that is, prior to stage 3. The specified hazards can only occur following confirmation that a hazard and risk analysis have been performed, and that the resultant recommendations have been implemented in the safety-technical system. This also means that deviations between the safety-technical system and the safety specifications have been determined and eliminated. Furthermore, all regulations, alarm plans, and plans for further evaluations must be presented, and the entire personnel must have been familiarized with the SIS.

The frequency of auditing, and the degree of independence between persons who perform a certain task, and persons who perform audits, must be specified. During auditing, the audit results must be documented and subsequently incorporated. For modifications of the safety-technical system, the procedure must be specified with regard to occasion, documentation, test, execution, and certification.

21.4.4 SIS Configuration Management

The SIS configuration management should contain regulations that specify the time of a formal test of the configuration, the procedure for identifying all parts, and a procedure for preventing the implementation of components which have not been certified.

21.5 Safety Lifecycle Requirements

All technical operations should be summarized in an organized flow chart, the safety lifecycle. For this, all phases of the cycle must be described, with all required inputs, the expected work results, and the tests to be performed. Actions that meet the requirements of this standard are assigned to phases.

Furthermore, safety planning must be in place for each phase, in order to ensure that the safety-technical system meets the safety requirements. Here, criteria, work techniques, measures and procedures are specified which examine whether

- the required safety of the SIS and the target values of the SIL are fulfilled in every operating mode;
- the SIL was mounted and activated flawlessly;
- the safety integrity of the safety-technical functions is maintained, after mounting and during operation;
- hazards can be controlled by the process applied during maintenance work.

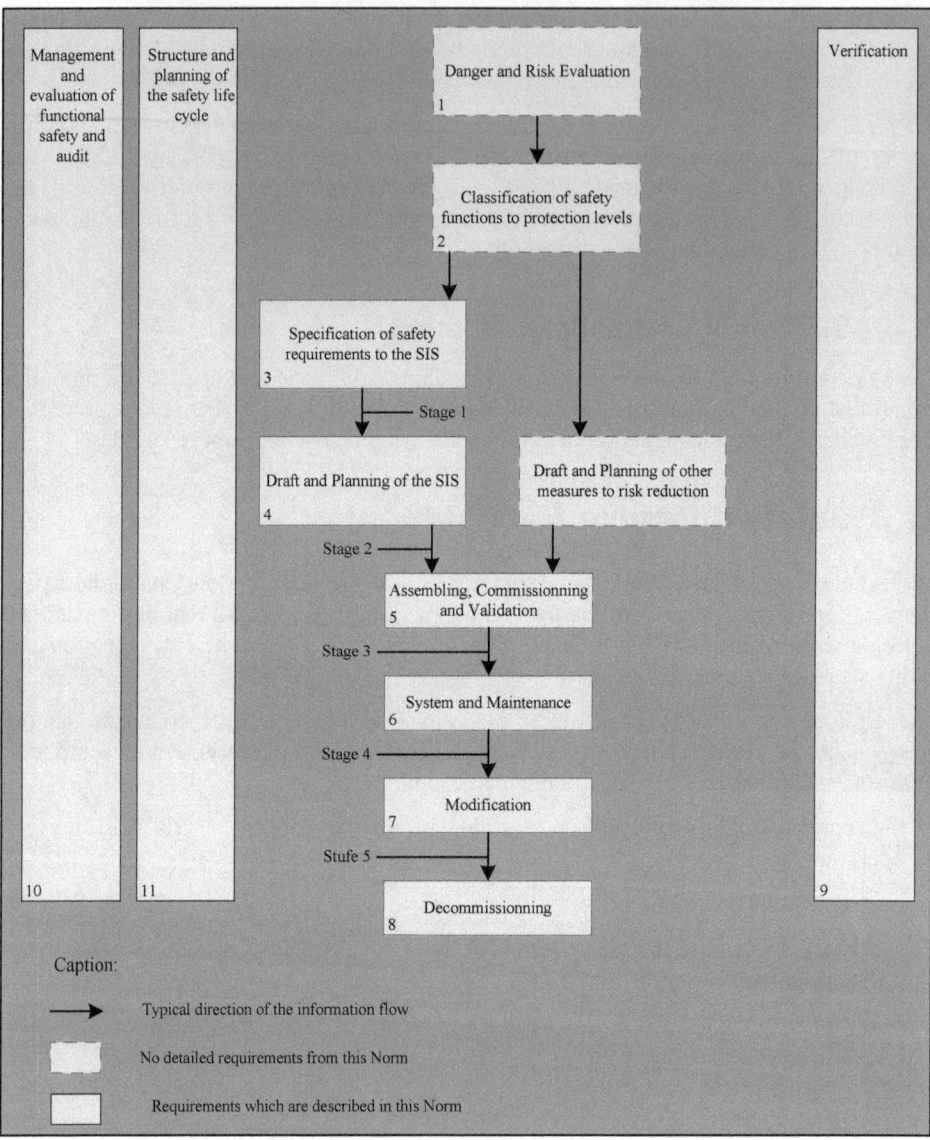

Figure 21.9: SIS safety lifecycle phases and steps in evaluating functional safety

21.5 Safety Lifecycle Requirements

Table 21.2: Overview of the safety life cycle of the SIS

Phase of Safety Life Cycle or Action		Goal	Requirements in segment	Specifications	Results
Box in Fig. 21.9	Title				
1	Hazard and risk evaluation	Specification of hazards and hazardous events through the process, respectively, the appropriate installations and chain of events which can lead to dangerous events, the corresponding process-technical risk, the requirements for risk reduction and the safety-technical functions which are required for risk reduction	8	Process-technical design, exposition, personal measures, safety goal	A description of the hazards, the required safety functions and the corresponding risk reduction
2	Assignment of safety functions to the protective level	Assignment of safety functions to the corresponding protective levels and, for each safety-technical function, assignment of the corresponding safety integrity level (SIL)	9	A description of required safety-technical functions including safety integrity requirements	A description of the assignment of safety requirements (see segment 21.9)
3	Specification of the safety requirements on the SIS	Definition of requirements on each SIS in the form of required safety-technical functions and the corresponding SIL for achieving the required functional safety	10	Description of the assignment of safety requirements (see 21.9)	Requirements on the safety-technical system and on software safety
4	Design and planning of the SIS	Design of an SIS which meets the requirements regarding the safety-technical functions and the safety integrity	11 and 12.4	Safey requirements on the SIS and the software safety	Design of the SIS in compliance with the safety requirements; planning of the SIS integration test
5	SIS installation, commissioning, and validation	Integration and test of the SIS. Validation that the SIS with its individual safety-technical functions and their safety integrity fulfills the specified safety requirements in every aspect	12.3, 14, 15	SIS-design, planning of the SIS integration test, SIS safety requirements, planning of the safety validation of the SIS	Fully functional SIS, according to design; results of the SIS integration test; results of the installation, commissioning and validation actions
6	SIS installation and maintenance	Ensuring that functional safety of the SIS is maintained during operation and maintenance work	16	SIS requirements; SIS design, planning for SIS operation and maintenance	Results of actions within the framework of the SIS operation and maintenance
7	Modifications of the SIS	Performing corrections, improvements, or adjustments of the SIS in order to achieve and maintain the required SIL	17	Revised SIS safety requirements	Results of SIS modification
8	Decommissioning	Ensure through appropriate tests and attention to the design structure that unaffected functions remain operational	18	Safety requirements and installation documents („as built")	Safety-technical function decommissioned

Phase of Safety Life Cycle or Action		Goal	Requirements in segment	Specifications	Results
Box in Fig. 21.9	Title				
9	SIS verification	Testing and evaluation of the results of a phase for correctness and consistency with regard to the products and standards, which are specified for this phase	7, 12.7	Phase-specific planning procedure for verification of the SIS	Results of SIS verification for each phase
10	Evaluation of functional safety of the SIS	Examination and evaluation of functional safety of the SIS	5	Plan for evaluation of functional safety of the SIS, SIS safety requirements	Result of the evaluation of functional safety of the SIS

21.6 Verification

21.6.1 Goal

With these requirements, evidence should be presented that the specifications of the verification plan are fulfilled in the results of individual steps of the safety lifecycle.

21.6.2 Requirements

In order to conform to the standard, the verification plan must contain all actions of the verification of individual steps. Herein, the time of verification, participating persons, methods applied, installations to be verified, and procedures, measures, and techniques must be documented.

21.7 Hazard Analysis and Risk Evaluation

21.7.1 Goal

The goal of hazard and risk analysis is the determination of hazards and events that can lead to dangers, by pointing out appropriate resources and a chain of events that can lead to a dangerous event. Furthermore, the process risk should be calculated by means of these events, measures and required safety functions for risk reduction should be specified, and safety-technical functions among the safety functions should be determined.

21.7.2 Requirements

The danger and risk analysis performed for the process must be described in detail as the result of dangerous conditions and their consequences, with specification of the probability of occurrence. In addition, the assumptions made during risk analysis, the demand and failure rates of devices, and the demands on methods and preparations for risk and danger reduction must be specified. Assigning safety functions to safety levels, and determining safety-technical functions among the safety functions, also are part of this analysis.

21.8 Allocation of Safety Functions to Protection Layers

21.8.1 Goal

Safety functions are allocated to different protection layers and safety-technical functions are specified with their appropriate safety integrity level.

21.8.2 Allocation Requirements

Until the end of allocation, safety functions must be allocated to protection layers, and the individual goals of risk reduction must be assigned to safety-technical functions. The assignment of the SIL to safety-technical functions in demand mode takes place according to Table 21.3, in continuous demand mode according to Table 21.4.

Table 21.3: Safety integrity level: Failure probability on demand

Safety integrity level (SIL)	Demand mode	
	Target value for medium failure probability	Target value for risk reduction
4	$\geq 10^{-5}$ to $< 10^{-4}$	> 10.000 to ≤ 100.000
3	$\geq 10^{-4}$ to $< 10^{-3}$	> 1.000 to ≤ 10.000
2	$\geq 10^{-3}$ to $< 10^{-2}$	> 100 to ≤ 1.000
1	$\geq 10^{-2}$ to $< 10^{-1}$	> 10 to ≤ 100

Table 21.4: Safety integrity level: Frequency of hazardous failures of the safety-technical function

Safety integrity level (SIL)	Continuous demand mode
	Target value for the frequency of hazardous failures per hour
4	$\geq 10^{-9}$ to $< 10^{-8}$
3	$\geq 10^{-8}$ to $< 10^{-7}$
2	$\geq 10^{-7}$ to $< 10^{-6}$
1	$\geq 10^{-6}$ to $< 10^{-5}$

21.8.3 Safety Integrity Level 4 Requirements

A function with a SIL greater than 4 must not be assigned to a SIS, although even an SIL of 4 is rare in the process industry. The construction of a safety-technical function with an SIL of 4 can take place if either the adherence of the threshold value of the failure probability for the SIL value of 4 is proven conclusively, or there are a number of experiences with the components used in the SIF, and the hardware failure rates of the components are known and justify the specified failure threshold value. It requires a very high and durable performance and very competent users and should therefore be avoided whenever possible. Another possibility is the reduction of the SIL, by improving intrinsic safety or by adding more protection layers.

21.8.4 Demands on Factory Devices Used as Protective Layers

As shown in Figure 21.10, factory devices can be counted among potential protection layers. When using a BPCS as protective level, the achievable risk reduction factor lies below 10. If a higher factor is required, the BPCS must conform to the standard.

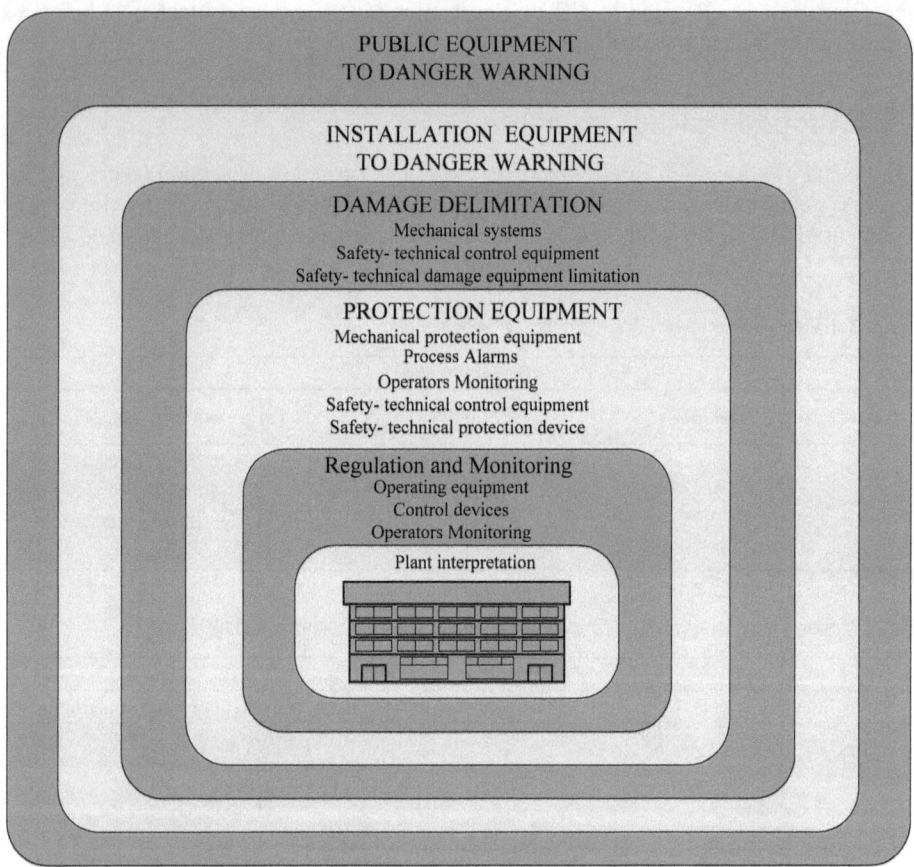

Figure 21.10: Typical methods of risk reduction in process engineering

21.8.5 Requirements for Failure Avoidance

During design of the protective layer it must be evaluated whether the probability of failures caused by the same event, of the same type, and of dependent failures in the protective layers is low enough, in comparison with the demands on overall system integrity of protective layers. Herewith, the independence, difference and physical separation of protective layers, and failures due to identical causes, must be taken into account.

21.9 Safety Specification of the SIS

21.9.1 Goal

This segment describes the safety demands on safety-technical functions. They are derived from the assignment of safety-technical functions and the requirements established during the safety planning phase.

21.9.2 SIS Safety Requirements

In the safety requirements, different points must be included, such as the description of all SIF with the definition of safe state, SIL, and operational modes of safety functions, the measuring signals of the SIS with their threshold values, the different requirements during error detection, deactivation, (re)installation, reversal, and re-examination of the SIS, threshold values of environmental conditions, measures for maintaining the safe state in the event of an error, repair time, functional coherence of the SIS etc.

Software specification must result from the selected architecture of the SIS and the general safety specifications.

21.10 SIS Design and Planning

21.10.1 Goal

It is the goal of this segment to design one or more SIS that fulfills the required safety-technical functions including the stated SIL.

21.10.2 General Requirements

The dependence, respectively, independence, between SIS and BPCS, and SIS and other protective layers, must be taken into account. If the BPCS does not conform to this standard, it must be separated and independent of the SIS.

If other safety-technical functions work within an SIS with differing SIL, then hardware and software must be treated as though they belong to the highest SIL. This becomes redundant if the safety-technical functions are independent from each other. If the SIS also contains safety-technical functions, then the entire hardware and software, which might have a negative influence on a safety-technical function, must be treated as part of the SIS and meet the highest SIL value. A component which is necessary as part of a safety-technical function must not be part of a function of the BPCS.

Once the process has been put into the safe state, it must remain in this state until resetting.

21.10.3 Demands on Safety Behavior upon Error Detection

If a dangerous error is detected in a subsystem which can handle simple hardware errors, the safe state must either be achieved or maintained, or the defective part must be repaired, while the process continues safely.

If a dangerous error is detected in a subsystem which is only operated for a SIF in demand mode, then here also the safe state must be achieved or maintained, or the defective component repaired, while the process continues safely. In doing so, risk reduction must be as low as if no error had been detected.

If a dangerous error is detected in a subsystem, which is operated for SIF with continuous demand, measures must be taken to achieve or maintain the safe state, such as for example reducing the load of the entire installation.

21.10.4 Demands on Hardware Error Tolerance

In safety-technical functions, sensors, logic controls, and actuators must have minimum hardware error tolerance.

Table 21.5 shows the required fault tolerance of PE logic systems.

Table 21.5: Minimum hardware fault tolerance of PE logic systems

SIL	SFF < 60 %	60 % < SFF < 90 %	SFF > 90 %
1	1	0	0
2	2	1	0
3	3	2	1
4	Special requirements apply, see IEC 61508		

Table 21.6 shows the required fault tolerance of all other subsystems, if the main failure type is directed into a safe state, or dangerous failures are recognized. Otherwise, error tolerance must be increased by an increment of 1. It can be decreased by an increment of 1 if the device was selected on the basis of an earlier application, or if only process-related parameters are set, or if protection for process-related parameters through jumper or password exists and the required SIL is smaller than 4.

Table 21.6: Minimum hardware error tolerance of sensors, actuators and non-programmable logic systems

SIL	Minimum hardware fault tolerance
1	0
2	1
3	2
4	Special requirements apply, see IEC 61508

Alternative requirements can also be used for error tolerance, following an evaluation according to IEC 61508-2.

21.10.5 Demands on the Selection of Components and Subsystems

Components or subsystems utilized as part of a SIS must be specified. Furthermore, demands on components that are integrated into the architecture of a safety-technical system and acceptance criteria are specified with regard to the safety-technical functions and the safety integrity for previously developed components.

For operationally proven components, examinations must be performed which contain evidence that a component is suitable for implementation in a safety-technical system.

21.10.6 Field Devices

During the selection and installation of field devices care must be taken that the number of errors through process and environmental conditions is as low as possible. On top of that, the integrity of input/output circuits must be examined according to the open-circuit principle and its energy supply. The entry and exit ways of a system must be connected with every field device via their own wiring, except in a case where several binary sensors located at the same entrance supervise the same process, or when an exit controls several final controlling devices, or when the overall safety of a bus connection fulfills the reliability requirements of the corresponding SIF. With intelligent sensors, an unintentional alteration must be prevented by write-protection, except if writing and reading are permitted following a safety examination.

21.10.7 Interfaces

The man-machine interfaces include:

- user interface
- maintenance and engineering interface
- communications interface

For the user interface, it is required that the user makes a choice or circumvents the system as rarely as possible during operation of the device. For user interference, protection again misuse must be put in place. A password or key switch must be built in, to protect against unauthorized use of a shunting switch. The SIS status information, which makes decisions on the adherence to the safety integrity level, must be shown as part of the user interface.

Changes to the application software of the SIS must be prevented by appropriate design of the user interface. The BPCS must write into specific variables of the SIS, in order to send safety information to the SIS. The devices used in this process should be able to confirm that data from the BPCS are transferred to and received by the SIS correctly. Therein, the safety functionality of the SIS must not be compromised.

The maintenance and engineering interface must be conceptualized such that during a failure the SIS is still capable of achieving a safe process state. This can be accomplished, for example, by separating one programming device from the SIS during normal operation. This interface has protected access to certain functions and therefore must not be used as user interface. Write/Read access must only be made available or blocked by a configuration or programming procedure at this interface, and by paying close attention to the safety rules and to the obligation of documentation.

For the communication interface, it is required that communication of the SIS with the BPCS and other devices must not influence the SIF. Any failure of this interface must not prevent the SIS from putting the process into a safe state. Furthermore, this interface must be sufficiently insensitive to electromagnetic rays and voltage fluctuations to prevent a dangerous SIS failure. This interface must also regulate communication among devices of varying electric potentials.

21.10.8 Maintenance and Test Device Requirements

Testing the entire SIS in one go, or in parts, must be permitted by SIS procedures. Testing must occur during operational mode, if the test procedure is to be performed more frequently than the powering-down of the system. The test devices must be integral components of the SIS design if a repeat test is necessary to check for hidden errors. Ultimately required test or bridging devices in the SIS must be taken into consideration in the maintenance and test requirements of safety specifications during SIS design, and the user must be informed by an alarm or in the operation manual. Forcing entries and exits into PE-SIS is forbidden in the application software, as part of the operating instructions and in maintenance, except under special rules and with additional reporting. Forcing must be demonstrated and alarmed correspondingly.

21.10.9 Failure Probability of Safety-technical Functions

This probability must have a maximum value equal to that of the failure threshold of the safety specification. It must be proven by calculation, where for each safety-technical function the failure rate due to errors and hardware failures with dangerous consequences, and the failure frequency due to common causes and due to human reactions must be taken into consideration.

21.11 Application Software Requirements

In this segment, three types of software are named: application software, firmware (included operation software), and software aids (tools for designing and testing application software). Furthermore, three language types for software construction: fixed program language (FPL), limited variability language (LVL), and full variability language (FVL). In this standard, only application software with FPL and LVL is described. The requirements described below are only valid up to SIL 3, and no distinction is made between SIL 1, 2, and 3. SIL 4 and FVL are discussed in IEC 61508.

21.11.1 Demands on the Safety Lifecycle of Application Software

The goals of this segment are the determination of actions and software aids designed for the creation of application software, and to secure a safe state by the use of an appropriate examination to achieve the functional safety goals of the application software.

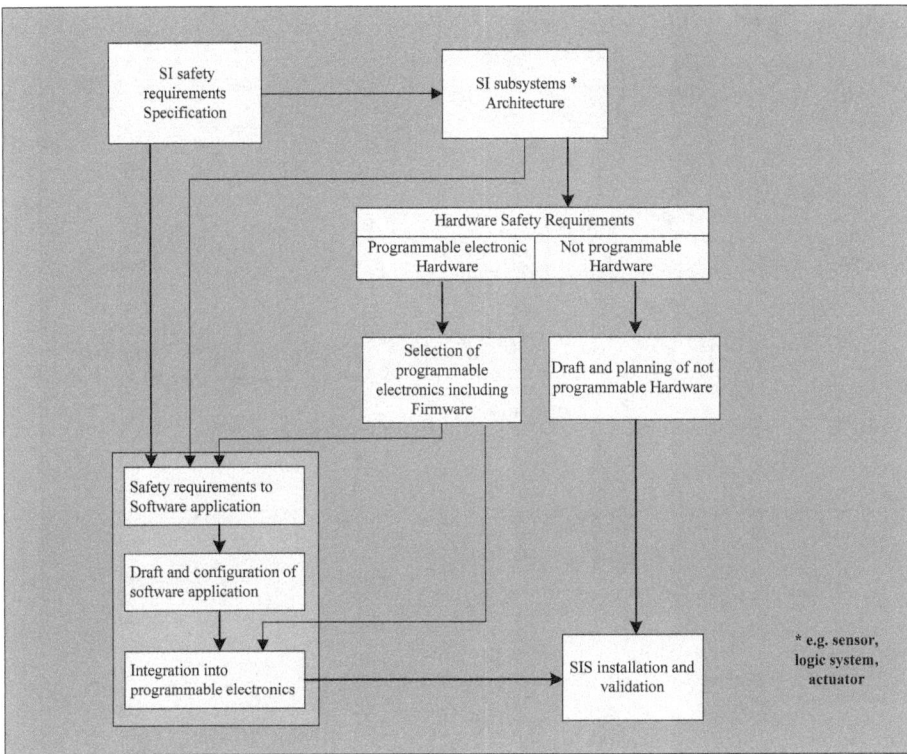

Figure 21.11: Safety lifecycle of application software and relationship to the SIS lifecycle

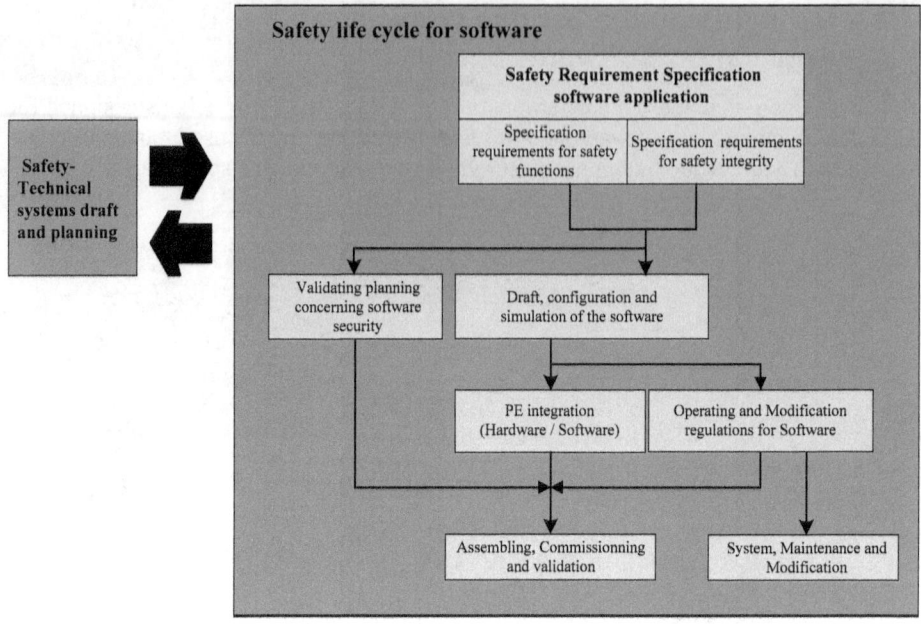

Figure 21.12: Safety lifecycle of the application software (in the implementation phase)

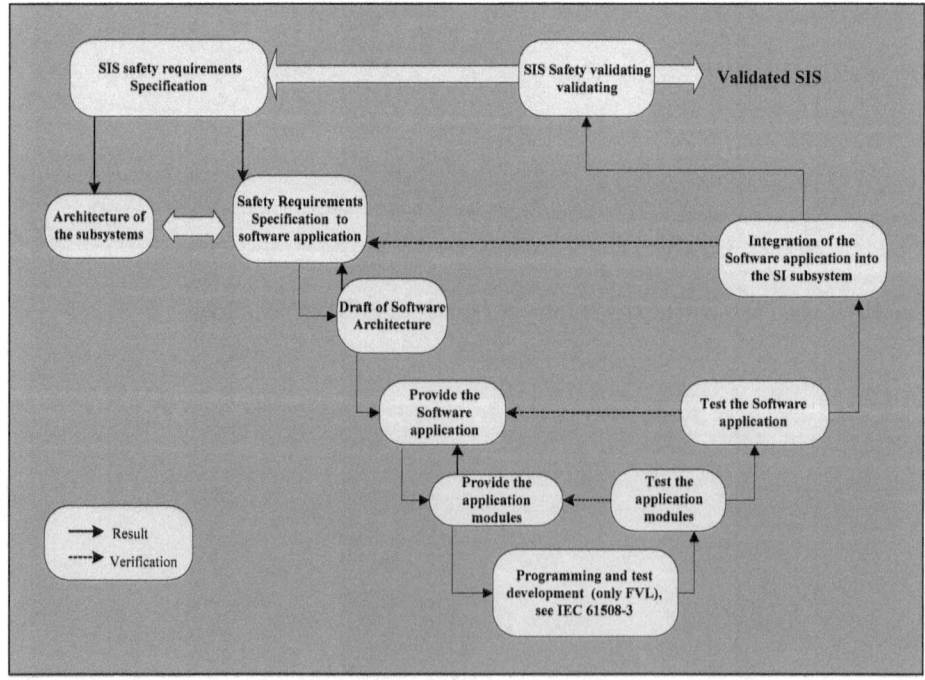

Figure 21.13: Software development process (V-model)

21.11 Application Software Requirements

Hereto, the specification of an appropriate safety lifecycle is required for the creation of application software. Furthermore, all actions, goals, required input information, results, demands on verification and responsible persons for each phase of the cycle must be described. In addition, the programmable electronic device for each SIF must have a suitable safety integrity.

Table 21.7: Instructions and limits regarding safety software

Language type	Applicable segments	Typical applications
Fixed program language (FPL)	Testing of data fields/manufacturers data Input of application data Integration test of the subsystem SIS validation test	Parameterizing within specified limits Alteration of functions not possible
Limited variability language (LVL)	Selection of the system and its tools Creation of application software Creation/selection of application modules Testing of application modules PE integration test SIS validation test	Combination of proven functions from a limited number of proven combinations, to create an application out of them
Full variability language (FVL)	Selection of system and language Design of the software architecture Design of the application software, module and code Test of the application module Test of the application software PE integration test SIS validation test	Development of complex functionalities out of elementary programming instructions

Table 21.8: Safety lifecycle of application software: overview

Safety lifecycle phase					
Number in Fig. 21.12	Title	Goals	Requirements according to segment	Required information	Required results
3.11.2	Specification of application software safety requirements	Specification of demands on the safety-technical functions for each SIS function which is required for the implementation of the desired safety-technical functions Specification of demands on software safety integrity for each safety-technical function of this SIS	3.11.2.2	Specification of SIS safety demands Safety manuals of the applied SIS SIS architecture	Specification of demands on the application software of the SIS Verification documentation
3.11.3	Validation planning for the safety of the application software	Designing a plan for the validation of application software	3.11.3.2	Specification of demands on the SIS application software	Validation plan for the safety of the SIS application software

Safety lifecycle phase			Require-ments according to segment	Required information	Required results
Number in Fig. 21.12	Title	Goals			
3.11.4	Design and creation of the application software	Architecture Design of a software structure which fulfills the safety specifications for the software Testing and evaluation of the software requirements, which result from the SIS hardware architecture	3.11.4.3	Specification of SIS software requirements Documentation on SIS hardware architecture	Description of system architecture Exemplary subdivision of the application software into process-appropriate subsystem and SIL values Exemplary specification of commonly used application components for pumps and valves Test specification for the software architecture and the integration of subsystems Documentation of verification
	Design and creation of the application software	Development tools and programming languages: Composition of a suitable selection of configuration, library, administration, simulation and testing tools for the entire safety lifecycle (software tools) Determination of rules for the creation of the application software	3.11.4.4	Specification of requirements on the SIS application software Description of system architecture SIS manuals Safety manuals for the SE logic systems applied	List of rules for the use of software tools Documentation of verification
3.11.4	Design and creation of application software	Design of application software and application components Creation of application software which fulfills the safety requirements of the application	3.11.4.5	Description of the planned system structure List of manuals and rules of the applied PES for the use of software tools	1) Application software (for example in functional component language or contact plan) 2) Simulation and integration test of the application programs 3) Specification of safety requirements of the application software for special applications 4) Documents for verification
3.11.4	Creation of application programs by the use of FVL	Creation and testing of the FVL application code: Implementation of program languages with unlimited variability, which fulfill the safety specifications for software	3.11.4.6, 3.11.4.7	Safety specification for application software for special applications	See IEC 61508-3

Safety lifecycle phase			Requirements according to segment	Required information	Required results
Number in Fig. 21.12	Title	Goals			
3.11.4	Design and creation of application software	Testing of application software: 1) Examination that the software safety requirements are fulfilled 2) Proof that all subsystems and systems of the application software work together flawlessly and fulfill their required functions and no unintended functions Following satisfactory test results, this can also be combined with the next phase (21.12.5)	3.11.4.6, 3.11.4.7 3.11.7	Specification for the simulation and the integration test of the application program (structure-oriented testing procedure) Specification for the integration test of software architecture	1) Results of the software tests 2) Verified and tested software system 3) Documentation of verification
3.11.5	Integration of programmable electronic (hardware and software)	Software integration into the target hardware	3.11.5.2	Test specification for the integration of hardware and software	Results of integration tests of hardware and software Verified software and hardware
3.11.3	SIS safety validation	Validation that the SIS including the safety application software conforms to safety requirements	3.11.3	Safety validation plan for software and SIS Validation planning of software and SIS with regard to safety	Validation result of software and SIS

21.11.2 Specification of Application Software Safety Requirements

This segment describes how the specification of application software safety requirements must be performed in order for the required safety-technical functions to be implemented such that they conform to the SIS architecture.

Table 21.9: Correlation between SIS hardware and software architecture

	Architecture of programmable SIS subsystems	
Hardware architecture	Software architecture (comprises firmware and application software)	
	Firmware	Application software
General and application-specific properties of hardware Examples are: Diagnostic tests Redundant processors Two-channel input/output cards	Examples are: Communication drivers Error routines Service programs	Examples are: Input/output functions Deduced functions (for example, sensor test, if not covered by firmware)

A safety specification must be documented for the application software of each SIS subsystem, which contains the safety requirements of the SIF, the requirements resulting from the SIS architecture, and the requirements which are necessary for safety planning.

In order to perform an evaluation of functional safety and achieve the required safety integrity, the specification must be conducted accordingly.

The specification requirements must be tested by the manufacturer of the application software with regard to clarity, consistency and comprehensibility, and potential flaws must be communicated to the creator of the SIS subsystem.

It must be possible to select suitable devices according to the specifications of the application software safety requirements. Therefore, functions must be described which elevate the process into a safe state and keep it there, functions which recognize, communicate and deal with errors in subsystems of the SIS, which perform regular tests of the SIF, and which facilitate safe alterations of the SIS.

Likewise with the SIL, each of the functions mentioned, the intersections with non-safety-related functions, and the processing capacity and response times of the system must be taken into account.

21.11.3 Validation Planning for Application Software Safety

A suitable software validation plan must be constructed which conforms to segment 15 of the IEC 61511-1.

21.11.4 Design and Construction of Application Software

A software architecture must be constructed which harmonizes with the hardware architecture and conforms to predetermined software safety requirements. Furthermore, testing and evaluation of the software requirements must be performed. Suitable tools must be selected for the creation of software. The application software should be designed or constructed in such a manner that it can be analyzed, verified, and safely modified and conforms to the safety requirements.

It is generally required that an application program in a program language with unlimited functional scope is constructed, tested, verified and validated according to IEC 61508-3. The development tools and the predetermined restrictions of the SIS subsystem concerned determine software design.

On the basis of properties of the selected design method and the application language (LVL or FPL), mastering the complexity, presentation, functional understanding of the application, technological limits and alterations can be simplified. The finished design must contain plausibility controls and tests of the data integrity; it can be tested, it permits safe changes, and must make sure that the safety-technical application software has minimal complexity and minimal size. The entire software is treated in such a manner as though it belongs to the highest SIL, if the application software comprises safety-technical functions of different SIL or non-safety-related functions. This demand is cancelled if the independence of the safety-technical functions with varying SIL is proven and documented. For previously developed components, their suitability with regard to the safety requirements of the application software must be proven. This must be done in a similar application that has the same functionality or the same criteria for verification and validation as the new software.

The architectural design of the application software must build on the required safety specifications of the SIS and conform to the design of the selected subsystem, its tools and its safety handbook. The methods and procedures used in the design of the application software should be listed and labeled on the basis of their selection. All procedures which lead to the maintenance of safety integrity of all data must be described and their application justified.

Suitable tools, a selection of program languages, configuration management, simulation and test programs must be selected. The availability of suitable tools must be taken into account, so that corresponding applications are facilitated at any given time. Suitable procedures must be specified for the use of tools. The application rules of the program language should contain instructions on good programming techniques, unsafe generic software properties, tests to detect configuration errors, and methods for program documentation. The software should be designed in a structured form, in order to achieve modularity and verifiability of functionality. Additionally, safe modifications, retraceability to the application functions and a description of the application functions with their corresponding limitations becomes then possible. For the production of application components, emphasis must be placed on insensitivity, in that all input variables and the global variables used for the input data can be tested for plausibility, all input and output intersections are fully defined, and the configuration, presence and accessibility of hardware and software components was tested.

The application software must be subjected to a review, in order to examine whether it corresponds to the specified design, the design rules and the requirements of safety validation planning. In order to ascertain that the input and output data are connected with the correct application logic, and that the intended functions work correctly, respectively, that unintended functions are not executed, the configuration of each input and each individual component must be simulated and tested.

It must be proven by examination that all building blocks and components, respectively, subsystems collaborate with the firmware in order to be able to execute the intended functions. The results of the integration test must contain the test results and the achievement or non-achievement of goals and criteria of the specification. If a test has a negative result, the reasons must be listed.

For each modification during the integration of application software, an influence analysis must be performed. It must contain a listing of all software components and the necessary design alteration with renewed verification.

21.11.5 Integration of the Application Software into the SIS Subsystem

It must be proven that the application software fulfills the safety specifications once it is used in conjunction with hardware and firmware of the SIS subsystem. In order to ascertain that the application software is compatible with hardware and firmware, the determination of integration tests must be performed as near as possible to the beginning of the safety lifecycle. Thereby, the safety-related functional requirements and capacity requirements are fulfilled. An influence analysis must be performed in the course of testing for each alteration. It must determine all subunits concerned and the steps to a renewed verification.

21.11.6 Procedure for Modification of Application Software

It must be ensured that the application software still fulfills the safety specifications after modification. During the implementation of a modification, the consequences for process safety must be taken into consideration, and the state of the software design must be analysed.

21.11.7 Verification of Application Software

It must be proven that the information is satisfactory and that all results of the safety lifecycle phases conform to the requirements. For each phase of the application software lifecycle, a verification plan must be performed. Moreover, the verification should concern itself with testability, legibility and retraceability. Non-safety functions and process intersections with integrated safety-related signals and functions should be verified with regard to the absence of interaction with safety functions, and the protection from interference with safety functions if disturbances occur in the non-safety functions.

21.12 Final Inspection

21.12.1 Goals

By examination of the logic system with the corresponding software prior to assembly, it can be determined whether the specified requirements are fulfilled. Furthermore, errors are easily recognized and can be fixed.

21.12.2 Recommendations

The necessity of a final examination should be documented already in the design phase of a project. During the planning process, the type of the necessary tests, the test cases, test descriptions and test dates should be specified. For this examination, a defined version of the logic control should be used. Thereby, the planning guidelines should be adhered to, in order to ensure logical correctness. The documentation of the results must contain the test cases performed with their test results, and a report on whether the goals have been achieved. Any errors which occurred must also be documented, analyzed and corrected correspondingly. With each alteration in the course of the final test, it is recommended to perform a safety analysis. Thereby, the extent of consequences on each safety-technical function, and the volume of examinations to be repeated, is determined.

21.13 SIS Assembly and Implementation

The goal is the assembly and implementation of the SIS according to specifications and plans for a subsequent validation.

During planning, the procedures of assembly and implementation must be specified and the time of their execution, and the persons responsible, must be specified. The installation of all components follows the design and implementation plan, and the implementation of the SIS must conform to planning. Furthermore, the test results of implementation and statements concerning the compliance with goals and test criteria from the design phase must be noted. Failure causes must be documented. A professionally competent person must evaluate potential deviations of the installation from the design. If the person does not detect any consequences for the safety, the next step is to adjust the plans to fit the status quo. Otherwise, the installation must be changed according to plans.

21.14 SIS Safety Validation

Through tests and examinations it should be validated that the implemented safety-technical system and corresponding safety-technical functions meet specified safety requirements.

In validation planning, all validation actions to be performed, the time of execution, the corresponding installations, responsible persons and the object of validation must be noted.

In addition, for the validation of application software the settings of validation, necessary input signals, expected output signals and further acceptance criteria are to be named. Validation actions include actions such as the examination of error-free behavior of the entire SIS in all operation modes, even with problems in the BPCS, the examination of communication between SIS and BPCS or other systems, the error-free function of by-passes, tractive bypasses and manual deactivating facilities, and the correct presentation of operation status and alarms. For software validation it is tested whether the required safety is maintained during normal status, under error conditions, in limited operation or with unspecified functions. If the expected results deviate, a decision must be made on whether the validation can be continued or whether it is necessary to implement a change.

Finally, to prevent hazards during operation, all stop-gaps applied during validation, forced values, and further testing measures, must be removed.

21.15 Operation and Maintenance of the SIS

21.15.1 Goals

During operation and maintenance, the required SIL and the required functional safety of each safety-technical function must be maintained.

21.15.2 Requirements

It is required to perform planning for the operation and maintenance of the SIS. It must contain routine procedures and procedures during special operational status, maintenance procedures, rules, means, methods, times, and persons responsible for operation and maintenance.

Operation and maintenance regulations must be constructed based on safety planning and contain the necessary routine measures to be performed in order to maintain functional safety, the necessary means and limitations to prevent unsafe conditions and to prevent consequences of dangerous events, maintenance regulations in the event of errors or failures, and documentation of results, demand rates, and system failures of the SIS. If deviations occur in the behavior of the SIS, an examination and, if necessary, a modification must be performed in order to maintain safety.

Following audits of the functional safety and examinations of the SIS, the operation and maintenance regulations are in need of revision. Any errors not identified during diagnosis must be uncovered with the help of written test regulations for each safety-technical function. These regulations describe the steps to be performed, such as the correct functioning of each sensor and actuator, the correct logic connection and the correct alarms and indicators.

Service personnel must be trained to understand and to operate the SIS. The education of maintenance personnel proceeds according to the maintenance of the full function of the SIS with the planned integrity.

21.15.3 Re-examination and Inspection

In order to find hidden errors, periodic re-examinations must be performed according to written instructions. Thereby, the entire SIS must be examined and any flaws encountered must be remedied in due course. The frequency of such examinations is calculated from the medium probability of a failure on demand, and can be re-specified from time to time. Following a modification of the application logic, a complete re-examination must be performed. It can be omitted only in exceptional cases, if the system has been properly examined. In order to ensure that there are no apparent flaws, every SIS must undergo a visual inspection at specific intervals. The performance of re-examinations and inspections must be documented, and must contain the date and the description of each examination, the specification of the system examined, the results, and the person who performed the examination.

21.16 SIS Modifications

21.16.1 Goals

Any changes made to a safety-technical system must be planned in great detail, tested, and authorized. It must be ensured that the SIL is maintained.

21.16.2 Requirements

Prior to a modification of the SIS, regulations must be specified and put in place. In these regulations, the procedure and the hazards of this work must be described. It must be analyzed whether the change will influence functional safety. Following the orderly authorization, the change can only be performed by qualified and appropriately trained personnel. Extensive documentation must be shown in the process.

21.17 Decommissioning of the SIS

Prior to decommissioning of the SIS, an accurate examination must take place and the required authorization must be acquired. At the same time, regulations for modification must be constructed and put in effect. These must describe the actions to be performed and the procedural manner for hazard determination. Furthermore, the influence of decommisioning on functional safety and all consequences on other units must be analyzed. The results of this analysis are to be documented in safety planning. The safety-technical function must continue to run during decomissioning.

21.18 Documentation Requirements

21.18.1 Goal

The purpose of documentation is that verification, validation and methods for evaluating functional safety and all phases of the safety lifecycle are conducted in the best possible manner.

21.18.2 Requirements

The documentation must describe the structure, the system and its use in a precise and easy to understand manner, and must furthermore be easy to access and maintain. For this purpose, a structure is needed which facilitates the search. The documentation must have an unambiguous name, which is also apparent from its content. In order to distinguish between different versions, a revision index is required. The documentation must be worked through, improved, tested, and accepted and must lead back to the standard. Moreover, it must always be kept up-to-date.

22 Terms and Definitions

22.1 Safety Systems

22.1.1 Risk

In safety engineering, the risk *(R)* of a technical procedure or a condition is calculated by the combination of frequency *(F)* of an event and the thereby possibly resulting extent of damage *(D)*:[198]

$$R = F \times D \qquad (22.1)$$

Risk is only rarely expressed quantitatively though, but mostly just described qualitatively.

22.1.2 Partial Risk

This refers to the part of a risk that is covered by a specific protection device.

22.1.3 Risk Limit

This is the biggest still acceptable risk of a specific technical procedure or condition (from: DIN VDE 31 000 Part 2/12.87). (The risk limit is mostly not only determined by objective factors but also very strongly by subjective factors. More on this in a later segment.)

22.1.4 Risk Parameters

Risk parameters provide a qualitative statement about damage extent and damage frequency. They are thereby essential for the determination of a requirement class.

22.1.5 Requirement Class

The requirement classes are a model subdivided into classes based on which the requirements on MSR installations are classified.

---

[198] *[DINa87]* DIN VDE 31000 part 2/12.87, *General principles for the safety-weighted design of technical products – Safety-technical terms – Basic terms*

22.1.6 Measures

Measures for implementation of safety-technical requirements.

22.1.7 Protection

Reduction of risk by measures. These should restrict[199]

- frequency of occurrence
- damage extent
- or both

22.1.8 Measurement and Control Protection Measures

Risk reduction measures that concern Measurement and Control devices.

22.1.9 MSR Protection Installation

MSR protection installations should supplement MSR operation installations. These can be functionally separated (process engineering installation) or form a unit (simple industrial machines). If they are separated, only the MSR protection installation is subjected to safety evaluation. If both form a unit, then naturally both must be examined together. The failure of MSR protection installations can result in personal injury, extensive damage to machines or apparatus or extensive production damage.

22.1.10 Undesired Event

Error condition of a unit under observation, which can lead to damage.

22.1.11 Error

Failure of compliance of at least one requirement of a required quality of a unit under observation.[200]

22.1.12 Redundancy

Presence of more than the necessary means for the implementation of scheduled tasks.[201]

22.1.13 Diverse Redundancy

Redundancy with dissimilar means, that is for example, a combination of electronic and pneumatic means. [202]

---

[199] *[DINa87]* DIN VDE 31000 Part 2/12.87, General principles for safety-weighted design of technical products – Terms of safety technique – Basic terms
[200] *[VDI-00]* VDI/VDE 3542 Leaflet 1-4, Safety-technical terms for automated systems
[201] *[VDI-00]* VDI/VDE 3542 Blatt 1-4, Sicherheitstechnische Begriffe für Automatisierungssysteme

22.1.14 Failsafe

Ability of a technical system to remain in a safe state upon occurrence of certain failures, or to immediate transfer to a different safe state.[203]

22.2 Dependability

Dependability requirements can be subdivided into requirements based on:

- Reliability
- Availability A
- Safety S
- Maintainability M

Reliability is a generic term which covers a multitude of concepts and measures. A system is reliable if one can justifiably trust in its ability to perform. The term reliability was originally introduced for computer systems with high reliability and safety requirements. Often, the term reliability is considered a synonym for reliability and safety. Reliability and safety definitions, as a rule, have been introduced for individual technological areas, such as for real-time computer systems, rail systems, airplanes, power plants, crane systems, medicinal-technical products, or tool machines. Frequently, the definitions date back to a time (1980 to 1990) where new areas such as mechatronics and new technologies such as the internet and mobile phones were still largely unknown. With regard to the conception and the operation of mechatronic units, the term reliability is defined as follows, in the style of listed definitions. Reliability is a measure for how reliable and safe a defined unit (component, system) fulfills its requirements over a certain period of time, with regard to all attributes. Depending on the application, new technological developments and new insights can (and must) this definition structure be broadened or even modified. The term reliability does justice to both the individual technological areas as well as the integration of individual areas - including the man-machine interaction. How far previous methods and procedures can be applied and/or joined, or must be newly developed, is considered relevant to research. The dependence of reliability, availability and safety of a unit of different influence magnitudes is signified in the definition "dependability". The all-encompassing view of reliability also means that previously conceptualized and built systems must, in certain intervals, undergo a reliability test (for example airplanes such as the Concorde, cable cars, or high speed trains), in order to sniff out safety risks. The human in the chain of operation can either be error-tolerant (for example through action based on experience) or can cause errors, which are often jointly responsible for catastrophies. The point is, to recognize these errors in reliability analyses and to prevent them. For example, the airplane and train catastrophies of recent times are always ascribed to "an unfortunate chain of events" – so read the official news, as a rule, which means, in plain English, a series of interdependent errors in which a human or humans was often involved. In the following segment, we go into more detail regarding some of these definitions of reliability.

---

[202] a. a. O.
[203] a. a. O.

22.2.1 Reliability

The reliability of R a system is the probability that a system will function during a specific time period within its specifications. If a system has a constant failure rate of λ failures/hour, then the reliability at time point t can be given as follows:

$$R(t) = e^{(-\lambda \cdot (t - t_0))}, \qquad (22.2)$$

where $(t - t_0)$ is given in hours. The reciprocal of the failure rate $1/\lambda$ = MTTF is the mean time until failure (Mean Time to Failure).

Reliability is a term which can be defined in different ways. The following segment will try to amplify the term „reliability".

Basic research and application research of reliability and maintenance of technical systems are not new, but nonetheless they have lost none of their actuality. Accidents with catastrophic consequences which have happened especially in recent years, outright demand an intensified attention to reliability and maintenance examinations of system structures as early as the project planning phase, and attention to the employment of special systems. The theory of reliability, a border area between the technical sciences and mathematics, even today commands extensive and verified knowledge and requires the application of methods from, among others, a variety of mathematical branches. In developing any system, one is actually concerned with how the system functions. The question as to why and how a system fails, the consequences of the failure, and the aspects of the production concepts, which influence failure probability, are usually not taught during education and training, mostly because one must first understand how a product works, before one can consider how it could fail. It is the developer's task to conceptualize and to service the product, so that a failure is delayed. Changeability and chance play an important role in the determination of the reliability of a system. The simplest principle of reliability is that in which a product is evaluated according to a specification, or according to a series of properties, and after passing the acceptance test is handed over to the customer. Once the customer accepts the product, he also accepts that a failure can occur after a certain period of time. This modus is often connected to a warranty or to a potential legal protection of the customer. He can demand a redress of the damages which have occurred within specified or reasonable time periods. This does, however, not yield a measure of the quality over a certain period of time, especially not following the expiration of the warranty period. Even within the warranty period, the customer can not make any further demands if the product fails once, twice, or several times, provided that the manufacturer repairs the product each time, as promised. More frequent failures will cause the manufacturer high warranty expenses, and the customer inconveniences. Outside of the warranty period, only the customer has to deal with trouble. In any case, the manufacturer loses his good reputation, which can have a negative impact on business. Reliability is a relatively new term, it originates from a new subject area which supplements quality control and is affiliated to quality science. In simple terms, reliability is the capacity of a material not to fail during use. It is often said that reliability corresponds to installation safety during a certain time period, but this definition is insufficient because it highlights the factor time, even though it does not clearly specify a measurable unit. Since reliability work was performed initially in the USA, the original American definition was also adopted at the onset: "Reliability is the probability that a unit does not fail during a defined period of time under given function and environmental conditions". Today, reliability of a component is defined as the ability of a unit under observation to perform the expected service under given conditions during a given period of time and following defined functional criteria. Since in German an immediate association of a probability to the term reliability is not possible, the IEC definition should be changed to: "A measure of reliability is, among others, the probability that a system performs satisfactorily under given conditions, during a given period of time".

According to DIN 40041, reliability is the ability of a unit under observation, within preset limits which are determined by the designated use, to satisfy requirements which are made on the behavior of its properties during a given period of time. Although this definition represents an information-rich capacity parameter, it has a disadvantage. For, on account of the necessity of specifying a given operational period for the unit under observation, the reliability of the unit under observation has a different value for each time interval. Therefore it is extremely useful to define different capacity parameters, which are not dependent on the functional period of time, such as, for example, the mean time between failures (MTBF) and the failure rate per hour (λ/[h]). Quoting a failure rate for a certain component though is pointless, unless the functional and environmental conditions which this failure rate refers to are given at the same time.

A unit under observation is understood as a random arrangement (component, module, system, device, installation) which is viewed as a unit within itself for the purpose of reliability examination.

22.2.2 Availability

The availability A of a technical system is the probability or the measure, that the system fulfills certain requirements within an agreed time frame, and is thereby a property of the system. It is a quality criteria of a system. Availability is a value between 0 and 1. In order to ensure a certain (minimum) availability of common central systems, known statistical error properties can be drawn on. Correspondingly, (sub)systems are installed redundantly, in order to minimize the probability of a total failure. Thereby the availability A is a function dependent on the time t. In systems with constant failure and repair rates, the following correlation exists between reliability (MTTF), serviceability (MTTR) and availability (A)

$$A = \frac{MTTF}{(MTTF + MTTR)} \qquad (22.3)$$

The operational period of a technical system can be limited due to regular maintenance and errors/damage, as well as repairs for their elimination. Availability is customarily given in percent. In computer systems, availability is measured as "duration of uptime UP per time unit" and is given in percent. The availability is then no longer indicated if the response time of a system exceeds a certain parameter. Typically, minutes, hours, days, months, quarters or years are used as time units. The reliability as a system property is thus documented in the contract (Service Level Agreement, SLA) between the system operator and the customer. Therein, the consequences of non-compliance of availability can be arranged. Depending on the agreement, availability has a major impact on the requirements regarding failure and serviceability of the system. A Service Level Agreement (SLA) is a service agreement between client and service provider, which configures recurring services for the client in control possibilities more transparently, such as the promised reaction time, scope, speed, and cost. It is characteristic for a SLA, that the service provider offers each relevant service parameter, unasked for, as different service levels, from which the client must choose, based on his economic aspects. With a classic service contract, the service provider does not offer these choices.

22.2.3 Safety

Safety S is the reliability with respect to critical failures. A critical failure is damaging, in contrast to a non-critical failure which can be labeled as good-natured. A critical failure can represent danger for material, environment or people. The costs of such a failure can be higher by several magnitudes than the normal operation costs of the system. Embedded systems find more and more entry even in safety-critical applications, such as for example ABS or airbag systems in vehicles. In many cases, a safety-critical system must be accepted through an independent certification agency.

22.2.4 Maintainability

Maintainability M is a measure of time which is necessary for bringing a system, following a failure, back into the „operative" state. Maintainability is measured through the probability M(t) that the system is operational again within the time t after the failure event.

22.3 Documentation of Failure Behavior

In order to materialize the failure behavior of components, there are a number of static procedures and functions which illustrate them. The following procedures:

- Density function $f(t)$, resp., failure density
- Failure probability, resp., distribution function $F(t)$
- Reliability $R(t)$, resp. survival probability and
- Failure rate $\lambda(t)$

are explained in more detail below.

22.3.1 Density Function, resp. Failure Density $f(t)$

For a simple graphic representation of the failure rates of components, the function of failure density is used. In general, failure density $f(t)$ can be calculated as follows:

$$f(t) = \frac{dF(t)}{dt} \tag{22.4}$$

Figure 22.1 shows the relative failures as a histogram of failure density. Figure 22.3.1 shows the approximation function of the failure density histogram.

22.3 Documentation of Failure Behavior

The x-coordinate is divided into intervals of time, so-called classes, in which the number of failures is determined. This ordinate value can be described as absolute or as relative frequency.

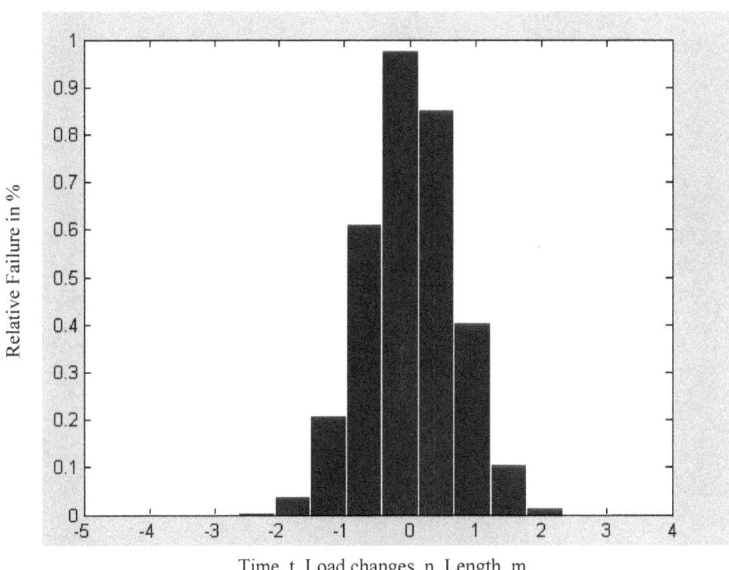

Figure 22.1: Failure density histogram

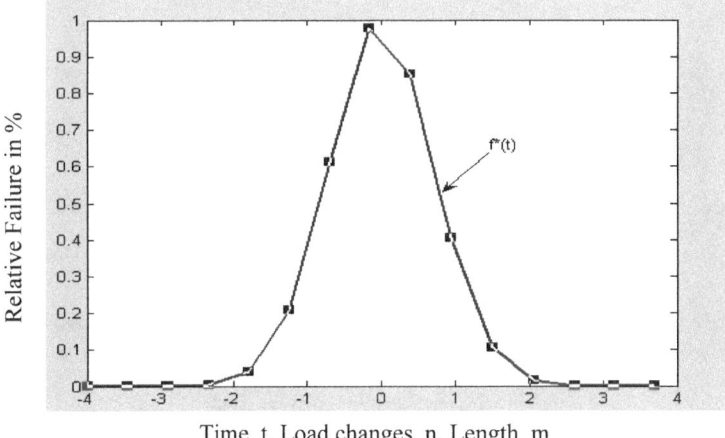

Figure 22.2: Approximation of failure density and density function $f^*(t)$

Different procedures exist for the mechanism of classification. For the transition limit of failures $n \rightarrow \infty$ the contour approaches the steady densitiy function $f(t)$.

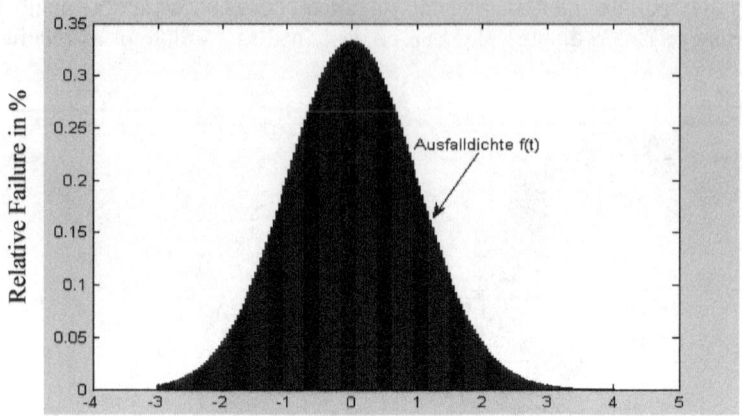

Figure 22.3: Failure density and density function f(t)

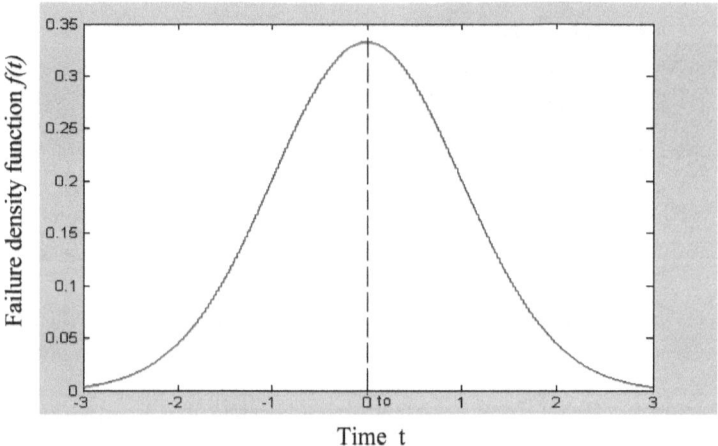

Figure 22.4: Failure density f(t) in Gaussian normal distribution

When using this density function one discusses (treats, approaches?) theoretical probabilities and leaves the experimentally determined densities behind. The transition limit $n \rightarrow \infty$ means, that all parts of a complete entity are tested, which is not realistic. On the basis of experimental, real data one can determine the empirical density function $f*(t)$ which deviates from the ideal density function $f(t)$. The formula for failure density $f(t)$ for the Gaussian normal distribution looks as follows

$$f(t) = \frac{1}{a} e^{-\frac{1}{b}(t-t_0)^2} \quad \text{with } a, b = \text{const.} \tag{22.5}$$

The corresponding graph in Figure 22.4 is almost identical to that in Figure 22.3.

22.3 Documentation of Failure Behavior

Next, the failure density $f(t)$ is examined in exponential form which calculates as follows

$$f(t) = a \cdot e^{-a \cdot t} \quad \text{with } a = \text{const.} \tag{22.6}$$

Figure 22.5 shows the density function in exponential form.

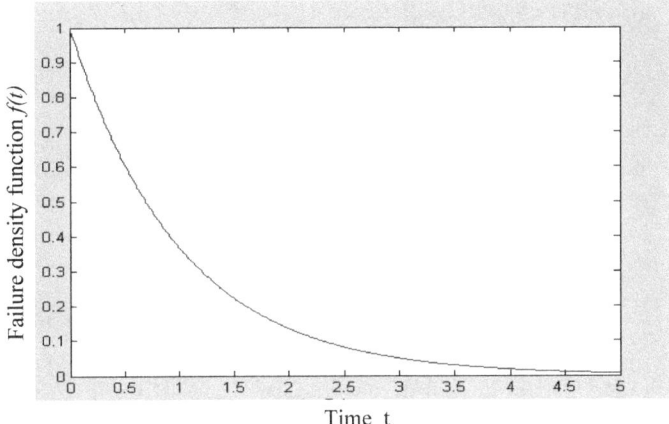

Figure 22.5: Density function in exponential form

Next, the Weibull distribution of this failure density is examined. The corresponding function looks as follows

$$f(t) = \frac{\beta}{\alpha} \cdot t^{(\beta-1)} e^{-(\frac{1}{\alpha}) \cdot t^{\beta}}, \text{ with } \alpha, \beta = \text{const.}, \beta = \text{form parameter} \tag{22.7}$$

22.3.2 Failure Probability, resp. Distribution Function F(t)

It can be seen, when using the cumulative frequency function, how many components in all have failed up to a certain time point. Herein, failures are cumulatively added with consecutive interval number. The result is an empirical distribution function $F^*(t)$. With the transition limit $n \to \infty$ one arrives again at the proper distribution function $F(t)$, which always begins at $F(t = 0) = 0$ and ends at $F(t \to \infty) = 1$ (Figure 22.6). The approximation of the failure probability histogram is shown in Figure 22.7.

The Gaussian distribution function results from the integral of the density function, or differently worded (restated), the density function results from derivation of the distribution function:

$$F(t) = \frac{1}{a} \int_{-\infty}^{t} e^{-\frac{1}{b}(t-t_0)^2} \cdot dt \quad \text{with } a, b = \text{const.} \tag{22.8}$$

From this follows the Gaussian distribution function of failure probability (see Figure 22.8).

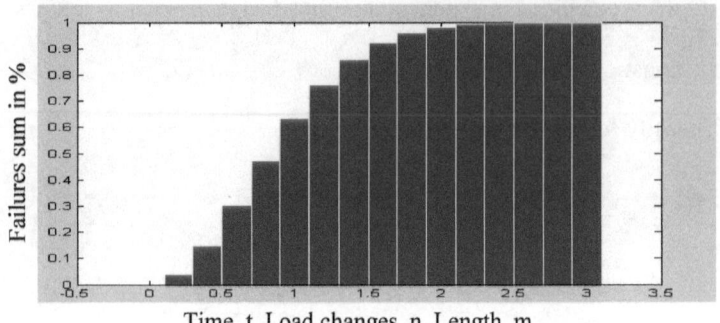

Figure 22.6: Failure histogram of failure probability

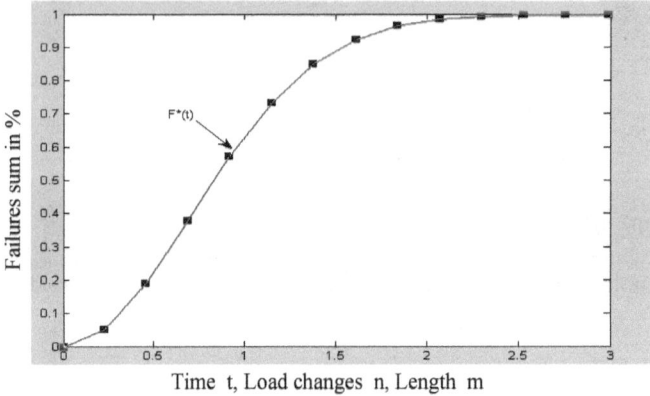

Figure 22.7: Approximation of the histogram to failure probability

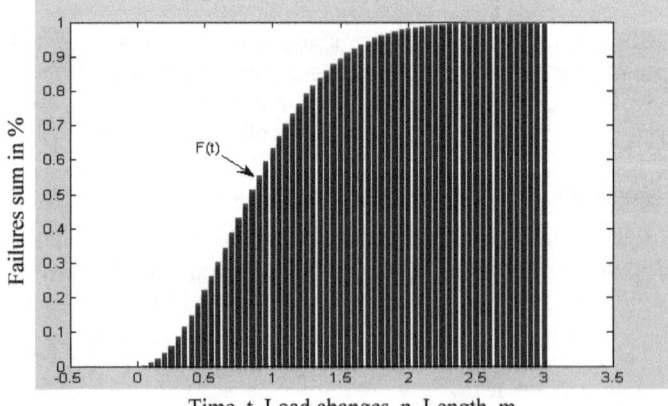

Figure 22.8: Gaussian distribution function of failure probability

22.3 Documentation of Failure Behavior

Now, we consider the failure probability in exponential form. Failure probability is calculated using the formula

$$F(t) = 1 - e^{-a \cdot t}, \text{ with } a = \text{const.} \tag{22.9}$$

Figure 22.9 shows the failure probability function in exponential form.

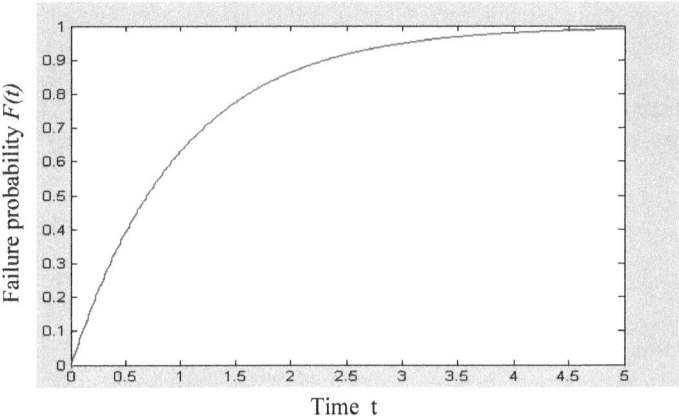

Figure 22.9: Failure probability function in exponential form

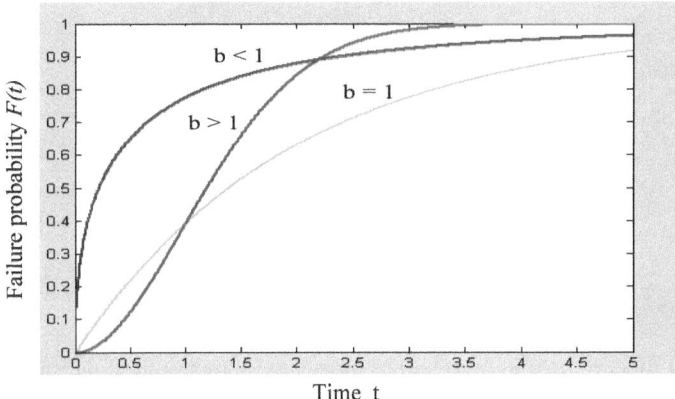

Figure 22.10: Failure probability function in Weibull distribution

Next, we examine the Weibull distribution of this failure probability. Figure 22.10 shows the failure probability $F(t)$ with a Weibull distribution. The function looks as follows:

$$F(t) = 1 - e^{-(\frac{1}{\alpha}) \cdot t^{\beta}}, \text{ with } \alpha, \beta = \text{const.}, \beta = \text{form parameter} \tag{22.10}$$

The distribution function $F(t)$ is called failure probability. It describes the sum of failures as a function of time.

22.3.3 Reliability, resp. Survival Probability $R(t)$

The survival probability describes the sum of operational units as a function of time. It can be expressed as a function of survival frequency. At a certain point in time t, the addition of the two cumulative frequencies, failures, and intact units, always adds up to 100%. Thereby, the survival probability is the complement of failure probability and is called reliability $R(t)$. It always starts with $R(t=0) = 100\%$, decreases monotonically and ends with $R(t \to \infty) = 0\%$, where all units have failed (Figure 22.11). The corresponding function is described as follows.

$$R(t) = 1 - \frac{1}{a}\int_{-\infty}^{t} e^{-\frac{1}{b}(t-t_0)^2} \cdot dt = 1 - F(t), \text{ with } a, b = \text{const.} \qquad (22.11)$$

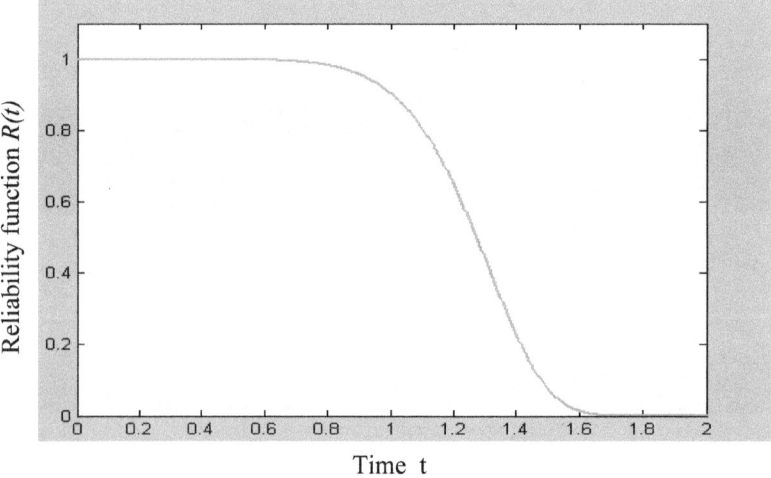

Figure 22.11: Reliability function $R(t)$ as Gaussian distribution function

To examine the reliability function $R(f)$ in exponential form, the following formula is used

$$R(t) = e^{-a \cdot t}, \text{ with } \alpha = \text{const.} \qquad (22.12)$$

Figure 22.12 shows the reliability function $R(f)$ in exponential form.

22.3 Documentation of Failure Behavior

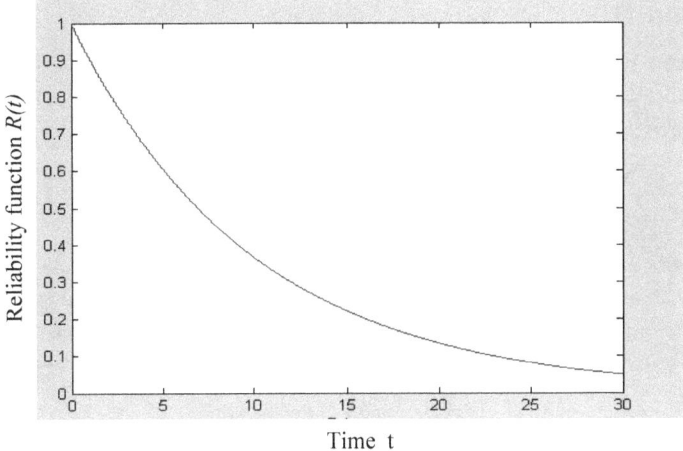

Figure 22.12: Reliability function $R(t)$ in exponential form

Next, we discuss the Weibull distribution of this reliability function. Figure 22.13 shows the reliability function $R(t)$ in Weibull distribution. The function looks as follows.

$$R(t) = e^{-(\frac{1}{a}) \cdot t^{\beta}}, \text{ with } \alpha, \beta = \text{constant}, \beta = \text{form parameter} \qquad (22.13)$$

Thereby, the qualitative definition of reliability can be quantified as:

Reliability is perceived as the probability that a unit, a component, or a system, does not fail during a defined period of time under given functional and environmental conditions.

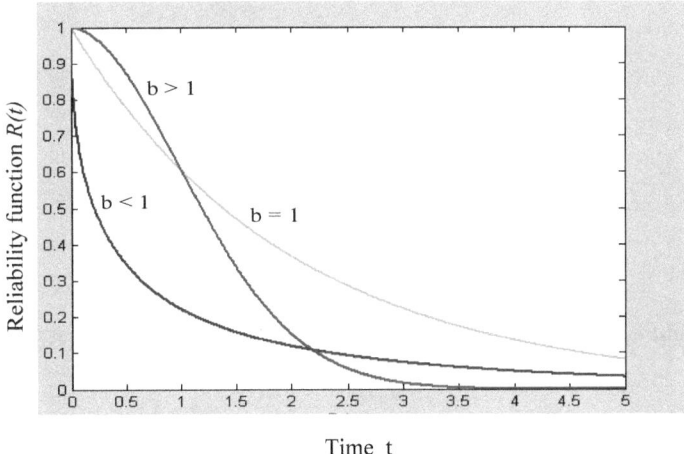

Figure 22.13: Reliability function $R(t)$ in Weibull distribution

22.3.4 Failure rate λ(t)

To describe the failure rate by means of a rate, the failures during a specified time t are referred to the sum of all still intact units. The failure rate can also be described as the quotient of failure density $f(t)$ and survival probability, resp. reliability:

$$\lambda(t) = \frac{\text{failure at time point t}}{\text{sum of faultless units at time point t}} = \frac{f(t)}{R(t)} \qquad (22.14)$$

The failure rate thus describes the risk of a unit, which is functional until time point t, to fail. With the failure rate an attempt is made to capture the entire failure behavior of a unit or of a system. Since the failure rate $\lambda(t)$ represents the quotient of the failure density $f(t)$ and the reliability function $R(f)$, the Gaussian normal distribution of this function $\lambda(t)$ looks as follows:

$$\lambda(t) = \frac{f(t)}{R(t)} = \frac{\frac{1}{a} \cdot e^{-\frac{1}{b}(t-t_0)^2}}{1 - \frac{1}{a} \int_{-\infty}^{t} e^{-\frac{1}{b}(t-t_0)^2} \cdot dt}, \text{ with } a, b = \text{const.} \qquad (22.15)$$

The Gaussian normal distribution of this function is shown in Figure 22.14.

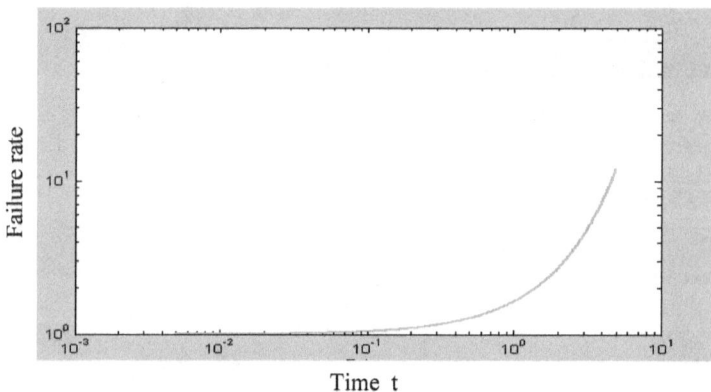

Figure 22.14: Failure rate in Gaussian normal distribution

Next, we examine the failure rate $\lambda(t)$ in exponential form. This is calculated according to the formula

$$\lambda(t) = a, \text{ with } a = \text{const.} \qquad (22.16)$$

Figure 22.15 shows the function of failure rate $\lambda(t)$ in exponential distribution.

22.3 Documentation of Failure Behavior

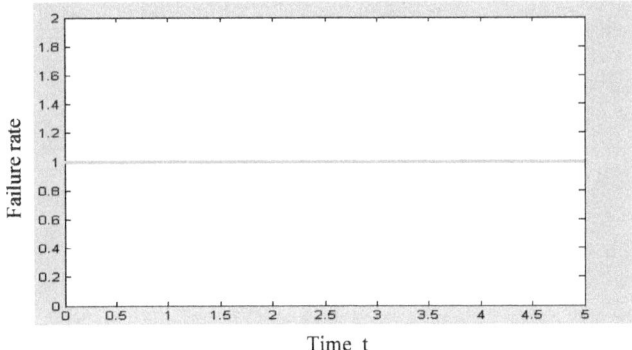

Figure 22.15: Failure rate λ(t) in exponential distribution

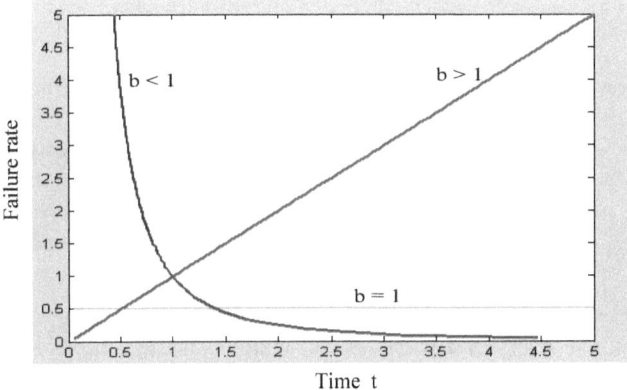

Figure 22.16: Failure rate λ(t) in Weibull distribution

Next, we examine the Weibull distribution of the function of failure rate. Figure 22.16 shows the failure rate $\lambda(t)$ in Weibull distribution. The formula of the function looks as follows

$$\lambda(t) = \frac{\beta}{a} \cdot t^{(\beta-1)}, \text{ with } \beta, a = \text{const.}, \beta = \text{form parameter} \quad (22.17)$$

The failure rate $\lambda(t)$ must be experimentally determined in most cases. Very often – especially with electronic components – the failure rate can be subdivided into three phases:

- Early failures with decreasing failure rate
- Accidental failures with constant failure rate
- Wearout failures with increasing failure rate.

This behavior is pictured in the so-called bathtub curve, see Figure 22.17.

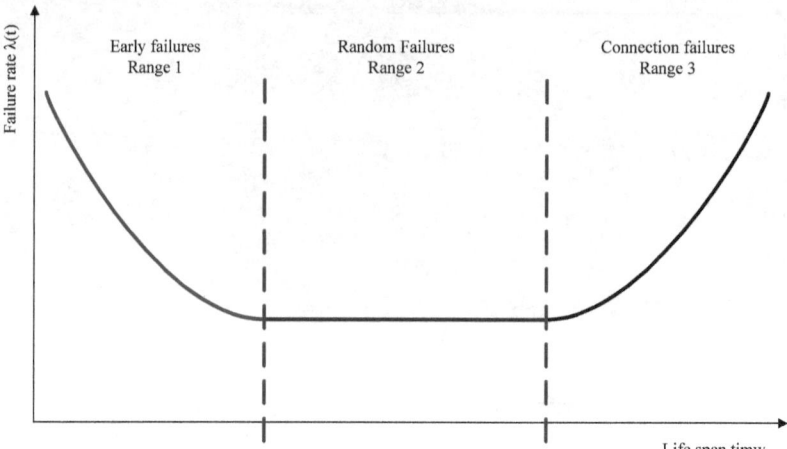

Figure 22.17: Bathtub curve

In the segment of early failures (segment 1), the failure rate steadily decreases with increasing life span. Early failures are caused by defective mounting, construction errors, material and fabrication flaws, and they are characterized by a relatively steeply decreasing failure rate during the startup phase. It can be eliminated by a „burn-in" (burning in, artificial aging effect), by test procedures during and after the manufacture, as well as by running in of all reliability-defining devices and components (for semiconductors about 168 hours at 125 °C).

Accidental failures are in the segment of constant failure rate (segment 2). They are, among others, caused by handling errors, dirt particles and maintenance errors, and are characterized by a constant failure rate λ. The assumption is made that the component was installed according to regulations, and that no outside overuse occurs. This time period describes the normal behavior of the components and provides the basis for the calculation of all reliability parameters.

The segment of wearout and fatigue failures (segment 3) begins with a steeply increasing failure rate. This phase is characterized by the fact that the failure rate increases again due to aging or wearout phenomena. It is synonymous with the end of the operational period.

Corresponding to the different failure causes in the various segments, different measures can be applied to increase reliability. Albeit, only segment 3 can be influenced with a constructive system design which can also captured by way of calculation.

22.3.5 Description of Failure Behavior by Examples

As was briefly shown above, mathematical distribution theories are used to describe failure behavior. We will discuss the most important distribution theories below and demonstrate some function examples.

The well-known bell curve (Gaussian normal distribution) is very rarely used in practice, because only one failure type can be described with its symmetric course. Exponential distribution is frequently used in electrical engineering. Essential characteristics are a

22.3 Documentation of Failure Behavior

monotonously decreasing density function and a constant failure rate λ as the sole parameter (Figure 22.18). This behavior is characteristic of electronic components. The formulas are listed below:

$f(t) = \lambda \cdot e^{-\lambda \cdot t}$ Density function (22.18)

$F(t) = 1 - e^{-\lambda \cdot t}$ Failure probability (22.19)

$R(t) = e^{-\lambda \cdot t}$ Reliability (22.20)

$\lambda = konst.$ Failure rate (22.21)

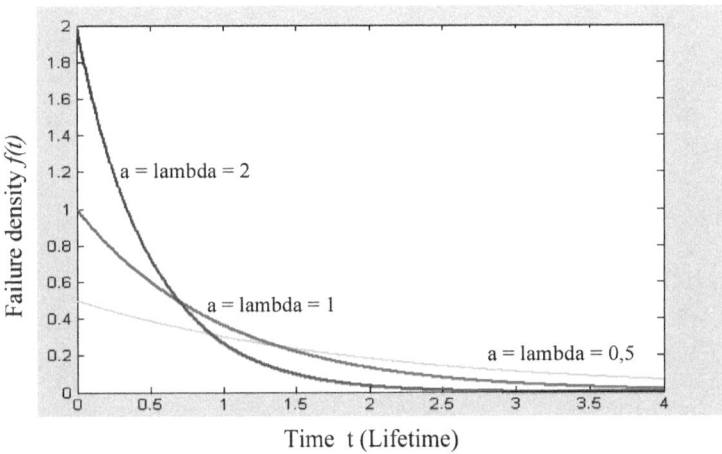

Figure 22.18: Failure density of exponential distribution

For the reciprocal of failure rate, the terms *MTTF* (Mean Time to Failure), resp. *MTBF* (Mean Time Between Failure), are discussed in more detail below.

Weibull distribution allows the most disparate failure types to be represented. The Weibull distribution can be applied as a two-parameter (characteristic life span T and form parameter b) or as three-parameter (additional failure-free time t_0) distribution. The form parameter b is a measure for the scattering of failure times and for the form of failure density. The characteristic life span T is the location parameter for the density function. With three-parameter distribution, the failure-free time t_0 determines at which point in time the failures begin. Thereby the curve is shifted by t_0 along the positive time axis. With two-parameter distribution, failures always start at $t = 0$. Thus, it represents a special case of the three-dimensional Weibull distribution with $t = 0$. The density function changes depending on the form parameter b. For values of $b < 1$, failures are described similar to the exponential distribution, for $b = 1$ the exponential distribution applies exactly. For values of $b > 1$, the density function starts with $f(t=0) = 0$, reaches a maximum with increasing life span, and then falls flat. For $b = 3.5$, the normal distribution is shown as an approximation. The formulas of the three-parameter Weibull distribution are given below, and those of the two-parameter distribution can be deduced directly from them by setting $t_0 = 0$.

Reliability:

$$R(t) = e^{-\left(\frac{t-t_0}{T-t_0}\right)^b} \tag{22.22}$$

Failure probability:

$$F(t) = 1 - e^{-\left(\frac{t-t_0}{T-t_0}\right)^b} \tag{22.23}$$

Density function:

$$f(t) = \frac{dF(t)}{dt} = \frac{b}{T-t_0} \cdot \left(\frac{t-t_0}{T-t_0}\right) \cdot e^{-\left(\frac{t-t_0}{T-t_0}\right)^b} \tag{22.24}$$

Failure rate:

$$\lambda(t) = \frac{f(t)}{R(t)} = \frac{b}{T-t_0} \cdot \left(\frac{t-t_0}{T-t_0}\right)^{b-1} \tag{22.25}$$

Conditions:

$t \geq t_0 \geq 0, T \geq t_0, b \geq 0$.

The failure and survival probability of the two-parameter Weibull distribution with a characteristic life span of $T = 1$ for different form parameters b is shown in Figures 22.19 and 22.20.

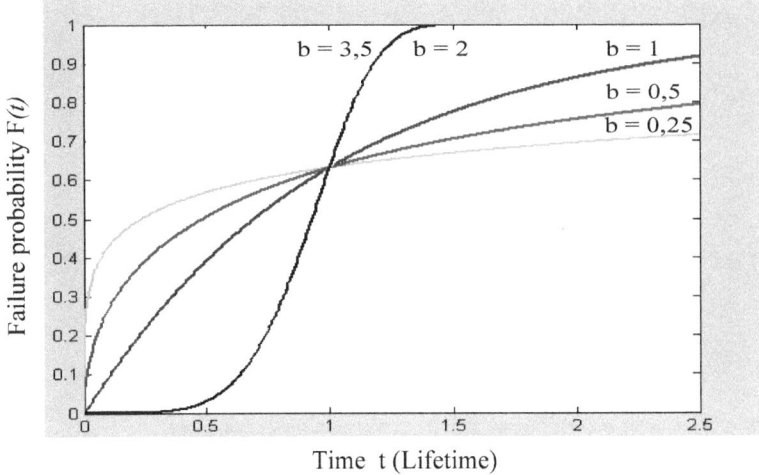

Figure 22.19: Failure probability of the Weibull distribution (T = 1, t₀ = 0)

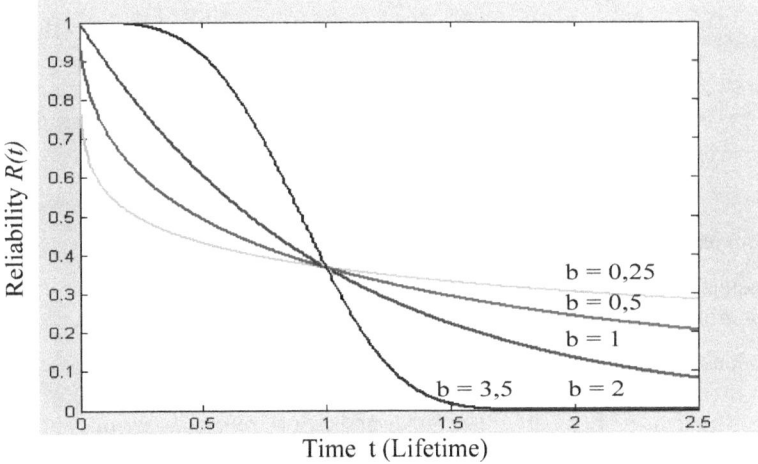

Figure 22.20: Reliability probability of the Weibull distribution (T = 1, t₀ = 0)

With varying values of the failure rate λ of the Weibull distribution, the three sections of the bathtub curve (see Figure 22.17) can be represented in correlation to the form parameter b. Early failures can be described for $b < 1$, because the failure rate decreases with increasing life span (section 1). For $b = 1$, the failure rate is constant and accidental failures can be shown (section 2). Wear out and aging failures (section 3) can be described with $b > 1$, because the failure rate increases distinctly with increasing life span. The failure rate, as well as the density function, for different values of the form parameter b are shown in Figure 22.21.

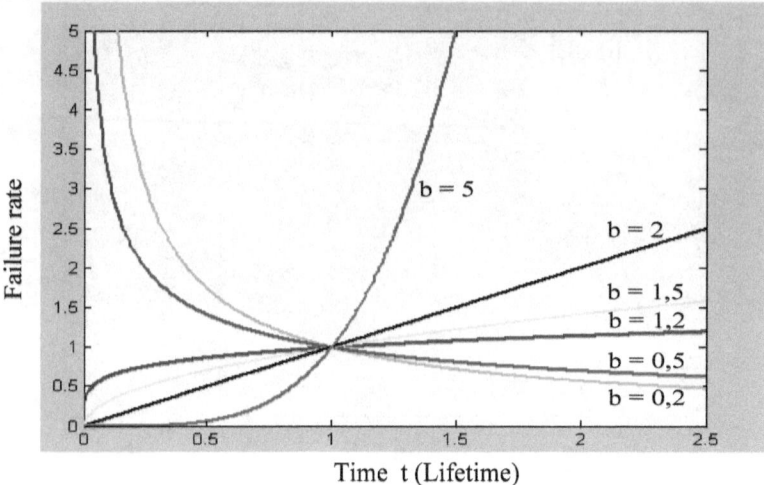

Figure 22.21: Failure rate of the Weibull distribution (T = 1, t₀ = 0)

22.3.6 Boolean Theory

The system failure behavior can be calculated by means of the Boolean model, proceeding from the failure behavior of individual components. The following restrictions apply to the application of this system theory:

- The system is „not repairable", that is, the first system failure ends the system lifecycle.
- The system elements can only exist in a state of „functional" or „failed",
- The system elements are „independent", that is, the failure behavior of a structural component is not influenced by the failure behavior of other structural components.

The reliability structure of this system is recognizable in the so-called reliability schematics, in which the effects of one component (a system component) on the whole system become apparent. The connections between entry E and exit A represent the possibilities of functionality of the system. For example, in a serial circuit, as shown in Figure 22.22, the failure of any one component leads to the failure of the entire system.

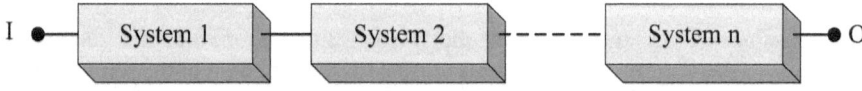

Figure 22.22: Clear serial structure of the Boolean model

The system can only function if all components are functional. The system reliability is calculated for n components according to the relationship

$$R_s(t) = R_1(t) \cdot R_2(t) \cdot \ldots \cdot R_n(t) = \prod_{i=1}^{n} R_i(t) \qquad (22.26)$$

22.4 Time Factor

In a parallel structure, the system only fails if all components fail.

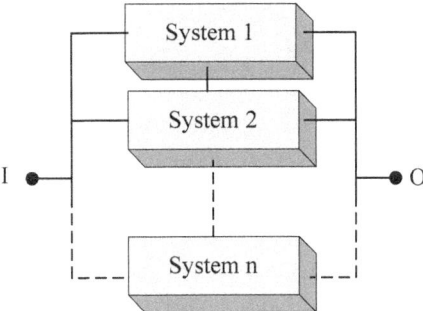

Figure 22.23: Clear parallel structure of the Boolean model

The relationship for the calculation of reliability of a clear parallel system with n individual components is shown in equation 22.26.

$$R_s(t) = 1 - (1 - R_1(t)) \cdot (1 - R_2(t)) \cdot \ldots \cdot (1 - R_n(t)) = 1 - \prod_{i=1}^{n}(1 - R_i(t)) \quad (22.27)$$

For reliability analysis of mixed forms, that is, combined or cross-linked structures, the system reliability can be calculated by stepwise application of the relationships of serial and parallel structure. Therein, substitute values are used. Further solution possibilities are the application of the disjunctive normal order, or of the minimal intersections, resp., success paths.

22.4 Time Factor

While for irreparable systems the reliability or the failure probability is considered, for repairable systems the availability or non-availability is examined. With endless repair times, availability corresponds to reliability. The Markov model determines the availability in reference to runtime. In the Markov method, repairable systems can be examined. Thereby, only those systems can be discussed whose system elements have constant failure and repair rates. The following terms and deduced formula symbols are applied in the Markov method:

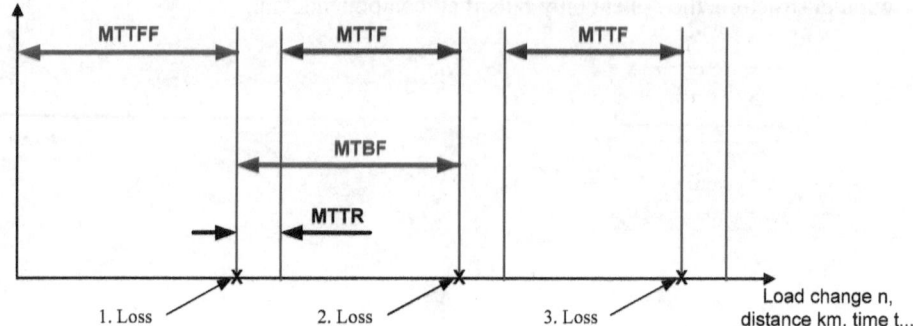

Figure 22.24: The terminology of time

MTTFF: Mean Time To First Failure. The mean lifetime until the first failure in repairable systems is called *MTTFF*.

MTTR: Mean Time To Repair. The mean time to repair, respectively, maintenance is called *MTTR*.

MTBF: Mean Time Between Failure. The mean value of failure-free times, or also mean failure distance, is called *MTBF*.

MTTF: Mean Time To Failure. The mean value of lifetime from the end of the repair time to the next failure is called *MTTF*.

These terms are explained in more detail below.

22.4.1 MTTF

MTTF stands for „Mean Time to Failure" and is the expected time to the first malfunction of a component of an instrument. It is a statistical value and is examined over a very long period of time and with a large number of units. In systems with constant error rate, the *MTTF* value is the reciprocal of the error rate. The error rate is given in errors/million hours. Thereby, an exponential distribution of errors is assumed. Technically, *MTBF* should only be used in connection with repairable instruments, whereas *MTTF* should only be used for irreparable instruments. Normally, *MTBF* is used though both for repairable as well as for irreparable instruments.

22.4.2 MTTFspurious

MTTFspurious stands for the mean estimated time which passes between two safe errors of components or of the system, which were caused accidentally.

22.4.3 MTBF

MTBF stands for „Mean Time Between Failures" and is the expected period of time between errors. It is measured in hours and is a statistical value. It is examined over a long period of time and with a large number of units. In systems with constant error rate, the

MTBF value is the reciprocal of the error rate. Technically, MTBF should only be used in connection with repairable instruments, whereas MTTF should only be used for irreparable instruments. Normally though, MTBF is used both for repairable instruments as well as for irreparable instruments. Figure 22.24 and equation 22.27 demonstrate the relationship between MTBF, MTTF, and MTTR.

$$MTBF = MTTF + MTTR \qquad (22.28)$$

22.4.4 MTTR

MTTR, „Mean Time To Repair", is the entire time needed for the performance of all maintenance repairs, divided by the total number of these repairs.

22.4.5 Example for Calculation of MTTF

With the given Weibull parameters b, T, t_0 and MTTR, the mean value of operating time MTTF can be calculated:

$$MTTF = E(t) = \int_0^\infty t \cdot f(t) \cdot dt = \int_0^\infty R(t) \cdot dt \qquad (22.29)$$

In three-parameter Weibull distribution according to equation 22.28 of the density function $f(t)$ with

$$f(t) = \frac{dF(t)}{dt} = \frac{b}{T-t_0} \cdot \left(\frac{t-t_0}{T-t_0}\right)^{b-1} \cdot e^{\left(\frac{t-t_0}{T-t_0}\right)^b} \qquad (22.30)$$

the expectancy value MTTF results, after a few conversions, in:

$$MTTF = E(t) = (T-t_0) \cdot \Gamma(1+\frac{1}{b}) + t_0, \qquad (22.31)$$

with Γ = Eulersche Gammafunktion

22.4.6 Continuous Availability

The continuous availability A_D results from the condition differential equations. Each element can have only two conditions, „functional" or „failed":

- Condition Z_0: The element is functional and in operative mode,
- Condition Z_1: The element has failed and is in repair mode.

The corresponding condition probabilities are labeled $P_0(t)$ and $P_1(t)$. The elements transform, with a certain transition probability, from one condition into the next. The sum of transition probabilities of an arrow leaving from a condition always has a value of 1:

$$P_0(t) + P_1(t) = 1 \qquad (22.32)$$

This prerequisite is called norming condition. At time $t = 0$, the observed element is functional and in mint condition. The starting conditions are:

$$P_0(t=0) = 0 \text{ and } P_1(t=0) = 1 \qquad (22.33)$$

The change of condition probabilities is constructed, in addition, from the transition probabilities, which result from multiplication of the condition probabilities and (with?) the corresponding transition rates. Thereby, arrows departing from a condition become negative, whereas arrows pointing towards a condition are shown with positive values (Figure 22.25). In the Markov graph, the transition rates of Z_0 to Z_1 are given as failure rate λ, and of Z_1 to Z_0 as repair rate μ.

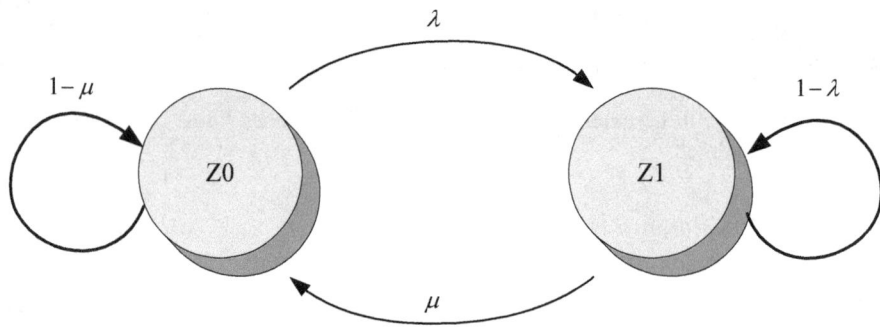

Figure 22.25: Markov model for a single element

In the case of the single element observed here, equations (22.33) and (22.34) result.

$$\frac{dP_0(t)}{dt} = -\lambda \cdot P_0(t) + \mu \cdot P_1(t), \qquad (22.34)$$

$$\frac{dP_1(t)}{dt} = -\mu \cdot P_1(t) + \lambda \cdot P_0(t). \qquad (22.35)$$

From these equations, the norming condition according to equation (22.31) and according to the starting conditions from equation (22.32), it follows for $P_0(t)$:

$$\frac{dP_0(t)}{dt} = -\lambda \cdot P_0(t) + \mu \cdot (1 - P_{01}(t)) \quad \Rightarrow \quad \frac{dP_0(t)}{dt} + (\lambda + \mu) \cdot P_0(t) = \mu. \qquad (22.36)$$

The solution of this inhomogeneous differential equation consists of a homogeneous and a particular part:

$$P_0(t) = P_{0H}(t) + P_{0P}(t). \qquad (22.37)$$

By separation of the variables, the homogeneous solution results:

$$\int \frac{dP_0(t)}{dt} = -\int (\lambda + \mu) dt \quad \Rightarrow \quad P_{0H}(t) = C \cdot e^{-(\lambda+\mu) \cdot t}. \qquad (22.38)$$

22.4 Time Factor

The particular solution results from a variation of the constants. The addition of the two parts of the solution results in the following equation for $P_0(t)$:

$$P_0 = \frac{\mu}{\mu + \lambda} + \frac{\lambda}{\mu + \lambda} \cdot e^{-(\lambda + \mu) \cdot t}. \tag{22.39}$$

Through the norming condition equation (22.31), the condition probability $P_1(t)$ is obtained. The condition probability $P_0(t)$, where the element resides in a functional condition at time point t, corresponds to the availability $A(f)$. Correspondingly, the non-availability $U(f)$ is defined as the complement of availability, $P_1(t)$:

$$A(t) = P_0(t), \quad U(t) = 1-A(t) = P_1(t). \tag{22.40}$$

The availability $A(f)$ converges for $t \to \infty$ towards the limiting value of the stationary solution. This value is called permanent availability A_D and is usually expressed by the previously introduced parameters (see Figure 22.24):

$$\text{Meantime of operation: } MTTF = \frac{1}{\lambda}, \tag{22.41}$$

$$\text{Meantime of repair: } MTTR = \frac{1}{\mu}. \tag{22.42}$$

The permanent availability A_D is therefore:

$$A_D = \lim_{t \to \infty} A(t) = \frac{\mu}{\lambda + \mu} = \frac{MTTF}{MTTF + MTTR} = \frac{1}{1 + \frac{MTTR}{MTTF}}. \tag{22.43}$$

It is only dependent on the quotient $MTTR/MTTF$. The bigger it gets, the more will permanent availability increase.

22.4.7 Downtime DT

Downtime *DT* is an expression that is used to characterize the time in which a computer system is not available, resp. not operational. Other terms for this would be failure time, even standstill, repair time, duration of disturbance, or cut-off time. One distinguishes between planned and unplanned downtime. A planned downtime is characterized by a directed and scheduled deactivation of the computer system. Downtime is often scheduled for the night, and users are informed of it in advance. Often, alternative systems are available. The planned downtime is used, for example, for hardware upgrades or moves. An unplanned downtime is a suddenly occurring disruption, such as a hard drive head-crash, or overheating. The unplanned downtime can lead to severe financial losses, because often the work of several days or weeks is destroyed. Backups are made as a means of countermeasure. In computer systems, the availability is measured in „length of downtime per year" and is given in percent. A system with 99% availability can fail a total of 3.6 times during one year, a system with 99.99% availability 52 minutes, and a system with 99.999% availability only 5 minutes.

22.4.8 Uptime *UT*

Uptime *UT* is an expression for the time in which a computer system runs and is operational. It can also be described as available operational time. Uptime is given counting back from the last system startup. Uptime is an indicator of the stability and safety of an operational system and its hardware. It determines the actual availability of the system. Computer service providers often advertise with uptime warranties, in order to attract customers. Thus, a 99.99% uptime corresponds to approximately 52 minutes of downtime in one year.

$$DT = \frac{t}{100\%} \cdot (100\% - UT)[h] \tag{22.44}$$

22.4.9 Mean Down Time *MDT*

MDT describes the mean time during which the system is not available. This comprises the time needed for determining the disruption, the time needed for mobilizing the service contractor, the time needed for procuring the defective material, and the repair time. Figure 22.26 shows the correlation between the individual times.

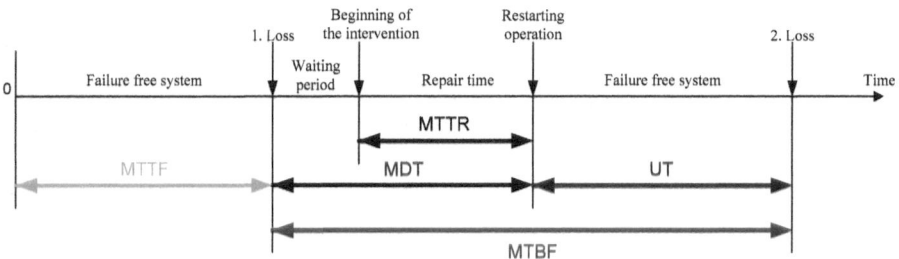

Figure 22.26: Relationship between the times

22.5 General Thoughts About Terms and Standards

Terms such as *PDF* (downtime/failure probability), *DC* (degree of diagnostic coverage), *CCF* (common cause failures), *MTBF* (mean time between failures), *MTTF* (mean time to failure/failure from an unknown cause), and *SIL* (safety integrity level, safety-related requirement level to the average probability of a failure) have now been integrated into the two standards and set new aspects. Thereby measures such as redundancy, serial connection of safety contacts, forced separation, cyclic test, short circuit recognition, etc. can be evaluated and classified according to their effectiveness and lifetime. For, what good is a test if it takes place at the wrong point in time or the test depth does not allow it to recognize errors? Apart from that, safety systems that are expected to fail shortly after implementation can now be classified accordingly. There are many arguments which speak for the IEC 61508. Even so, the IEC 61508 is not meant for direct application, but rather is designed as a template (model) for the creation of safety standards. Therefore, a variety of sector standards are currently emerging, for example the IEC 61511 is designed for the

22.5 General Thoughts About Terms and Standards

process engineering sector, the IEC 61513 for the power plant sector, and the IEC 62061 for the machine sector. This „accumulated load" (cumulative charge, load) of functional safety has just a small disadvantage: the IEC 61508 and its sector standards have become too academic, a fact that does not make their implementation any easier. Life is breathed back into regulation EN 954-1. The IEC 61508 has not succeeded in supplanting the EN 954-1, whose critics have always found fault with the missing probabilistic view of safety. As prEN ISO 13849-1 it now takes into account all safety aspects which it had been lacking until now, namely in a form that is applicable for mechanical engineers. It is a non-electric standard created under ISO (International Organization for Standardization) and DKE (Deutsche Kommission Elektrotechnik Elektronik Informationstechnik im DIN und VDE). The prEN ISO 13849-1 has, in contrast to all these other norms, an applicable concept and, in contrast to the IEC norms, generally includes even non-electric hazards, like those found in pneumatics, hydraulics and mechanics. Once one is familiar with the procedure, the advantages of this norm are readily apparent. Users must adjust themselves especially upon achieving a risk category, because now the performance level PL is the goal. The categories still continue to exist though. As before, they determine the structure of the safety circles, yet they can be freely chosen. The higher the PFD value, the higher the probability of a dangerous failure. The diagnostic coverage degree DC is the partial reduction of the probability of dangerous hardware failures, on the basis of the application of automatic diagnostic tests. The diagnostic coverage degree is calculated with the following equation:

$$DC = \frac{\sum \lambda_{DD}}{\sum \lambda_{DTotal}} \qquad (22.45)$$

λ_{DD} is about the probability of recognized (known?), hazardous failures, the λ_{DTotal} is about the probability of all dangerous failures. The diagnostic coverage degree is subdivided into layers. To be accurate, it concerns the depth of testing and the detection of errors. There are errors which are relatively easy to detect, and errors which can only be found with great effort. Furthermore, errors are possible which are harmless in their effect and can actually not be classified as errors. In conjunction with other „harmless" errors though they can lead to a dangerous condition. Additionally, practical experience has shown that even redundant systems can have early dangerous failures. As a rule, the causes are not systematic errors but early failures. These are usually not traced back to wearout or excessive age, but rather to a manufacturing error, in which case all components with the same production date and the same production process are concerned. This is a case of an error of common cause. If this is not recognized and if such components are applied in safety circuits, then redundancy-caused duplication does not change the failure probability fundamentally. For, why should a defective component in conduit A be more reliable than in the identically constructed conduit B (Figure 22.27). There really is only one measure of controlling failures of common cause: diversity. It can be achieved by redundancy with differing components, techniques, or principles. Attention must be paid to the fact that both conduits have approximately the same quality. It is a fact that every system fails if it is operated long enough, or if it is excessively complicated and extravagantly designed. The time to the dangerous failure is identified with the value $MTTF$. The more components are present, the more can fail.

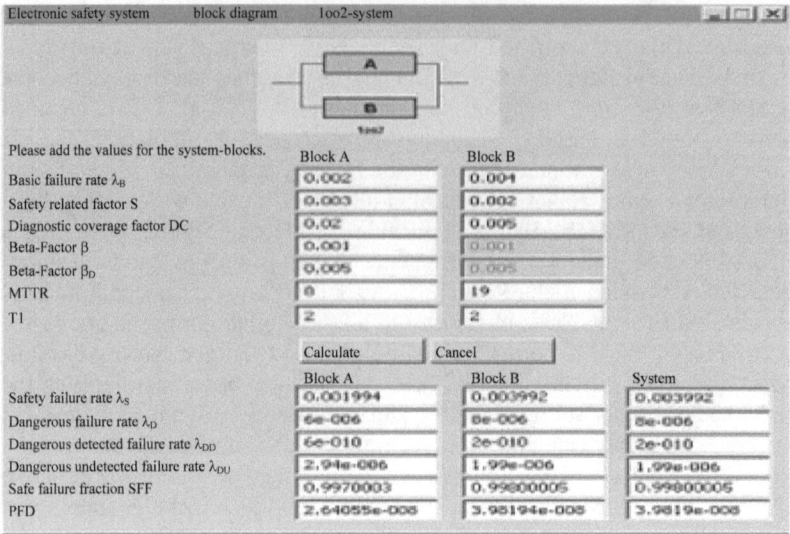

Figure 22.27: Redundant 1oo2-system

One can even imply that, in the course of its life phases, a system can change from an extremely safe system to an unsafe, dangerous system. There are two possibilities of counteracting this phenomenon:

- constructing the system qualitatively of such high quality that the dangerous failure only occurs after the end of the operational period, or
- testing the system regularly in order to ensure that the required functions are still safeguarded for the next demand.

The *MTTF* is approximately the time that passes until about 63 % of the applied components have failed. Some of these terms are described in more detail below.

22.5.1 Degree of Diagnostic Coverage *DC*

The *DC* factor (Diagnostic Coverage Factor) is a measure for how good a safety-relevant system detects an error. To determine the mean value of the entire system, there is also the mean degree of diagnostic coverage DC_{avg}. The mean degree of diagnostic coverage DC_{avg} is subdivided into four classes, that is, the classes „none", „low", „medium", and „high". The value range of the classes is shown in Table 22.1, according to ISO 13849-1.

22.5 General Thoughts About Terms and Standards

Table 22.1: Degree of diagnostic coverage DC factor

Label	Value range diagnostic coverage DC
none	$DC < 60\%$
low	$60\% \leq DC < 90\%$
medium	$90\% \leq DC < 99\%$
high	$99\% \leq DC$

Example:

There are n components with n different *MTTF*- and *DC*-values (with $n = 1, 2, \ldots 3$). The mean degree of diagnostic coverage DC_{avg} is:

$$DC_{avg} = \frac{\dfrac{DC_1}{MTTF_1} + \dfrac{DC_2}{MTTF_2} + \ldots \dfrac{DC_n}{MTTF_n}}{\dfrac{1}{MTTF_1} + \dfrac{1}{MTTF_2} \ldots \dfrac{1}{MTTF_n}} \qquad (22.46)$$

With:

- $MTTF_1$ = Mean value of the operational period without a dangerous error of the first component
- $MTTF_2$ = Mean value of the operational period without a dangerous failure of the second component
- $MTTF_n$ = Mean value of the operational period without a dangerous failure of the nth component
- DC_1 = Degree of diagnostic coverage of the first component
- DC_2 = Degree of diagnostic coverage of the second component
- DC_n = Degree of diagnostic coverage of the nth component

22.5.2 Common Cause Failure CCF

Common Cause Failure CCF describes the failure which is the result of one or more events that cause the simultaneous failure of one or more separate conduits in a multiconduit system and lead to a system failure. This failure rate is weighted by the β-factor in the PFD_{avg}-determination (Average Probability of Failure on Demand) of a multiconduit system. PFD_{avg} stands for the average probability of a safety function error on demand.

In general, the following can be written for the calculation of an error f with common cause

$$f = \lambda_D \cdot \beta \qquad (22.47)$$

λ_D stands for the failure rate of dangerous errors (D = dangerous) and β for β-factor for the common error.

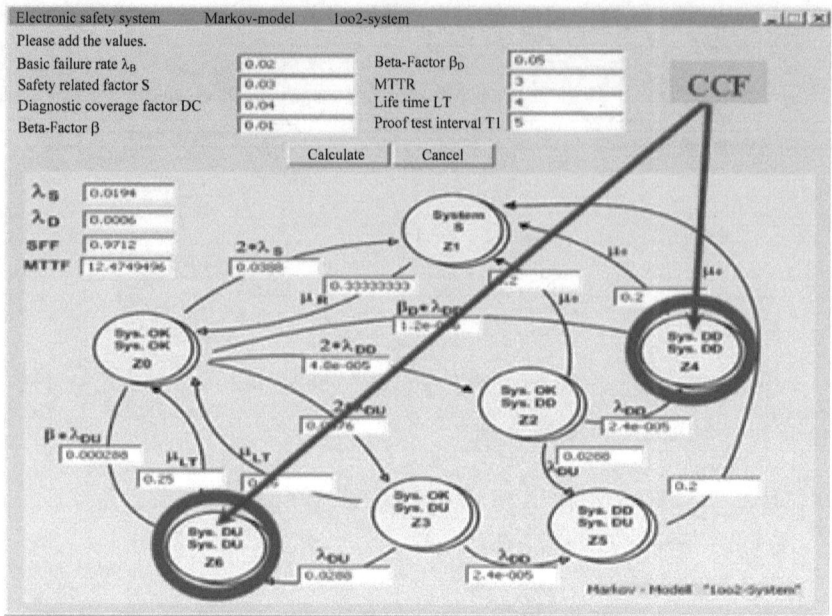

Figure 22.28: Common Cause Failure of a 1oo2-Markov system

One distinguishes between two β-factors, the β-factor for the dangerous, undetected error λ_{DU} (DU = dangerous undetected) condition Z6 in the Markov model of the 1oo2-system, and the β_D-factor for the dangerous, detected λ_{DD} (DD = dangerous detected) condition Z4 in the Markov model of the 1oo2-system (Figure 22.28). Dangerous, undetected errors λ_{DU} are not recognized by the diagnostic tests, whereas the dangerous, detected errors λ_{DD} are diagnosed.

22.5.3 Probability of Failure on Demand PFD

PFD stands for the probability of an error on demand of a safety function. This is the probability that a system fails in the moment it is needed. The smaller the *PFD*-value, the better the system.

PFD$_{avg}$ specifies the average probability with which a safety system fails just at the moment when it is needed. This value refers to a selectable period of time, typically per year. A *PFD$_{avg}$* of, for example, $3 \cdot 10^{-3}$ therefore means that it is highly probable that the safety function will fail once in 333 years at the very moment when it is needed. This does not mean though that the system will operate without failure for 333 years. The safety-critical failure can already occur after just one year and then not for 332 years - this is what the probability calculation implies. The calculation of *PFD$_{avg}$* of components occurs in a rather elaborate analytical procedure, the so-called FMEDA (Failure Mode Effects and Detectability Analysis), in which is analyzed down to the individual components, what happens with what error and how it can be detected. Possible errors in components can be classified into five different error types, which have different consequences on the *PFD$_{avg}$*.

22.5 General Thoughts About Terms and Standards

Table 22.2: Error types in safety systems

Error type	Harmless failures	Dangerous failures
Detected failures	λ_{SD} (Safe Detected)	λ_{DD} (Dangerous Detected)
Undetected failures	λ_{SU} (Safe Undetected)	λ_{DU} (Dangerous Undetected)
Failure of components which are not part of the protection layer	λ_{NP} (not part)	

22.5.4 Failure Rates

The failure rate λ is subdivided into the safe failures λ_S and the dangerous failures λ_D, see Figure 22.29.

$$\lambda = \lambda_S + \lambda_D \tag{22.48}$$

Thereby the safe failures are subdivided into the safe, detected failures λ_{SD} and into the safe, undetected failures λ_{SU}, see equation (22.49). The same goes for the dangerous failures, for these are subdivided into dangerous, detected failures λ_{DD} and dangerous, undetected failures λ_{DU}.

$$\lambda_S = \lambda_{SU} + \lambda_{SD} \tag{22.49}$$

$$\lambda_D = \lambda_{DU} + \lambda_{DD} \tag{22.50}$$

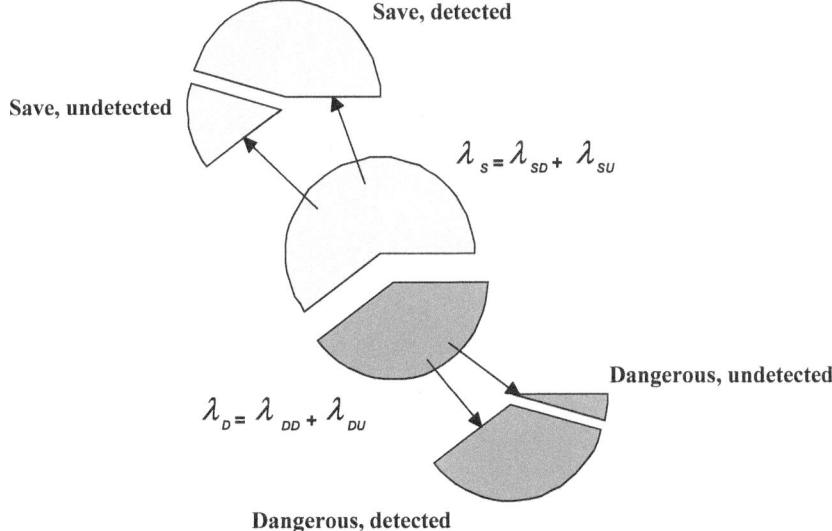

Figure 22.29: Failure rates

These different failure rates (see Table 22.2 and Figure 22.29) are required for the construction of a Markov model, for example for a 1oo2-system. 1oo2-system stands for „1 out of 2"-system and means, that in this case at least one system must function, in order to allow the system to correctly perform its safety function on demand. Figure 22.30 demonstrates a possible Markov model for a 1oo2-system.

In condition Z_0 both systems are operational, that is both systems are ok. Next, one has to ask to which possible conditions the condition Z_0 can convert directly or indirectly. The condition Z_0 can convert into the following conditions:

The system can, with an error transition rate of $2 \cdot \lambda_S$ convert into condition Z_1, in which the system is „safe" (S). The value „2" before the λ_S stands for two conduits, respectively, systems. λ_S stands for the „safe failures" (S = Safe). These are failures that are safe and known. They represent no problem for the system. λ_S is subdivided into two safe failures, that is into „safe, detected failures" λ_{SD} and „safe, not detected failures" λ_{SU}. For λ_S it can be written:

$$\lambda_S = \lambda_{SD} + \lambda_{SU} \tag{22.51}$$

Through the transition rate μ_R (R = repair) the system can return again to condition Z_0

$$\mu_R = \frac{1}{\tau_R} \quad \text{with } \tau_R = \text{repair time} \tag{22.52}$$

The system can convert with an error transition rate of $2 \cdot \lambda_{DD}$ to condition Z_2, that is, that one subsystem is operational and the other in condition „dangerous, detected failure" λ_{DD}. This case is not disturbing because the failure of the second system, while dangerous, is known. Therefore, through the transition rates μ_R and μ_0 the condition Z_2 can transfer through the safe condition Z_1 into condition Z_0.

22.5 General Thoughts About Terms and Standards

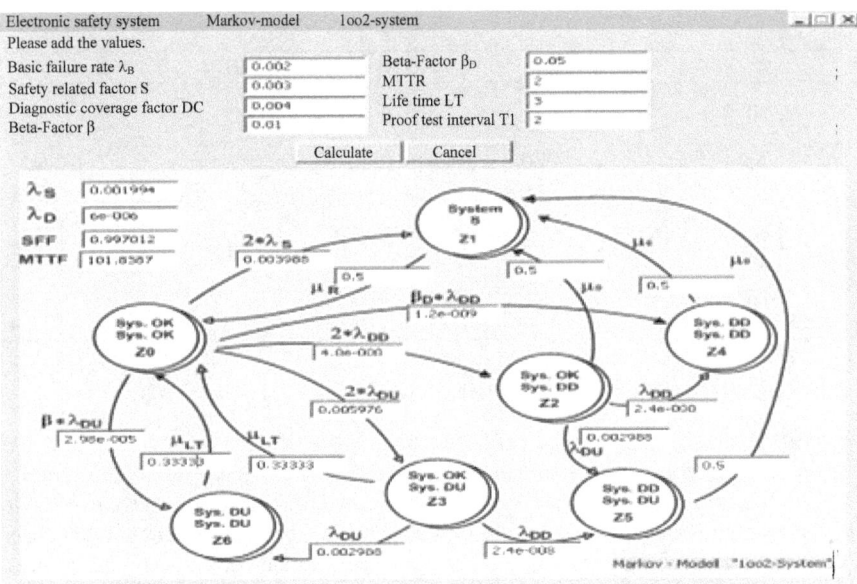

Figure 22.30: Failure rates of a 1oo2-Markov system

$$\mu_0 = \frac{1}{\tau_{Test}} \quad \text{with } \tau_{Test} = \text{test interval} \qquad (22.53)$$

With a failure transition rate of $2 \cdot \lambda_{DU}$, the system can transfer into condition Z_3 where one subsystem is functioning „ok" and the other has a dangerous, not detected error with λ_{DU}. The second subsystem has an error which can not be repaired. One must therefore wait until this second subsystem has reached its lifetime τ_{LT} (Life Time). Only once the subsystem has exceeded its lifetime, it can be exchanged and, possibly, repaired. With a transition rate μ_{LT}, the system can then transfer to condition Z_0.

$$\mu_{LT} = \frac{1}{\tau_{LT}} \left[\frac{1}{s} \right] \quad \text{with } \tau_{LT} = \text{lifetime} \qquad (22.54)$$

There is also the possibility to reach condition Z_5, with a transition rate of μ_0, and the safe condition Z_1, transition rat μ_R, to reach condition Z_0.

In summary it can be said that in the „low demand" systems examined here, only the dangerous, non-detected error λ_{DU} plays a major role, namely as refers to a defined time interval which is called T_1 (Proof Test Interval). It is the goal of these tests, to detect and to eliminate the dangerous error that could lead to failure of the safety function, through tests. In the reverse, incidentally, this means that a change of the time intervals for testing will result in a change of the failure probability on demand. Every driver knows this procedure when he presents his car to the TÜV every two years. Naturally, an annual or even biannual TÜV would increase the safety of the vehicle, but this would also mean an increase in cost. Sometimes though the shortening of the test time T_1 is the only method for achieving a required SIL. The calculated value PFD_{avg} allows the assignment of the device to an SIL.

Table 22.3: Failure probability and achievable SIL

PFD_{avg}	SIL
$\geq 10^{-2} .... < 10^{-1}$	SIL 1
$\geq 10^{-3} .... < 10^{-2}$	SIL 2
$\geq 10^{-4} .... < 10^{-3}$	SIL 3
$\geq 10^{-5} .... < 10^{-4}$	SIL 4

22.5.5 Risk, Damage, and Danger

The three terms damage, danger, and risk are very closely connected with each other. Damage is the impairment of a system caused by an event or a circumstance, or an undesired change in the original state of an object. The result of danger in the system leads to damage of this system. Danger is a situation or a state which can lead to a negative consequence. This negative consequence can concern persons, objects, conditions, or animals. A danger exists if a person can be spatially and temporally in contact with a injury-causing factor. Several dangerous conditions can interact. Dangers and the effects of dangers should be averted by technical, organizational and personal measures. Dangers which can not be eliminated by technical means are called residual risk. The dangers inherent in an actual workplace are determined and protocolled during the danger evaluation, for example through the specialist on workplace safety, on order by the employer. The employer decides on what measures should be taken.

Normally, „risk" (S) is often equivalent to danger. Frequently, more than one risk occurs at the same time, and one problem in risk evaluation is that there is more than one possible event. Often there is a tendency of neglecting essential risk and only examining the closest one. If the risk is so small that it is considered negligible, it is often called „residual risk". Misjudging risks can always lead to very negative consequences (catastrophies).

- Risk is a combination of the probability and the severity of potential injury or harm to a person's health in a dangerous situation (according to DIN EN 292-1). Risk is an expression of probability which quantitatively summarizes the expected frequency H of the event leading to damage, and the expected damage extent S upon the occurrence of the event (according to VDI 2542-2).

Risk is defined as the product of occurrence probability of an undesired event in a specified period of time and the associated potential damage (= effect).

$$\text{Risk} = \text{Occurrence probability} \bullet \text{Effect} \qquad (22.55)$$

The system of risk management implies that, if only 20% of the entire system are impaired by events, respectively, risks, they cause 80% of the problems. In order to prevent these 20%, the underlying causes of disturbances must be identified as early as possible and fixed as fast as possible.

22.5.6 Hazard Rate

The Hazard Rate H (coincidence, risk) is an element of the resting time analysis in statistics. It indicates a value for the probability that a specified „exit event" occurs during a specified time span. This is also called momentary inclination towards change of condition. With an underlying continuous and non-discrete time axis, the hazard value $h(t)$ does not meet a true probability, because, upon entry of the event (or exit through an event different from the wanted one), the remaining time of the reference period is „given away". The true probability for the entry of an event for a single object therefore results from integration over the associated density function $f(t)$, which results from the $h(t)*P$ (at time point t, the object is still under the risk for the first entry of the event). If no „concurring events" exist (that is, every object for which the event has not yet taken place, is still under risk), this corresponds to $f(t) = h(t) * (1-F(t))$, where $F(t)$ is the distribution function for the events. The Hazard function looks like this.

$$H(t) = \frac{f(t)}{1 - F(t)} \qquad (22.56)$$

22.5.7 Safety Integrity Level SIL

The definition of „safe state" is an important component of a safety concept. Depending on the application, a variety of different observations are possible. In a process in which a fluid is heated the safe state can be achieved, for example, through switching off all electric circuits including heating, so that the fluid will not boil over. For an error condition in an airplane, on the other hand, everything possible should be done to maintain all systems functional. By means of this example it becomes clear that the energy-free condition of a system is quite easy to achieve and should therefore be preferred for safety functions. Safe states in automatization components can, for example, „hold the last position", „switch off hardware" or „drive safely down". The definition of safe state is an important component of the test reports of components with SIL classification. SIL itself is an abbreviation of Safety Integrity Level and has, in the meantime, become synonymous with Functional Safety.

The achievable Safety Integrity Level is determined by the following safety-technical parameters:

- mean probability of dangerous failures of a safety function on demand (PFD_{avg})
- hardware error tolerance (HTF) and
- percentage of harmless failures (SFF).

The following table shows the dependence of the „Safety Integrity Level" (SIL) on the mean probability of dangerous failures of a safety function of the entire safety-related system" (PFD_{avg}). Thereby, the „low demand mode" is examined, that is, the demand rate on the safety-related system is maximal only once in a year. One of four discrete steps for specification of the requirements for safety integrity of the safety functions, which are assigned to the safety-technical system, where the safety integrity level 4 is the highest degree of safety integrity, and safety integrity level 1 the lowest.

SIL „only" defines a measure for the safety-related capacity or reliability of an electronic or electric control device, and has little stand-alone significance. Since the SIL is determined and examined differently from the AK (demand class) of DIN V 19250, a direct comparison is not so easy. One criterion is that the AK 3 system approximately corresponds to a SIL 1 system, or a AK 5 system to a SIL 3.

For the SIL, as opposed to the AK determination, the evaluation of the safety chain, the so-called SIS, is given priority. With safety instrumentation SIS (Safety Instrumented System), a combination of sensors, logic units and actuators, abnormal operating conditions are detected and the installation is again brought to a safe state, without a person's intervention. The SIS has a predetermined safety function, a defined reliability, and must work independently from other protection and safety systems. International standards such as IEC 61508 and IEC 61511 are, by and by, pushing back national, respectively, factory standards, in the course of globalization. Furthermore, installation managers want to be able to document the application of certified components. Due to the fact that more and more installations are evaluated more strictly with regard to their safety-critical importance, due to new regulations and standards, the topic of SIS grows in importance. As a rule, there are two methods for instrumentation components in safety circles, to prove conformity according to IEC 61511. These are:

- Operational reliability is proven through the long-term and successful use of safety components, which require a whole series of organizational and test methods by operators or device manufacturers, or

- a certificate of safety for each component of the safety installation, according to IEC 61508.

Table 22.4: Relationship between IEC 61511, DIN 19250, VDI/VDE 2180

IEC 61511	DIN V 19250	VDI/VDE 2180
SIL 1	AK 1	Risk area I
	AK 2	(low risk)
	AK 3	
SIL 2	AK 4	Risk area I
SIL 3	AK 5	Risk area II
	AK 6	(higher risk)
SIL 4	AK 7	Not covered solely by PLT protection devices
	AK 8	

While the SIS comprises the entire safety circle, the Safety Integrity Level (SIL) is a performance criterion which is determined for the devices of the safety installation. The usual subdivision of the mean probability of dangerous failures of a safety function on demand (PFD$_{avg}$), referring to the subsystems, can be seen in Figure 22.31.

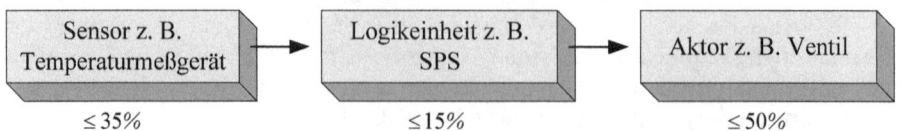

Figure 22.31: Subsystems of the safety function

22.5 General Thoughts About Terms and Standards

The SIS „Safety Instrumented System" is created from one or more safety chains. Generally, safety systems are classified, according to IEC 61508, as „low demand" and „high demand" systems. Hereby is defined how frequently during operation of an installation or machine the safety function is activated. If it is expected that the safety function is called upon several times a day or more, then one speaks of a „high demand" system. Examples for this are found in applications of machine engineering, when, for example, a light barrier must trigger the safety function upon intervention of a user, and the user continuously works at the machine. In process automation, it must be assumed that the safety function is only triggered very rarely, typically once a year or less. This is termed a „low demand" system. Examples for this are, for example, fire alarms or emergency fail safes. A statement concerning the SIL of individual components is not sufficient for the projection of safety systems. While in the past, the safety chain could achieve exactly the lowest AK of the individual components, with the SIL a calculation must occur on the basis of the failure probabilities. For this, the value PFD_{avg} has a central significance.

Two additional parameters describe the safety quality of a device. The *SFF* (Safe Failure Fraction) states, how big the percentage of harmless errors is compared to the total number of possible errors. A harmless error is defined as an error which is relevant for safety but which is either identified or transfers the system into a safe state. A simple example for this is a device fuse, which will transfer the device into a safe out-state upon over voltage of the device - as long as this is the safe state. An *SFF* of, for example, 90%, means that only 10% of possible errors in a safety installation can not be recognized and lead to a dangerous condition. The second parameter which is relevant here, is the *HFT* (Hardware Failure Fraction). With the *HFT*, the redundancy tolerance of an instrument, respectively, a system, is described. Systems without redundancy for which therefore the safety function is no longer guaranteed after a failure, have a $HFT = 0$. With simple redundancy, the $HFT = 1$, and with double redundancy the $HFT = 2$. The connection of the two parameters SFF and *HFT* consequently results in the SIL of an instrument. Herein, a distinction is made between simple type A instruments, where all errors are known and described (see Table 22.5), and more complex type B instruments, where not all errors are known and described, such as is the case for most microprocessor systems, respectively, software (see Table 22.6).

Table 22.5: Correlation between SFF and HFT in simple instruments

	HFT (Hardware Failure Tolerance)		
SFF (Safe Failure Fraction)	0	1	2
<60 %	SIL 1	SIL 2	SIL 3
60 ... 90 %	SIL 2	SIL 3	SIL 4
90 ... 99 %	SIL 3	SIL 4	SIL 4
>99 %	SIL 3	SIL 4	SIL 4

Table 22.6: Correlation of SFF and HFT in complex instruments

SFF (Safe Failure Fraction)	HFT (Hardware Failure Tolerance)		
	0	1	2
<60 %	----	SIL 1	SIL 2
60 ... 90 %	SIL 1	SIL 2	SIL 3
90 ... 99 %	SIL 2	SIL 3	SIL 4
>99 %	SIL 3	SIL 4	SIL 4

In the IEC 61511, the term of „operational reliability" is defined.

The term „operational reliability" actually originates from the field of hardware, especially with respect to components. The evaluation whether software – either an individual module or, for example, a complete compiler – is operationally reliable, draws on three criteria:

- Time period of application at least two years
- High number of installations
- Different applications

To fulfill these criteria the specification must not be changed. Furthermore, no or only negligible malfunctions must occur. The application of operationally tested software is - like with hardware - a measure to increase availability and reliability of the entire system. For some software products, meeting the above-mentioned three criteria does not necessarily mean that the product is „operationally verified". Thus, an operating system by a well-known firm from Redmond is not necessarily „operationally verified" software, because the occurring errors are often very rare and immaterial. The number of installations and the number of different applications may well be higher than that of other manufacturers, but nonetheless this product should not be considered „operationally verified".

If the manufacturer can prove the operational reliability of his components or systems, then the achievable SIS can be raised. This evidence is based on the number of instruments available on the market, and the systematic evaluation of safety-relevant errors of these instruments during their operation. This proof of operational stability is the only effective possibility, for more complex system which are frequently equipped with microprocessors, to perform a SIL evaluation. From the two, possibly differing, SIL which result from the PFD_{avg} and the SFF and HFT, the lowest value is accepted of the SIL of the instrument or system. Different from, for example, the explosion protection, there is not certification duty for Functional Safety. According to IEC 61508, a manufacturers test is sufficient for SIL 2 applications, although this must be performed by an independent department within the enterprise. From SIL 3 on, the norm recommends that an outside enterprise perform the analysis.

22.6 Process Control Technique PLT

The guideline VDI/VDE 2180 is in reference to the conception, planning, implementation and operation of process control techniques (PLT) for securing process engineering installations. Safety is of the utmost importance for this technique PLT. Therefore, the primary goal is always to organize the process in such a way that it is safe within itself. Wherever that is not feasible, additional measures are required to reduce the residual risk, which is inherent in the process, to acceptable levels. This can be achieved by means of process control techniques (PLT), if they are suitable for the special task. PLT measures and installations are suitable for tasks of securing an installation, if they were specially designed for it. In order to minimize the risk by means of PLT, measures are needed when errors, or disturbances in the PLT-operation of supervisory installations can lead to a dangerous event or dangerous state of the installation, and if the resultant risk is unacceptably high. In this case it is necessary to minimize the probability for the occurrence of a dangerous event by suitable protection measures, or to minimize the extent of damage. This can be performed by PLT protection devices, if these fulfill the safety requirements.

22.7 Performance Level PL

Performance Level PL stands for the evaluation of safety-technical quality of the implementation of the safety function and is subdivided into steps a through e. The following table 22.7 shows the relationship between Performance Level PL, Safety Integrity Level SIL, and the average probability of an error on demand of the safety function PFD_{avg} per hour.

Table 22.7: Correlation between PL, SIL and PFD_{avg} per hour

PL	SIL	PFD_{avg} per hour (=x)
a	no special requirements	$10^{-5} \leq x < 10^{-4}$
b	1	$3 \cdot 10^{-6} \leq x < 10^{-5}$
c	1	$10^{-6} \leq x < 3 \cdot 10^{-6}$
d	2	$10^{-7} \leq x < 10^{-6}$
e	3	$10^{-8} \leq x < 10^{-7}$

Literature

[5-INT98] 5. Internationales Symposium, *Programmierbare Systeme für sicherheitsgerichtete Anwendungen*, Tagungsband, Köln 1998.

[ABRA79] Abraham, J., *An improved algorithm for network reliability*, IEEE Trans. R28, 1979, pp. 58-61.

[ABRA86] Abramovici, M.; Kulikowski, J. J.; Menon, P. R.; Miller, D. T., *SMART and FASR: Test generation for VLSI scandesign circuits*, IEEE Design and Test of Computers, vol 3, 1986 Aug, pp. 43 - 54.

[AICH93] Guidelines for Safe Automation of Chemical Processes, Center for Chemical Process Safety/AIChE, 1993.

[AFR-87] AFR 122-9 Report, *The Nuclear Security Design Certification Program for Nuclear Weapon System Software and Firmware*, Department of the Air Force, 1987.

[AFSD] Air Force Space Division Handbook.

[ANDE90] Anderson, T., Lee, P. A., *Fault Tolerance: Principles and Practice*, Springer Verlag, 2nd edn. New York: 1990.

[AONI05] Aonix, *Software through Pictures*, www.aonix.de/stp.html, Stand: 31.10.2005.

[ARPa96] SAE-ARP4754, *Certification Considerations for Highly-Integrated Or Complex Aircraft Systems*, SAE International, see www.sae.org, 1996.

[ARPb96] SAE-ARP4761, *Guidelines and Methods for Conducting the Safety Assessment Process on Civil Airborne Systems and Equipment*, SAE International, see www.sae.org, 1996.

[ARNE87] Arnett, D. J., *A high performance solution for in-vehicle networking – controller area network (CAN)*, SAE 870823, 1987.

[ASAN] Asanek, V., Sweating, M., Ward, J., *Cost-Effective Microsatellites*, University of Surrey, Guildford.

[ASHE59] Ashenhurst, R., *The decomposition of switching functions*, Lab. Harvard 29, 1959, pp. 74-116.

[AVIZ83] Avizienis, A., *Framework for a taxonomy of faulttolerance attributes*. 10 Int. Symposium on computer architecture SIGARCH Newsletter 11, No. 3, 1983.

[AVIZ85]   Avizienis, A., *The N-version approach to fault-Tolerant software.* IEEE Trans. Software Eng., 11(12) 1985, pp. 1491-1501.

[AFCI83]   AFCIQ, *Données de Fiabilité en Stockage des Composants Electroniques*, Association Francaise pour le contrôle industriel et la qualité, 1983.

[ANSI01]   ANSI/ISA-91.00.01-2001, *Identification of Emergency Shutdown Systems and Controls That Are Critical to Maintaining Safety in Process Industries*, Research Triangle Park, NC, ISA, 2001.

[ANSI04]   ANSI/ ISA-84.00.01-2004, Parts 1-3 (IEC 61511-1 to 3 Mod). *Functional Safety: Safety Instrumented Systems for the Process Industry Sector*, 2004.

[ANSI96]   ANSI/ISA-84.01-96, *Application of Safety Instrumented Systems for the Process Industries*, Research Triangle Park, NC, ISA, 1996

[API-01]   API RP 14C-2001, *Analysis, Design, Installation, and Testing of Basic Surface Safety Systems for Offshore Production Platforms*, 2001.

[API-95]   API RP 554-1995, *Process Instrumentation and Control*, 1995.

[APIa95]   American Petroleum Institute, *API Recommended Practice 752 - Management of Hazards Associated with Location of Process Plant Buildings*, Washington, DC, American Petroleum Institute, 1995.

[APIb95]   American Petroleum Institute, *Management of Hazards Associated with Location of Process Plant Buildings*, API RP-752, Washington, DC, API, 1995.

[BACA93]   Baca, A., *Examples of Monte Carlo Methods in Reliability Estimation Based on Reduction of Prior Informations* IEEE Transactions on Reliability, Vol.42, No.4, 1993.

[BALL88]   Ball, M.; Provan, J., *Disjoint products and efficient computation of reliability*, Oper. Res. 36, 1988, pp. 703-715.

[BALL98]   Ball, D. J.; Floyd, Peter J, *Societal Risks: A Report Prepared for the Health and Safety Executive*, London, HM Stationery Office, 1998.

[BANE90]   Banerjee, P., Abraham J. A., *Characterisation and testing of physical failures in MOS logic circuits.* IEEE Design and Test, 1(3) 1990, pp. 76-86.

[BANS82]   Bansal, V. K., Misra K. B., Jain M. P, *Minimal pathset an minimal cutsets using search technique.* Microelectronics and Reliability, 22(6) 1982, 1067-1075.

[BARJ]   Barjenbruch, U., *Meßgenauigkeit und Zuverlässigkeit* Skriptum zur Vorlesung an der Universität Kassel FB Elektonik.

[BARL75]   Barlow, R.; Fussell, J.; Singpurwalla, N., Reliability and Fault Tree Analysis; *theoretical and applied aspects of system reliability and safety assessment*, Philadelphia, Pennsylvania, SIAM, 1975.

[BARL98]   Barlow, R. E., *Engineering Reliability* ASA-SIAM, 1998.

[BARN93]   Barnes, J. G. P., *Highlights of Ada 9X.* In Ada Yearbook 1993 (Loftus C., ed.). Amsterdam: IOS Press, 1993.

[BATE99]   Bate, J., Burns, A., *Real-Time Systems – Multidisciplinary Engineering* IEEE Prog.- Softw. Vol. 146 No. 2, April 1999.

[BAUE78]   Bauer, H., *Wahrscheinlichkeitstheorie und Grundzüge der Maßtheorie-DeGruyter,* Berlin, 1978.

[BCS-95]   BCS; *Standard oft Software Component Testing.* Working Draft 3.0 British Computers Society Specialist Interest Group in Software Testing (BCS SIGIST) (standard in preparation) 1995.

[BECa82]   Becker, A.; Camarinopoulos, L., *Delay times in fault tree analysis*, Microelectron, Reliab. 22, 1982, pp. 819-836.

[BECb82]   Beckman, L. V., *How Reliable Is Your Safety System* Chemical Engineering, 1982.

[BECK93]   Becker, W. J., et al., *Analoge und digitale Messverfahren* Script zur Vorlesung, Universität Kassel, 2. Auflage 1993.

[BECK98]   Becker, W. J., (Hrsg.) *Handbuch elektrische Messtechnik2*. Auflage 2001, Hüthig Verlag 1998.

[BECK90]   Becker, W. J., et al., *Leistungsfähigkeit objektorientierter Messgeräteprogrammierung am Beispiel der messtechnischen Charakterisierung eines Ultraschall-Interferometers zur Dichte Messung* MessLab 90, 2. Kongressmesse.

[BECK92]   Becker, W. J., et al., *Objektorientierte Meßdaten-Erfassungsmessen + prüfen* 28, S. 14 und 16 – 18, 1992.

[BECK95]   Beckman, L. V., *Match Redundant System Architectures with Safety Requirements* Chemical Engineering Progress, December 1995.

[BECK96]   Becker, W. J. et al., *Die Meß- und Automatisierungspraxis* Band 1, S.38-44, Expert-Verlag, 1996.

[BEIC88]   Beichelt, F., *Zuverlässigkeit strukturierter Systeme*, Berlin, VEB Verlag Technik, 1988.

[BEIC93]   Beichelt, F., *Zuverlässigkeits- und Instandhaltungstheorie*, Stuttgart, Teubner, 1993.

[BEIC95]   Beichelt, F., *Stochastik für Ingenieure: Eine Einführung in die Wahrscheinlichkeitstheorie und mathematische Statistik; mit zahlreichen Beispielen und Übungsaufgaben,* Teubner, Stuttgart, 1995.

[BEIZ90]   Beizer, B., *Software Testing Techniques* 2nd Edition Thomson Computer Press 1990.

[BELL93]   Bell, R., Reinert, D., *Risk and system integrity concepts for safety related control systems Microprocessors and Microsystems,* 17(1), pp. 3-15, 1993.

[BELL95]   Bellcore (Bell Communication Research), *Reliability Prediction Procedure for Electronic Equipment,* Livingston, NJ, Bellcore, 4nd Ed. TR –332, 1995.

[BELK97]   Belke, James C. *Recurring Causes of Recent Chemical Accidents*. U.S Environmental Protection Agency - Chemical Emergency Preparedness and Prevention Office, 1997.

[BEND88]   Bendell, T., *An overview of collection, analysis, and application of reliability data in the process industries*, IEEE Trans. R37, 1988, pp. 132-137.

[BENG76]   Bengiamin, N. N., Bowen, B. A.; Schenk, K. F., *An efficient algorithm for reducing the complexity of computation in fault tree analysis*, IEEE Trans. Nuclear Science, vol NS-23, 1976 Oct, pp. 1442 – 1446.

[BENN75]   Bennets, R. G., *On the analysis of fault trees, IEEE Transactions on Reliability,* vol R-24, 1975 Aug, pp. 175 – 185.

[BEIZ90]   Beizer, B., *Software Testing Techniques*, 2nd Edition, Thomson Computer Press, 1990.

[BIA-87]   BIA-Handbuch, Fehlerliste für elektrische Bauelemente bei Prüfung unterstellter Fehlerarten 7. Lfg. VI/87, Berufsgenossenschaftliches Institut für Arbeitsschutz.

[BIAN90]   Bianchini, R., Goodwin, K., Nydick, D., *Practical Application and Implementation of Distributed System-Level Diagnosis Theory,* Digest 20th Int'l Symp. Fault-Tolerant Computing, pp. 332-339, 1990.

[BILL83]   Billinton, R. Allan, R. N. *Reliability Evaluation of Engineering Systems,* Concepts and Techniques New York, Plenum Press, 1983.

[BIRD96]   Bird, Frank E., George L. Germain; Bird F.E. Jr., *Practical Loss Control Leadership,* International Loss Control Institute, 1996.

[BIRN92]   Birnbaum, H., et al., *Lehr- und Übungsbuch Mathematik* IVFachbuchverlag Leipzig – Köln 1992.

[BIRO91]   Birolini, A., *Qualität und Zuverlässigkeit technischer Systeme*, 3. Auflage, Berlin, Springer, 1991.

[BIRO96]   Birolini, A., *Reliability engineering: Cooperation between University and Industry at the ETH* Zürich, Quality Eng., 8 (1996) 4. pp. 659 – 674.

[BIRO97]   Birolini, A., *Zuverlässigkeit von Geräten und Systemen,* Springer-Verlag, 4. Auflage, 1997.

[BISH85]   Bishop, P. G., Esp D. G., Barnes, M., Humphreys, P., Dahll, G., Lhti, J., *PODS-a project on diverse software*. IEEE Trans. Software Eng.,12(9) 1985, 929-40.

[BITT77]   Bitter, P., *Technische Zuverlässigkeit Springer*, Verlag Berlin, 1977.

[BALZ92]   Balzert, H., *Die Entwicklung von Software-Systemen*, Spektrum-Verlag, 1992.

[BALZ96]   Balzert, H., *Lehrbuch der Software-Technik* Spektrum Akademischer Verlag Heidelberg, Berlin, Oxford, 1996.

[BALZ97]   Balzert, H., *Lehrbuch der Software-Technik*, Spektrum-Verlag, 1997.

[BOEH73]	Boehm, B., Software and its Impact: *A Quantitative Assessment,* Datamation, 05/1973, 48-59.
[BOEH79]	Boehm, B., *Software Engineering,* MIT Press, 1979.
[BOEH81]	Boehm, B., *Software Engineering Economics*, Prentice-Hall, 1981.
[BOEH88]	Boehm, B., *A Spiral Model of Software Development and Enhancement*, IEEE Computer, 5/1988, 61-72.
[BOOL58]	Boole, G., *The Laws of Thought*, London, Macmillan 1854, Neudruck: New York, Dover 1958.
[BÖRC00]	Börcsök, J. *Risikobewertung und Qualitätssicherung bei Prozessrechnern I. und II.,* Scriptum zur Vorlesung Universität Kassel 2000.
[BÖRC04]	Börcsök, J., *Elektronische Sicherheitssysteme: Hardwarekonzepte, Modelle und Berechnung*, Heidelberg, Hüthigverlag, 2004.
[BÖRC92]	Börcsök, J. *Echtzeitbetriebssysteme,* Scriptum zur Vorlesung Universität Kassel, 1992.
[BÖRC93]	Börcsök, J., *Prozessrechner,* Scriptum zur Vorlesung Universität Kassel, 1993.
[BÖRC97]	Börcsök, J., *Prozessrechner und -automation*, Hannover, Verlag Heinz Heise, 1997.
[BÖRC98]	Börcsök, J., *Echtzeitrechnerarchitekturen I. und II.,* Scriptum zur Vorlesung Universität Kassel, 1998.
[BÖRC99]	Börcsök, J., *Moderne Programmiertechniken für die Automatisierungstechnik nach IEC 61131-3* Scriptum zur Vorlesung, Fachhochschule Darmstadt, 1999.
[BRAA00]	Braasch, W., *Workshop on Risk Assessment*, The Third International Conference on Quality Management 2000, Internet-Recherche.
[BRAU91]	Braun, I., *Approximative Fehler-Baum-Analyse mittels unvollständiger Shannon-Zerlegung der Redundanz-Struktur-Funktion*, Hagen, Fern-Universität, Informatik-Bericht 115, 1991.
[BRIT97]	British Telecom Rel. HDBK HRD5 and Italtel Rel. Pred. HDBK IRPHB93. IEC 61709, Elec. Components *Reliability-Reference-Condition for Failure Rates and Stress Models Conversions,* in press, 1997.
[BRÖH03]	Bröhl, A.-P.; Dröschel, W., *Das V-Modell*, München, R. Oldenbourg Verlag, 2003.
[BRÖH95]	Bröhl, A.-P., Dröschel, W., Das *V-Modell,* Oldenburg Verlag, 1995.
[BRON97]	Bronstein, I. N.; et al., *Taschenbuch der Mathematik* Verlag Harri Deutsch, 1997.
[BROW76]	Brown, D., Fault tree analysis, *Systems Analysis and Design for Safety*, Prentice-Hall, 1976, pp. 152-193.

[BRYA92]   Bryant, R., *Symbolic Boolean manipulation with ordered binary decision diagrams,* ACM Computing Surveys, vol. 24, 1992, pp. 293 – 318.

[BSI-98]   BS 7925-2, *Software testing. Software component testing,* London, BSI Group, British Standards Publishing Limited (BSPL), 1998.

[CAP-84]   Cap, F., *Einführung in die Plasmaphysik* Vieweg-Verlag Wiesbaden, 1984.

[CARR85]   Carré, B.A., Debney C., *SPADE-Pascal* Southampton: Program Validation Limited, 1985.

[CARR90]   Carré, B. A., Jennings T. J., Maclennan F. J., Farrow P. F., Garnsworthy, J. R, *SPARK – The SPADE Ada Kernel, 3rd edn.* Southampton: Program Validation Limited, 1990.

[CARR95]   Carrasco J. A., *Improving availability bounds using the failure distance concept,* Dependable Computing and Fault-Tolerant Systems, vol 5, 1995, pp. 479 – 497, Springer-Verlag.

[CARR96]   Carrasco J. A., Escribá J., Calderón A., *Efficient exploration of availability models guided by failure distances,* ACM Performance Evaluation Review, vol 24, 1996 no. 1, May, pp. 242 – 251.

[CASS05]   CaSSPack (Control and Safety System Modeling Package). L&M Engineering (Kingwood, TX). Web site: www.landmengineering.com (Retrieved 6/28/2005 from source).

[CCPS89]   Center for Chemical Process Safety, *Guidelines for Process Equipment Reliability Data, with Data Tables,* New York, NY American Institute of Chemical Engineers Center for Chemical Process Safety, 1989.

[CCPS93]   Center for Chemical Process Safety, *Guidelines for Safe Automation of Chemical Processes,* New York, NY, American Institute of Chemical Engineers Center for Chemical Process Safety, 1993.

[CCPS96]   Center for Chemical Process Safety, *Guidelines for Chemical Process Quantitative Risk Analysis,* New York, NY, American Institute of Chemical Engineers Center for Chemical Process Safety, 1996.

[CCPS01]   Center for Chemical Process Safety, *Guidelines for Hazard Evaluation Techniques,* New York, NY, American Institute of Chemical Engineers Center for Chemical Process Safety, 2001.

[CCPS01a]   Center for Chemical Process Safety, *Layer of Protection Analysis, Simplified Process Risk Assessment,* New York, NY, American Institute of Chemical Engineers Center for Chemical Process Safety, 2001.

[CHAO87]   Chao, Anne, Huwang, L. C., *A Modified Monte Carlo Technique for Confidence Limits of System Reliability Using Pass-Fail Data* IEEE Transactions on Reliability, Vol. R-16, No. 1, 1987.

[CHAT75]   Chatterjee, P., *Modularization of fault trees: A method to reduce the cost of analysis,* Reliability and Fault Tree Analysis, 1975, pp. 101 – 126, SIAM.

[CHEN]	Chen, L., Avizienis A., *N-version programming: a fault tolerant approach to reliability of software operation*. In Proc. 8th Ann. Int. Conf. on Fault-Tolerant Computing, Toulouse, France, 21-23 June, pp. 3-9. New York: IEEE.
[CHWA81]	Chwa, K. Y., Hakimi, S. L., *Schemes for Fault-Tolerant Computing: A Comparison of Modularly Redundant and t-Diagnosable Systems* Information and Control, vol. 49, pp. 212-238, 1981.
[CLUT92]	Clutterbuck, W. J., *Drive-report: review of current tools and techniques for the development of safety-critical software. Software in Safety-Related Systems* (Wichmann B.A., ed.), pp. 145-75. Chichester: John Wiley, 1992.
[CNET93]	CNET, RDF 93, *Recueil de Données de Fiabilité des Composants Electroniques, Lannion,* CENT, 1993 British Telecom Rel. HDBK HRDS and Italtel Rel. Pred. HDBK IRPHB93.
[COBL98]	Coble, W.M. *Control Systems Safety Evaluation and Reliability.* Second edition. ISA, 1998.
[CODE]	Codes and Standards – ANSI/ISA S84.01 and Draft IEC 61508, *How This Standards Will Affect Your Business.*
[COLB87]	Colbourn, C., *The Combinatorics of Network Reliability*, Oxford, University Press, 1987.
[CONT95]	CONTESSE, *CONTESSE Test-Handbook* Glasgow, BEeSEMA, 1995.
[COOK92]	Cook, A., *Certification of digital systems in commercial avionics applications.Safety of Computer Control Systems 1992* (SAFECOMP'92), pp. 179-84. Oxford: Pergamon Press.
[COUD94]	Coudert, O.; Madre, J., MetaPrime, *An interactive fault-tree analyzer*, IEEE Trans. R43, 1994, pp. 121-127.
[COUD95]	Coudert, O., Madre, C., *MeaPrime: An interactive fault-tree analyzer,* IEEE Transactions on Reliability, vol. 43, pp. 121-127, Mar. 1994.
[COVE83]	Covello, V.T., The Perception of Technological Risks: A Literature Review, *Technological Forecasting and Social Change 23,* 285-97, 1983.
[CRIS85]	Cristian, F., Aghili, H., Strong, R., *Atomic Broadcast: From Simple Message Diffusion to Byzantine Agreement* Digest 15th Int'l Symp. Fault-Tolerant Computing, pp. 200-206, 1985.
[CRIS89]	Cristian, F., *Probabilistic clock synchronization*, Distrib Comput. 3, pp. 146-158, 1989.
[CRIS91]	Cristian, F., *Understanding fault-tolerant distributed systems* ACM. 34 (2), pp. 56-78, Feb. 1991.
[CRIS95]	Cristian, F., Schmuck, F., *Agreeing on processor-group membership in asynchronous distributed systems* 1995 Technical Report CSE95-428. Dept of Computer Science and Engineering. University of California. San Diego. La Jolla. CA.

[CRIS96]   Cristian, F., *Synchronous and asynchronous group communication* ACM. 39 (4), pp. 88-97, April 1996.

[CRIS99]   Cristian, F., Fetzer, C., *The timed asynchronous distributed system model* IEEE Trans. Parallel Distrib. Syst., June 1999.

[CULL91]   Cullyer, W. J.; Goodenough S. J.; Wichmann, B. A., *The choice of computer languages for use in safety-critical systems,* Software Eng. J., 6(2), 1991, 51-58.

[CULL94]   Cullyer, W. J., Storey, N., *Tools and techniques for testing of safety critical software,* Computing and Control Engineering Journal, 5(5): 239-- 244, October 1994.

[CUMM94]   Cummings, D., Alkalaj, L, Checkpoint / *Rollback in a Distributed System using Coarse-Grained Dataflow* Digest 24th Int'l Symp. Fault-Tolerant Computing, pp. 424-433, 1994.

[CURR89]   Currie, I. F., *NewSpeak: a reliable programming language.* High Integrity Software (Sennett C., ed.), London Pitman, 1989.

[DAHB84]   Dahbura, A. T.; Masson, G. M., *An $O(n^{2.5})$ Fault Identification Algorithm for Diagnosable Systems* IEEE Trans. Computers, vol. 33, no. 6, pp. 486-492, June 1984.

[DAHB88]   Dahbura, A. T., *System-Level Diagnosis: A Perspective for The Third Decade,* Concurrent Computation: Algorithms, Architectures, Technologies Plenum, 1988.

[DANN97]   Dann, P. H., *Global positioning system (gps) time dissemination for real-time applications*, Real-Time Syst. J. 12. (1), Kluwer Acad. Publishers, pp. 9-40, Jan. 1997.

[DEFE]   Defence Standard 00-56, *Safety Management Requirements for Defense Systems.*

[DEH-96]   Deh, K., Limnios N., *A new algorithm for fault trees prime implicant computations,* ESREL´96, Crete (Greece), 1996 Jun, pp. 1085-1090.

[DEMI78]   DeMillo, R. A., Lipton, R. J., Sayward, F. G., *Hints on data selection: help for the practising programmer.* IEEE Computer, 9(4) 1978, 34-41.

[DEUT98]   Deutsche Gesellschaft für Qualität e.V., *Zuverlässigkeit komplexer Systeme aus Hardware und Software* Berlin 1998, Beuth Verlag GmbH.

[DENK98]   Denker, M., Woyczynski, W. A., Ycart, B., *Introductory statistics and random phenomena: uncertainty, complexity and chaotic behavior in engineering and science* Birkhäuser, Boston, Basel, Berlin, 1998.

[DIN]   DIN IEC 65A/180/CD, *Kapitel 7.4.4, (VDE 0801 Teil 2)*, Beuth Verlag Berlin.

[DINa85]   DIN 25419, *Ereignisablaufanalyse; Verfahren, graphische Symbole und Auswertung*, Beuth Verlag Berlin, 1985.

[DINa87]   DIN VDE 31000 Teil 2:1987-12, *Allgemeine Leitsätze für das sicherheitsgerechte Gestalten technischer Erzeugnisse – Begriffe der Sicherheitstechnik – Grundbegriffe*, Beuth Verlag Berlin, 1987.

[DINa90]   DIN 25448, *Ausfalleffektanalyse,* Beuth Verlag Berlin, Ausgabe Mai 1990.

[DINb90]   DIN 40041, 1990-12, *Zuverlässigkeit, Begriffe*, Beuth Verlag Berlin, 1990.

[DINc90]   DIN 25424, 1990-04, *Fehlerbaumanalyse; Handrechenverfahren zur Auswertung eines Fehlerbaumes*, Beuth Verlag Berlin, 1990.

[DINa94]   DIN V VDE 0801, *Änderung A1 vom 16.01.94, S. 46.* Beuth Verlag Berlin.

[DINb94]   DIN V 19250, *Leittechnik: Grundlegende Sicherheitsbetrachtungen für MSR-Schutzeinrichtungen*, Berlin, Deutsche Elektrotechnische Kommission im DIN und VDE (DKE), 1994.

[DINc94]   DIN 19250, *Grundlegende Sicherheitsbetrachtungen für MSR-Schutzeinrichtungen*, Beuth Verlag Berlin, 1994.

[DINd94]   DIN EN ISO 8402:1994, *Qualitätsmanagement und Qualitätssicherung Begriffe*, zurückgezogene Norm, ersetzt durch DIN EN ISO 9000, see [ISOa05].

[DINa95]   DIN V 19251, *Leittechnik; MSR-Schutzeinrichtungen; Anforderungen und Maßnahmen zur gesicherten Funktion*, Beuth Verlag Berlin, 1995.

[DINa96]   DIN EN 954-1, *Sicherheit von Maschinen - Sicherheitsbezogene Teile von Steuerungen – Teil 1: Allgemeine Gestaltungsleitsätze*; Deutsche Fassung EN 954-1:1996, Beuth Verlag Berlin, 1996.

[DINa98]   DIN V 19250, *Grundlegende Sicherheitsbetrachtungen für MSR-Schutzeinrichtungen,* Beuth Verlag Berlin 1998.

[DINb98]   DIN IEC 65A/255/*CDV VDE 0801,* Beuth Verlag Berlin, 1998.

[DINc98]   DIN V VDE 0801, Teil 2, *Funktionale Sicherheit sicherheitsbezogener elektrischer/elektronischer/programmierbarer elektronischer Systeme (E/E/PES), Teil 2: Anforderungen an sicherheitsbezogene elektrische/elektronische/programmierbare elektronische Systeme* (IEC 65A/254/CDV:1998), S.27f, August 1998.

[DINa00]   DIN EN ISO 14971, *Medizinprodukte - Anwendung des Risikomanagements auf Medizinprodukte*, Beuth Verlag Berlin, 2001.

[DINb00]   DIN EN 954-1, Beiblatt 1, *Sicherheit von Maschinen – Sicherheitsbezogene Teile von Steuerungen – Teil 100: Leitfaden für Benutzung und Anwendung der EN 954-1:1996*; Deutsche Fassung CR 954-100:1999, Beuth Verlag.

[DINa03]   DIN IEC 61511, Teil 1 bis 3, (VDE 0810 Teil 1), *Funktionale Sicherheit: Sicherheitstechnische Systeme für die Prozessindustrie*, Frankfurt am Main, Deutsche Elektrotechnische Kommission im DIN und VDE (DKE), 2003.

[DINb03]   DIN EN ISO 14971/A1, *Medizinprodukte - Anwendung des Risikomanagements auf Medizinprodukte - Änderung 1: Begründung der Anforderungen*, Beuth Verlag Berlin, 2003.

[DINf]   DIN V VDE 0801/A1, *Grundsätze für Rechner in Systemen mit Sicherheitsaufgaben*, Beuth Verlag.

[DINg]   DIN EN61131-3, *Speicherprogrammierbare Steuerungen*, Beuth Verlag.

[DINj]   DIN 31004, *Zuverlässigkeit, Begriffe*.

[DO-92]   DO178B, RTCA, Inc. 1992, *Fehlerliste für elektrische Bauelemente – Bei der Prüfung unterstellte Fehlerarten*, BIA- Handbuch 7. Lfg. VI/87.

[DOD-92]   DoD, *Military Standardization Handbook: Reliability Prediction of Electronic Equipment*. United States Department of Defence MIL-HDBK-21F 1992.

[DOD-96]   DoD, *Defense Standard 00-56, Safety-Management Requirements for Defense Systems*, Department of Defense Standard, 1996.

[DOLE82]   Dolev, D., *The Byzantine Generals Strike Again Algorithms*, vol. 3, pp. 14-30, 1982.

[DOUG99]   Douglas, M., et al., *The Broadcast Comparison Model for On-Line Fault Diagnosis in Multicomputer Systems:Theory and Implementation* IEEE TRANSACTIONS ON COMPUTERS, VOL. 48, NO. 5, May 1999.

[DOWa97]   Dowell, A. M., III, *Layer of Protection Analysis: A New PHA Tool, After HAZOP, Before Fault Tree Analysis*, Presented at Center for Chemical Process Safety International Conference and Workshop on Risk Analysis in Process Safety, Atlanta, GA, October 21, 1997, American Institute of Chemical Engineers Center for Chemical Process Safety, New York, NY, 1997.

[DOWa98]   Dowell, A. M., III, *Layer of Protection Analysis for Determining Safety Integrity Level*, ISA Transaction 37, 155-166, 1998.

[DOWb98]   Dowell, A. M., III, *Layer of Protection Analysis and Inherently Safety Processes*, Process Safety Progress 18, no. 4, 1999.

[DOWa02]   Dowell, A. M., III, Hendershot D. C., *Simplified Risk Analysis - Layer of Protection Analysis (LOPA)*, AIChE 2002 National Meeting, unpublished paper, Copyright Rohm and Haas Company, 2002.

[DUKE86]   Duke, Geoff R. *Calculation of Optimum Proof Test Intervals for Maximum Availability*, Quality & Reliability Engineering International, Volume 2, pp. 153-158, 1986.

[DUNN93]   Dunn, R. H., *Software Qualität, Konzepte und Pläne*, Carl Hanser Verlag, 1993.

[DUTU96]   Dutuit, Y., Rauzy, A., *A linear-time algorithm to find modules of fault trees*, IEEE Trans. Reliability, vol 45, 1996 Sep, pp. 422-425.

[EDWA86]   Edwards, C.H.; Penney, D.E., *Calculus and Analytical Geometry*, 2d ed. Englewood Cliffs, NJ, Prentice-Hall, 1986.

[ELDR59]   Eldred, R. C. *Test routines based on symbolic logic statements*. JACM, 33-66, 1959.

[EN-a01]	EN 61508-1:2001, deutsche Fassung, *Funktionale Sicherheit sicherheitsbezogener elektrischer/elektronischer/programmierbarer elektronischer Systeme – Teil 1: Allgemeine Anforderungen*, Frankfurt am Main, Berlin, DKE, Deutsche Kommission Elektrotechnik Elektronik Informationstechnik im DIN und VDE, VDE Verlag GmbH, Berlin, 2001.
[EN-b01]	EN 61508-2:2001, deutsche Fassung, *Funktionale Sicherheit sicherheitsbezogener elektrischer/elektronischer/programmierbarer elektronischer Systeme – Teil 2: Anforderungen an sicherheitsbezogene elektrische/elektronische/programmierbare elektronische Systeme*, Frankfurt am Main, Berlin, DKE, Deutsche Kommission Elektrotechnik Elektronik Informationstechnik im DIN und VDE, VDE Verlag GmbH, Berlin, 2001.
[EN-c01]	EN 61508-3:2001, deutsche Fassung, *Funktionale Sicherheit sicherheitsbezogener elektrischer/elektronischer/programmierbarer elektronischer Systeme – Teil 3: Anforderungen an Software*, Frankfurt am Main, Berlin, DKE, Deutsche Kommission Elektrotechnik Elektronik Informationstechnik im DIN und VDE, VDE Verlag GmbH, Berlin, 2001.
[EN-d01]	EN 61508-4:2001, deutsche Fassung, *Funktionale Sicherheit sicherheitsbezogener elektrischer/elektronischer/programmierbarer elektronischer Systeme – Teil 4: Begriffe und Abkürzungen*, Frankfurt am Main, Berlin, DKE, Deutsche Kommission Elektrotechnik Elektronik Informationstechnik im DIN und VDE, VDE Verlag GmbH, Berlin, 2001.
[EN-e01]	EN 61508-5:2001, deutsche Fassung, *Funktionale Sicherheit sicherheitsbezogener elektrischer/elektronischer/programmierbarer elektronischer Systeme – Teil 5: Beispiele zur Ermittlung der Stufe der Sicherheitsintegrität (safety integrity level)*, Frankfurt am Main, Berlin, DKE, Deutsche Kommission Elektrotechnik Elektronik Informationstechnik im DIN und VDE, VDE Verlag GmbH, Berlin, 2001.
[EN-f01]	EN 61508-6:2001, deutsche Fassung, *Funktionale Sicherheit sicherheitsbezogener elektrischer/elektronischer/programmierbarer elektronischer Systeme – Teil 6: Anwendungsrichtlinien für IEC 61508-2 und IEC 61508-3*, Frankfurt am Main, Berlin, DKE, Deutsche Kommission Elektrotechnik Elektronik Informationstechnik im DIN und VDE, VDE Verlag GmbH, Berlin, 2001.
[EN-g01]	EN 61508-7:2001, deutsche Fassung, *Funktionale Sicherheit sicherheitsbezogener elektrischer/elektronischer/programmierbarer elektronischer Systeme – Teil 7: Anwendungshinweise über Verfahren und Maßnahmen*, Frankfurt am Main, Berlin, DKE, Deutsche Kommission Elektrotechnik Elektronik Informationstechnik im DIN und VDE, VDE Verlag GmbH, Berlin, 2001.
[EN-90]	EN ISO 9000, *Quality management systems*, ISO, International Organization for Standardization, 2005.
[EPA-99]	Environmental Protection Agency, *Risk Management Program Guidance for Offsite Consequence Analysis*, Washington, DC, United States Environmental Protection Agency, 1999.

[ER-A95]   Er, A. *Entwicklung eines Modells zur optimierten Nutzung des Wissenspotentials einer Prozess-FMEA,* Fortschrittsberichte VDI, Reihe 20, Nr. 176, 1995.

[ERMA75]   Ermakov, S. M., *Die Monte-Carlo-Methode und verwandte Fragen* Oldenbourg, München, 1975.

[ESSA99]   Essamed, D, Aralt, J., Powell, D., Padre, A., *A protocol for asymmetrie duplex redundancy* Proceedings of the Seventh IFIP International Working Conference on Dependable Computing for Critical Applications. January 1999. San Jose USA.

[EXID03]   *Safety Equipment Reliability Handbook.* exida.com, 2003.

[FAGA86]   Fagan, M. E., *Advances in Software Inspections* IEEE Tranaction on software engineering, Vol. SE-12, No. 7, July 1986 pp.744-751.

[FAIR85]   Fairley, R.E., *Software Engineering Concepts*, McGraw-Hill, 1985.

[FAR-93]   FAR 25.1309, *Federation Aviation Regulation 25.1309: Equipment, Systems and Installations.* USA: Office of the Federal Register National Archives and Records Administration 1993.

[FEMA90]   Federal Emergency Management Agency, Department of Transportation, Environmental Protection Agency, *Handbook of Chemical Hazard Analysis Procedures*, Washington, DC, Federal Emergency Management Agency, 1990.

[FETZ]   Fetzer, C., Cristian, F., *A highly available local leader service* IEEE Trans. Softw. Eng.

[FETa96]   Fetzer, C., *Fail-aware clock synchronization.* 138. Technical Report Time-Service. Dagstuhl- Seminar-Report: Mach 1996.

[FETb96]   Fetzer, C., Cristian, F., *Fail-awareness in timed asynchronous systems* Proceedings of the 15th ACM Symposium on Principles of Distributed Computing, pp. 314-321 Philadelphia, May 1996.

[FETa97]   Fetzer, C., Cristian, F., *Fail-awareness: An approach to construct fail safe applications* In Proceedings of the 27th Annual International Symposium on Fault-Tolerant Computing, Seattle, June 1997.

[FETb97]   Fetzer, C., Cristian, F., A fail-aware membership service Proceedings of the 16th Symposium on Reliable Distributed Systems, pp. 157-164, October 1997.

[FETc97]   Fetzer, C., Cristian, F., *Fortress A system to support fail-aware real-time applications.* IEEE Workshop on Middleware for Distributed Real-Time Systems and Services, San Francisco, Dec. 1997.

[FETZ98]   Fetzer, C., *The message classification model In Proceeding of the 17th AMC Symposium on Principles of Distributed Computing*, Puerto Vallarta Mexico, June 1998.

[FETZ99]	Fetzer, C. Cristian, F. *Building fault-tolerant hardware clocks Proceedings of the Seventh IFIP International Working Conference on Dependable Computing for Critical Applications,* Jan 1999, pp. 59-78, San Jose USA.
[FINK99]	Finkelstein, M. S., *Multiple Availlability on Stochastic Demand* IEEE TRANSACTIONS ON RELIABILITY, VOL. 48, NO. 1, March 1999.
[FISH96]	Fishman, G. S., *Monte Carlo, Concepts, Algorithms and Applications* Springer Verlag, Berlin, 1996.
[FLEM85]	Fleming, R. E., Josselyn, J. V., Dolny, L. J., DeHoff, R. L, *Complex System RMA and T, Using Markov Modells* Proccedings of Annual Reliability and Maintainability Conference, New York, IEEE, 1985.
[FOAT99]	Foata, D., Fuchs, A.,*Wahrscheinlichkeitsrechnung* Birkhäuser, Basel, Boston, Berlin, 1999.
[FRUE83]	Fruehwirth, R. Regler, M. *Monte-Carlo-Methoden: eine Einführung* Bibliograph. Inst., Mannheim, 1983.
[FRÜH00]	Frühauf, K.; Ludewig, J.; Sandmayr, H., *Software Prüfung*, ETH Zürich, vdf Hochschulverlag AG, 2000.
[FRÜH95]	Frühauf, K., et al., *Software-Projektmanagement und -Qualitätssicherung* B.G. Teubner Verlag 1995.
[FUJI83]	Fujiwara, H., Shimano, T., *On the acceleration of test generation algorithms,* IEEE Trans. Computers, vol. C-32, 1983 Dec, pp. 1137-1144.
[FUSS72]	Fussell, J. B., Vesely, W. E. *A new methodology for obtaining cut sets for fault trees,* Trans. American Nuclear Society, vol. 15, 1972 Jun, pp. 262-263.
[FUSS74]	Fussell, J.; Powers, G.; Bennets, R., *Fault trees –a state of the art discussion*, IEEE Trans. R23, 1974, pp. 51-55.
[G3-I95]	G3-IQSE, Bericht des IQSE, *component failure assumtions,* 08.02.95.
[GAED77]	Gaede, K. W., *Zuverlässigkeit - Mathematische Modelle* München 1977, Carl Hanser Verlag.
[GAED96]	Gaedke, K. *Ein netzlistenbasierendes Verfahren zur Zuverlässigkeitsanalyse fehlertoleranter VLSI-Schaltkreise* VDI-Verlag Reihe 9 Nr. 238, 1996.
[GALL97]	Gall, H., Kemp, K., *Wirksamkeit von zeitlichen und logischen Programmlaufüberwachungen beim Betrieb von Rechnersystemen,* Dortmund / Berlin 1997, Bundesanstalt für Arbeitsschutz und Arbeitsmedizin.
[GAMM95]	Gamma, E.; Helm, R.; Johnson, R.; Vlissides, J., Design Patterns: *Elements of Reusable Object-Oriented Software,* Addison-Wesley, 1995.
[GERT94]	Gertman, David I.; Blackman, Harold S., *Human Reliability and Safety Analysis Data Handbook*, New York: Wiley-Interscience, 1994.
[GLAS80]	Glasstone, S.; Dolan, P.J., *The Effects of Nuclear Weapons*, 3d ed. Tonbridge Wells, UK, Castle House Publishers, 1980.

[GNED]   Gnedenko, B. V., Roßberg, H. -J., *Einführung in die Wahrscheinlichkeitstheorie* Akademie-Verlag, Berlin.

[GOBL01]   Goble, William M.; Scharpf, Eric W., 61508 Overview Report, *Safety Lifecycle Report*, <http://www.exida.com>, February 2001.

[GOBL88]   Goble, William M., *Control Systems: Safety, Evaluation, and Reliability*, Research Triangle Park, NC, ISA, 1988.

[GOBL90]   Goble, W. M. *Evaluating Control Systems Reliability*, Instrument Society of Amerika, 1990.

[GOBL92]   Goble, W. M., *Evaluating Control Systems Reliability*, ISA, 1992.

[GOBL95]   Goble, W. M., *Safety of Programmable Electronic Systems – Critical Issues, Diagnostic and Common Cause Strength* Proceedings of the IchemE Sympossium, Rugby, U.K. Institution of Chemical Engineers, June 1995.

[GOBL96]   Goble, W. M., *Using PLCs for Safety Applications* Hydrocarb Processing, June 1996.

[GOBL98]   Goble, William M., *Control Systems Safety Evaluation and Reliability*, Research Triangle Park, NC, ISA, 1998.

[GOEL81]   Goel, P., *An implicit enumeration algorithm to generate tests for combinational logic circuits*, IEEE Trans. Computers, vol C-30, 1981 Mar, pp. 215-222.

[GOND86]   Gondran, M., Launch meeting of the European Safety and Reliability Association held in Brussels, Belgium, October 1986

[GÖRK78]   Görke, W. (Hrsg), *Zuverlässigkeit von Rechensystemen* Fachberichte und Referate Bd. 9 Oldenbourg Verlag 1978.

[GÖRK68]   Görke, W., *Probleme der Zuverlässigkeit elektronischer Schaltungen* Habilitationsschrift, TU Karlsruhe, 1968.

[GÖRK84]   Görke, W., *Was ist Fehlertoleranz?* Elektronische Rechenanlagen, 26. Jahrgang 1984 Nachrichtenteschnische Gesellschaft im VDE (NTG) Hrsg., Heft 1, S.29-31.

[GÖRK92]   Görke, W., *Fehlertoleranz in Rechnern* Interner Bericht Nr. 28/92 Universität Karlsruhe, 1992.

[GOLD79]   Goldstein, L. H., *Controllability / observability analysis of digital circuits*, IEEE Trans. Circuits and Systems, vol CAS-26, 1979 Sep, pp. 685-693.

[GOLD85]   Goldsack, S., *Ada for Specification and Design – Possibilities and Limitations*. Cambridge: Cambridge University Press, 1985.

[GORI94]   Goring, A., *Methods and techniques of improving the safety classification of programmable logic controller safety systems*. Technology and Assessment of Safety-Critical Systems (Redmill F. and Anderson T., eds), pp. 21-30. London: Springer-Verlag 1994.

[GOWL04] Gowland, R., *Practical Experience of applying Layer of Protection Analysis For Safety Instrumented Systems (SIS) to comply with IEC 61511*, aus Internet, 2004.

[GRAM90] Grams, T., *Denkfallen und Programmierfehler* Berlin 1990, Springer Verlag.

[GREE94] Greenway, A., *A user's perspective of programmable logic controllers (PLCs) in safety-related applications.* Technologiy and Assessment of Safety-Critical Systems (Redmill F. and Anderson T., eds), pp. 1-20. London: Springer Verlag 1994.

[GRUH92] Gruhn, Paul and A. Rentcome. "Safety Control System Design: Are All the Bases Covered?" *Control,* July 1992.

[GRUH98] Gruhn, Paul; Cheddie, Harry L., *Safety Shutdown Systems: Design, Analysis, and Justification*, Research Triangle Park, NC, ISA, 1998.

[GUY-91] Guy, C. G., *Errors, reliability and redundancy.* Digital Systems Reference Book (Holdsworth B. and Martin G.R., eds.), ch. 1.16. Oxford: Butterworth-Heinemann, 1991.

[HADZ93] Hadzilacos, V., Toueg, S., *Fault-Tolerant Broadcasts and Related Problems* Distributed Systems, Mullender, S,, chapter 5. ACM Press, 1993. Hill, I:D: (1972). Wouldn't it be nice if we could write computer programs in ordinary English – or would it? Comp. Bull., 16(6), pp. 306-312.

[HAGS83] Hagstrom, J., *Using the decomposition tree of a network in reliability computation*, IEEE Trans. R32, 1983, pp. 71-78.

[HALA99] Halang, W. A., Programmierung auf den verschiedenen Sicherheitsniveaus nach IEC 61508 Fern Universität Hagen bzw. in *Sicherheit und Zuverlässigkeit software-basierter Systeme*, F. Saglietti and W. Goerigk (Eds.), pp. 101 - 116, Garching: Institut für Sicherheitstechnologie 1999, ISBN 3-00-004872-3.

[HARI87] Hariri, S.; Raghavendra, C., SYREL, *a symbolic reliability algorithm, based on path and cutest methods*, IEEE Trans. C36, 1987, pp. 1224-1232.

[HARR82] Harris, L. N., Dale, C. J., *Approaches to software reliability prediction.* Proc. Ann. Reliability an Maintainability Symp., Los Angeles, CA, 26-28 January, pp. 167-75, New York: IEEE 1982.

[HARV76] Harvey, B.H.(chairman), *First Report of the Advisory Committee on Major Hazards*, London, HM Stationery Office, 1976.

[HEID89] Heidtmann, K., *Smaller sums of disjoint products by subproduct inversion*, IEEE Trans. R38, 1989, pp. 305-311.

[HEIL93] Heiler, G3-IEA03, Institut für Qualität und Sicherheit in Elektronik, 2/93.

[HELL] Heller, W. D., *Wahrscheinlichkeitsrechnung: mit vollständig gelösten Aufgaben* Birkhäuser, Basel.

[HENG78] Hengartner, W., Theodorescu, R., *Einführung in die Monte-Carlo-Methode* Hanser, München,1978.

[HENL92]   Henley, E. J., Kumamoto, H., *Probabilistic Risk Assessment - Reliability engeneering, design and analysis,* New York, IEEE 1992.

[HENN95]   Hennings, W. Kuznetsov, N., *FAMOCUTN & CUTQN: Programs for fast analysis of lerge fault trees with replicated & negated gates,* IEEE Trans. Reliability, vol. 44, 1995 Sep., pp. 368 – 376.

[HENS78]   Hentschke, S., Überlagerung, Berechnung und Auswertbarkeit höherer Momente von Zufallsverkehr, in *Mitteilungen der Standard Elektrik Lorenz AG*, Stuttgart, 1978.

[HENS81]   Hentschke, S., Predective process overload control strategies for SPC switching system, in *Ntz. Archiv*, Bd. 3 (1981) H.5, pp. 121-126.

[HENS88]   Hentschke, S., *Grundzüge der Digitaltechnik,* Wiesbaden, B. G. Teubner Verlag, 1988.

[HENS96]   Hentschke, S., Digital stochastic magnetic-field detection, *Sensor and Actuators,* A 57, 1996, pp. 1-8.

[HENT88]   Hentschke, S., *Grundzüge der Digitaltechnik* B.G. Teubner Verlag, 1988.

[HENT78]   Hentschke, S., *Überlagerung, Berechnung und Auswertbarkeit höherer Momente von Zufallsverkehr* Mitteilungen der Standard Elektrik Lorenz AG Stuttgart 1978.

[HENT81]   Hentschke, S., *Predective process overload control strategies for SPC switching system* Ntz. Archiv Bd. 3 (1981) H.5, pp. 121-126.

[HENT96]   Hentschke, S., *Digital stochastic magnetic-field detection* Sensor and Actuators A 57 (1996), pp. 1-8.

[HERM96]   Hermanns, O., *Schuba M., Performance investigations of the ip multitest architecture* Compu Netw. ISDN Syst. 28, pp. 429-439, 1996.

[HILL72]   Hill, I. D., *Wouldn't it be nice if we could write computer programs in ordinary English or would it?* Comp. Bull., 16(6), pp. 306-12, 1972.

[HILT95]   Hiltunen, M., *Membership and System Diagnosis* Proc. 14th Symp. Raliable Distributed Systems, pp. 208-217, 1995.

[HORN74]   Horning, J. J., Lauer, H. C., Melliar-Smith, P. M.; et al., *A program structure for error detection and recovery.* Proc. Int. Symp. on Operating Systems, Rocquencourt, France, 23-25 April, pp. 171-87, Berlin Springer-Verlag, 1974.

[HOSS84]   Hosseini, S., Kuhl, J., Reddy, S., *A Diagnosis Algorithm for Distributed Computing Systems with Dynamic Failure and Repair* IEEE Trans Computers, vol. 33, no. 3, pp. 223-233, Mar. 1984.

[HSEa87]   HSE (1987), *Programmable Electronic Systems in Safety-Related Applications* Vol. 2: General Technical Guidelines. London: Her Majesty's Stationery Office, 1987.

[HSEb87]   UK Health and Safety Executive, *Programmable Electronic Systems in Safety-Related Applications, Part 1: An Introductory Guide,* Sheffield, UK, UK Health an Safety Executive, 1987.

[HSE-95]   UK Health & Safety Executive, *Out of Control: Why control systems go wrong and how to prevent failure*, UK Health & Safety Executive, 1995.

[HSE-96]   Health and Safety Executive, *The Setting of Safety Standards: A Report by an Interdepartmental Group of External Advisors*, London, HM Stationery Office, 1996.

[HUMM88]   Hummel, R. A., *Automatic Fault Injection for Digital Systems Proceedings of the Annual Reliability and Maintainability Symposium* New York, IEEE, 1988.

[HWAN81]   Hwang, C.; Tillmann, F.; Lee, M., *System reliability evaluation techniques for complex/large systems - review*, IEEE Trans. R30, 1981, pp. 416-423.

[IBM-05]   IBM, *Rational Software*, www.306.ibm.com/software/rational/, Stand: 31.10.2005.

[IEC]   IEC Norm 271, Geneva: International Electrotechnical Commission.

[IECa]   IEC 61508, *International Standard 61508 Functional Safety: Safety-Related System*, Geneva: International Electrotechnical Commission.

[IEC-86]   DIN IEC 880, *Software for computers in the safety systems of nuclear power stations*, 1986, als DIN 1987.

[IECa00]   IEC 61508, *International Standard 61508: Functional safety of electrical/electronic/programmable electronic safety-related systems*, Geneva, International Electrotechnical Commission, 2000.

[IECb00]   IEC 61511, *Functional Safety: Safety Instrumented Systems for the Process Industry Sector*, Geneva, International Electrotechnical Commission, 2000.

[IECa90]   IEC 60050-191 Ed. 1.0 b: 1990, *International Electrotechnical Vocabulary, Chapter 191: Dependability and quality of service*, 1990.

[IECa99]   IEC 61508-1:1998 + Corrigendum 1999, *International Standard: 61508 Functional safety of electrical/electronic/programmable electronic safety-related systems, Part 1: General requirements*, Geneva, International Electrotechnical Commission, 1999.

[IECb01]   IEC 61508-2:2001, *International Standard: 61508 Functional safety of electrical/electronic/programmable electronic safety-related systems, Part 2: Requirements for electrical/electronic/programmable electronic safety-related systems*, Geneva, International Electrotechnical Commission, 2001.

[IECc99]   IEC 61508-3:1998 + Corrigendum 1999, *International Standard: 61508 Functional safety of electrical/electronic/programmable electronic safety-related systems, Part 3: Software requirements*, Geneva, International Electrotechnical Commission, 1999.

[IECd99]   IEC 61508-4:1998 + Corrigendum 1999, *International Standard: 61508 Functional safety of electrical/electronic/programmable electronic safety-related systems, Part 4: Definitions and abbreviations*, Geneva, International Electrotechnical Commission, 1999.

[IECe99]   IEC 61508-5:1998 + Corrigendum 1999, *International Standard: 61508 Functional safety of electrical/electronic/programmable electronic safety-related systems, Part 5: Examples of methods for the determination of safety integrity level*, Geneva, International Electrotechnical Commission, 1999.

[IECf00]   IEC 61508-6:2000, *International Standard: 61508 Functional safety of electrical/electronic/programmable electronic safety-related systems, Part 6: Guidelines on the application of IEC 61508-2 and IEC 61508-3*, Geneva, International Electrotechnical Commission, 2000.

[IECg00]   IEC 61508-7:2000, *International Standard: 61508 Functional safety of electrical/electronic/programmable electronic safety-related systems, Part 7: Overview of techniques and measures*, Geneva, International Electrotechnical Commission, 2000.

[IEC-02]   IEC 65A/324/FDIS 2002, *Functional safety: Safety Instrumented Systems for the process industry sector*, Geneva, International Electrotechnical Commission, 2002.

[IEC-03]   IEC 61131, *Programmable Controllers*, Geneva, International Electrotechnical Commission, 2003.

[IEC-91]   IEC 1078, *International Standard 1078 Analysis Techniques for Dependability – Reliability Block Diagram Method.* Geneva: International Electrotechnical Commission, (also available in the UK as BS 5760: Reliability of Systems, Equipment and Components. Part 9: Guide to Block Diagram Technique, 1992, British Standards Institution), 1991.

[IEC-97]   IEC 61709, *Electronic Components Reliability-Reference-Condition for Failure Rates and Stress Models Conversions,* in Press 1997.

[IEC-98]   International Electrotechnical Commission, *Functional Safety of Electrical / Electronic / Programmable Electronic Safety-Related Systems*, IEC 61508, Geneva, International Electrotechnical Commission, 1998.

[IEC-99]   International Electrotechnical Commission, IEC draft standard 61511, Part 3, *Guidelines in the Application of Hazard and Risk Analysis*, Geneva, IEC, 1999.

[IEEE91]   IEEE 603-1991. *Standard Criteria for Safety Systems for Nuclear Power Generating Stations*, 1991.

[IEEE80]   IEEE-Std 493, *Recommended Practise for the Design of Reliable Industrial and Commercial Power Systems,* 500-1984, Reliability Data for Nuclear-Power Generation Stations, 1980.

[IEEE84]   IEEE 500-1984, *Equipment Reliability Data for Nuclear-Power Generating Stations.*

[IEEE92]   IEEE Spectrum, *Faults and failures,* Febraury 1992.

[IEEE95]   IEEE, *Safety-Related Systems Postgraduate Qualifications Syllabus Proposals,* Stevenage: Institution of Electrical Engineers, 1995.

[INST]   Institut für Mathematik; *Mathematische Lehrbücher und Monographien,* Akademie-Verlag, Berlin.

[INST91]	Institut für Qualität und Sicherheit in der Elektronik (IQSE), SAA-IQSE-S/02; *Anweisungen zur Prüfung festverdrahteter Elektronik-Systeme,* Version 0.5.EN, 08.07.91.
[IQSE98]	IQSE (Bericht des Instituts für Qualität und Sicherheit in der Elektronik), *Component Failure Assumptions* G3-IQSE, 08.02.98.
[ISA-02]	ISA-TR84.00.02-2002, Parts 1-5. *Safety Instrumented Functions (SIF) – Safety Integrity Level (SIL) Evaluation Techniques,* 2002.
[ISA-96]	ISA-84.01-1996. *Application of Safety Instrumented Systems for the Process Industries,* 1996.
[ISOa99]	ISO/IEC Guide 51:1999, *Safety aspects – Guidelines for their inclusion in standards,* 1999.
[ISOa00]	ISO 9001:2000, *Quality management systems – Requirements,* ISO, International Organization for Standardization, 2000.
[ISOb00]	ISO 9004:2000, *Quality management systems – Guidelines for performance improvements,* ISO, International Organization for Standardization, 2000.
[ISOa05]	ISO 9000:2005, *Quality management systems – Fundamentals and vocabulary,* ISO, International Organization for Standardization, 2005.
[ITEM05]	ITEM Software (Irvine, CA), 1998. Web site: www.itemsoft.com (retrieved 6/28/2005 from source).
[JACK83]	Jackson, M., *System Design,* Englewood Cliffs, NY Prentice-Hall, 1983.
[JAGE86]	Jager, R., *Integrated Diagnostics – Extension of Testability* Proceedings of the Annual Reliability and Maintainability Symposium New York, IEEE, 1986.
[JAR-94]	JAR 1994, *Joint aviation requirements 25.1309* Equipment: Systems and installations Cheltenham: Civial aviation authority.
[JECK05]	Jeckle, M., *Unified Modeling Language (UML) Tools,* www.jeckle.de/umltools.htm, Stand: 31.10.2005.
[JOHN94]	Johnson, D. A., *Automatic Fault Insertion* INTECH, Research, Triangle Park, NC: ISA, November 1994.
[JOSS86]	Josselyn, J. V., Fleming, R. E., Frenster, J. A., DeHOFF, R. L., *Application of Markov Models for RMA Assessment* Proceedings of the Annual Reliability and Maintainability Conference New York, IEEE, 1986.
[KBST05]	KBSt, *V-Modell XT Release 1.1 Gesamtumfang,* www.kbst.bund.de/V-Modell/-,293/V-Modell-XT.htm, Stand: 31.10.2005.
[KIM-89]	Kim, Chul, *MTBF of a Complex Binary Coherent System* IEEE Transactions on Reliability, Vol.38, No.4, 1989.
[KIM-92]	Kim, C., Lee, H. K., *A Monte Carlo Simulation Algorithm for finding MTBF,* IEEE Transactions on Reliability, Vol.41, No.2, 1992.
[KLET93]	Kletz, T., *Lessons From Disaster: How Organizations Have no Memory and Accidents,* Gulf Publishing Co, 1993.

[KLET95]   Kletz, T., *Computer Control and Human Error*, Gulf Publishing Co., 1995.

[KNIG86]   Knight, J. C., Leveson, N. G., *An experimental evaluation of the assumption of independence in multiversion programming*, IEEE Trans. Software Eng., 12(1), 1986, pp. 96-109.

[KOHA78]   Kohavi, Z., *Switching and Finite State Automata Theory*. New York: McGraw-Hill, 1978.

[KOHD89]   Kohda, T., Henley, E. J., Inoue, K., *Finding modules in fault trees*, IEEE Trans. Reliability, vol 38, 1989 Jun, pp. 165 – 176.

[KOHL73]   Kohlas, J., *Relative Werte und Monte Carlo Analyse von Kosten- oder Erloes-Strukturen auf Markoff-Ketten*, Phil. Diss., Zürich, 1973.

[KOLM]   Kolmogorov, A. N., *Mathematics of the 19th century* Birkhäuser, Basel, Stuttgart.

[KRAE75]   Kraemer, U., Rackwitz, R., Grasser, E., *Monte-Carlo-Studie zur Zuverlässigkeit von durchlaufenden Stahlbetondecken in Bürogebäuden* Technische Universität München, 1975.

[KREY65]   Kreyszig, E., *Statistische Methoden und ihre Anwendungen* Vandenhoeck & Ruprecht, Göttingen, 1965.

[KREY67]   Kreyszig, E., *Advanced Engineering Mathematics*, Chapter 4.4, New York John Wiley & Sons, 1967.

[KÜFN95]   Küfner, H.; Schneeweiss, W., *Petrinetze zur Modellierung von Wartungsstrategien*, Proc. 4. Fachtagung „Entwurf komplexer Automatisierungssystem", 1995, S. 296-313.

[KUEF97]   Kuefner, H.; Schneeweiss, W., *Qualifying the fault tolerance of multiple bus-based systems*, Proc. ESREL, 1997, pp. 2242-2254.

[KUHL80]   Kuhl, J. G., Reddy, S. M., *Distributed Fault-Tolerance for Large Multiprocessor Systems,* Proc. Seventh Int'l Symp. Computer Architecture, pp. 23-30, 1980.

[KUHN04]   Kuhn, *Risk Matrix as a Tool for Risk Assessment in the Chemical Process Industry*, Workshop Bieleschweig 4, BASF, Braunschweig, 2004.

[KUMA93]   Kumamoto, H., Henley, E. J., *Probabilistic Risk Assessment and Management for Engineers and Scientists,* New York, IEEE 1993.

[KUTT81]   Kutter, R., *Ein Beitrag zum Zuverlässigkeitsnachweis für mechanische Bauteile mit hohem Sicherheitsanspruch durch Anwendung gewichteter Monte-Carlo-Simulation, GHS,* Wuppertal, 1981.

[LAMP77]   Lampson, B. W.; Horning, J. J.; London, R. L.; Mitchell, J. G.; Popek, G. L., *Report on the programming language Euclid.* ACM SIGPLAN Notices, 12(2) 1977.

[LAPR92]   Laprie, J., Dependability, *Basic Concepts and Terminology*, in English, French, German, Italien and Japanese, Wien/New York, Springer, 1992.

[LAUa81] Lauber, R., *Zuverlässigkeit und Sicherheit in der Prozessautomatisierung* Tagung Prozessrechner 1981, München.

[LAUb81] Lauber, R., *Zuverlässigkeit und Sicherheit in Prozessrechner gesteuerten technischen Anlagen*, Tagung Technische Zuverlässigkeit 1981, Nürnberg.

[LAUB82] Lauber, R., *Verfahren zur Erfüllung von Sicherheitsanforderungen bei der Anwendung von Mikrorechnertechnik im Verkehrswesen*, 13. verkehrswissenschaftliche Tagung Dresden 1982.

[LEE-85] Lee, W.; Grosh, D.; Tillmann, F.; Lie, C., *Fault tree analysis, methods and applications – a review*, IEEE Trans R34, 1985, pp. 194-203.

[LEES92] Lees, F. P., *Loss Prevention for the Process Industries*, London, Butterworth and Heinemann, 1992.

[LEIT95] Leitch, R. D., *Reliability Analysis for Engineers, An Introduction*, Oxford Science Publications, 1995.

[LEVE86] Leveson, N. G., *Software safety: why, what, and how*. ACM Comp. Surv., 18(2) 1986, 25-69.

[LEVE91] Leveson, N. G., *Software safety in embedded computer system*. Comm ACM, 34(2), 34-46, 1991.

[LEVE95] Leveson, N. G., *Safeware: System Safety and Computers. Reading*, Addison-Wesley 1995.

[LEWI90] Lewis, B. F., Bunker, R. L., *MAX: An Advanced Parallel Computer for Space Applications*, Proc. Second Int'l Symp. Space Information Systems, Sept. 1990.

[LEWI92] Lewis, B. F., et al., *COSMOS Multicomputer Operating System and Development Environment Functional Specification*, NASA Technical Memorandum, Caltech, JPL, Aug. 1992.

[LEWI96] Lewis, E. E., *Introduction to Reliability Engineering 2nd edn.* New York: John Wiley 1996.

[LIGG90] Liggesmayer, P., *Modultest und Modulverfikation – State of the art* BI-Wissenschaftsverlag Mannheim, Wien, Zürick, 1990.

[LIGG96] Liggesmayer, P., *Die Prüfung von objektorientierten Systemen* OBJEKTspectrum No. 6, November/Dezember 1996, S. 68-78.

[LIMN86] Limnios, N., Ziani, R., *An algorithm for reducing cut sets in fault-tree analysis*, IEEE Trans Reliability, vol R-35, 1986 Dec, pp. 559 – 561.

[LIMN91] Limnios, N., *Arbres de défaillance*, Paris, Hermes, 1991.

[LITT93] Littlewood, B., Strigini, L., *Validation of ultrahigh dependability for software based Systems.* Comm. ACM, 36(11) 1993, 69-80.

[MAEN81] Maeng, J., Malek, M., *A Comparison Connection Assignment for Self-Diagnosis of Multiprocessor Systems* Digest 11th Int'l Symp. Fault Tolerant Computing, pp. 173-175, 1981.

[MANI99]   Manian, R.; Coppit, D.; Sullivan, K.; Bechta Dugan, J., *Bridging the gap between systems and dynamic fault tree models*, Proc. Ann. Reliab. & Maintainab. Symp., 1999, pp. 105-111.

[MARa00]   Marszal, E.M., *Notes of Confidential Survey and Literature Search on Tolerable Risk Guidelines and Third Party Liability Settlements and Judgments*, Columbus, OH, Exida, 2000.

[MARb00]   Marszal, E.M.; Fuller, B.A.; Shah, J.N., Utilization of Risk Based Criteria for Safety Integrity Level Selection, *Process Safety Progress* 43, no. 34, 75, 2000.

[MARS02]   Marszal, Edard M.; Scharpf, Eric W., *Safety Integrity Level Selection: Systematic Methods Including Layer of Protection Analysis*, ISA, 2002.

[MAKI73]   Maki, D. P., Thompson, M., *Mathematical Models and Applications*, Englewood Cliffs, NJ, Prentice-Hall, 1973.

[MALE80]   Malek, M., *A Comparison Connection Assignment for Diagnosis of Multiprocessor Systems*, Proc. Seventh Int'l Symp. Computer Architecture, pp. 31-36, 1980.

[MARS02]   Marszal, E., Scharpf, E., *Safety Integrity Level Selection, Systematic Methods Including Layer of Protection Analysis*, ISA – The Instrumentation, Systems, and Automation Society, 2002.

[MATH99]   *Mathcad, Benutzerhandbuch,* MathSoft Inc. Cambride, MA, 1999.

[MAUR00]   Mauri, G., *Integrating Safety Analysis Techniques, Supporting Identification of Common Cause Failures*, Dissertationsarbeit, University of York, Department of Computer Science, September 2000.

[MCDE92]   McDermid, J., *Education and training for safety-critical systems practitioners* Software in Safety-Related Systems (Wichmann B.A., ed.), pp. 177-207, Chichester: Wiley, John 1992.

[MEIL91]   Meilir, P. -J., *Praktisches DV-Projektmanagement*, Hanser Verlag, 1991.

[MEUS82]   Meusemann, B., *Ursachen für Hardwarefehler und -ausfälle*, In Seminar: Zuverlässigkeit und sichere Rechnersysteme, TÜV Rheinland e.V. Köln, 1982.

[MERC90]   Mercedes-Benz-AG; *Fehler-Möglichkeits- und Einfluss-Analyse (FMEA)*, Leitfaden zur Anwendung, Stuttgart, 1990.

[MEYE84]   Meyer, G. G. L., *A Diagnosis Algorithm for the BGM System Level Fault Model*, IEEE Trans. Computers vol. 33, no. 8, pp. 756-758, Aug. 1984.

[MEYE90]   Meyer, B., *Objektorientierte Softwareentwicklung*, München, Hanser Fachbuchverlag, 1990.

[MEYE97]   Meyers, Robert E., *Handbook of Petroleum Refining Processes*, 2d ed. New York, McGraw-Hill, 1997.

[MIL]   MIL HDBK-217, *Reliability Prediction of Elec. Equip.*, Ed. F, 1991, Not. 2, 1995.

[MILa] MIL Standard 882D.

[MISR89] Misra, K.; Weber, G., *A new method for fuzzi fault tree analysis*, Microelectron., Reliab. 29, 1989, pp. 195-216.

[MOCK95] Mock, R., *Methoden zur Datenhandhabung in Zuverlässigkeitssystem*, Hochschulverlag AG an der ETH Zürich, 1995.

[MOD-91] MoD; *Interim Defence Standard 00-55 The Prourement of Safety Critical Software in Defence Equipment*, Glasgow: Directorate of Standardiziation, 1991.

[MOHR] Mohr, S., Montenegro S., *Analysen der Anlagen Sicherheit und Gefahren (Hazard-Analyse)* GMD-FIRST, Frauenhofer Institute for Comptuter Architecture and Software Technology.

[MOSL93] Mosleh A., Fleming K., Parry G., Paula H., Warledge D., Rasmusson D., *Procedures for Analysis of Common-Cause Failures in Probabilistic Safety Analysis*, NUREG/CR-5801, Vol. 1., Office of Nuclear Regulatory Research, Washington, DC, 1993.

[MUDA95] Mudan, K.S.; Shah, J.N.; Myers, P.M., *Financial Risk Assessment: A Uniform Approach to Manage Liabilities*, New York, American Institute of Chemical Engineers, 1995.

[MUDA96] Mudan, K.S.; Croce, P., *NFPA/SFPE Handbook of Fire Protection Engineering*, section 2, chapter 4, Batterymarch, MA, National Fire Protection Association, 1996.

[MÜLL03] Müller, F., *Entwicklungshelfer*, iX, 06/2003, 82-90.

[MUSA87] Musa, J. D., Iannino, A., Okumoto, K., *Reliability Prediction and Measurement*, McGraw-Hill London, 1987.

[NAKA79] Nakashima, K., Hattori, Y., *An efficient bottom-up algorithm for enumerating minimal cut sets of fault trees,* IEEE Trans. Reliability, vol R-28, 1979 Dec, pp. 353 – 357.

[NARA86] Narasimhan, J., Nakajima, K., *An Algorithm for Determining the Fault Diagnosability of a System* IEEE Trans. Computers, vol. 35, no. 11, pp. 1, 004-1,008, Nov. 1986.

[NEUM95] Neumann, Peter. G., *Computer Related Risk*, Addison-Wesley, 1995.

[NEUM96] Neumann, P. G., *On hierarchical design of computer systems for critical applications* IEEE Trans.Software Eng., 12(9) 1996, 905-20.

[NFPA01] NFPA 85-2001, *Boiler and Combustion Systems Hazards Code*, 2001.

[NIED94] Niedermeier, A., *Fehlertoleranz durch Kombination statischer und dynamischer Redundanz*, VDI-Verlag Reihe 10 Nr. 327, 1994.

[NTG82] NTG 3004, *Zuverlässigkeitsbegriffe im Hinblick auf komplexe Software und Hardware*, Entwurf einer NTG Empfehlung, Nachrichtentechnische Zeitung Nr. 35 1982, S.327-333.

[NTT-85] NTT, Standard Rel. Tables for Semicond. Dev.,Tokyo; Nippon Telegr. and Tel. Corp., 1985.

[NURE81] U.S. Nuclear Regulatory Comission, NUREG-0492, *Fault Tree Handbook*, 1981.

[O'CO88] O'Connor, P., *Undue faith in US Mil-Hdbk-217 for reliability prediction*, IEEE Trans. R37, 1988, pp. 468.

[O'CO90] O'Connor, P. D. T, *Zuverlässigkeitstechnik; Grundlagen und Anwendung*, VCH Verlagsgesellschaft Weinheim, 1990.

[ODEH94] Odeh, K.; Limnios, N., *Comparaison de principales methods d'evaluation probabiliste des arbres de défaillance*, Proc. ESREL, 1994, pp. 513-518.

[OEST05] Oesterreich, B., *Analyse und Design mit UML 2*, Oldenbourg Verlag, 2005.

[OGJ-90] *Oil and Gas Journal*, Aug. 27, 1990.

[ORED92] OREDA-92, *Offshore Reliability Data Handbook*, 2d ed. Hovik: Det Norske Veritas, 1992, 1997, 2002.

[ORLa83] Orlowski, S., Sibbertsen, W., *Statistik I*, Demmig Verlag, 1983.

[ORLb83] Orlowski, S. Sibbertsen, W., *Statistik II*, Demmig Verlag, 1983.

[OSHA93] Occupational Safety and Health Administration, *Process Safety Management (Pamphlet OSHA 3132)*, Washington, DC, U.S. Department of Labor, Occupational Safety and Health Administration, 1993.

[OTTI99] OTTI Technologie-Kolleg, *Seminarunterlagen zum FMEA-Intensivseminar* Regensburg, 02.1999.

[PAGE88] Page-Jones, M., *The Practical Guide to Structured Systems Designs*, Prentice-Hall, 1988.

[PAGE89] Page, L. B., Perry, J. E., *A Model for System Reliability with Common-Cause Failures*, IEEE Transactions on Reliability, Vol. 38, No 4, October 1989.

[PAN-88] Pan, Z. -J., Tai, Y. -C., *Variance Importance of System Components by Monte Carlo*, IEEE Transactions on Reliability, Vol.37, No.4, 1988.

[PAPA90] Paula H. M., Parry G. W., *A Cause Defence Approach to the Understanding and Analysis of Common Cause Failures*, NUREG/CR-5460, Vol. 1., Office of Nuclear Regulatory Research, Washington, DC, 1990.

[PAPO84] Papoulis, A., *Probability, Random Variables, and Stochastic Processes* New York, McGraw-Hill, 1984.

[PAUL90] Paula, H. M., Roberts, M. W., Battle, R. E., *Reliability Performance of Fault Tolerant Digital Control Systems* Proceeding of the 24th Annual Loss Prevention Sympossium, New York American Institute of Chemical Engineers, August 1990.

[PAVE93] Pavey, D. J., Winsborrow, L. A., *Demonstrating equivalence of source code and PROM contents,* Computer Journal, 36(7) 1993, 654-67.

[PECH88]   Pecht, M.; Kang, W., *A critique of Mil HDBK-217 reliability prediction method*, IEEE Trans. R37, 1988, pp. 453-457.

[PETE86]   Peters, O.; Meyna, A. (Hrsg.), *Handbuch der Sicherheitstechnik*, München, Hanser, 1986.

[PETE91]   Peters, Max S.; Timmerhaus, K. D., *Plant Design and Economics for Chemical Engineers*, 4th ed., New York, McGraw-Hill, 1991.

[PFLU92]   Pfluegl, M., Blough, D., *Communication Protocols for Fault-Tolerant Clock Synchronization in Not-Completely-Connected Networks*, Proc. 11th Symp. Reliable Distributed Systems, pp. 130-137, 1992.

[PLC-03]   Webseite unter www.plcopen.org.

[POST80]   Postel, J., *User datagram protocol 1980*. Technical Report RFC768 USC/information Sciences Institute.

[POTO93]   Potocki de Montalk, J. P., *Computer software in civil aircraft*, Microproc. Microsys., 17(1) 1993, 17-23.

[POWE88]   Powell, D., et al., *The Delta-4 Approach to Dependability in Open Distributed Computing Systems* Digest 18th Int'l Symp. Fault-Tolerant Computing, pp. 246-251, 1988.

[POWE92]   Powell, D., *Failure mode assumptions and assumption coverage*, Proceedings of the 22nd International Symposium on Fault-Tolerant Computing Systems, 1992, pp. 386-395.

[PREP67]   Preparata, F. P., Metze, G., Chien, R. T., *On the Connection Assignment Problem of Diagnosable Systems*, IEEE Trans. Electronic Computers, vol. 16, pp. 848-854, Dec. 1967.

[PRES97]   Pressman, R., *Software Engineering*, WCB/McGraw-Hill, 1997.

[PYLE91]   Pyle, I. C., *Developing Safety Systems. A Guide Using Ada*, Hemel Hempstead: Prentice-Hall 1991.

[QSAA91]   QSAA IQSE – S/02, *Anweisungen zur Prüfung festverdrahteter Elektronik-Systeme*, Institut für Qualität und Sicherheit in der Elektronik (IQSE), Version 0.5.EN, 08. Juli1991, Draft, IEC 1508 (Functional Safety, Safety-related systems).

[RAC-90]   RAC, NONOP-1, *Nonoperating Reliability Data*, 1992; *NPRD-95: Nonelectronic Parts Reliability Data*, 1995; TR-89-177, *VHSIC/VHSIC-like Reliabilty Modeling*, 1989; TR-90-72, *Reliability Analysis* Assessment of Advanced Technologies, 1990.

[RASM78]   Rasmuson, D. M., Marshall, N. H., *FATRAM – A core efficient cut-set algorithm*, IEEE Trans. Reliability, vol. R-27, 1978 Oct, pp. 250-153.

[RAUZ93]   Rauzy, A., *New algorithms for fault trees analysis*, Reliability Engineering and System Safety, vol. 40, 1993, pp. 203-211.

[RAZO62]   Razovsky, Igor, *Reliability Theory And Practice* Prentice-Hall, Englewood Cliffs, New Jersey, 1962.

[REAS97]   Reason, J.T. *Managing the Risks of Organizational Accidents*, Ashgate, 1997.

[REIN77]   Reinschke, K., *Aufstellen von Zuverlässigkeitsersatzschaltungen und Fehlerbäumen*, Berlin, VEB Verlag Technik, 1977.

[REIN87]   Reinschke, K.; Usakov, I., *Zuverlässigkeitsstrukturen*, Berlin, Verlag Technik, 1987.

[REIS86]   Reisig, W., *Petrinetze*, 2. Auflage, Berlin, Springer, 1986.

[RICH]   Richter, K., *Statistische Stichprobenverfahren*, Frako, Kondensatoren- und Apparatebau GmbH, Tenningen.

[ROBE99]   Robert, C. P., Casella, G., *Monte Carlo; Statistical Methods*, Springer Verlag, Berlin, 1999.

[ROMA96]   Romanski, G, *Safety critical software handbook*, Thomson Software Products, Safety Mangement Requirements for Defense Systems (Def Stan 00-56), 2. Auflage, 1996.

[ROSE75]   Rosenthal, A., *A computer scientist looks at reliability computations*, Reliability and Fault Tree Analysis, 1975, pp. 133-152, SIAM.

[ROSE80]   Rosenthal, A., *Decomposition methods for fault tree analysis*, IEEE Trans. Reliability, vol R-29, 1980 Jun, pp. 136-138.

[ROWE94]   Rowe, R., *Safety-critical systems computer language survey results. Internet comp.* Software-eng newsgroup, November 1994.

[ROYA92]   Royal Society; *Risk analysis, perception and management*, London, Royal Society 1992.

[RSRE87]   RSRE 1987; *The Official Handbook of MASCOT, Version 3*, Malvern, Computer Division, RSRE 1987.

[RTCA92]   RTCA/EUROCAE 1992; *Software Considerations in Airborn Systems and Equipment Certification. RTCA/DO-178B; EUROCAE/ED-12B*, Washington: Radio Technical Commission for Aeronautics. Paris: European Organisation for Civil Aviation Electronics, 1992.

[SACH72]   Sachs, L., *Statistische Auswertungsmethoden*, Springer, Berlin, Heidelberg 1972.

[SALI92]   Salietti, F., et al., *Software-Diversität für Steuerungen mit Sicherheitsverantwortung*, Schriftenreihe der Bundesanstalt für Arbeitsschutz Dortmund 1992.

[SCHA00]   Scharpf, Eric W.; Goble, William M, *Implementing IEC 61508 in the Process Industries*, Presented at SCS2000, Melbourne, Australia, November 2000.

[SCHI91]   Schier, D., *Network Reliability and Algebraic Structures*, Oxford University Press, 1991.

[SCHL83]	Schlichting, R., Schneider, F., *Fail-Stop Processors: An Approach to Designing Fault-Tolerant Computing Systems, ACM Trans. Computing Systems,* vol. 1, pp. 222-238, Aug. 1983.
[SCHN74]	Schneeweiss, W., *Zufallsprozesse in dynamischen Systemen*, Heidelberg, Springer, 1974.
[SCHN76]	Schneeweiss, W., *"Sicherheit" und "Schutz" oder die zuverlässigste Sprachverwirrung*, RTP (atp) 18, 1976, S. 201.
[SCHN78]	Schneider, C., *Fehlerbaumanalyse von periodisch inspizierbaren Systemen mit Hilfe von Monte-Carlo-Methoden* Univ. Diss., Karlsruhe, 1978.
[SCHN81]	Schneeweiss, W., *Use of a fault tree with delayed inputs*, IEEE Trans. R30, 1981, pp. 339-344.
[SCHN85]	Schneeweiss, W., *Grundbegriffe der Graphentheorie*, Heidelberg, Hüthig, 1985.
[SCHN86]	Schnieder, E., *Prozessinformatik*, Braunschweig, Vieweg, 1986.
[SCHa87]	Schneeweiss, W., *Approximate fault-tree analysis with prescribed accuracy*, IEEE Trans. R36, 1987, pp. 250-254.
[SCHb87]	Schneeweiss, W., *SyRePa'86, ein neues Programmsystem zur Bestimmung von Verfügbarkeit und MTBF von reparierbaren technischen Systemen*, Automatisierungstechnik (at) 354, 1987, pp. 144-147.
[SCHN88]	Schneeweiss, W.; Schulte, M., *SyRePa'87-A package of programs for systems reliability evaluation*, Hagen, FernUniversität, Informatik-Bericht 76, 1988.
[SCHa89]	Schneeweiss, W., *Boolean Functions with Engineering Applications and Computer Programs*, Berlin, Springer, 1989.
[SCHb89]	Schneeweiss, W., *Fault tree synthesis via backtracking*, Proc. RELIABILITY'89, 1989, 4C/1/1-4C/1/9.
[SCHN90]	Schneeweiss, W., *Fault tree evaluation for many sets of input data*, IEEE Trans. R39, 1990, pp. 296-300.
[SCHa91]	Schneeweiss, W., *A review of recent work on fault trees*, Proc. RELECTRONIC'91, 1991, pp. 328-337.
[SCHb91]	Schneeweiss, W.; Nießen-Gillhaus, U., *Fehlerbaum-Auswertung auf der Basis von zwei Versionen der Shannon-Zerlegung*, Proc. TTZ, IGT-Fachberichte 116, Berlin, VDE-Verlag, 1991, S. 31-41.
[SCHa92]	Schneeweiss, W., Zuverlässigkeits-Technik, *Von den Komponenten zum System*, Köln, Datakontext-Verlag, 1992.
[SCHb92]	Schneeweiss, W., *Approximate fault-tree analysis without cut sets*, Proc. Annual Reliability and Maintability Symposium 1992, New York, IEEE Press, 1992, pp. 370-375.
[SCHc92]	Schneeweiss, W., *Reliability theory for large 'linear' systems with helping neigbors*, IEEE Trans. R41, 1992, pp. 343-351.

[SCHd92]   Schneeweiss, W. G., *Usefulness of MTTF of an-Independent Case in Other Cases*, IEEE Transactions on Reliability, Vol.2, No.4, 1992.

[SCHN93]   Schneeweiss, W., *Calculating MTBF for modularized fault trees*, Proc. Annual Reliability and Maintainability Symposium 1993, New York, IEEE Press, 1993, pp. 206-213.

[SCHa95]   Schneeweiss, W., *Fehlerbaum-Analysen mit System-Reliability-Paket*, SyRePa Proc. Tagung "Technische Zuverlässigkeit", VDI-Bericht 1239, 1995, S. 275-289.

[SCHb95]   Schneeweiss, W., *Modularization of fault trees for determining mean life of non-repaired electronics systems*, Proc. RELECTRONIC'95, 1995, pp. 85-94.

[SCHc95]   Schneeweiss, W., *On the mean number of disconnected nodes of a network*, Diagnostic et Surete de Fonctionnement 5, 1995, pp. 107-119.

[SCHd95]   Schneeweiss, W., *Recursive analysis of fault trees of big k-out-of-n systems with non-identical components*, Proc 1ist ISSAT Int. Conf. On Reliab. & Quality in Design, 1995, pp. 70-73.

[SCHN96]   Schneeweiss, W., *A tighter upper bound for system unavailability*, Proc ESREL'96, Berlin, Springer, 1996, pp. 63-68.

[SCHa97]   Schneeweiss, W.; Buschhorn, P., *Mincut-based fault tree analysis revisited and extended for calculating MTBF*, Rel. Engg. & System Safety 57, 1997, pp. 121-127.

[SCHb97]   Schneeweiss, W., *Calculating mean system-failure frequency with prescribed accuracy*, IEEE Trans. R46, 1997, pp. 201-207

[SCHN99]   Schneeweiss, W., *Petri Nets for Reliability modeling*, Hagen, LiLoLe-Verlag, 1999.

[SCHÜ98]   Schürger, K., *Wahrscheinlichkeitstheorie*, Oldenbourg, München, Wien, 1998.

[SCHW90]   Schwab, A. J., *Elektromagnetische Verträglichkeit [EMV]*, Springer, Berlin, Heidelberg, 1990.

[SEMA71]   Semanders, S. N., *ELRAFT: A computer program for efficient logic reduction analysis of fault trees"*, IEEE Trans. Nuclear Sience, vol NS-18, 1971 Feb, pp. 481 – 487.

[SEMI99]   Seminarunterlagen zum FMEA-Intensivseminar des OTTI Technologie-Kolleg, 25.02. – 26.02.99 in Regensburg.

[SENG92]   Sengupta, A., Dahbura, A. T., *On Self-Diagnosable Multiprocessor Systems: Diagnosis by the Comparison Approach*, IEEE Trans. Computers, vol. 41, no. 11, pp. 1386-1396, Nov. 1992.

[SFKa04]   Störfallkomission, SFK-GS-41, *Risikomanagement im Rahmen der Störfall-Verordnung*, Bericht, 21.04.2004.

[SHAN89]	Shanmugam; Ramalingam; Richards; Dale, O., *On Estimating the Mean Time To Failure With Unknown Censoring*, IEEE Transactions on Reliability, Vol.38, No.3, 1989.
[SHOO68]	Shooman, Martin L., *Probabilistic reliability: an engineering approach* McGraw-Hill, New York, 1968.
[SIEM91]	Siemens, SN 29 500 Teil 1, *Ausfallraten Bauelemente,* München, Siemens 1991, bzw. DIN 40039 (1988 E).
[SIEW82]	Siewiorek, D. P., Swarz, R. S., *The Theory and Practice of Reliable System Design*, Bedford, MA. Digital Press 1982.
[SIEW90]	Siewiorek, D. P., *Fault Tolerance in Commercial Computers,* Computer, July 1990.
[SIL-05]	SILver^{TM} (SIL Verification). exida.com (Sellersville, PA) Web site: www.exida.com (Retrieved 6/28/2005 from source).
[SINN96]	Sinnamon, R., Andrews, J. D., *Fault tree analysis and binary decision diagrams,* Proc. Ann. Reliability and Maintainability Symp, pp. 215-222, 1996.
[SIRJ88]	Sirjaev, A. N., *Wahrscheinlichkeit*, VEB Deutscher Verlag der Wissenschaften Berlin, 1988.
[SMIT93]	Smith, David J., *Reliability, Maintainability, and Risk: Practical Methods for Engineers,* 4th edition, Butterworth-Heinemann, 1993 (Note: 5th[1997] and 6th [2001] editions of this book are also available).
[SMIT97]	Smith, David J., *Reliability, Maintainability, and Risk: Practical Methods for Engineers,* 5th edition, Butterworth-Heinemann, 1997.
[SOBO85]	Sobol, M., *Die Monte-Carlo-Methode*, Verlag der Wissenschaften, 1985.
[SÖDE93]	Söder, G., *Modellierung, Simulation und Optimierung von Nachrichtensystemen*, Springer Verlag, 1993.
[SOHa91]	Soh, S.; Rai, S., CAREL, *computer aided reliability evaluator for distributed computing networks*, IEEE Trans. Parallel Distr. Syst. 2, 1991, pp. 199-213.
[SOHb91]	Soh, S.; Rai, S., *Experimental results on preprocessing of paths/cuts terms in sum of disjoint products technique*, IEEE Trans. R42, 1993, pp. 24-33.
[SOMM87]	Sommerville, I., *Software Engineering*, Addison-Wesley, 1987.
[STAR87]	STARTS Purchasers´Group; *The STARTS guide: Vol. 1,2nd edn.* Manchester National Computing Centre Publications 1987.
[STAR89]	STARTS Purchasers´Group; *The STARTS Purchasers´Handbook: Software Tools for Application to Large Real Time Systems, 2nd edn.* Manchester: National Computing Centre Publications, 1989.
[STOL87]	Stoll, J. J., *Anwendungsorientierte Techniken zur Fehlertoleranz in hierarchisch verteilten Realzeitsystemen*, Dissertation an der Universität der Bundeswehr, 1987.

[STÖR70] Störmer, H., *Mathematische Theorie der Zuverlässigkeit*, München, Oldenbourg, 1970.

[STOR88] Storm, R., *Wahrscheinlichkeitsrechnung, mathematische Statistik und statistische Qualitätskontrolle*, Leipziger Fachbuchverlag, 1988.

[STOR95] Storey, N., *Safety trough training.* Proc. Second Safety Trough Quality Conference, Cape Canaveral, Fl, 23-25 October 1995, pp. 261-70.

[STOR96] Storey, N., *Safety Critical Computer Systems*, Addison Wesley, 1996.

[SULL84] Sullivan, G., *A Polynomial time Algorithm for Fault Diagnosability*, Proc. 25th Ann. Symp. Foundations of Computer Sience, pp. 148-156, 1984.

[SUMM99] Summers, A. E.; Ford K. A.; Raney G., *Estimation and Evaluation of Common Cause Failures in SIS*, Premier Consulting + Engineering, Triconex Corporation, published in Chemical Engineering Progress, November 1999.

[SUNE97] Suné, V., Carrasco, J. A., *A method for the Computation of reliability bounds for non-repearable fault-tolerant systems*, Proc. 5th Int'l Symp. On Modeling, Analysis and Simulation of Computers and Telecommunication Systems (MASCOTS'97), 1997 Jan, pp. 221-228. Haifa.

[TAYL94] Taylor, J. R., *Safety assessment of control systems – the impact of computer control*, Israel Institute for Petroleum and Energy Conference on Process Safety Management held in Tel Aviv, Israel, October 1994.

[THAa00] Thaller, G. E., *Softwaremetriken, einsetzen - bewerten – messen,* Verlag Technik, Berlin, 2000.

[THAb00] Thaler, G. E, *Softwaredesign und Implementierung* Verlag Technik 2000.

[TIEZ98] Tiezema, R., *Common Cause effects on safety rated PLCs*, Yokogawa, Apeldoorn, Niederlande, Mai 1998.

[TIEZ03] Tiezema, R., *Risk Reduction in the Process Industry, Proof testing*, Yokogawa, Apeldoorn, Niederlande, März 2003.

[TIWA75] Tiwari, R.; Verma, M., *On the analysis of fault treest*, IEEE Trans. R24, 1975, pp. 194-203.

[TOGE06] Together 2006 für Eclipse, www.borland.de/together/index.html, Stand: 31.10.2005.

[TRAU] Trautwein, W. M., *Ein Beitrag zur Funktionsüberwachung und Fehlertoleranz in mikroprogrammierbaren Prozessoren*, Dissertation Universität Duisburg.

[UKOO99] *Guidelines for Instrument-Based Protective Systems*, U.K. Offshore Operators Association, 1999.

[UNVE92] *Unveröffentlichte Arbeitskreisunterlagen,* Kapitel "FMEA, VDA Nr. 4, neu", Verband der Automobilindustrie e.V., 26.08.1992.

[USFR92] U.S. Federal Register, 29 CFR Part 1910.119., *Process Safety Management of Highly Hazardous Chemicals*, U.S.Federal Register, Feb. 24, 1992.

[USMI00] US MIL-STD-882D-2000, *Standard Practice for System Safety*.

[VDE-95] VDE 0801, Teil 4, *Definitionen, Entwurf*, September 1995, entspricht IEC 61508, Part 4, 06.1995.

[VDI-93] VDI-GIS (Hrsg.), *Software-Zuverlässigkeit: Grundlagen, konstruktive Maßnahmen, Nachweisverfahren*. VDI-GIS (Gemeinschaftsausschuß Industrielle Systemtechnik) VDI-Verlag, 1993.

[VDI-91] VDI Bildungswerk, *Die FMEA, Seminarunterlagen zum FMEA-Moderatoren-Seminar*, Stuttgart, 1991.

[VDI-00] VDI/VDE 3542 Blatt 1-4, *Sicherheitstechnische Begriffe für Automatisierungssysteme*, 2000.

[VELT06] Velten-Philipp, W.; Houtermans, M. J. M., *The Effect of Diagnostic and Periodic Testing on the Reliability of Safety Systems*, TÜV Industrie Service GmbH, 2006.

[VEMU99] Vemuri, K.; Bechta Dugan, J.; Sullivan, K., *A design language for automatic synthesis of fault trees*, Proc. Ann. Reliab. & Maintainab. Symp., 1999, pp. 91-96.

[VERB86] Verband der Automobilindustrie e.V., *Sicherung der Qualitätskontrolle vor Serieneinsatz*, Schriftenreihe "Qualitätskontrolle in der Automobilindustrie", Band 4, 1986.

[VESE81] Vesely, W.; Goldberg, F.; Roberts, N.; Haasl, D., *Fault Tree Handbook*, Washington, NUREG-0492, 1981.

[VILL91] Villemeur, A., *Reliability, availability, maintainability and saftey assessment*, Vol. 1, (trans. A. Cartier and M.-C. Lartisien), Wiley, Chichester, 1991.

[WALL01] Wallmüller, E. *Software-Qualitätsmanagement in der Praxis*, München, Hanser Fachbuchverlag, 2001.

[WALT94] Walter, C., Suri, N., Hugue, M., *Continual On-Line Diagnosis of Hybrid Faults* Proc. Fourth IFIP Working Conf. Dependable Computing for Critical Applications, pp. 233-249, 1994.

[WANG94] Wang, H., Blough, D., Alkalaj, L., *Analysis and Experimental Evaluation of Comparison-Based System-Level Diagnosis for Multiprocessor Systems* Gigest 24th IEEE Int'l Symp. Fault-Tolerant Computing, pp. 55-64, 1994.

[WANG95] Wang, H., *Practical Comparison-Based Fault Diagnosis in Multiprocessor Systems* PhD dissertation, Dept. of Electrical and Computer Eng., Univ. of California, Irvine, 1995.

[WARD85] Ward, P. T., et al., *Structured development of real-time systems*, Prentice Hall, 1985.

[WEIL04] Weilkiens, T.; Schröder, C., Praxisbericht: *Erste Projekt-Erfahrungen mit der UML 2.0*, Objekt-Spektrum 3/2004.

[WHET94]   Whetton, C., *Maintainability and its influence on system safety*. In Technology and Assesment of Safety-Critical Systems (Redmill, F. and Anderson, T., eds), pp. 31-54. London, Springer-Verlag, 1994.

[WICH89]   Wichmann, B. A., *Insecurities in the Ada Programming Language*, NPL Report DITC 137/89. Teddington, National Physical Laboratory, 1989.

[WICH94]   Wichmann, B. A., *Producing critical systems – the Ada 9X solution*. In Technolog. and Assessment of Safety-Critical Systems, Redmill F. and Anderson T., eds, pp. 194-203, London, Springer-Verlag, 1994.

[WICH95]   Wichmann, B. A., Canning, A. A., Clutterbuck, D. L., et al., *Industrial perspective of static analysis*, Software Eng. J., 10(2) 1995, pp. 69-75.

[WILS85]   Wilson, J. M., *Modularizing and minimizing fault trees*, IEEE Trans. Reliability, vol R-34, 1985 Oct, pp. 320 – 322.

[WILS90]   Wilson, J., *An improved minimizing algorithm for sum of disjoint products*, IEEE Trans. R39, 1990, pp. 42-45.

[WITT95]   Witter, A., *Entwicklung eines Modells zur optimierten Nutzung des Wissenspotentials einer Prozess-FMEA*, Fortschrittsberichte VDI, Reihe 20, Nr. 176, 1995.

[WLOK95]   Wloka, D., *Softwaretechnologie*, Skriptum zur Vorlesung Universität Kassel, 1995.

[WLOK90]   Wloka, D., *Robotersimulation Teil 1, 2 und 3,* Springer Verlag, 1990.

[WOLF93]   Wolf, B., *Fehler-Baum-Synthese für Zusammenhangs-Probleme in stochastischen Graphen*, Hagen, FernUniversität, Informatik-Bericht 148, 1993.

[WOOD86]   Wood, K., *Factoring algorithms for computing k-terminal network reliability*, IEEE Trans. R35, 1986, pp. 269-278.

[YOUN]   Young, E., *Structured design: Fundamentals of a disciplin of computer program and system design.*

[YOUR79]   Yourdon E., Constantine, L., *Structured Design: Fundamentals of a Disciplin of Computer Program and Systems Design, Englewood Cliffs,* Prentice-Hall, 1979.

[YOUR89]   Yourdon E., *Structured Walkthroughs*, Prentice-Hall, 1989.

[ZIO-02]   Zio, E., *Common Cause Failures, An Analysis Methodology and examples*, Internetrecherche, April 2002.

Index

1

1oo1-system 150, 260
1oo2-„Single-Board-System" 264
1oo2-architecture 152, 156, 158
1oo2D-architecture 158, 272
1oo2-system 150, 225
1oo3-system 150

2

2oo2-system 265
2oo3-architecture 158
2oo3-system 156, 268
2oo4-architecture 158

A

Able to function 68
Absorbing state 175, 260, 261, 266
Actuator 192
ALARP 390
Alterable memory 351
Application software 438
Architectural design 285, 297
Architecture 292, 301
Automation industry 167
Availability 73, 168, 172, 265
Average
  - failure probability 241
  - Repair Time 71, 72
  - time to repair 177
Average value 72
  - of the failure free working time. 168
Avoidance 93

B

Backward search 82
Basic
  - failure rate λ 255

  - Parameter-Model 218
  - Process Control System 133
Behavior of a system over time 172
Beta Factor Model 210
Binomial Failure Rate Model 220
Black box 295, 301, 323
Bottom-up 83
  - integration 326
BPCS 411
Bridge structure 140

C

CASE tools 304
CCF 200, 206
Change in state 169
Circuit block diagram 145
Combination without repetition 150
Common Cause Failure ... 173, 175, 199,
  263, 270
Common Mode Failure 205
Communication and control 155
Complement
  - of the reliability-function 70
Component 104, 108
  - intersections 302
Computer
  - simulation 173
  - system 24
Condition
  - analysis 169
  - description 171
  - probabilities 173
  - transition 173
Constant random variable 70
Control system 411
CPU-block 256
CRC-algorithm 351
Customer requirement specifications 293

Cut set 200

D

Damage 49, 53
Danger analysis 82, 183
Dangerous 54
  - detectable failure 263, 270
  - detected failure 258, 272
  - error 159
  - failure 265, 412
  - malfunction 389
  - state 256
  - undetected failure 258, 270
  - undetected failure 263
  -undetected failure 272
DC 61
Degree of diagnostic coverage . 351, 390
Dependability 53
Dependent failure 412
Design 292
  - classes 302
  - pattern 296
Determination of reliability 160
Deviation 389
DGQ 27
Diagnose coverage factor 61
Diagnosis connection 274
Diagnostic
  - coverage 236
  - installation 272
  - switch 272
  - system 174
  - test 234
DIN 29
DIN EN 801 150
DIN V 19250 .... 34, 89
DIN V VDE 0801 35
Directional arrow 167
Distinct system conditions 169
Distribution
  - model 302
  - of probability of failure 166
  - of the condition 169
  - of the reliability 166
Diversity 388
Division 122
Documentation 376
DoD 27

E

E/E/PES 247, 374, 386
Early phase 59, 74
Effect 67
  - analysis 389
Energy-free state 264, 274
Energyless condition 157, 158
Entrance/exit system 151
ERRF 135
Error 319, 389
  - avoidance 321, 389
  - of the quartet configuration 160
  - probability 148
  - tolerance 321
Error tree analysis 369, 386
ESD 373
EUC 378
Event 122, 260, 261
  - initial 122
  - safe 261
Event tree 368, 378, 382
  - analysis 121
  - linkage 213
  - symbol 123
Expectancy-value 70
Exponential distribution ... 148, 169, 257
Exponential distribution of failure ... 143
Exposure time 92
Extent of Damage 92

F

Failure .. 53, 70, 105, 155, 187, 199, 389
  - behavior 159, 161
  - condition 168
  - correction procedure 175
  - criteria 108
  - dangerous detected 60
  - dangerous undetected 60
  - density 119
  - detected 256
  - diagnosis 258
  - diagnostic 175
  - due to a common cause 188
  - due to interruption 159
  - free condition 173

- mechanism 159
- model 187
- probability 145, 231
- propagation 83
- undetected 256
Failure In Time 59, 74
Failure probability 265
- of a 1oo1-system 257
Failure rate 71, 144, 146, 174, 177
- λ_D 257
- average 74
Failure/repair process 169
Failure-free state 258, 274
Failure-mechanism function 68
Fault
- Propagation Model 85
- tolerance 199, 256, 259
- tolerating control system 168
Fault tree 105, 109
- analysis 103
Field devices 431
Final inspection 440
Fine tuning design 285
FIT 59
- value 59, 74
Forward Engineering 304
Forward search 82
Frames 305
FTA 103
Functional
- binding lines 122
- element 104
- part of a system 104
- Safety 50
- safety management 354
- unit 104
Functioning string 150

G

Goal control 282

H

Halstead 318
Hardware
- diagnosis connection 273
- failure 24, 396
- fault tolerance 251

- fault-tolerance 190
Harmless malfunction 389
Hazard 47
Hazardous event 48
Hazardous situation 48
HAZOP 183
High availability 152
Homomorphism 315
Human error 389

I

I/O block 256
IEC 28
IEC 61131 42
IEC 61508..36, 150, 193, 233, 341, 373, 406
IEC 61511 .... 39, 193, 342, 409
Independent parameters 146
Initial state 171
Input 150
Inspection 315
Installation 195
Integration 325
Intended use/purpose 49
Interaction model 303
Inversion 181
ISA TR 84.02 44
Item 104

K

Kinds of redundancies 145

L

Late phase 59
Layer model 301
Level of difficulty 318
Life Cycle Phase 185
Lifetime 175
Lines of Code 317
Logic function 414
Logic solver 192
Logical system 386
LOPA 127
Lowering the expenses of material ... 166

M

MacLaurin progression 257

Maintenance 195
Majority decision 268
Marginal case failure...................... 161
Markov chains homogeneous........... 169
Markov model
   - of a 2oo3-"single board system" 271
Markov Model........................ 167, 177
Markov Processes 170
Markov-model
   - of a 1oo1-architecture 258
   - of a 1oo2-architecture 264
   - of a 1oo2D-"single board system"
   274
   - of a 1oo2D-architecture 274
   - of a 2oo2"Single-Board System"266
Markov-Model
   - Model Calculation...................... 173
Matrix element 177
Matrix I - Q 180
Matrix M 171
Matrix N 181
Matrix Q 179
McCabe-measure............................. 317
MDA 305
Mean 71
Mean Time Between Failures 72
Mean Time To Failure 70, 168
Measurement and control................ 373
        - safety equipment........................... 89
        - technology.................................... 89
Memory-Less-Systems..................... 173
Milestone... 285
Minimal cut sets 114
Minimal system 256
MISRA 28
mixed systems 148
Modul 415
MSP430.. 367
MTBF.. 72, 310
MTTF .. 70, 71, 145, 148, 161, 162, 164,
   165, 168, 175, 180, 181, 231, 259,
   260, 261, 264, 266, 274
   - calculation.................................. 180
MTTF-calculation
   - of the 1oo2D-system 274
MTTF-term
   - for a 1oo2D-system..................... 276
   - for a 1oo2-system 264
   - for a 2oo3-system 271
MTTF-value
   - for a 2oo2-system 267
   - of a 1oo1-system........................ 262
MTTR 310
   - of distribution function 72
Multiple redundancies..................... 154
Multy Greek Letter Model 219
MVC 297

N

Naming convention 298
NASA.. 28
No absorbing condition................... 179
Nonconformity.................................. 53
Non-redundancy system configuration
   166
Non-Shock-Model 217
Not able to function 68

O

Object flow 303
Object orientation............................ 296
Object oriented design 301
Online diagnostic 172
Operating error.................................. 24
Operating mode............................... 388
Operation 195
Operational availability................... 143
Organisation.................................... 281
OR-junction 265
Output 150
Outside-in integration 327
Over dimensioning.......................... 143

P

Parallel connection.......................... 265
Path 114
Pattern of behavior.......................... 297
Performance indicators 235
Period under consideration 69
Permission... 67
PES... 386, 416
PFD 168, 187, 232
PFD$_{avg}$-equation
   - of a 1oo1-system................ 257, 258

- of a 1oo2D system 273
- of a 1oo2D-system..................... 273
- of a 2oo2-system........................ 265
PFD_{avg}-formula
- for a 1oo2-system 263
PFD_{avg}-value
- for a 2oo3-system 270
PFD-fault tree
- of a 1oo2-architecture................ 262
Phase diagram 167
Pipe to pipe approach....................... 345
P-matrix 260
Point-Availability............................. 168
Positive random variable 71
Probability
- analysis 84
- density 260
- of occurrence 94
- of survival................................... 69
Probability block diagram................ 146
Probability condition........................ 168
Probability density 172
Probability function 170
Probability of failur.......................... 146
Probability of failure
- by demand of a safety function.. 168
- on demand 168
Procedural structures........................ 302
Procedure 49, 104
Process 49
- automation system 67, 68
- chart 111
Process industry 167, 183
Processing unit................................. 150
Processor unit................................... 350
Production design 297
Productivity...................................... 281
Program volume............................... 318
Programmable electronic 416
Programming convention................. 297
Programming effort.......................... 319
Project planning 289
Proof 49
- of safety 67
Proof test.. 229
- expansion factor......................... 244
- frequency 242

- interval..232
Protection...54
- layer..427
- layer analysis 129
- levels...132
Prototype...286

Q

Q-matrix 261
Quality 281
- control...308
- management................................307
- management system....................307
- plan 307
- planning......................................307
- politics 307
- testing 308
Quartet configuration.........................158

R

Random
- variable 70, 143
Random -
value...71
Read only memory.............................350
Redundancy 145, 388
Redundancy circuits..................150, 165
Redundancy concepts 165
Redundancy of components..............146
Redundancy of systems 146
Redundant
systems..161
Redundant architecture 146
Redundant connected subsystems.....151
Redundant safe processing unit 152
Redundant secure processing unit 155
Redundant system 169
Redundant system configuration 166
Redundantly connected blocks 158
Relation...122
Reliability 53, 67, 143, 145, 154, 155,
161, 168, 172, 314
- analysis 367
- block diagram 114, 370, 375
- Block Diagram Analysis.............137
- function.....................69, 70, 73, 231
- matrix..261

- measures 310
- models 143
- of a 1oo1-system 257
- of the redundantly connected blocks with two failure types 158
- procedure 167
- values from the Markov Model . 168
Reliability block diagram 145, 169
Reliability function 143, 144, 146
Reliability functions 169
Reliability increase 158
Reliability of failure of the redundant circuits 161
Reliabiltiy
- function 148
Repair 195
Repair rate 172, 175
- μ_R 172
Repair time 72, 167
Repairable system 167
Requirement classes 96
Residual risk 49, 52
Resources 108, 294
Reverse engineering 304
Review 315
Revision 422
risk
- graph 372
Risk 49, 52, 54, 67, 90
- analysis 52, 81
- control 52
- evaluation 52, 426
- Evaluation Analysis 183
- management 52
- parameter 91
- rate 241
- reduction 90, 240, 255
Risk Assessment 183
Risk graph 89, 94, 97, 99, 367, 376, 381
Round trip-engineering 304
RTCA 27
RTCA DO 178B 45

S

SAE ARP 4754 56
SAE ARP 4761 56
safe

- detectable error 174
Safe
- state 174, 256, 260, 261, 264, 276
- State 48
- undetectable error 174
Safe - state 268
Safe condition 158
Safe failure 174
Safe state 179
Safety 49, 52, 54, 67, 143, 145, 385
- analysis 32
- dimensions from the Markov Model 168
- function 268, 353, 388
- instrumented function 185
- integrity 388
- life cycle 343
- life cycle 423
- lifecycle 377, 389, 419
- loop 362
- management 376
- oriented function 346
- related system 31
- relevant-process 67
- technical control system 418
- technical function 414, 415, 418
- technical system 418
- validation 193
Safety consideration 168
Safety factor 143
Safety procedure 167
Safety related
- process and automation industry 146
- system 386
Safety related automation technology 152
Safety related industrial automation. 155
Safety related system-model 173
Safety related systems 148
Safety system 183
Scale types 316
Self diagnosis 258
Sequence 122
Serial-parallel combination 158
Series architecture 143
Series parallel system 149
Series system 149

Severity level 53
SFF 189, 235, 355, 361
Shock-Model 217
Short circuit failure 159
Safety Life Cycle 423
SIF 183, 240
SIL .. 185, 342, 347, 353, 377, 379, 388, 392, 393, 394, 419
Single configuration 166
Single element 165
Single-Controller-System 256
SIS 193, 206, 341, 405, 429
    - configuration 423
    - safety validation 441
Software 277
    - architecture 293
    - development 278, 283, 284
    - engineering 280
    - evolution 279
    - inspection 313
    - lifecycle 389
    - measures 316
    - quality 306, 315
    - reliability 309, 316
    - safety lifecycle 375
Sources of failures 199
Specification of requirements 289
Spiral model 288
SPS 416
SRS 195
Standard 25
Standard architecture for safety related systems 154
Startup 195
State 122, 167
    - DU 260
Statistical Analysis 84
Status models 303
Structural design 296
Sub system 104
Subject classes 300
Subsystem model 302
Sure
    - failure 159
System 104
    - analysis 105, 107
    - architecture 189

    - availability 167
    - in operation 266
    - operational 261
    - safety 167
    - system configurations for industrial use 150
    - without redundancy 145
    reliability 165
System condition 169
System error free 168
System failure 145
System function 150
System in operation 168, 179
System process 300
System reliability 145, 148
System safety 145
System state 173
Systematic failure 389
Systems with mixed parallel series combinations 149

T

Test 303
Test interval 174, 259
Testing facility 390
Three component system 215
Threshold risk 54
Time dependent Markov model 173
Time interval 169
TMR 156
Tolerable risk 48
Tolerance of failure 166
Top-Down 83
Top-Down integration 326
Top-Down-Algorithm 118
Transition
    - curve 167, 171
    - matrix 170
    - matrix *P* 176
    - probability 174, 175, 176
    - rate 169, 175, 258, 260, 264, 270, 274
Transition matrix 259
    - for a 1oo1-system 259
Transmission rate 274
Triple module redundancy 156

U

UML 299, 306
Unavailability 119
Unit matrix I 171
Useful life phase 59
User interface 415
Utilization phase 74

V

Validation ... 54, 195, 289, 312, 314, 390
Verification . 53, 54, 289, 312, 379, 389, 426
V-model 55, 194, 288, 434
Voting circuit 268
Voting-unit 156

W

Walkthrough 315
Waterfall model 287
Wearout phase 74
Weibull Distribution 74
White box 324
Working condition 168
Working environment 108

α

α-factor model 219

β

β-factor 177, 221, 222
  - method 221
  - model 218